Ecology and Evolution of Cooperative Breeding in Birds

Cooperative breeders are species in which more than a pair of individuals assist in the production of young. Cooperative breeding is found in only a few hundred bird species worldwide, and understanding this often strikingly altruistic behavior has remained an important challenge in behavioral ecology for over 30 years. This book highlights the theoretical, empirical, and technical advances that have taken place in the field of cooperative breeding research since the publication of the seminal work *Cooperative Breeding in Birds: Long-Term Studies of Behavior and Ecology* (Cambridge, 1990). Organized conceptually, this book pays special attention to ways in which cooperative breeders have proved fertile subjects for testing modern approaches to classic evolutionary problems including those of sexual selection, sex-ratio manipulation, life-history evolution, partitioning of reproduction, and incest avoidance. It will be of interest to both students and researchers in the fields of behavior and ecology.

WALTER KOENIG is Research Zoologist at the Museum of Vertebrate Zoology, and Adjunct Professor in the Department of Integrative Biology, at the University of California, Berkeley. He has studied the phenomenon of cooperative breeding for over 25 years. He is the coeditor of *Cooperative Breeding in Birds: Long-Term Studies of Ecology and Behavior* with P. B. Stacey, and author or coauthor of over 150 technical and popular articles on behavioral and population ecology.

JANIS DICKINSON is Associate Research Zoologist at the Museum of Vertebrate Zoology at the University of California, Berkeley. She became interested in cooperative breeding in birds in 1988 after studying sperm competition and mating behavior in insects in New York, Texas, and Arizona. She has used the western bluebird as a model system to study sex allocation, sperm competition, parental care, cooperative breeding, and dispersal.

Ecology and Evolution of Cooperative Breeding in Birds

Edited by
Walter D. Koenig
Janis L. Dickinson
Hastings Reservation and Museum of Vertebrate Zoology,
University of California, Berkeley

GOVERNORS STATE UNIVERSITY
UNIVERSITY PARK
IL 60466

PUBLISHED BY THE PRESS SYNDICATE OF THE UNIVERSITY OF CAMBRIDGE
The Pitt Building, Trumpington Street, Cambridge, United Kingdom

CAMBRIDGE UNIVERSITY PRESS
The Edinburgh Building, Cambridge, CB2 2RU, UK
40 West 20th Street, New York, NY 10011–4211, USA
477 Williamstown Road, Port Melbourne, VIC 3207, Australia
Ruiz de Alarcón 13, 28014 Madrid, Spain
Dock House, The Waterfront, Cape Town 8001, South Africa

http://www.cambridge.org

© Cambridge University Press 2004

This book is in copyright. Subject to statutory exception and
to the provisions of relevant collective licensing agreements,
no reproduction of any part may take place without
the written permission of Cambridge University Press.

First published 2004

Printed in the United Kingdom at the University Press, Cambridge

Typeface Ehrhardt 9.5/12pt.   *System* LATEX 2$_\varepsilon$ [TB]

A catalog record for this book is available from the British Library

Library of Congress Cataloging in Publication data
Ecology and evolution of cooperative breeding in birds / edited by Walter D. Koenig and Janis L. Dickinson.
  p. cm.
Includes bibliographical references.
ISBN 0 521 82271 8 (hardback) – ISBN 0 521 53099 7 (paperback)
1. Birds – Behavior – Evolution.   2. Birds – Ecology.   3. Cooperative breeding in animals.   I. Koenig, Walter D., 1950–
II. Dickinson, Janis L., 1955–
QL698.3.E36   2004
598.156 – dc22   2003055589

ISBN 0 521 82271 8 hardback
ISBN 0 521 53099 7 paperback

QL 698.3 .E36 2004

Ecology and evolution of
cooperative breeding in

To the "founding fathers" of the modern study of cooperative breeding
Jerram Brown, Steve Emlen, Frank Pitelka, Ian Rowley, and Glen Woolfenden

and our favorite pseudo-pensioner
Nick Davies

whose influence, both through their own work and through that of their students and associates, is evident on every page of this book

Contents

List of contributors viii

Introduction 1
WALTER D. KOENIG AND JANIS L. DICKINSON

1 **Evolutionary origins** 5
J. DAVID LIGON AND D. BRENT BURT

2 **Delayed dispersal** 35
JAN EKMAN, JANIS L. DICKINSON, BEN J. HATCHWELL, AND MICHAEL GRIESSER

3 **Fitness consequences of helping** 48
JANIS L. DICKINSON AND BEN J. HATCHWELL

4 **Parental care, load-lightening, and costs** 67
ROBERT G. HEINSOHN

5 **Mating systems and sexual conflict** 81
ANDREW COCKBURN

6 **Sex-ratio manipulation** 102
JAN KOMDEUR

7 **Physiological ecology** 117
MORNÉ A. DU PLESSIS

8 **Endocrinology** 128
STEPHAN J. SCHOECH, S. JAMES REYNOLDS, AND RAOUL K. BOUGHTON

9 **Incest and incest avoidance** 142
WALTER D. KOENIG AND JOSEPH HAYDOCK

10 **Reproductive skew** 157
ROBERT D. MAGRATH, RUFUS A. JOHNSTONE, AND ROBERT G. HEINSOHN

11 **Joint laying systems** 177
SANDRA L. VEHRENCAMP AND JAMES S. QUINN

12 **Conservation biology** 197
JEFFREY R. WALTERS, CAREN B. COOPER, SUSAN J. DANIELS, GILBERTO PASINELLI, AND KARIN SCHIEGG

13 **Mammals: comparisons and contrasts** 210
ANDREW F. RUSSELL

14 **Summary** 228
STEPHEN PRUETT-JONES

Names of bird and mammal species mentioned in the text 239
References 242
Taxonomic index 279
Subject index 290

Contributors

RAOUL K. BOUGHTON
Department of Biology
University of Memphis
Memphis, TN 38152-3540, USA

D. BRENT BURT
Department of Biology
Stephen F. Austin State University
Nacogdoches, TX 75962, USA

ANDREW COCKBURN
Evolutionary Ecology Group
School of Botany and Zoology
Australian National University
Canberra ACT 0200, Australia

CAREN B. COOPER
Cornell Laboratory of Ornithology
159 Sapsucker Woods Road
Ithaca, NY 14850, USA

SUSAN J. DANIELS
Department of Biology
Virginia Polytechnic Institute and State University
Blacksburg, VA 24061–0406, USA

JANIS L. DICKINSON
Hastings Reservation
University of California, Berkeley
38601 E. Carmel Valley Rd.
Carmel Valley, CA 93924, USA

MORNÉ A. DU PLESSIS
Percy FitzPatrick Institute
University of Cape Town
Private Bag, Rondebosch 7701, South Africa

JAN EKMAN
Evolutionary Biology Centre/Population Biology
Uppsala University
Norbyv. 18D, SE-752 36 Uppsala, Sweden

MICHAEL GRIESSER
Evolutionary Biology Centre/Population Biology
Uppsala University
Norbyv. 18D, SE-752 36 Uppsala, Sweden

BEN J. HATCHWELL
Department of Animal and Plant Sciences
University of Sheffield
Sheffield S10 2TN, UK

JOSEPH HAYDOCK
Department of Biology
Gonzaga University
Spokane, WA 99258, USA

ROBERT G. HEINSOHN
Centre for Resource and Environmental Studies
Australian National University
Canberra ACT 0200, Australia

RUFUS A. JOHNSTONE
Department of Zoology
University of Cambridge
Cambridge CB2 3EJ, UK

WALTER D. KOENIG
Hastings Reservation
University of California, Berkeley
38601 E. Carmel Valley Rd.
Carmel Valley, CA 93924, USA

JAN KOMDEUR
Zoological Laboratory
University of Groningen
PO Box 14, 9750 AA Haren
The Netherlands

J. DAVID LIGON
Department of Biology
University of New Mexico
Albuquerque, NM 87131, USA

ROBERT D. MAGRATH
School of Botany and Zoology
Australian National University
Canberra ACT 0200, Australia

GILBERTO PASINELLI
Zoological Institute
University of Zurich
Winterhurerstr. 190, CH-8057 Zurich, Switzerland

STEPHEN PRUETT-JONES
Department of Ecology and Evolution
1101 E. 57th Street
University of Chicago, Chicago, IL 60637, USA

JAMES S. QUINN
Department of Biology
McMaster University
Hamilton, Ontario L8S 4K1, Canada

S. JAMES REYNOLDS
School of Biosciences
The University of Birmingham
Edgbaston, Birmingham B15 2TT, UK

ANDREW F. RUSSELL
Department of Animal and Plant Sciences
University of Sheffield
Sheffield S10 2TN, UK

KARIN SCHIEGG
Zoological Institute
University of Zurich
Winterhurerstr. 190, CH-8057 Zurich, Switzerland

STEPHAN J. SCHOECH
Department of Biology
University of Memphis
Memphis, TN 38152-3540, USA

SANDRA L. VEHRENCAMP
Department of Neurobiology and Behavior
Cornell University
Ithaca, NY 14850, USA

JEFFREY R. WALTERS
Department of Biology
Virginia Polytechnic Institute and State University
Blacksburg, VA 24061-0406, USA

Introduction

WALTER D. KOENIG AND JANIS L. DICKINSON
University of California, Berkeley

Cooperative breeding continues to engender considerable interest among behavioral ecologists. However, the players and issues have changed dramatically since the publication of the first *Cooperative Breeding in Birds* volume (Stacey and Koenig 1990a). Back then, a series of long-term demographic studies were coming to fruition, opening the door for a synthetic volume that would "search for common themes and patterns" while illustrating "the great diversity that exists among cooperatively breeding birds" (Stacey and Koenig 1990b). At the time it appeared that the "common themes and patterns" would outstrip the "great diversity" and that a general understanding of the main issues raised by the phenomenon of cooperative breeding was about to be achieved (Emlen 1997a).

Such optimism concerning a general answer to the paradox of helping behavior was quickly dismissed (Cockburn 1998), and it has continued to elude our grasp. Instead, new theoretical approaches and studies have emerged to reinvigorate the field. Three stand out in particular. First is DNA fingerprinting, which was just getting started in the late 1980s and was only minimally represented in the 1990 volume. Multilocus minisatellite fingerprinting and its descendant, microsatellite fingerprinting, provided the long-sought-after ability to determine parentage and estimate relatedness. Fingerprinting allowed those who were continuing long-term studies or who had been fortunate enough to collect and save blood samples either to confirm prior inferences regarding patterns of parentage (as in Florida scrub-jays and acorn woodpeckers: Quinn *et al.* 1999; Dickinson *et al.* 1995; Haydock *et al.* 2001) or to turn all prior inference on its head (as in the splendid fairy-wren: Brooker *et al.* 1990). This latter case was particularly dramatic, since it made what was already a perplexing mating system (Rowley *et al.* 1986) even more extraordinary. More importantly, it raised questions about all other presumptions about paternity based on banding of cooperative breeders: no longer would it be possible to assume that the mating system of a cooperative breeder would necessarily bear close correspondence to the demographically observed social unit.

Unfortunately, inferring parentage (as opposed to performing paternity exclusion analyses) is still not easy in birds, particularly in cooperative breeders, where potential sires (or dams) are close relatives. Consequently, the number of studies of cooperative breeders with unambiguous data on parentage is still relatively small. However, the conclusion from studies performed thus far, discussed by Cockburn (Chapter 5), is clear: diversity rules. Explaining this diversity remains a challenge, and is likely to become even more difficult as additional data on other species become available.

Second has been the consistent failure of attempts to predict the occurrence of cooperative breeding based on ecological features or life-history characteristics (Dow 1980; Yom-Tov 1987; Brown 1987; Ford *et al.* 1988; Du Plessis *et al.* 1995; Cockburn 1998). This is not to say that ecological factors are unimportant (Chapter 3), or that cooperative breeders do not share a variety of ecological and life-history characteristics (Chapter 14). However, many of the characteristics shared by cooperative breeders, such as year-round residency, prolonged dependence of offspring, and even ecological constraints on dispersal, are found in many non-cooperative breeders as well. In other words, we can often do a reasonable job of answering the question of why a particular species is a cooperative breeder, but we continue to be abject failures at offering a convincing explanation for why many other species are *not* cooperative breeders.

The third, and perhaps the most important, factor generating renewed excitement in the field of

---

Ecology and Evolution of Cooperative Breeding in Birds, ed. W. D. Koenig and J. L. Dickinson. Published by Cambridge University Press.
© Cambridge University Press 2004.

cooperative breeding has been the new generation of field studies that began yielding important results in the 1990s. Notable among these was work on the Seychelles warbler, the long-tailed tit, the Siberian jay, and the onslaught of work on various Australian cooperative breeders seemingly competing to be designated "most bizarre," including the inimitable fairy-wrens, the white-winged chough, the white-browed scrubwren, noisy and bell miners, the eclectus parrot, and more. These systems simply cannot be assimilated into prior frameworks concerning the evolution of cooperative breeding based on work summarized in Stacey and Koenig (1990a).

The bottom line is that we have more questions, and fewer answers, to the central questions in the field of cooperative breeding than we did a decade ago. Furthermore, the field has progressed conceptually as well as empirically, leading to novel ways of analyzing new genetic and old demographic data. As the genetic data and their interpretations are not yet available for many of the newer studies, we felt that a thematic volume based on major concepts and issues was more timely than a follow-up compilation focused on individual species. The current volume is the result of this effort.

Several of these theoretical issues are addressed explicitly. A good example is Jamieson's (1989, 1991) "unselected hypothesis," which was just gathering steam (and controversy) as Stacey and Koenig (1990a) went to press. Although hammered at the level of functional consequences (Koenig and Mumme 1990; Emlen et al. 1991; Ligon and Stacey 1991), it has returned, stronger than ever, at the level of evolutionary, or phylogenetic, origins, and is discussed in detail by Ligon and Burt (Chapter 1).

At least two conceptual issues addressed here owe much of their recent development to advances in molecular biology similar to those that now allow determination of parentage. The first is the problem of sex allocation, an area poised for an explosion now that sexing techniques in birds have become relatively cheap and easy. Although research exploiting this breakthrough is still young, cooperative breeders are positioned to play a key role in testing hypotheses for sex allocation, an area that has continued to interest and befuddle workers ever since Fisher (1930) laid down the theoretical foundation that currently defines the field. Progress in this area is summarized by Komdeur (Chapter 6).

The second is how reproduction is partitioned among individuals within social groups. This may or may not be an issue among the "simpler" cooperative breeders in which groups consist of pairs with non-breeding helpers that are constrained in their reproductive activities by incest avoidance (Chapter 9). However, things become considerably more complicated in species in which groups contain more than one potential breeder of one or both sexes. In fact, even describing such systems can be a challenge.

Compare three groups of acorn woodpeckers, each of which contains one breeder female and two males. In group 1, male 1 is an unrelated immigrant from elsewhere that bred with the female the previous year and produced one surviving male offspring that stayed in the natal group and became male 2. In the other two groups, the two males are brothers that immigrated into the group together. All groups breed. In group 1, male 1 sires all the young, since male 2, the helper, is constrained from breeding by incest avoidance (Chapter 9). In groups 2 and 3, neither male is constrained by incest avoidance and both mate-guard and attempt to mate with the female. In group 2 only male 1 is successful in siring young in the nest, whereas in group 3 there is multiple paternity and both males successfully sire offspring. Group 1 is a standard cooperatively breeding group with a single non-breeding helper male, while group 3 is a cooperatively polyandrous group with two cobreeder males. But where does group 2 fit in?

Both males in group 2 were potential mates of the female, even though one failed to sire any offspring. In terms of his genetic contribution, this unsuccessful male is equivalent to the non-breeding helper in group 1, since neither sired any offspring in the nest. Both are related to the nestlings indirectly through male 1 (to which both male 2s are genetically related).

The two males do, however, differ in two ways: relatedness to the chicks, which is higher for the non-breeding helper since he is also related to the nestlings through the breeder female, and copulatory access to the female, which the potential cobreeder may have had even though he was not successful in siring offspring. Unless the potential cobreeder has perfect information regarding his paternity in the nest, his behavior toward the nestlings should be influenced by the possibility that he may have sired at least some offspring (even if he did not).

In contrast, the non-breeding helper has been exposed to strong selection to avoid engaging in reproductive activities with his mother because of incest avoidance, and his treatment of nestlings should not be affected by his mating access.

Such complexities continue to result in considerable differences in the field. This starts immediately with the definition of cooperative breeding, defined inclusively by Cockburn (Chapter 5) to include all three hypothetical groups, but more exclusively by Ligon and Burt (Chapter 1) to include only groups containing non-breeding helpers. This latter definition clearly eliminates our hypothetical group 3; how it deals with the problem of group 2 is less clear.

In any case, cooperative breeders in which groups contain more than one potential breeder raise the theoretically important issue of how reproduction is partitioned. This field of "reproductive skew" was originally developed by Vehrencamp (1979, 1983a, 1983b) well before Stacey and Koenig (1990a). However, relatively little could be done empirically with skew theory until methods of determining parentage were developed. Availability of parentage data led to an explosion of interest, both empirically and theoretically. The impact of reproductive skew theory on our understanding of cooperative breeding systems is addressed extensively by Magrath et al. (Chapter 10) and by Vehrencamp and Quinn (Chapter 11), who focus more generally on joint nesting systems.

Other chapters presented here focus on issues that were controversial in Stacey and Koenig (1990a) and have remained so since. Why, in cooperative breeders, do helpers delay dispersal? And why, once dispersal is delayed, do they help? A general answer to the first of these questions once appeared to be within our grasp. This answer involved "ecological constraints," which were poised as a major factor in the evolution of cooperative breeding despite some controversy (Stacey and Ligon 1987, 1991). Although "ecological constraints" are clearly important in many cooperative breeding species, non-complementary alternatives have since surfaced, including nepotism and other "benefits of philopatry" that appear to be particularly important in species with delayed dispersal and no helping behavior. Ekman et al. (Chapter 2) bring us up to date on this important issue.

But what about helping behavior itself? At the time of Stacey and Koenig (1990a), the major issue was the importance of kin selection (indirect fitness benefits), brought to the forefront because the vast majority of cooperative breeding systems are family-based. Yet direct fitness benefits may be far more important than previously suspected, an hypothesis explored by Heinsohn (Chapter 4). Still there is debate over the relative importance of direct and indirect benefits and the quality of evidence for various costs and benefits of helping behavior that have been addressed over the years, as evidenced by the different viewpoints taken by Heinsohn (Chapter 4) as compared to Dickinson and Hatchwell (Chapter 3).

A long-standing issue that is revisited in this volume is that of incest, which is a potential problem due to the high relatedness among group members in most cooperative breeders. Does this result in rampant inbreeding, or at least a higher incidence of incest than in non-cooperative species? Although controversy remains, recent studies, many making use of molecular techniques to determine parentage, have in general presented a unified front supporting a central role of incest avoidance as a determinant of reproductive roles in cooperative breeding societies. The saga leading to this conclusion, along with a discussion of studies and investigators challenging this interpretation, is discussed by Koenig and Haydock (Chapter 9).

One of the more important ways that the study of cooperative breeding has diversified since Stacey and Koenig (1990a) has been its expansion into questions directed at levels of analysis other than that of ultimate fitness consequences. Besides evolutionary origins, discussed by Ligon and Burt (Chapter 1), the role of physiological constraints in cooperative breeders is summarized by Du Plessis (Chapter 7), while the hormonal correlates of cooperative breeding are reviewed by Schoech et al. (Chapter 8). The latter, in particular, offer several excellent examples in which physiological traits are modified to facilitate helping behavior, a finding that counters the original "unselected hypothesis": regardless of how it originated, helping behavior is clearly under strong selection in many species and is correlated with numerous physiological adaptations.

Two additional issues, largely ignored in Stacey and Koenig (1990a), are covered in detail here. First, Walters et al. (Chapter 12) discuss reasons why cooperative breeders are of particular interest to the emerging field of conservation biology and how these species are faring relative to non-cooperative breeders in the face of expanding threats of habitat loss and fragmentation. As they point out, many cooperative breeders exhibit traits that potentially make them uniquely vulnerable to such threats, including philopatry, small population size, and specific habitat requirements. On the other hand, populations of cooperative breeders typically contain relatively large numbers of "extra" adults in the form of nonbreeding helpers, which can in some cases buffer against the effects of demographic stochasticity. Whether these and other life-history characteristics make cooperatively breeding species more or less vulnerable to habitat loss and fragmentation is an important issue that Walters *et al.* discuss for the first time.

Second is work that has been done on mammals. Although the chapters in Stacey and Koenig (1990a) were restricted to avian systems, studies in other taxa have contributed significantly to our understanding of cooperative breeding, to the extent that a parallel volume devoted to mammalian cooperative breeding was published several years later (Solomon and French 1997). Acknowledging these contributions, we enlisted Russell (Chapter 13), one of the few workers to have experience in both avian and mammalian cooperative systems, to discuss ways in which study of the latter has contributed, both theoretically and empirically, to our understanding of cooperative breeding in general.

We conclude with a summary by Pruett-Jones (Chapter 14), who generates a series of 13 synthetic statements about cooperative breeding with which all workers in the field, or at least the majority, can agree. Although not the synthesis that seemed so close back in 1990, his chapter offers as close to a set of common patterns among cooperative breeders as has ever been conceived, leaving considerable hope that a general understanding of this phenomenon may exist after all, despite the ever greater diversity being discovered in such systems.

We did not start out with the goal of either excluding contributors to Stacey and Koenig (1990a) or highlighting younger workers. However, many of the new ideas and data that have continued to draw attention to the field have come from a new generation of investigators, as evidenced by the relatively low overlap between the two volumes, which share only four authors in common. This high proportion of "new blood" is part of what has kept the field of cooperative breeding dynamic and active. It has also helped generate new controversies, many of which are highlighted in the chapters presented here. Our hope is that these chapters, and the alternative viewpoints they present, will provide yet another generation of students with the same kind of excitement and inspiration that we experienced when first discovering this field.

1 · Evolutionary origins

J. DAVID LIGON
University of New Mexico

D. BRENT BURT
Stephen F. Austin State University

Cooperative breeding (hereafter often abbreviated as CB) is an umbrella label that includes a diverse array of mating and social systems (Ligon 1999). For example, Brown (1987) lists 13 separate categories of CB (see also Chapter 5). The variability in the forms of CB is due to differences in both the strength and the forms of selection on helping behaviors, mating strategies, and other aspects of group living. Here we follow the commonly employed definition of avian cooperative breeding, which is that it involves the existence of social units composed of two or more breeding birds, plus one or more (often presumed) non-breeding "helpers-at-the-nest" (Brown 1987; Edwards and Naeem 1993). It is the feeding of young birds by the helpers – also referred to as alloparental behavior – that characterizes cooperative breeding and that has made it of singular interest.

For most of the history of CB studies, researchers have sought ecological factors that might have promoted the evolutionary development of CB. This search has met with limited success, in part because ecological and climatic considerations, in themselves, offer little predictive power beyond the fact that north-temperate-zone species are unlikely to be cooperative breeders (Heinsohn *et al.* 1990; Mumme 1992a; Cockburn 1996). Even in tropical and subtropical areas, where cooperative breeders occur most frequently, one typically cannot offer a good guess, based solely on environmental conditions, as to whether or not a given species will prove to exhibit CB. The only factor that does provide good predictive power is whether the species in question has cooperatively breeding relatives. This suggests that phylogenetic history may be a critical consideration in any attempt to address the origins and, to a lesser extent, the maintenance of cooperative breeding.

IDENTIFYING COOPERATIVE BREEDING AND THE ISSUE OF HOMOLOGY

Some writers have lumped a wide array of social and genetic mating systems under the label of cooperative breeding (Brown 1987; Hartley and Davies 1994; Arnold and Owens 1998, 1999). This is understandable to the extent that the social and sexual relationships among members of a group are often not well known. In some cases, individuals that first were assumed to be non-breeding helpers have, with the use of molecular techniques, been shown to breed, albeit rarely (Rabenold *et al.* 1990; Haydock *et al.* 1996). This dichotomy between actual non-breeding helpers (usually the offspring of one or both members of the breeding pair) and would-be breeders is clearly seen in pied kingfishers. In this species, "primary" helpers typically are offspring of the nesting pair and they do not attempt to mate with a parent. In contrast, "secondary" helpers are unmated, unrelated males that may, depending on circumstances, form a pair bond with the breeding female at a later date (Reyer 1990). Both primary and secondary helpers deliver food to nestlings.

In other social mating systems, all members of a social unit are breeders or potential breeders; the "goal" for each group member is actual parentage. For example, in dunnock groups all members are actual or hopeful breeders (Davies 1990). There are no non-breeding "helpers," even though a beta male may not have sired any offspring during a particular nesting attempt (see also Chapter 5). The term polygynandry more accurately labels the dunnock's unusually variable social-mating system than does cooperative breeding.

In still other cases, both non-breeding helpers and breeders or would-be breeders occur in the same social

---

Ecology and Evolution of Cooperative Breeding in Birds, ed. W. D. Koenig and J. L. Dickinson. Published by Cambridge University Press.
© Cambridge University Press 2004.

unit (Haydock et al. 1996). If there has been no selection to preferentially feed one's own chicks, one could argue that the role played by the prospective breeders is no different than the role of the true helpers, despite the fact that any selective benefits may differ: both feed nestlings that are not their own offspring.

The rule we follow here for including a given species is that non-breeding helpers occur within a social unit beyond the primary pair, irrespective of the presence or absence of potential breeders. This approach is weakened by the scanty knowledge we have of genetic parentage in most species that appear to breed cooperatively. We feel that this weakness is offset, however, by obtaining a clearer focus on the phenomenon of interest here, the feeding of chicks by individuals that have little or no possibility of parentage within the brood they are provisioning.

Another important point relates to the issue of homology. Is the CB reported for an ecologically and taxonomically diverse array of species homologous? In other words, is CB across different species and lineages derived from a common ancestor, or has it appeared *de novo* in different lineages? This is one of the most interesting and difficult questions we attempt to address in this chapter. We argue below that for altricial groups, the answer ultimately depends on whether or not altriciality evolved one or more times. If the altriciality of the groups we consider is derived from a common ancestor, then it would be appropriate to view the concomitant intense parental care shared by these groups as homologous.

Conversely, if it could be shown that altriciality evolved separately from precocity in two or more of these lineages (the coraciiform and passeriform birds, for example), one might argue that the associated parental care exhibited by these two groups reflects analogy rather than homology. In either case, we argue that the intense parental care associated with altricial lineages predisposed individuals to alloparental care, given close proximity of non-breeders and begging young. In other words, altriciality and alloparental care evolved essentially in concert, but alloparental care (excluding the hosts of social parasites) is normally unexpressed in descendant lineages in which individuals typically have no close contact with young birds that are not their own offspring.

EVOLUTIONARY ORIGIN VERSUS EVOLUTIONARY MAINTENANCE

The issue of its evolutionary origin has been largely ignored for most of the modern history of the study of CB. Rather, the level of analysis (Sherman 1988) on which most students of this phenomenon focused was the current adaptive significance of CB, sometimes assuming that the environmental factors promoting or maintaining CB in the particular species they studied also accounted for its evolutionary origin.

The appearance of a number of publications that considered phylogenetic history (Russell 1989; Peterson and Burt 1992; Edwards and Naeem 1993; Ligon 1993, 1999; Farley 1995; Burt 1996; Cockburn 1996) clearly demonstrated the importance of distinguishing between evolutionary origins of CB and current maintenance of this trait. Why is this important? First, identifying the patterns of CB evolution provides us with opportunities for further study. For example, are certain environmental, behavioral or life history features associated with the origin or expression of CB? Second, when the ecological correlates associated with CB change, do we see a subsequent loss of CB? If so, this pattern implies that specific ecological factors play an important role in the maintenance of CB. Alternatively, if transitions from CB to non-CB do not occur under different ecological conditions, three interpretations are possible: (1) specific ecological settings are not a primary factor in the maintenance of CB as an adaptive social system, (2) CB is adaptive in different ways in a variety of ecological circumstances, or (3) CB is not adaptive in at least some of the species exhibiting it (Ligon and Stacey 1989).

An evolutionary framework also provides a fresh perspective on the interaction between the two most widely recognized aspects of CB, delayed dispersal and helping behavior. For example, life-history characteristics associated with delayed dispersal have recently been identified as important in the origins of CB (Arnold and Owens 1998). However, alloparental care may initially have been nothing more than a response to the stimuli of begging nestlings (Jamieson and Craig 1987a; Jamieson 1989). In such cases, although the breeding system fits the definition of CB, at this initial evolutionary stage CB as a "trait" is simply an epiphenomenon of delayed dispersal. When alloparental care subsequently became adaptive in certain group-living lineages, the various forms of CB could be labeled as "exaptations" (Gould and Vrba 1982). That is, delayed dispersal was the original, adaptive response to particular ecological or physiological circumstances that provided the opportunity for alloparental behavior among related individuals, but subsequent benefits associated with helping behaviors give CB a new exaptive role.

The initial evolution of intense parental care associated with production of altricial young, together with group living, was the raw material for the subsequent adaptive development of CB. Ecological factors over the tens of millions of years from the early Tertiary to the present have modified this behavior in many ways, including, for a majority of altricial lineages, the absence of strong alloparental tendencies, or at least the absence of the regular expression of the behavior. However, in the ancestors of other species, those recognized today as regular or frequent cooperative breeders, the feeding of nestlings by non-parents set the stage for the development of a whole suite of adaptive modifications associated with CB, many of which are treated in this volume.

EVOLUTIONARY ORIGINS OF COOPERATIVE BREEDING

The origins of altriciality

Because the initial appearance of intense parent care, including parental feeding, must have been critically linked to the altricial mode of chick development, we first consider the origins of altriciality. Traditionally, the usual assumption has been that among birds as a whole precocity was the evolutionary precursor of altriciality (Gill 1995). However, Starck and Ricklefs (1998; Ricklefs and Starck 1998) mapped chick developmental mode onto the phylogeny of Sibley and Ahlquist (1990) and concluded that altricial development is probably ancestral for the infraclass Neoaves (which includes all modern bird lineages) except the ratite–anseriform clade and the turniciform lineage (Fig. 1.1). Ricklefs and Starck (1998) suggested that within the Neoaves precocity has re-evolved in both the superorder Strigimorphae and the common ancestor of the orders Gruiformes and Ciconiiformes. Altriciality has then again re-evolved numerous times within the Ciconiiformes. These multiple evolutionary transitions probably account for the variability along the altriciality–precocity spectrum among living groups of Neoaves. For purposes of this chapter, the key point is that while altriciality may or may not be ancestral in birds as a whole, it probably is ancestral in all but the two most basal major avian lineages (Fig. 1.1).

Did types of birds likely to produce altricial young exist during the early history of modern birds? Avian evolution during the Paleocene and early Eocene apparently was explosive, with most modern types except the passerines appearing in the fossil record between the end of the Cretaceous and the lower Eocene, a period of only about 13 million years (Feduccia 1996). Many small arboreal or aerial species existed by this time (Mayr 2000, 2001), which strongly suggests that fully altricial young had evolved even earlier. This is because chicks of such species probably could not have been sufficiently precocial and mobile at hatching to accompany their parents as they foraged. In fact, the altricial condition may have initially evolved in response to the development of arboreal and aerial lifestyles of small, actively feeding lineages (Ricklefs and Starck 1998). In short, specialized parental care, including the delivery of food to the mouths of nestlings, was a key requisite for the evolution of altricial young and, based on the types of birds present at that time, probably was already well developed by the early Tertiary.

In summary, the analyses of Starck and Ricklefs lead to the conclusion that altriciality is ancient and, by implication, that intense parental care of helpless young is also an ancient adaptation. Finally, in support of this point, a number of altricial groups (the parvclass Coraciae, including coraciiforms, galbuliforms, bucerotiforms, upupiforms, trogoniforms, as well as the piciforms and the coliiforms) are among the oldest neoavian lineages with living descendants (Sibley and Ahlquist 1990). Some of these groups contain a number of species that breed cooperatively (Fig. 1.1).

We used the concentrated changes test (Maddison 1990) in MacClade 4.0 (Maddison and Maddison 2000) to test our assertion that altriciality influences the evolution of CB. Multiple equally parsimonious reconstructions between breeding system states were found and optimization options were utilized to demonstrate the

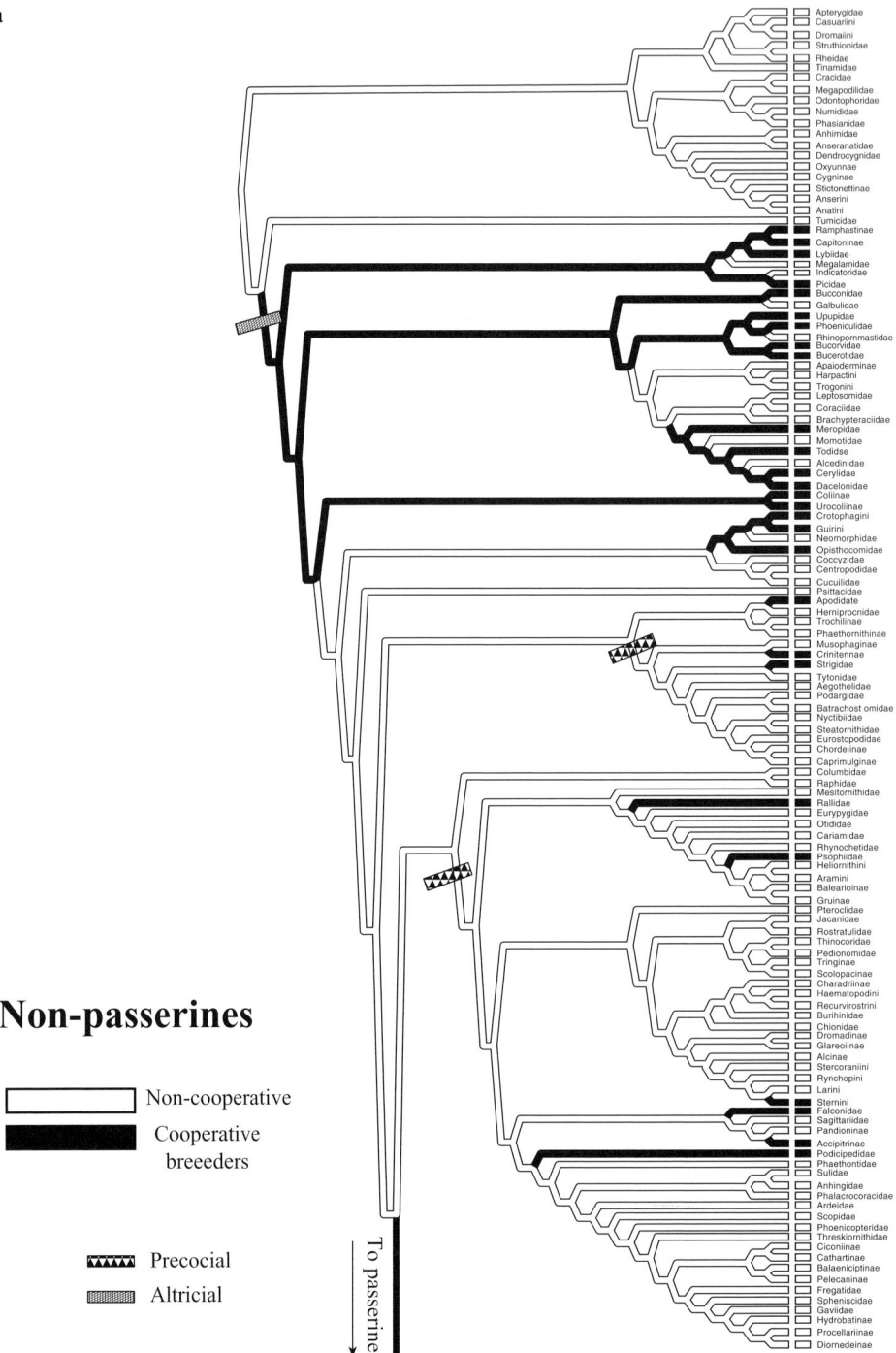

Figure 1.1. Reconstructed evolutionary transitions in breeding systems and developmental modes between lineages on the phylogeny of Sibley and Ahlquist (1990). Reconstruction on the (a) non-passerine portion and (b) passerine portion of the avian tree. Assuming this ACCTRAN reconstruction, transitions from non-CB to CB occur more frequently ($P = 0.10$) on altricial branches than one would expect at random (concentrated changes test, Maddison 1990). The DELTRAN reconstruction (not shown) indicates a highly significant concentration of CB gains on altricial branches ($P = 0.01$).

Figure 1.1. (cont.)

range of reconstructions that placed transitions from non-CB to CB either as close to the base of the tree as possible (ACCTRAN option, Fig 1.1) or as close to the tips of the tree as possible (DELTRAN option, not shown).

The ACCTRAN reconstruction of breeding system evolution (Fig. 1.1) shows 28 gains of CB (non-CB to CB) and 20 losses of CB (CB to non-CB). Of the 28 gains, 20 are found in altricial lineages. The DELTRAN reconstruction produces 38 gains of CB and 12 losses of CB, with 30 of these gains found in altricial lineages. The concentrated changes test suggests transitions from non-CB to CB lineages occur more frequently in lineages having altricial development than one would expect if these traits evolved independently. That is, our reconstructions show more gains of CB in altricial lineages than would be expected if breeding system and developmental mode were evolving randomly relative to each other.

This test does not include likely transitions between developmental traits within the Ciconiiformes. However, given that a number of these lineages have likely evolved altriciality and that we currently have them reconstructed as precocial, our tests are conservative. The evolutionary pattern verified by these tests, that there is concordance between altriciality and CB, is not surprising. However, we feel the evolutionary relationship between CB and altricial development has often been underappreciated by past researchers.

Origins of sociality in certain lineages of cooperative breeders

A key aspect of the scenario we present here is that cooperative breeding may initially have arisen more or less incidentally in response to the evolution of altricial young and the existence of factors favoring group living. In the prior section we discussed the relationship between altriciality and the concomitant intense parental care that it demands. Here we offer some suggestions concerning the initial factors leading to social living.

In terms of percentage of species of a particular lineage exhibiting this behavior, CB is most prevalent in certain families of rather small, primarily arboreal groups of birds within the ancient orders Coliiformes, Coraciiformes, Upupiformes, Bucerotiformes, and Piciformes (Fig. 1.1).

What traits might have predisposed these ancient and primarily or exclusively tropical birds to live in groups and possibly subsequently to become cooperative breeders? Physiological limitations of one sort or another may be a primary factor that led to group-living in some of these groups, at least some of which have unusually low basal metabolic rates. Energy conservation by cavity- and group-roosting green woodhoopoes is one example (Boix-Hinzen and Lovegrove 1998). Individuals of the highly social speckled mousebird also have been shown to benefit greatly both by group-clustering during the day and by group-roosting at night (McKechnie and Lovegrove 2001a, 2001b). An apparently similar relationship between group-roosting, social living, and CB can be seen in certain groups of cooperatively breeding passeriform birds, including the family Pomatostomidae and the genera *Turdoides*, *Campylorhynchus*, and *Daphoenositta*. Further discussion of the potential interplay between physiological limitations and the evolution of cooperative breeding can be found in Chapter 5.

To summarize, for cooperative breeders in several tropical non-passerine groups, one response to low nocturnal temperatures, or, in colies, even low diurnal temperatures, is social roosting or clustering in order to conserve energy. Although data are few, it is becoming increasingly clear that in some groups of tropical and subtropical birds the correlation between latitude and cooperative breeding (Brown 1974) is strongly affected by the relationship between social living and the behavioral and physiological characters affecting energy balance.

Origins of alloparental behavior

What factors promoted the initial appearance of cooperative breeding in the ancestors of today's cooperative breeders? Addressing this question requires consideration of the evolution and care of altricial chicks. Here we modify and extend the arguments of Jamieson and Craig (1987a), Jamieson (1989, 1991) and Ligon and Stacey (1989, 1991) concerning the origins of alloparental feeding.

We envision two likely evolutionary routes to CB. In the first, alloparental care is initially an epiphenomenon of delayed dispersal (generalized feeding response plus access to the stimuli of begging chicks), with subsequent modifications due to one or more of several possible selective advantages. This route is most applicable to territorial species living in situations where ecological conditions of one or more kinds either provide benefits to natal philopatry or limit the option of immediate dispersal, or both (Stacey and Ligon 1987,1991; Koenig et al. 1992; Komdeur 1992).

The second evolutionary route is initiated with the occasional, originally non-adaptive, feeding of chicks by non-parents, similar to that seen today in species often referred to as rare, opportunistic, facultative, or irregular cooperative breeders. Often these "pseudohelpers" will be failed breeders, stimulated by the sound of nearby begging offspring. This route may be particularly relevant for colonial species, such as the bee-eaters we discuss below. Note that in this second route, CB first appears due to the adaptive nature of parental care, without reliance on ecological factors that promote delayed dispersal.

Over time, in the diverging lineages of altricial birds, natural selection (1) largely eliminated this non-discriminating tendency to feed young birds or (2) did not completely eliminate this generalized tendency, as seen in contemporary irregular cooperative breeders and in the numerous cases of interspecific feeding, including the hosts of social parasites, or (3) favored adaptive refinements which led to the diversity of sophisticated CB social systems present today. In the latter two scenarios, the essential precursor of CB as exhibited by living species was the "hard-wired" feeding of chicks that necessarily coevolved with the development of altricial young. The importance of this hard-wired feeding response cannot be overemphasized as the primary contributing factor explaining the initial evolution of allofeeding in CB systems. This response also likely accounts for the allofeeding behavior seen outside CB systems. For example, interspecific feeding is surprisingly common (Shy 1982; Skutch 1999), and the feeding of nestlings by non-parents may occur as a freak event in almost any kind of altricial bird (Eltzroth and Robertson 1984; Welty and Baptista 1988, Fig. 17-7). The occurrence of interspecific feedings of chicks provides perhaps the best evidence that proximity to begging chicks can stimulate or trigger alloparental feeding in species that normally do not exhibit such behavior. Brood parasites and their hosts offer additional examples.

When "helping" behavior involving two different species is observed, it is clearly non-adaptive to the individuals providing the help. In contrast, when the feeding of chicks by a conspecific is recorded, an adaptive explanation may be too readily invoked, even in species in which an observation of intraspecific alloparental feeding has been reported only once or twice. Thus, here we confine our attention to avian taxonomic groups with altricial young and in which apparently non-breeding helpers are believed to be a regular aspect of the biology of one or more species.

EVOLUTIONARY MAINTENANCE OF COOPERATIVE BREEDING

Some authors have made a distinction between CB in which helping is frequent or regular and those in which is it rare or irregular. We believe that recognizing this dichotomy makes the history and current significance of CB more amenable to study. Here we briefly present our views of the origins and current adaptive significance (or lack of it) of the two. We begin with regular CB because this includes nearly all well-studied CB species.

Regular cooperative breeders

A few frequent or regular CB species such as white-winged choughs (Heinsohn 1991c) are "obligate" in that simple pairs never breed successfully. Chough groups composed of fewer than seven individuals cannot bring a young bird through its first winter. In most regular CB species, however, both simple pairs and cooperatively breeding groups occur in the same population, and sometimes at about equal frequencies. Thus the labels "frequent" or "regular" CB do not imply that unassisted pairs never breed successfully. Rather, they indicate that CB is a common aspect of the biology of the species in question.

In many cases, simple pairs make up a minority of the reproductively successful social units, and they tend to be less successful by various measures and for various specific reasons than are larger groups. Not surprisingly, attempts to ascertain the adaptive significance of frequent CB in individual species almost always leads to identification of benefits associated with this behavior (Stacey and Koenig 1990a). For this and other forms of group living to persist, there must be benefits to sociality, if not to alloparental care per se, that override its costs (Alexander 1974), and enterprising investigators can identify many of those benefits. Until recently, an unfortunate effect of this approach was to reinforce the notion that identification of specific benefits, which vary from species to species, can explain the origin of CB in species exhibiting those benefits.

Irregular cooperative breeding

Irregular CB includes species in which an extra bird has been reported to feed conspecific nestlings on one or two occasions. In contrast to regular CB, selection has not operated in a directional way to refine this behavioral syndrome. Rather, the *absence* of strong directional selection either promoting or eliminating it has allowed alloparental feeding to persist over evolutionary time as a rare and possibly aberrant behavior.

Like regular CB, irregular CB appeared first as an unselected epiphenomenon based on rare or occasional alloparental care of altricial chicks. Examples include the hooded warbler and ovenbird (Tarof and Stutchbury 1996, King *et al.* 2000). In most such cases, too little is known to warrant detailed interpretation, but it is likely that the extra bird is often a male attracted to the female member of the pair. Other than possible parentage (which falls outside the phenomena considered here), one would be hard-pressed to come up with a convincing adaptive explanation for most cases of irregular CB (Ligon and Stacey 1989, 1991).

ECOLOGY AND COOPERATIVE BREEDING: PROBLEMS OF CAUSE AND EFFECTS

Over the past several decades, a number of workers have attempted to identify the general ecological factors that promoted the evolutionary development of CB. However, no consensus has been reached (Cockburn 1996). For example, Ford *et al.* (1988) suggest that in Australia CB is favored by aseasonal environments, while Du Plessis *et al.* (1995) conclude that in South Africa regular CB is associated with seasonal environments. Both probably are correct about the correlation between climate and CB in Australia and South Africa respectively, but we doubt that these correlations are causal. Rather, both environments, which are relatively benign, permit the retention of CB that had evolved earlier, but in most cases they probably are not the basis for the origins of CB among contemporary species in either region.

Arnold and Owens (1998, 1999) recently produced a comparative analysis of the relationship between life-history characteristics and CB, concluding that CB developed primarily as a result of decreased annual mortality associated with living in warm, stable environments. These lineages then evolved increased sedentariness, which leads to saturation of breeding habitats and additional reductions in population turnover.

Arnold and Owens' studies illustrate one of the most frequent and long-standing problems in the field of CB, namely, the issue of cause and effect. For example, they argue that certain environmental features, including a stable and warm climate, may lead to the evolution of CB in certain lineages, via lowered mortality. We do not question the existence of a correlation between the frequency of CB and a relatively mild, aseasonal climate: this has been recognized for many years (Rowley 1968, 1976). However, we are dubious about a causal relationship between low mortality and the original development of CB (Poiani and Jermin 1994).

It is widely recognized that social living in general is often a response to predation pressures (Alexander 1974). We believe that this is often the case for cooperative breeders, and thus that the group-living aspect of CB leads to lowered mortality, rather than the other way around (Stacey and Ligon 1987). Many sorts of adaptive features, including several related to deterrence of predation, are associated with cooperatively breeding species. As one example, sophisticated sentinel behavior is a well-documented benefit of group living (Gaston 1977; McGowan and Woolfenden 1989; Hailman *et al.* 1994). Other aspects of the biology of cooperatively breeding species may also reduce their mortality as compared to their non-cooperative relatives (Noske 1991).

In short, we suggest that adaptive behaviors associated with group living, such as sentinel behavior, cause lowered mortality, rather than lower mortality promoting group living. Moreover, tropical and subtropical species often exhibit low mortality relative to temperate species, whether or not they are cooperative breeders (Fry 1980; Rowley and Russell 1997). Straightening out the potential circularity of this issue is admittedly not an easy task.

With regard to the relationship between environmental or climatic factors and CB, we suggest that CB appeared in several lineages in the early Tertiary, when the climate of most of the planet was warm and aseasonal as compared to today's world, and that in some lineages the ancient trait of CB has persisted over time in geographic regions which probably retained comparatively benign environments throughout the Cenozoic. In short, the issue is: do warm, stable environments somehow promote the repeated evolution of CB among living species, as suggested by Arnold and Owens (1999) and some earlier authors, or have they merely permitted its retention and adaptive refinement in some of the lineages occupying such habitats? We believe that employing this perspective will contribute to the understanding of CB in many, but not all, lineages. As discussed below, It also seems clear that CB has evolved, or become re-expressed from a retained ancestral condition, in one or several closely related species of basically non-cooperative lineages.

The scenario suggested here may provide an explanation for the difficulty in identifying ecological factors that favor the evolutionary development of CB. Rather than assuming that particular ecological or demographic factors promoted the origin of this phenomenon, we reiterate that the key characteristic of cooperative breeding – the feeding of non-offspring – developed as an epiphenomenon of evolved parental care for altricial nestlings, channeled by factors that affected dispersal. This may have occurred early in the diversification of altricial lineages in the more climatically benign world of the Lower Tertiary. In addition, the inclination, or more likely the opportunity, to feed non-offspring was almost completely lost in many, but not all, altricial lineages. This includes a large majority of the passerine parvorder Passerida, which is one of today's largest avian groups. Given this general pattern, we suggest that to better understand the relationship between CB and ecological factors, in most cases it would be more profitable to examine the ecological contexts in which CB systems are lost than to attempt to identify one or more environmental correlates responsible for the evolutionary origin of CB.

In summary, despite considerable effort, the goal of identifying ecological factors that predictably correlate with CB has remained elusive (Heinsohn *et al.* 1990; Cockburn 1996). The ideas offered here about the evolutionary origins of CB may contribute to the resolution of this difficulty.

PHYLOGENETIC PATTERNS AND COOPERATIVE BREEDING

Sibley and Ahlquist (1985) made the revolutionary discovery that oscine passerines of Australia belong to one of two major groups, or parvorders. One of these, their parvorder Corvida, had undergone its earliest evolutionary radiation in Australia (the "old endemics"), while the other, the parvorder Passerida (Sibley and Ahlquist 1990), is thought to be a more recent arrival from Eurasia. Following up on this discovery, Russell (1989) pointed out that in Australia only one of these two major passerine radiations contains any cooperatively breeding species at all. While CB occurs relatively frequently in the Corvida, it is totally absent in the Australian Passerida. This dichotomy provides striking evidence to counter the hypothesis that environmental factors are sufficient to account for the relative frequency of CB on that continent (Rowley 1965, 1968; Harrison 1969; Ford *et al.* 1988).

Edwards and Naeem (1993) published the next breakthrough in linking CB to evolutionary history. These authors analyzed the occurrence and distribution of cooperative breeding in 71 polytypic genera that contained at least one cooperative breeder, and compared its incidence in each genus with a random distribution among these genera. They did not deal with the various forms of cooperative breeding; therefore their analysis included both species in which it is regular and those in which it is irregular. Edwards and Naeem's results indicate that the most parsimonious assumption is that cooperative breeding in several lineages arose prior to many of the speciation events that occurred within those lineages.

Appendix 1.1 lists avian taxa containing one or more cooperative breeders, along with the frequency of CB, as currently known. Species in which helping has been recorded but that do not meet our definition of cooperative breeding are listed in Appendices 1.2 and 1.3. Fig. 1.1 reconstructs the most parsimonious pattern of evolutionary transitions between CB and non-CB states on the phylogeny of Sibley and Ahlquist (1990). This optimization assumption is congruent with our hypothesis of an ancient origin of alloparental care.

In contrast to what we propose here, however, this reconstruction does not indicate a single ancient origin of CB congruent with the evolution of altriciality. Two aspects of our data and methodology contribute to this pattern. First, our current state of knowledge of the breeding systems of the majority of avian species is so poor as to significantly bias our data toward non-CB species. That is, species with unknown breeding systems are assumed to be non-CB. As data become available, we are confident that additional lineages will be shown also to have a propensity to breed cooperatively. With this additional information the continuity of CB across altricial lineages will become evident. Second, our analyses utilize a highly resolved phylogeny, with terminal taxa representing families, subfamilies and even tribes. This approach allows examination of more detailed patterns of evolutionary transitions, but may accentuate the problems outlined in our previous point. An analysis using only families as terminal taxa pulls many more lineages together as CB. However, the more detailed phylogeny is preferred in that it likely will give a more realistic picture of breeding-system evolution once ecological data catch up to our evolutionary hypotheses. Finally, our restrictive definition of CB makes our analyses conservative since we have listed many species as non-CB that were considered CB by previous authors. This definition has reduced the number of lineages considered CB in our analyses by eleven. Given these methodological and data limitations, what can we learn concerning CB evolution in non-passerine and passerine lineages with our current state of knowledge?

Cooperative breeding in non-passerine birds

Cooperative breeding is relatively common in certain non-passerine groups, typically families or genera (Fig. 1.1). Two points stand out. First, these are thought to be among the most ancient of all living neoavian birds (Sibley and Ahlquist 1990). Second, nearly all members of most of these groups are currently restricted to tropical regions. These ancient lineages support the hypothesis that CB, along with altriciality, appeared early in the Neoaves, and that since the early Cenozoic CB has been either retained or lost in different lineages, rather than having evolved *de novo* in certain species within these groups.

Cooperative breeding in passerine birds

Cooperative breeding appears to be the basal condition in the order Passeriformes (Fig. 1.1). Within the passerines, the single most striking relationship between the presence of CB and phylogeny is seen in the parvorder Corvida (Russell 1989; Cockburn 1996; Appendix 1.1). In his review and analysis of the distribution of CB in the Corvida, Cockburn (1996) makes several important points, with which we largely agree:

1. The extent of CB in the Corvida has previously been "spectacularly underestimated."
2. Within the Corvida, the proportion of clades originating outside Australia that contain at least one cooperative species is similar to the pattern in Australia–Papua New Guinea. Thus, there probably is no special environmental factor that has led to the relatively frequent occurrence of CB in Australia.
3. Contrary to earlier claims, there is no obvious relationship between CB and habitat type.
4. Pairs-only mating systems may often have been derived from CB systems.
5. Neither the habitat saturation model nor any other environmentally based model can apply to many Australian CB, which occur in a wide variety of densities and habitats yet are cooperative breeders throughout their range.

Cooperative breeding also appears in several lineages of the other major oscine group, the parvorder Passerida (Appendix 1.1, Fig. 1.1). Unlike the Corvida, no family in this large assemblage is composed either entirely or primarily of cooperative breeders. However, our parsimony reconstruction indicates that CB is ancestral in the Passerida as well. This behavior certainly is well developed in certain taxa. Good examples include the genus *Turdoides* (babblers, subfamily Sylviinae, tribe Timaliini), of which 28 of 29 species may breed cooperatively (Gaston 1977), and the genus *Campylorhynchus* (wrens, subfamily Troglodytinae), within which up to 12 of 13 species may breed cooperatively (Farley 1995).

PRIMITIVE VERSUS DERIVED COOPERATIVE BREEDING

Cooperative breeding appears to be the primitive condition

Some of the best known species of cooperative breeders belong to larger taxonomic groups in which most or all species breed cooperatively (Appendix 1.1). This suggests that CB was present in the common ancestor of such lineages and has been retained in their living descendants. Because closely related species often have similar ecologies, it can be difficult to ascertain whether the CB exhibited by two or more such species is due to a shared cooperatively breeding ancestor or to selection for similar modes of life. Perhaps most likely, both factors are usually involved. Thus, groups containing closely related CB species that occupy widely differing habitats could provide insights into the relative importance of phylogenetic history versus adaptive response to environmental variables in both the evolution and maintenance of CB.

In this section we consider two groups, the Upupiformes and the Meropidae. In both these polytypic taxa the majority of species breed cooperatively and there is considerable variation among species in the kinds of habitat occupied.

Woodhoopoes, family Phoeniculidae

Traditionally, the woodhoopoes (*Phoeniculus*) and scimitarbills (*Rhinopomastus*) have been placed in one family, the Phoeniculidae, which today is confined to sub-Saharan Africa. However, based on their DNA-hybridization studies, Sibley and Ahlquist (1990) determined that these two genera should be separated at the family level, the Phoeniculidae and the Rhinopomastidae. The available evidence indicates that at least four of the five species of *Phoeniculus* are cooperative breeders, while the social system of the forest woodhoopoe is unknown, and even its placement in *Phoeniculus* rather than *Rhinopomastus* is uncertain (Ligon 2001). In contrast, none of the three species of *Rhinopomastus* exhibit this trait (Ligon and Davidson 1988; Ligon 2001).

In reviewing the habitat types occupied by each species of *Phoeniculus* and *Rhinopomastus*, three main points stand out. First, the four species of *Phoeniculus* that breed cooperatively occupy habitats ranging from high montane rain forest (white-headed woodhoopoe) to low, hot desert (black-billed woodhoopoe). Second, in some cases, a species of *Phoeniculus* is broadly sympatric with one or two species of *Rhinopomastus*, yet no species of the latter genus exhibits CB. Third, the range of one species, the green (or red-billed) woodhoopoe, is huge, covering most of sub-Saharan Africa. Despite the great diversity in habitats occupied, this species breeds cooperatively throughout this vast area. In short, CB was retained in the radiation of *Phoeniculus* into all major habitat types of sub-Saharan Africa that contain trees.

The Messelirrisoridae, fossil upupiform birds from the middle Eocene (about 49 million years ago) form the sister group of the hoopoes (Upupidae) and woodhoopoes and scimitarbills (Mayr 2000). Messelirrisorids were very small, apparently arboreal, perching birds. Because CB occurs in hoopoes (Upupidae), the sister taxon of the woodhoopoes and scimitarbills, it appears that CB appeared early in this group and apparently was completely lost in the branch leading to the living species of *Rhinopomastus* (Fig. 1.2).

Bee-eaters, family Meropidae

Of the 26 species in the family Meropidae, 17 species are known or likely to be cooperative breeders. Five species apparently are not. At least one species shows variation among populations in its breeding system. The breeding systems of the remaining three species are unknown. Reconstructions on six alternative trees (Burt 1996) indicate one of two basic patterns of breeding system evolution (Fig. 1.3a). As in the upupiforms, cooperative breeding is an ancient trait that evolved either before the diversification of the entire family or before the diversification of the genus *Merops*.

Merops bee-eaters are widely distributed over the paleotropics and southern Eurasia. Throughout this range they occupy a variety of habitats including tropical forests, grasslands, marshes, savanna woodlands, semi-desert, and cultivated areas. They also vary greatly in their nesting substrates, from flat ground to cliff banks, and differ in their migratory behavior. These traits generally show no correlation with breeding systems within the family (Burt 1996), with the exception of colonial versus solitary nesting. Species that nest in solitary-only situations have evolved non-CB breeding from a CB ancestor more often than one would expect if degree of sociality and breeding system evolved and were maintained independently of each other (Fig. 1.3b).

Given the ecological diversity of extant bee-eaters and the extreme age of the family, determining the factors responsible for the origin of CB in this group is unrealistic. However, the general pattern of evolutionary stasis, with CB retained in the majority of lineages, can be explained in one of two ways. First, individuals in colonial species may be more likely to exhibit alloparental care simply because they are in close association with begging young. This association of non-breeders and begging young is absent in solitary nesting species, where non-CB may be more likely to evolve. Alternatively, CB may be adaptive in certain colonial species and natural selection may maintain helping behavior (Emlen 1990; Jones *et al.* 1991). The adaptive or non-adaptive nature of alloparental care in other colonial species requires additional research. Finally, as we argue with other avian groups, perhaps the best way to study the potential adaptive nature of helping in

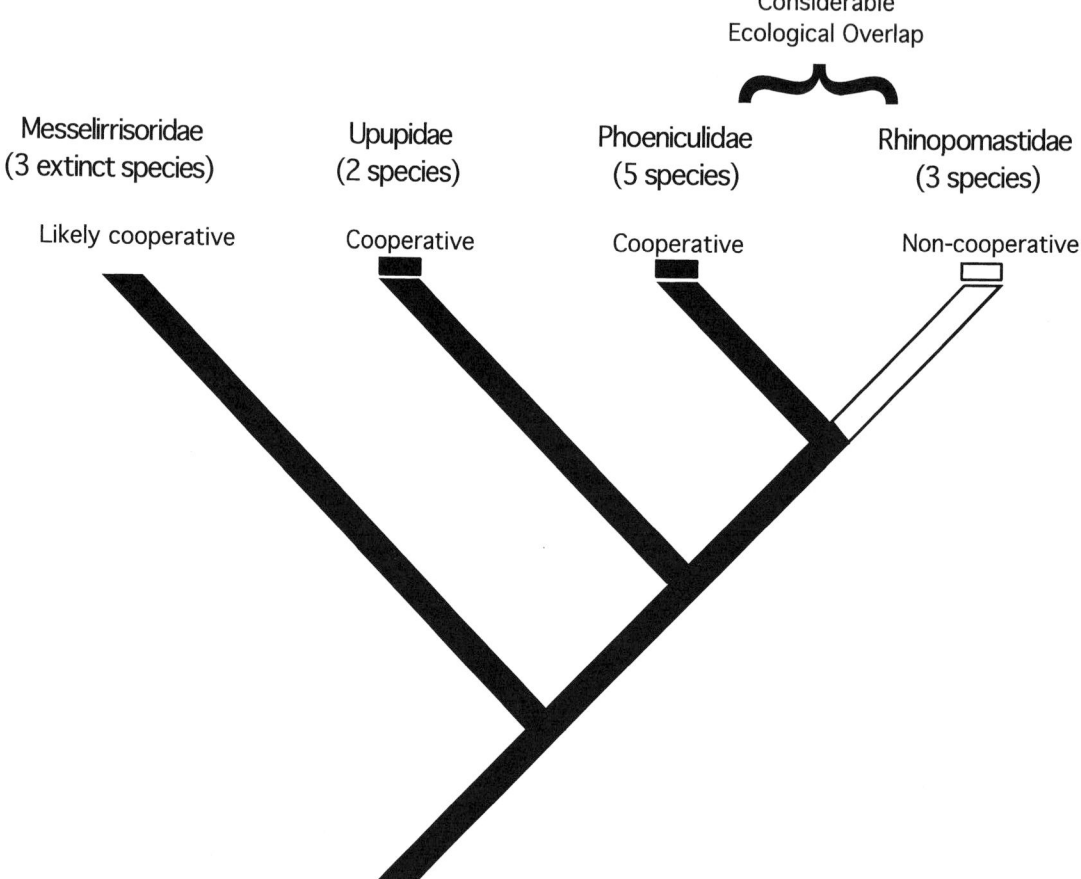

Figure 1.2. Phylogeny of upupiform birds based on information from Sibley and Ahlquist (1990) and Mayr (2000). Cooperative breeding likely evolved in the lineage that subsequently gave rise to this group since the hoopoes (Upupidae) and woodhoopoes (Phoeniculidae) breed cooperatively. Cooperative breeding is retained in at least some of the woodhoopoes despite lack of a clearly defined adaptive benefit in contemporary populations. Additionally, woodhoopoe and scimitarbill (Rhinopomastidae) species show considerable geographic overlap, suggesting that their current ecologies have minimal effect on either the maintenance of CB in the former or the lack of CB in the latter.

bee-eaters is to study the ecology of lineages that have lost CB.

Cooperative breeding is the derived condition

In other cases, cooperative breeding occurs in only one or few species of a polytypic genus. This suggests that cooperative breeding is a derived condition, and that CB species arose relatively recently from non-cooperative ancestors. Our use of the term "derived" does not necessarily indicate independent evolution of CB. Instead, we suggest that the derived state of alloparental care in these species is simply a re-expression of a trait that evolved deep within the avian tree. Two examples from North America are the red-cockaded woodpecker and the brown-headed and pygmy nuthatches.

Red-cockaded woodpecker
The genus *Picoides* contains 11 species of which only one, the red-cockaded woodpecker, breeds cooperatively. Red-cockaded woodpeckers depend on a single, self-constructed, critical resource, namely cavities excavated in living pine trees that are used for roosting and nesting (Ligon 1970; Walters 1990; Conner *et al.* 2001).

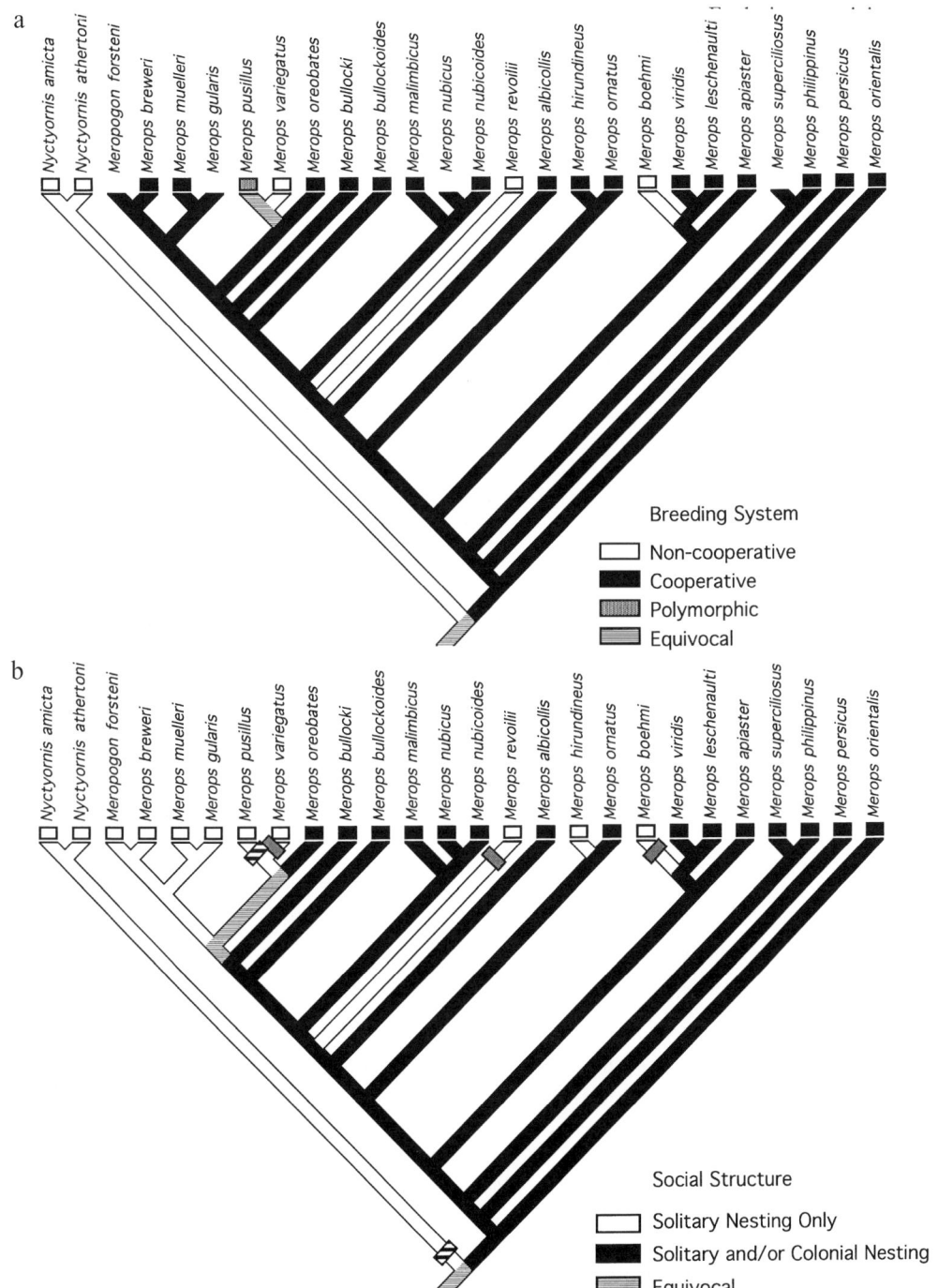

Figure 1.3. (a) One of six alternative phylogenies for bee-eaters from Burt (1996) with the most parsimonious pattern of breeding-system evolution reconstructed. Cooperative breeding is either the basal state in the family or evolved before the diversification of the genus *Merops*. Subsequent reversals to non-CB are seen in three to five lineages. (b) Degree of sociality reconstructed on the same phylogeny. Some species show plasticity in nesting both colonially and solitarily. Others apparently nest only solitarily. Evolutionary transitions from CB to non-CB occur significantly more often in solitary-only lineages than expected if the traits were evolving in an uncorrelated manner. The probability of three reversal events occurring in solitary-only lineages (gray bars) is 0.046 and the probability of this occurring five times (two additional dashed bars) is 0.014 (concentrated changes test, Maddison 1990).

The valuable cavities set the stage for delayed dispersal and CB. Because this species is the only member of its genus to exhibit delayed dispersal, and because the benefits associated with natal philopatry are well understood (Walters *et al.* 1992a, 1992b), we can confidently conclude that delayed dispersal set the stage for cooperative breeding as a derived trait.

The environmental factor that promoted delayed dispersal by ancestral red-cockaded woodpeckers was occupancy of open, fire-maintained pine forest, where, prior to fire suppression by humans, dead trees were rare. Occupation of open pine savannas required the excavation of cavities in fire-resistant living trees. Such cavities take much time to excavate and thus are extremely costly to construct. Cavities are passed from one generation to the next for as long as the tree remains alive. In good habitat, each occupied territory typically contains from two to several cavities. The critical variation in territory quality, which is related to factors associated with the presence, number, and quality of cavities, appears to be the basis for delayed dispersal by many of the young males produced in a given territory.

In contrast, natal philopatry is unknown in the congeneric downy and hairy woodpeckers, which do not breed cooperatively, and excavate cavities relatively quickly and easily in dead wood, which is relatively common in the habitats they primarily occupy. Thus for these species there are no major or unusual benefits to philopatry (Ligon 1999).

Brown-headed and pygmy nuthatches
The genus *Sitta* contains 24 species, of which only two are known to be cooperative breeders, the brown-headed nuthatch of pine forests of the southeastern USA and the pygmy nuthatch of pine forests of the western USA and Mexico. In these two very closely related forms, as in the red-cockaded woodpecker, it appears that CB is derived from non-CB ancestors. Unlike the situation for the red-cockaded woodpecker, however, there is no obvious relationship between the ecology or life-history traits of these nuthatches and CB. For example, these two are more likely to excavate their own cavities, making them less of a limiting resource, than the two other North American species of *Sitta*. To date, no convincing adaptive benefit of helping behavior per se has been demonstrated for either the brown-headed or the pygmy nuthatch (Sydeman 1989).

Recent transitions
One recurring and especially instructive pattern is the development of a non-cooperative population or species from a largely cooperative lineage: that is, cases where non-cooperative breeding appears to be the derived state. Strictly speaking, the origins of CB in those lineages where it appears to be basal cannot be studied. For example, we may never be able to identify the specific factors that led to the development of cooperative breeding in the New World jays, the great majority of which exhibit this social system in one form or another. In contrast, it might be possible to identify important factors that led to the loss of CB in scrub-jays of western North America (Peterson and Burt 1992). Here we consider scrub-jays and two other cases in which non-CB appears to the derived condition.

Western scrub-jay
Cooperative breeding occurs in most species of New World jays, including all six of the genera found in Mexico, plus the pinyon jay. In *Aphelocoma*, all populations of two of five species, the Mexican and unicolored jays, exhibit CB. In contrast, the scrub-jay group, recently recognized as three closely related species, shows a more variable pattern. While the Florida scrub-jay is renowned as the archetypal cooperative breeder (Woolfenden and Fitzpatrick 1984), a helper has never been recorded in the western and island scrub-jays of the western United States. Finally, the situation in southern Mexico is especially intriguing in that non-breeding individuals of the western scrub-jay subspecies *sumichrasti* feed fledglings, help in nest and territory defense, and attempt to feed nestlings, but usually are deterred from doing so by dominant group members, presumably the parents (Burt and Peterson 1993).

Although they did not pursue the point, Woolfenden and Fitzpatrick (1984) suggested that the non-cooperative system of western scrub-jays could have been derived from a cooperatively breeding ancestor. A hint that this might be the case was mentioned by Ligon (1985), who reported that captive juvenile western scrub-jays had a propensity to feed fledgling Mexican and pinyon jays, as well as other young scrub-jays. The fact that the now geographically isolated Florida scrub-jay is a cooperative breeder, together with the observation that scrub-jays of southern Mexico show CB tendencies, supports the idea that CB was indeed the

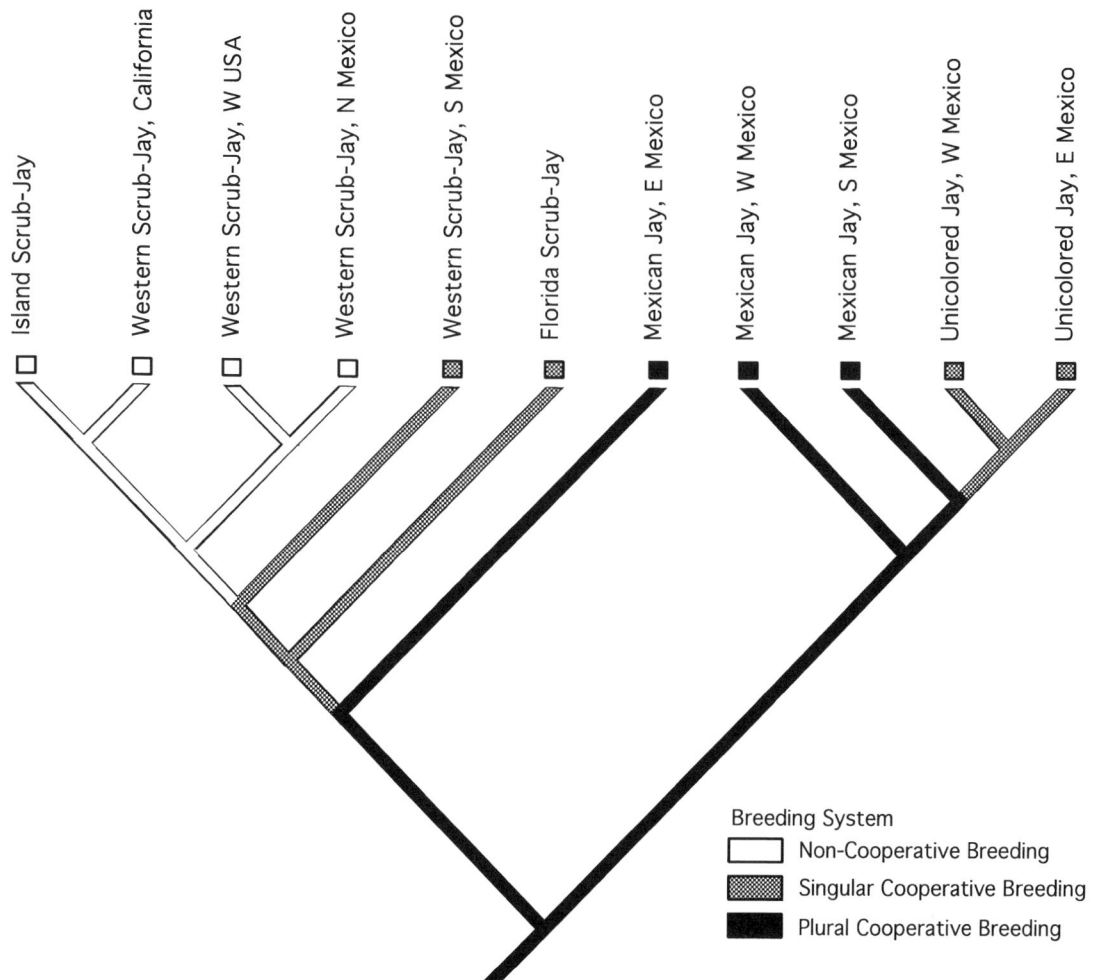

Figure 1.4. *Aphelocoma* phylogeny with CB pattern reconstructed, redrawn from Peterson and Burt (1992). Cooperative breeding is the ancestral state for the genus with a reduction in sociality in two derived lineages. The instability and unpredictability of habitats in the western USA may make delayed dispersal, and therefore helping, impractical. Alternatively, the loss of cooperative breeding may have been a necessary step before the western scrub-jay could expand its range across the western USA.

ancestral condition for scrub-jays of western North America (Fig. 1.4) (Peterson and Burt 1992). Assuming that this is true, what factors might have promoted the loss of this behavior in western US populations?

One possibility is that environmental unpredictably and strong seasonality, particularly as they affect food supply, have led to the loss of CB in western scrub-jays. Although pairs of these jay are territorial throughout the year, their territories often are incapable of fully supporting even two birds over the annual cycle. In the southwestern USA, for example, established adult jays often leave their territories to obtain food elsewhere. Over this species' range, serious food shortages come about as a result of several factors, including failure of mast crops, prolonged snow cover, or drought during the spring and summer. In short, the instability and unpredictability of their environment, as it affects the food supply, often makes it impossible for a pair of western scrub-jays to occupy a territory that meets their needs throughout the year, much less the needs of a larger social unit.

An alternative scenario is that the suite of traits that characterize cooperative breeding in many jays may have been lost in some Mexican populations of scrub-jays, as evidenced by the apparently non-functional tendency to help at the nest (Burt and Peterson 1993). This suggests that cooperative breeding may first have been selected against in an environment with less drastic seasonality than that of the western USA. Loss of cooperative breeding in Mexico might have been a necessary requisite for successful colonization of more rigorous climates. That is, the ancestors of contemporary western scrub-jays probably could not have moved into the seasonally cold environments of what is now the western USA if cooperative breeding was an important aspect of their biology. Edwards (1986) discusses ecological factors that may have determined the abrupt northern boundary of the range of the congeneric and cooperatively breeding Mexican jay in the southwestern United States.

Cactus wren
The tropical genus *Campylorynchus* contains 13 species, 12 of which are certainly or probably cooperative breeders (Selander 1964; Edwards and Naeem 1993; Rabenold 1990; Farley 1995), suggesting that in this genus CB is a primitive trait. The different species occupy a variety of habitats, ranging from hot low desert to cool montane cloud or rain forest. All species construct enclosed roost and breeding nests, and probably in all species except one, the cactus wren, group members typically roost together. The cactus wren, a derived species within the genus (Selander 1964), is also the only species that does not exhibit regular CB anywhere within its large range, which extends from south-central Mexico north into the southwestern US (Farley 1995).

At the southern end of the species' range in southern Mexico, territories are extremely small, producing a population density similar to that of their cooperatively breeding congeners. That is, it is possible that the resource base could support CB groups that occupied larger territories. However, strong seasonality, with periods of wet and dry, could make year-round occupancy of a territory by a group of southern cactus wrens impossible. Second, unlike the other species of *Campylorhynchus*, southern cactus wrens roost alone, as do their northern relatives.

In considering the genus as a whole, both of these traits are puzzling. Farley (1995) speculated that the absence of CB and the trait of solitary roosting may reflect high levels of predation, due both to low placement of roost nests and to the possibly high densities of predators, especially snakes, in this hot, subtropical/tropical environment. Whatever the specific factors were, we emphasize that loss of CB in the cactus wren apparently occurred in the tropical portion of its current range, prior to its range expansion into more temperate climes.

The key point is that the initial evolutionary shift from CB to non-CB may have had nothing to do with colonization and occupancy of a more temperate environment. Rather, the apparently derived behaviors of CB and solitary roosting evolved in the tropics and may have made it possible for cactus wrens subsequently to expand their range northward from southern Mexico into more temperate regions. Three traits exhibited by the more northern wrens apparently are directly associated with colonization of much cooler and more seasonal environments (Farley 1995):

(1) In the northernmost portion of their range, some individuals may migrate.
(2) Northern cactus wrens are larger than their southern counterparts, in accordance with Bergmann's rule.
(3) Roost nests of northern cactus wrens are significantly heavier and denser and have thicker walls than the nests of southern cactus wrens, suggesting adaptation to thermal challenges presented by seasonally cold nocturnal environments.

The cactus wren, and possibly the western scrub-jay, make the interesting and perhaps counterintuitive point that loss of CB in subtropical populations may have made it possible for subsequent expansion into temperate environments. That is, CB may limit the geographic ranges of certain lineages to benign, usually tropical or subtropical climes. This is a different scenario than the more frequent suggestion that temperate environments prevent the evolutionary origin and development of CB (Brown 1974).

Lanius *shrikes*
With regard to the issue of loss of CB and its distribution within a monophyletic lineage, laniid shrikes present an especially interesting variation on the general pattern shown by the jays and wrens. Although CB seems to be the ancestral or primitive state in the Laniidae, most of the 30 extant members of the group are not cooperative breeders (Zack 1995; Ligon 1999). In Africa, two *Lanius* species breed cooperatively, as do the four species in two other African genera. None of the remaining 24 species

of *Lanius* is thought to breed cooperatively. Of these, nine occur in sub-Saharan Africa and 15 outside Africa.

The difference between *Lanius* and the patterns of cooperative and non-cooperative breeding shown by the genera *Aphelocoma* and *Campylorhynchus* is that a lot of speciation has taken place in the first genus, producing many non-CB forms. At present, we do not know whether CB was lost in a single common ancestor of all of the non-CB species, or whether it was lost multiple times. In either case, loss of CB probably was associated with the colonization of temperate environments and subsequent speciation there.

LIMITATIONS OF PHYLOGENETIC ANALYSES

We have attempted to clarify the benefits of comparative evolutionary analyses for a more complete understanding of CB. However, comparative methods are hampered by several factors. One assumption of these methods is that we have an accurate understanding of the relatedness of the taxa involved in our analyses. That is, our phylogeny accurately reflects evolutionary history. Simulation studies (Hillis *et al.* 1994; Wiens and Servedio 1998) have corroborated the accuracy of several methods of phylogenetic reconstruction, as have studies using organisms with well-understood evolutionary histories (Hillis *et al.* 1994; Russo *et al.* 1996). Most of these methods appear to be quite reliable under various evolutionary conditions. Additionally, new metrics are continually being developed that help indicate the reliability of specific phylogenies and their associated data sets (Sanderson and Kim 2000). For these reasons, we believe that any uncertainties associated with well-supported evolutionary hypotheses are no more problematic than uncertainties associated with hypotheses derived from any other biological field.

Recent phylogenetic analyses using mitochondrial and nuclear DNA sequence data have provided additional support for many of the branching patterns in the Sibley and Ahlquist tree used in our study (Mindell *et al.* 1999; Johnson 2001; Barker *et al.* 2002; Edwards and Boles 2002; Ericson *et al.* 2002). However, these phylogenetic analyses question both the placement of certain lineages, including the passeriform birds, and the sister-group relationship between the Corvida and Passerida. Additional studies will be required to identify how these tree topology changes would change the specific patterns of CB evolution. However, our initial analyses indicate that the patterns are qualitatively similar and would not alter our primary conclusions.

A potentially more problematic issue is our ability to accurately reconstruct patterns of breeding system evolution on any given phylogeny. Character-state assignments become less reliable as one moves from the terminal tips to the base of a phylogeny. In other words, the older the ancestors, the more likely we are to inaccurately predict their behavior (Maddison 1995; Garland *et al.* 1999; Schultz *et al.* 1996). There is strong evidence that CB is very old. Given this deep or ancient origin, the specific branch where CB was derived in any particular avian lineage will usually be uncertain.

This complicates any attempt to determine specific ecological or environmental correlates associated with the origins of CB. Again, we advocate an alternative approach to this problem: study correlates associated with the more recent losses of CB in various lineages (Peterson and Burt 1992). It is in the analyses of such systems that we are more likely to gain an understanding of ecological factors associated with the maintenance and loss of CB.

Another issue of concern is whether CB systems in different lineages are homologous (Edwards and Naeem 1994). Homology as used here is "similarity due to inheritance from a common ancestor" (Edwards and Naeem 1994). We believe that there is a reasonable expectation of homology among closely related lineages. How far this homology extends can then be determined only by mapping CB on a phylogeny. This procedure provides an objective means of identification of the origins and losses of helping behavior.

We have addressed different scenarios of CB evolution under the assumption that helping behavior had its origins as a simple by-product of misplaced parental care associated with delayed dispersal or colonial living in lineages with altricial young. If this is true, then the genetic basis for helping behavior is much older than previously appreciated. Descendants of these lineages then may simply express the trait "helping behavior" when delayed dispersal is seen. In other words, the basic elements of alloparental care have not evolved independently in most avian lineages: helping behavior simply re-emerges in a variety of appropriate ecological situations, including many, but not all, of those that promote long-term natal philopatry. Edwards and Naeem (1993: 772) state: "If the diversity of mating systems and 'routes' toward CB in particular groups . . . is too great for them to be considered 'homologous' . . . then phylogenetic analysis of such systems would be inappropriate." We disagree. Alloparental behavior can be ancestral for the group in question and CB may be adaptive in substantially different ways in different lineages. That is, the behavior is malleable to different ecologies and social systems. Alternatively, CB may be retained as a holdover independent of the ecologies or social systems of descendant lineages due to evolutionary stasis (Burt 2001). In either case the derived differences in the ecologies or social systems of each lineage do not negate the homology of the basic alloparental tendencies that evolved in the ancestor. Again, we suggest a different question for studies of this issue: why do some altricial lineages with delayed dispersal fail to express alloparental care (Stacey and Ligon 1991; Ekman *et al.* 2001a)?

Last, the most fundamental limitation to a more complete understanding of the evolution of cooperative breeding is a lack of basic natural-history data for the majority of avian species. Some might even argue that our analysis, like that of Edwards and Naeem (1993), is too preliminary given the current state of our data (McLennan and Brooks 1993). We do not take such a negative view, but agree that reanalyses will be necessary as more information becomes available.

ACKNOWLEDGMENTS

We thank Rebecca Kimball, Peter Stacey, Andrew Cockburn, Priscilla Coulter and Walt Koenig for their constructive comments.

APPENDIX 1.1 COOPERATIVELY BREEDING BIRDS, FOLLOWING THE CLASSIFICATION OF SIBLEY AND MONROE (1990, 1993)

Family, subfamily, or tribe (N cooperative breeders/N total species in taxon)	Common name	Scientific name	Reference
Picidae (10/215)	Red-cockaded woodpecker	*Picoides borealis*	Brown 1987
	Yellow-tufted woodpecker	*Melanerpes cruentatus*	Brown 1987
	Acorn woodpecker	*M. formicivorus*	Brown 1987
	White woodpecker	*M. candidus*	Winkler *et al.* 1995
	Golden-naped woodpecker	*M. chrysauchen*	Winkler *et al.* 1995
	Yellow-fronted woodpecker	*M. flavifrons*	Winkler *et al.* 1995
	White-fronted woodpecker	*M. cactorum*	Winkler *et al.* 1995
	Hispaniolan woodpecker	*M. striatus*	Winkler *et al.* 1995
	Ground woodpecker	*Geocolaptes olivaceus*	Winkler *et al.* 1995
	Campo flicker	*Colaptes campestris*	Winkler *et al.* 1995
Lybiidae (24/42)	Naked-faced barbet	*Gymnobucco calvus*	Short and Horne 1988 (possible)
	Bristle-nosed barbet	*G. peli*	Short and Horne 1988 (possible)
	Grey-throated barbet	*G. bonapartei*	Short and Horne 1988
	White-eared barbet	*Stactolaema leucotis*	Brown 1987
	Whyte's barbet	*S. whytii*	Short and Horne 1988
	Anchieta's barbet	*S. anchietae*	Short and Horne 1988
	Green barbet	*S. olivacea*	Short and Horne 1988
	Spot-flanked barbet	*Tricholaema lacrymosa*	Short and Horne 1988 (likely)
	Pied barbet	*T. leucomelas*	Skutch 1999
	Black-throated barbet	*T. melanocephala*	Short and Horne 1988 (likely)

Appendix 1.1. (cont.)

Family, subfamily, or tribe (N cooperative breeders/N total species in taxon)	Common name	Scientific name	Reference
	Vieillot's barbet	*Lybius vielloti*	Short and Horne 1988 (likely)
	Black-collared barbet	*L. torquatus*	Brown 1987
	Chaplin's barbet	*L. chaplini*	Short and Horne 1988
	Red-faced barbet	*L. rubrifacies*	Short and Horne 1988 (likely)
	Black-billed barbet	*L. guifsobalito*	Short and Horne 1988
	Brown-breasted barbet	*L. melanopterus*	Short and Horne 1988 (likely)
	Black-backed barbet	*L. minor*	Short and Horne 1988
	Double-toothed barbet	*L. bidentatus*	Short and Horne 1988
	Bearded barbet	*L. dubius*	Short and Horne 1988 (likely)
	Black-breasted barbet	*L. rolleti*	Short and Horne 1988 (likely)
	White-headed barbet	*L. leucocephalus*	Grimes 1976
	D'Arnaud's barbet	*Trachyphonus darnaudii*	Brown 1987
	Red-and-yellow barbet	*T. erythrocephalus*	Brown 1987
	Yellow-breasted barbet	*T. margaritatus*	Short and Horne 1988 (likely)
Ramphastidae			
Capitoninae (1/14)	Toucan barbet	*Semnornis ramphastinus*	Restrepo and Mondragón 1998
Ramphastinae (2/41)	Collared aracari	*Pteroglossus torquatus*	Brown 1987
	Fiery-billed aracari	*P. frantzii*	Stiles and Skutch 1989
Galbulidae (3/18)	Three-toed jacamar	*Jacamaralcyon tridactyla*	Tobias *et al.* 2002
	Chestnut jacamar	*Galbalcyrhynchus purusianus*	Tobias *et al.* 2002
	Rufous-tailed jacamar	*Galbula ruficauda*	Langham *et al.* 2003
Bucconidae (1/1)	White-fronted nunbird	*Monasa morphoeus*	Brown 1987
Bucerotidae (13/54)	White-crowned hornbill	*Aceros comatus*	Brown 1987
	Bushy-crested hornbill	*Anorrhinus galeritus*	Brown 1987
	Assam hornbill	*A. austeni*	Kemp 2001
	Brown hornbill	*A. tickelli*	Witmer 1993
	Black-and-white-casqued hornbill	*Ceratogymna subcylindricus*	Kemp 2001 (possible)
	Trumpeter hornbill	*C. bucinator*	Du Plessis *et al.* 1995
	Black-casqued hornbill	*C. atrata*	Kemp 2001 (possible)
	Rufous hornbill	*Buceros hydrocorax*	Witmer 1993
	Rhinoceros hornbill	*B. rhinoceros*	Kemp 2001 (probable)
	Red-billed dwarf hornbill	*Tockus camurus*	Kemp 2001 (probable)
	Sulawesi hornbill	*Penelopides exarhatus*	Kemp 2001

(cont.)

Appendix 1.1. (cont.)

Family, subfamily, or tribe (N cooperative breeders/N total species in taxon)	Common name	Scientific name	Reference
	Tarictic hornbill	*P. panini*	Kemp 2001 (possible)
	Luzon hornbill	*P. manillae*	Kemp 2001
Bucorvidae (1/2)	Southern ground-hornbill	*Bucorvus leadbeateri*	Brown 1987
Upupidae (1/2)	African hoopoe	*Upupa africana*	Brown 1987
Phoeniculidae (4/5)	Green woodhoopoe	*Phoeniculus purpureus*	Brown 1987
	White-headed woodhoopoe	*P. bollei*	Ligon 2001
	Black-billed woodhoopoe	*P. somaliensis*	Ligon 2001
	Violet woodhoopoe	*P. damarensis*	Ligon 2001
Todidae (1/5)	Puerto Rican tody	*Todus mexicanus*	Brown 1987
Halcyonidae (5/61)	Blue-winged kookaburra	*Dacelo leachii*	Brown 1987
	Laughing kookaburra	*D. novaeguineae*	Brown 1987
	Striped kingfisher	*Halcyon chelicuti*	Brown 1987
	Forest kingfisher	*Todirhamphus macleayii*	Brown 1987
	Buff-breasted paradise-kingfisher	*Tanysiptera sylvia*	Brown 1987
Cerylidae (1/9)	Pied kingfisher	*Ceryle rudis*	Brown 1987
Meropidae (17/26)	Black-headed bee-eater	*Merops breweri*	Burt 1996 (possible)
	Blue-headed bee-eater	*M. muelleri*	Burt 1996 (possible)
	White-throated bee-eater	*M. albicollis*	Brown 1987
	European bee-eater	*M. apiaster*	Brown 1987
	Red-throated bee-eater	*M. bulocki*	Brown 1987
	White-fronted bee-eater	*M. bullockoides*	Brown 1987
	Swallow-tailed bee-eater	*M. hirundineus*	Du Plessis *et al.* 1995
	Chestnut-headed bee-eater	*M. leschenaulti*	Burt 2002
	Carmine bee-eater	*M. nubicus*	Brown 1987
	Cinnamon-chested bee-eater	*M. oreobates*	Burt 1996
	Little green bee-eater	*M. orientalis*	Burt 2002
	Rainbow bee-eater	*M. ornatus*	Brown 1987
	Blue-tailed bee-eater	*M. philippinus*	Burt 2002
	Little bee-eater	*M. pusillus*	Burt 1996
	Blue-cheeked bee-eater	*M. persicus*	Kossenko and Fry 1998
	Rosy bee-eater	*M. malimbicus*	Skutch 1999
	Blue-throated bee-eater	*M. viridis*	Burt 1996
Coliidae			
Coliinae (4/4)	White-backed mousebird	*Colius colius*	Du Plessis *et al.* 1995
	Speckled mousebird	*C. striatus*	Brown 1987
	Red-backed mousebird	*C. castanotus*	Decoux 1988a
	White-headed mousebird	*C. leucocephalus*	Decoux 1988a (probable)
Urocoliinae (2/2)	Red-faced mousebird	*Urocolius indicus*	Grimes 1976
	Blue-naped mousebird	*U. macrourus*	Decoux 1988a (possible)
Opisthocomidae (1/1)	Hoatzin	*Opisthocomus hoazin*	Brown 1987

Appendix 1.1. (cont.)

Family, subfamily, or tribe (N cooperative breeders/N total species in taxon)	Common name	Scientific name	Reference
Crotophagidae			
Crotophagini (3/3)	Smooth-billed ani	*Crotophaga ani*	Brown 1987
	Greater ani	*C. major*	Brown 1987
	Groove-billed ani	*C. sulcirostris*	Brown 1987
Guirini (1/1)	Guira cuckoo	*Guira guira*	Brown 1987
Apodidae (10/99)	Mottled swift	*Tachymarptis aequatorialis*	Grimes 1976
	Alpine swift	*T. melba*	Grimes 1976
	Horus swift	*Apus horus*	Grimes 1976
	Ashy-tailed swift	*Chaetura andrei*	Brown 1987
	Short-tailed swift	*C. brachyura*	Brown 1987
	Chimney swift	*C. pelagica*	Brown 1987
	Vaux's swift	*C. vauxi*	Brown 1987
	White-rumped swiftlet	*Collocalia spodiopygius*	Clarke 1995
	Bat-like spinetail	*Neafrapus boehmi*	Grimes 1976
	Cassin's spinetail	*N. cassini*	Grimes 1976
Musophagidae			
Criniferinae (1/6)	Grey go-away-bird	*Corythaixoides concolor*	Du Plessis *et al.* 1995
Strigidae (1/161)	Verreaux's eagle-owl	*Bubo lacteus*	Du Plessis *et al.* 1995
Psophiidae (1/3)	Pale-winged trumpeter	*Psophia leucoptera*	Sherman 1995a
Rallidae (7/142)	Red-knobbed coot	*Fulica cristata*	Brown 1987
	Giant coot	*F. gigantea*	Skutch 1999
	Common moorhen	*Gallinula chloropus*	Brown 1987
	Dusky moorhen	*G. tenebrosa*	Brown 1987
	Purple gallinule	*Porphyrula martinica*	Brown 1987
	Pukeko	*Porphyrio porphyrio*	Brown 1987
	Black crake	*Amaurornis flavirostra*	Brown 1987
Laridae			
Larinae			
Sternini (1/45)	Arctic tern	*Sterna paradisaea*	Brown 1987
Accipitridae			
Accipitrinae (3/239)	Mississippi kite	*Ictinia mississippiensis*	Brown 1987
	Harris's hawk	*Parabuteo unicinctus*	Brown 1987
	Bateleur	*Terathopius ecaudatus*	Brown 1987
Falconidae (1/63)	Peregrine falcon	*Falco peregrinus*	Brown 1987
Podicipedidae (1/21)	Australasian grebe	*Tachybaptus novaehollandiae*	Brown 1987
Acanthisittidae (1/4)	Rifleman	*Acanthisitta chloris*	Cockburn 1996
Tyrannidae			
Tyranninae (2/340)	White-bearded flycatcher	*Phelpsia inornata*	Brown 1987
	Rusty-margined flycatcher	*Myiozetetes cayanensis*	Brown 1987
Cotinginae (1/69)	Purple-throated fruitcrow	*Querula purpurata*	Brown 1987

(cont.)

Appendix 1.1. (cont.)

Family, subfamily, or tribe (N cooperative breeders/N total species in taxon)	Common name	Scientific name	Reference
Furnariidae			
Furnariinae (3/231)	Plain thornbird	*Phacellodomus rufifrons*	Skutch 1999
	Rufous cacholote	*Pseudoseisura cristata*	Zimmer and Whittaker 2000
	Lark-like brushrunner	*Coryphistera alaudina*	Fraga 1979
Climacteridae (4/7)	Red-browed treecreeper	*Climacteris erythrops*	Brown 1987
	Black-tailed treecreeper	*C. melanura*	Brown 1987
	Brown treecreeper	*C. picumnus*	Brown 1987
	Rufous treecreeper	*C. rufa*	Brown 1987
Maluridae			
Malurinae			
Malurini (15/15)	Orange-crowned fairy-wren	*Clytomyias insignis*	Rowley and Russell 1997
	Wallace's fairy-wren	*Sipodotus wallacii*	Rowley and Russell 1997
	Broad-billed fairy-wren	*Malurus grayi*	Rowley and Russell 1997
	Campbell's fairy-wren	*M. campbelli*	Rowley and Russell 1997
	White-shouldered fairy-wren	*M. alboscapulatus*	Rowley and Russell 1997
	Red-backed fairy-wren	*M. melanocephalus*	Rowley and Russell 1997
	Lovely fairy-wren	*M. amabilis*	Rowley and Russell 1997
	Purple-crowned fairy-wren	*M. coronatus*	Rowley and Russell 1997
	Emperor fairy-wren	*M. cyanocephalus*	Rowley and Russell 1997
	Superb fairy-wren	*M. cyaneus*	Brown 1987
	Red-winged fairy-wren	*M. elegans*	Brown 1987
	Variegated fairy-wren	*M. lamberti*	Brown 1987
	White-winged fairy-wren	*M. leucopterus*	Brown 1987
	Blue-breasted fairy-wren	*M. pulcherrimus*	Brown 1987
	Splendid fairy-wren	*M. splendens*	Brown 1987
Amytornithinae (8/8)	Grey grasswren	*Amytornis barbatus*	Rowley and Russell 1997 (possible)
	White-throated grasswren	*A. woodwardi*	Rowley and Russell 1997
	Carpentarian grasswren	*A. dorotheae*	Rowley and Russell 1997
	Striated grasswren	*A. striatus*	Rowley and Russell 1997 (possible)
	Eyrean grasswren	*A. goyderi*	Rowley and Russell 1997 (possible)
	Dusky grasswren	*A. purnelli*	Rowley and Russell 1997 (possible)
	Thick-billed grasswren	*A. textiles*	Rowley and Russell 1997 (possible)
	Black grasswren	*A. housei*	Rowley and Russell 1997 (possible)

Appendix 1.1. (cont.)

Family, subfamily, or tribe (N cooperative breeders/N total species in taxon)	Common name	Scientific name	Reference
Meliphagidae (24/182)	Red wattlebird	*Anthochaera carunculata*	Clarke 1995
	Little wattlebird	*A. lunulata*	Brown 1987
	Rufous-throated honeyeater	*Conopophilia rufogularis*	Brown 1987
	Yellow-tufted honeyeater	*Lichenostomus melanops*	Brown 1987
	White-plumed honeyeater	*L. penicillatus*	Brown 1987
	Varied honeyeater	*L. versicolor*	Cockburn 1996
	White-throated honeyeater	*Melithreptus albogularis*	Brown 1987
	Brown-headed honeyeater	*M. brevirostris*	Brown 1987
	Black-chinned honeyeater	*M. gularis*	Clarke 1995
	Black-headed honeyeater	*M. affinis*	Clarke 1995
	White-naped honeyeater	*M. lunatus*	Brown 1987
	Golden-backed honeyeater	*M. laetior*	Skutch 1999
	Strong-billed honeyeater	*M. validirostris*	Cockburn 1996
	White-lined honeyeater	*M. albilineata*	Cockburn 1996
	New Holland honeyeater	*Phylidonyris novaehollandiae*	Brown 1987
	Striped honeyeater	*P. lanceolata*	Brown 1987
	Blue-faced honeyeater	*Entomyzon cyanotis*	Clarke 1995
	Black-eared miner	*Manorina melanotis*	Clarke 1995
	Yellow-throated miner	*M. flavigula*	Brown 1987
	Noisy miner	*M. melanocephala*	Brown 1987
	Bell miner	*M. melanophrys*	Brown 1987
	Stitchbird	*Notiomystis cincta*	Cockburn 1996
	Little friarbird	*Philemon citreogularis*	Brown 1987
	White-fronted chat	*Ephthianura albifrons*	Cockburn 1996
Pardalotidae Acanthizinae Sericornithini (2/26)	Large-billed scrubwren	*Sericornis magnirostris*	Brown 1987
	Speckled warbler	*Chthonicola sagittatus*	Cockburn 1996
Acanthizini (11/35)	Yellow-rumped thornbill	*Acanthiza chrysorrhoa*	Brown 1987
	Striated thornbill	*A. lineata*	Brown 1987
	Yellow thornbill	*A. nana*	Brown 1987
	Buff-rumped thornbill	*A. reguloides*	Brown 1987
	Chestnut-rumped thornbill	*A. uropygialis*	Brown 1987
	Papuan thornbill	*A. murina*	Cockburn 1996
	Yellow-bellied gerygone	*Gerygone chrysogaster*	Brown 1987
	Brown gerygone	*G. mouki*	Brown 1987
	Banded whiteface	*Aphelocephala nigricincta*	Cockburn 1996
	Southern whiteface	*A. leucopis*	Cockburn 1996
	Weebill	*Smicrornis brevirostris*	Brown 1987

(cont.)

Appendix 1.1. (cont.)

Family, subfamily, or tribe (N cooperative breeders/N total species in taxon)	Common name	Scientific name	Reference
Petroicidae (3/46)	Yellow robin	*Eopsaltria australis*	Brown 1987
	White-breasted robin	*E. georgiana*	Brown 1987
	Grey-breasted robin	*E. griseogularis*	Brown 1987
Orthonychidae (1/2)	Logrunner	*Orthonyx temminckii*	Brown 1987
Pomatostomidae (5/5)	Hall's babbler	*Pomatostomus halli*	Brown 1987
	New Guinea babbler	*P. isidorei*	Brown 1987
	Chestnut-crowned babbler	*P. ruficeps*	Smith 1992
	White-browed babbler	*P. superciliosus*	Brown 1987
	Grey-crowned babbler	*P. temporalis*	Brown 1987
Laniidae (6/30)	Yellow-billed shrike	*Corvinella corvina*	Brown 1987
	Magpie shrike	*C. melanoleuca*	Brown 1987
	White-crowned shrike	*Eurocephalus anguitimens*	Brown 1987
	White-rumped shrike	*E. rueppelli*	Zack 1995
	Grey-backed fiscal shrike	*Lanius excubitoroides*	Brown 1987
	Long-tailed fiscal shrike	*L. cabanisi*	Zack 1995
Corvidae			
Cinclosomatinae (1/15)	Cinnamon quail-thrush	*Cinclosoma cinnamomeum*	Brown 1987
Corcoracinae (2/2)	White-winged chough	*Corcorax melanorhamphos*	Brown 1987
	Apostlebird	*Struthidea cinerea*	Brown 1987
Pachycephalinae			
Neosittini (2/2)	Varied sitella	*Daphoenositta chrysoptera*	Brown 1987
	Black sitella	*D. miranda*	Cockburn 1996
Mohouini (2/3)	Whitehead	*Mohoua albicilla*	Cockburn 1996
	Yellowhead	*M. ochrocephala*	Cockburn 1996
Falcunculini (1/3)	Crested shrike-tit	*Falcunculus frontatus*	Brown 1987
Corvinae			
Corvini (26/117)	Florida scrub-jay	*Aphelocoma coerulescens*	Brown 1987
	Mexican jay	*A. ultramarina*	Brown 1987
	Unicolored jay	*A. unicolor*	Burt and Peterson 1993
	Western scrub-jay	*A. californica*	Burt and Peterson 1993
	White-throated magpie-jay	*Calocitta formosa*	Skutch 1999
	Black-throated magpie-jay	*C. colliei*	Brown 1987
	Azure-winged magpie	*Cyanopica cyana*	Brown 1987
	Formosan magpie	*Urocissa caerulea*	Cockburn 1996
	American crow	*Corvus brachyrhynchos*	Brown 1987
	Northwestern crow	*C. caurinus*	Brown 1987
	Carrion crow	*C. corone*	Cockburn 1996
	Violaceous jay	*Cyanocorax violaceus*	Cockburn 1996
	Curl-crested jay	*C. cristatellus*	Cockburn 1996

Appendix 1.1. (cont.)

Family, subfamily, or tribe (N cooperative breeders/N total species in taxon)	Common name	Scientific name	Reference
	Black-chested jay	*C. affinis*	Cockburn 1996
	Beechey jay	*C. beecheii*	Brown 1987
	Tufted jay	*C. dickeyi*	Brown 1987
	Bushy-crested jay	*C. melanocyaneus*	Brown 1987
	San Blas jay	*C. sanblasianus*	Brown 1987
	Green jay	*C. yncas*	Brown 1987
	Yucatan jay	*C. yucatanicus*	Brown 1987
	Pinyon jay	*Gymnorhinus cyanocephalus*	Brown 1987
	Gray jay	*Perisoreus canadensis*	Waite and Strickland 1997
	Siberian jay	*P. infaustus*	Cockburn 1996
	Brown jay	*Psilorhinus morio*	Brown 1987
	Piapiac	*Ptilostomus afer*	Brown 1987
	Stresemann's bush-crow	*Zavattariornis stresemanni*	Brown 1987
Artamini (9/24)	Black-faced woodswallow	*Artamus cinereus*	Brown 1987
	Dusky woodswallow	*A. cyanopterus*	Brown 1987
	White-breasted woodswallow	*A. leucorynchus*	Brown 1987
	Little woodswallow	*A. minor*	Brown 1987
	Great woodswallow	*A. maximus*	Cockburn 1996
	Australian magpie	*Gymnorhina tibicen*	Brown 1987
	Pied butcherbird	*Cracticus nigrogularis*	Brown 1987
	Grey butcherbird	*C. torquatus*	Brown 1987
	Hooded butcherbird	*C. cassicus*	Cockburn 1996
Oriolini (2/111)	Ground cuckoo-shrike	*Coracina maxima*	Brown 1987
	Green figbird	*Sphecotheres viridis*	Clarke 1995
Dicrurinae			
Dicrurini (1/24)	Black drongo	*Dicrurus macrocercus*	Brown 1987
Monarchini (4/98)	Chestnut-capped flycatcher	*Erythrocercus mccalli*	Brown 1987
	African blue-flycatcher	*Elminia longicauda*	Brown 1987
	African paradise-flycatcher	*Terpsiphone viridis*	Cockburn 1996
	Magpie-lark	*Grallina cyanoleuca*	Clarke 1995
Malaconotinae			
Malaconotini (1/48)	Black-backed puffback	*Dryoscopus cubla*	Cockburn 1996
Vangini (8/58)	Chabert's vanga	*Leptopterus chabert*	Brown 1987
	White-spotted wattle-eye	*Platysteira tonsa*	Brown 1987
	Black-throated wattle-eye	*P. peltata*	Cockburn 1996
	Yellow-crested helmetshrike	*Prionops alberti*	Cockburn 1996

(cont.)

Appendix 1.1. (cont.)

Family, subfamily, or tribe (N cooperative breeders/N total species in taxon)	Common name	Scientific name	Reference
	White helmetshrike	*P. plumatus*	Brown 1987
	Retz's helmetshrike	*P. retzii*	Brown 1987
	Chestnut-fronted helmetshrike	*P. scopifrons*	Brown 1987
	Rufous vanga	*Schetba rufa*	Cockburn 1996
Picathartidae (1/4)	Rufous rockjumper	*Chaetops frenatus*	Brown 1987
Muscicapidae			
Turdinae (2/179)	Western bluebird	*Sialia mexicana*	Cockburn 1996
	Eastern bluebird	*S. sialis*	Gowaty and Plissner 1998
Muscicapinae			
Muscicapini (3/115)	Pale flycatcher	*Bradornis pallidus*	Brown 1987
	African forest-flycatcher	*Fraseria ocreata*	Brown 1987
	Abyssinian slaty-flycatcher	*Dioptrornis chocolatinus*	Brown 1987
Saxicolini (2/155)	Schalow's wheatear	*Oenanthe lugubris*	Brown 1987
	Anteater-chat	*Myrmecocichla aethiops*	Brown 1987
Sturnidae			
Sturnini (10/114)	Yellow-billed oxpecker	*Buphagus africanus*	Brown 1987
	Red-billed oxpecker	*B. erythrorhynchus*	Brown 1987
	Golden-breasted starling	*Cosmopsarus regius*	Brown 1987
	Red-shouldered glossy-starling	*Lamprotornis nitens*	Brown 1987
	Long-tailed glossy-starling	*L. caudatus*	Wilkinson 1988
	Chestnut-bellied starling	*L. pulcher*	Brown 1987
	Superb starling	*L. superbus*	Brown 1987
	African pied starling	*Spreo bicolor*	Brown 1987
	Fischer's starling	*S. fischeri*	Skutch 1999
	Violet-backed starling	*Cinnyricinclus leucogaster*	Cockburn 1996
Mimini (6/34)	Hood mockingbird	*Nesomimus macdonaldi*	Brown 1987
	Galápagos mockingbird	*N. parvulus*	Brown 1987
	Charles mockingbird	*N. trifasciatus*	Brown 1987
	Chalk-browed mockingbird	*Mimus saturninus*	Fraga 1979
	Tropical mockingbird	*M. gilvus*	Fraga 1979
	Long-tailed mockingbird	*M. longicaudatus*	Fraga 1979 (possible)
Sittidae			
Sittinae (2/24)	Brown-headed nuthatch	*Sitta pusilla*	Brown 1987
	Pygmy nuthatch	*S. pygmaea*	Brown 1987
Certhiidae			
Troglodytinae (16/75)	Fasciated wren	*Campylorhynchus fasciatus*	Brown 1987

Appendix 1.1. (cont.)

Family, subfamily, or tribe (N cooperative breeders/N total species in taxon)	Common name	Scientific name	Reference
	Bicolored wren	*C. griseus*	Brown 1987
	Boucard's wren	*C. jocosus*	Brown 1987
	Gray-barred wren	*C. megalopterus*	Brown 1987
	Stripe-backed wren	*C. nuchalis*	Brown 1987
	Band-backed wren	*C. zonatus*	Brown 1987
	Rufous-naped wren	*C. rufinucha*	Farley 1995
	Thrush-like wren	*C. turdinus*	Farley 1995
	Yucatan wren	*C. yucatanicus*	Farley 1995 (probable)
	Spotted wren	*C. gularis*	Farley 1995 (probable)
	White-headed wren	*C. albobrunneus*	Farley 1995 (probable)
	Giant wren	*C. chiapensis*	Farley 1995 (probable)
	Sepia-brown wren	*Cinnycerthia peruana*	J. W. Fitzpatrick pers. comm. (probable)
	Musician wren	*Cyphorhinus ardus*	J. W. Fitzpatrick pers. comm. (probable)
	Black-capped donacobius	*Donacobius atricapillus*	Brown 1987
	Banded wren	*Thryothorus pleurostictus*	Brown 1987
Paridae			
Remizinae (1/12)	Tit-hylia	*Pholidornis rushiae*	Brown 1987
Parinae (2/53)	Bridled titmouse	*Baeolophus wollweberi*	Nocedal and Ficken 1998
	Black tit	*Parus niger*	Brown 1987
Aegithalidae (2/8)	Long-tailed tit	*Aegithalos caudatus*	Brown 1987
	Bushtit	*Psaltriparus minimus*	Brown 1987
Hirundinidae			
Hirundininae (3/87)	Barn swallow	*Hirundo rustica*	Brown 1987
	Tree swallow	*Tachycineta bicolor*	Skutch 1999 (probable)
	Northern house-martin	*Delichon urbica*	Skutch 1999 (probable)
Pycnonotidae 2/137	Spotted greenbul	*Ixonotus guttatus*	Brown 1987
	Swamp greenbul	*Thescelocichla leucopleura*	Brown 1987
Zosteropidae (2/96)	Seychelles grey white-eye	*Zosterops modesta*	Brown 1987
	Mascarene grey white-eye	*Z. borbonicus*	Skutch 1999
Sylviidae			
Acrocephalinae (5/221)	Pitcairn reed-warbler	*Acrocephalus vaughani*	Cockburn 1996
	Seychelles warbler	*A. sechellensis*	Brown 1987
	Asian stubtail	*Urosphena squameiceps*	Brown 1987
	Senegal eremomela	*Eremomela pusilla*	Brown 1987
	Greencap eremomela	*E. scotops*	Du Plessis *et al.* 1995
Sylviinae			
Timaliini (13/233)	Yellow-eyed babbler	*Chrysomma sinense*	Brown 1987
	Yellow-billed babbler	*Turdoides affinis*	Brown 1987

(cont.)

Appendix 1.1. (cont.)

Family, subfamily, or tribe (N cooperative breeders/N total species in taxon)	Common name	Scientific name	Reference
	Common babbler	*T. caudatus*	Brown 1987
	Striated babbler	*T. earlei*	Brown 1987
	Arrowmarked babbler	*T. jardineii*	Brown 1987
	Large grey babbler	*T. malcolmi*	Brown 1987
	Black-lored babbler	*T. melanops*	Brown 1987
	Brown babbler	*T. plebejus*	Brown 1987
	Blackcap babbler	*T. reinwardtii*	Brown 1987
	Arabian babbler	*T. squamiceps*	Brown 1987
	Jungle babbler	*T. striatus*	Brown 1987
	Southern pied babbler	*T. bicolor*	Du Plessis *et al.* 1995
	Taiwan yuhina	*Yuhina brunneiceps*	Brown 1987
Passeridae			
Passerinae (1/36)	House sparrow	*Passer domesticus*	Brown 1987
Motacillinae (1/65)	Cape wagtail	*Motacilla capensis*	Brown 1987
Prunellinae (1/13)	Alpine accentor	*Prunella collaris*	Brown 1987
Ploceinae (3/117)	White-browed sparrow-weaver	*Plocepasser mahali*	Brown 1987
	Sociable weaver	*Philetairus socius*	Brown 1987
	Grey-headed social-weaver	*Pseudonigrita arnaudi*	Brown 1987
Fringillidae			
Emberizinae			
Thraupini (8/413)	Speckled tanager	*Tangara guttata*	Brown 1987
	Plain-colored tanager	*T. inornata*	Brown 1987
	Golden-hooded tanager	*T. larvata*	Brown 1987
	Turquoise tanager	*T. mexicana*	Brown 1987
	Dusky-faced tanager	*Mitrospingus cassinii*	Skutch 1999
	Medium ground-finch	*Geospiza fortis*	Brown 1987
	Common cactus-finch	*G. scandens*	Brown 1987
	Thick-billed euphonia	*Euphonia laniirostris*	Skutch 1999 (probable)
Cardinalini (1/42)	Black-faced grosbeak	*Caryothraustes poliogaster*	Skutch 1999
Icterini (5/97)	Bobolink	*Dolichonyx oryzivorus*	Brown 1987
	Bay-winged cowbird	*Molothrus badius*	Brown 1987
	Austral blackbird	*Curaeus curaeus*	Brown 1987
	Bolivian blackbird	*Oreospar bolivianus*	Skutch 1999 (probable)
	Brown-and-yellow marshbird	*Pseudoleistes virescens*	Brown 1987

APPENDIX 1.2 SPECIES NOT CONSIDERED COOPERATIVE BREEDERS DUE TO DIRECT BREEDING OPTIONS, BUT CONSIDERED SO IN OTHER PAPERS

Family, subfamily, or tribe	Common name	Scientific name	Reference
Struthionidae	Ostrich	*Struthio camelus*	Brown 1987
Rheidae	Greater rhea	*Rhea americana*	Codenotti and Alvarez 1997
Anseranatidae	Magpie goose	*Anseranas semipalmata*	Brown 1987
Rallidae	Tasmanian native hen	*Gallinula mortierii*	Brown 1987
Charadriidae			
Charadriinae	Southern lapwing	*Vanellus chilensis*	Brown 1987
Accipitridae			
Accipitrinae	Galápagos hawk	*Buteo galapagoensis*	Brown 1987
	Pale chanting goshawk	*Melierax canorus*	Malan *et al.* 1997
Falconidae	Merlin	*Falco columbarius*	James and Oliphant 1986
Pardalotidae			
Acanthizinae			
Sericornithini	White-browed scrubwren	*Sericornis frontalis*	Brown 1987
Petroicidae	Hooded robin	*Melanodryas cucullata*	Brown 1987
Sylviidae			
Acrocephalinae	Moustached warbler	*Acrocephalus melanopogon*	Fessl *et al.* 1996
Passeridae			
Prunellinae	Dunnock	*Prunella modularis*	Brown 1987
Fringillidae			
Emberizinae			
Emberizini	Smith's longspur	*Calcarius pictus*	Cockburn 1996

APPENDIX 1.3 SPECIES NOT CONSIDERED COOPERATIVE BREEDERS BECAUSE HELPING IS NOT A REGULAR EVENT OR IS LIKELY ACCIDENTAL, MISDIRECTED CARE

Family, subfamily, or tribe	Common name	Scientific name	Reference
Picidae	Middle-spotted woodpecker	*Dendrocopos medius*	Winkler *et al.* 1995
Psittacidae	Eclectus parrot	*Eclectus roratus*	Arnold and Owens 1998
Musophagidae			
Criniferinae	White-bellied go-away-bird	*Corythaixoides leucogaster*	Brosset and Fry 1988
Laridae			
Larinae			
Stercorariini	South polar skua	*Catharacta maccormicki*	Brown 1987

(cont.)

Appendix 1.3. (cont.)

Family, subfamily, or tribe	Common name	Scientific name	Reference
Accipitridae			
Accipitrinae	Cooper's hawk	*Accipiter cooperii*	Boal and Spaulding 2000
	Red-tailed hawk	*Buteo jamaicensis*	Boal and Spaulding 2000
	Swainson's hawk	*B. swainsoni*	Boal and Spaulding 2000
	Bald eagle	*Haliaeetus leucocephalus*	Boal and Spaulding 2000
Falconidae	American kestrel	*Falco sparverius*	Wegner 1976
Podicipedidae	Horned grebe	*Podiceps auritus*	Brown 1987
Scopidae	Hammerkop	*Scopus umbretta*	Du Plessis *et al.* 1995
Pardalotidae			
Pardalotinae	Striated pardalote	*Pardalotus striatus*	Brown 1987
Muscicapidae			
Turdinae	Groundscraper thrush	*Psophocichla litsipsirupa*	Du Plessis *et al.* 1995
Muscicapinae			
Muscicapini	Mariqua flycatcher	*Bradornis mariquensis*	Du Plessis *et al.* 1995
Saxicolini	European robin	*Erithacus rubecula*	Brown 1987
Certhiidae			
Troglodytinae	Cactus wren	*Campylorhynchus brunneicapillus*	Anderson and Anderson 1972
Paridae			
Parinae	Tufted titmouse	*Baeolophus bicolor*	Brown 1987
Hirundinidae			
Hirundininae	Blue swallow	*Hirundo atrocaerulea*	Du Plessis *et al.* 1995
	Brown-chested martin	*Phaeoprogne tapera*	Fraga 1979
Alaudidae	Spike-heeled lark	*Chersomanes albofasciata*	Du Plessis *et al.* 1995
Fringillidae			
Emberizinae			
Emberizini	Stripe-headed sparrow	*Aimophila ruficauda*	Brown 1987
	Chestnut-eared bunting	*Emberiza fucata*	Brown 1987
Parulini	Ovenbird	*Seiurus aurocapillus*	King *et al.* 2000
	Hooded warbler	*Wilsonia citrina*	Tarof and Stutchbury 1996
Cardinalini	Northern cardinal	*Cardinalis cardinalis*	Brown 1987

2 · Delayed dispersal

JAN EKMAN
Uppsala University

BEN J. HATCHWELL
University of Sheffield

JANIS L. DICKINSON
University of California, Berkeley

MICHAEL GRIESSER
Uppsala University

Cooperatively breeding birds are species in which social groups comprise at least three individuals that share parental care at a single nest. Although same-sex group members are sometimes non-relatives (Davies 1992), the majority of cooperative breeders exhibit delayed dispersal of offspring, which subsequently forgo reproduction and become non-reproductive helpers at the nests of parents or other close relatives. Group formation, social interactions, and reproduction within groups are characterized by both cooperation and competition among family members (Mumme 1997). In many cases, helpers derive indirect fitness benefits by increasing the productivity of their parents' nest. However, because helping at the nest only partially compensates helpers for failing to breed independently, helpers usually pay a net fitness cost by helping instead of breeding on their own (Stacey and Koenig 1990a).

Because most helping appears to represent a "best of a bad job" strategy, rather than an adaptive peak, it has become clear that we can understand the evolution of kin-based helping only by investigating why offspring forgo personal reproduction in the first place, and the associated question of why offspring, once they postpone personal reproduction, remain on their natal territories (Emlen 1982a). The decision to delay dispersal should be the key to the formation of family units, hence the aim of this chapter is to explore the selective basis of delayed dispersal. The selective factors favoring postponement of reproduction and helping are discussed in Chapter 3.

The route leading to cooperative breeding in multi-generational family groups involves a series of decisions. Mature offspring that become helpers have usually postponed independent reproduction, delayed dispersal to remain on their natal territories, and helped (Brown 1987). Theoretically, the decisions to postpone, delay, and help could have a common or related cause, but this is not necessarily the case. Individuals may postpone reproduction and float or they may delay dispersal without ever helping (Koenig et al. 1992). Similarly, natal philopatry may set the scene for kin-directed interactions throughout life rather than acting as a precursor to dispersal and independent breeding. For example, helpers may not postpone personal reproduction and may move back and forth between their parents' nest and their own nest within a day (Dickinson and Akre 1998), or they may switch to helping following the failure of their own breeding attempt (MacColl and Hatchwell 2002).

Dispersal is normally considered delayed when the offspring remain on their natal territories after they are competent to reproduce. This definition obscures much of the interesting variation in avian dispersal strategies and ignores the continuous nature of variation in timing of departure from the natal territory. A delay that extends into a bird's second year of life, when it is sexually mature, may not be very different from a delay that terminates just prior to the age of first reproduction. Increased understanding of the evolution and maintenance of delayed dispersal may come from a broader approach that seeks to explain continuous variation in the timing and modes of dispersal. The scope of this approach allows us to investigate the selective factors that favor prolonged association with parents and the natal site beyond the fledgling stage.

The value of the natal site for young birds that delay dispersal has received considerable attention (Brown 1969; Emlen 1982a; Stacey and Ligon 1987; Zack 1990), while changes in the social environment as a consequence of natal philopatry have been neglected until recently (Ekman et al. 2001a). One goal of this chapter is to integrate these two sets of ideas into a more comprehensive framework for studying delayed dispersal.

Ecology and Evolution of Cooperative Breeding in Birds, ed. W. D. Koenig and J. L. Dickinson. Published by Cambridge University Press.
© Cambridge University Press 2004.

The opportunity for prolonged interaction with relatives may lead to nepotism, which should be considered when developing hypotheses for why young birds should stay at home. Nepotism is considered of wide importance in species without cooperative brood care (Sherman 1985), but in studies of cooperatively breeding birds it has tended to be overshadowed by issues of habitat quality and consideration of the inclusive fitness benefits of helping.

The hypothesis that natal philopatry is favored by "extended parental investment," in which parents promote offspring fitness through prolonged brood care, is best considered as a non-mutually exclusive alternative to the hypothesis that delayed dispersal is favored by benefits associated with the quality or familiarity of the natal site (Ligon 1981; Brown and Brown 1984; Ekman and Rosander 1992; Ekman et al. 2001a). Prolonged brood care is not only an issue for species in which young remain on their natal territories, it is also relevant when young birds maintain close association with their relatives after they have left the natal territory to become independent breeders. In some species with helpers, young maintain social connections with their relatives long after they have dispersed to breed nearby (Curry and Grant 1989; Dickinson and Akre 1998; Hatchwell et al. 2001a; Kraaijeveld and Dickinson 2001; Russell and Hatchwell 2001). Such species may provide special insight into the importance of prolonged association with parents versus benefits of associating with the natal site.

PHYLOGENETIC CONSIDERATIONS

Phylogenetic analysis alone can distinguish evolutionary loss of helping from cases in which it never evolved in the first place. This question is not trivial, and a similar problem exists with identifying species in which delayed dispersal has either not arisen or has disappeared. Non-cooperative breeders that are derived from cooperative breeders should vary in the extent to which they have retained delayed dispersal. The corvids are one such lineage with representatives that breed singularly (Carmen 2004), delay dispersal without helping (Gayou 1986; Birkhead 1991; Ekman et al. 1994), delay dispersal, but usually help only during the fledgling stage (Burt and Peterson 1993), and have helpers-at-the-nest of either or both sexes (Woolfenden and Fitzpatrick 1984; Caffrey 2000; Baglione et al. 2002a, 2002b).

Phylogenetically controlled comparison of species that vary in whether or not they have delayed dispersal could ultimately point to life-history or ecological factors that favor retention of young. The focus of this chapter is the current functional utility of delayed dispersal, which should be addressed separately from the question of evolutionary origins (Reeve and Sherman 1993). Nevertheless, life-history traits are highly conserved in the adaptive radiation of birds (Owens and Bennett 1995), and given that such traits are likely to be associated with cooperative breeding (Arnold and Owens 1998; Hatchwell and Komdeur 2000), phylogenetic factors should be taken into account whenever demographic or life-history components are considered (Chapter 1).

Therefore, although we do not consider phylogeny further, it is important to investigate phylogenetic explanations for dispersal strategies because, while information on evolutionary origins cannot provide a rigorous test of current function, it is still critical for understanding the evolution of a trait. On the other hand, the response to transfer of carrion crow chicks from a non-cooperative to a cooperative population indicates that the phylogenetic legacy does not commit the species to a specific social behavior. Rather, delayed dispersal is a plastic response to local ecological or social conditions (Baglione et al. 2002a).

THE UNCOUPLING OF DELAYED DISPERSAL FROM COOPERATIVE BREEDING

Studies of delayed dispersal in birds have usually focused on cooperative breeders, giving the impression that the two behaviors are inextricably linked. While it is certainly true that delayed dispersal is a permissive factor allowing offspring to help their parents, it does not necessarily follow that delayed dispersal is maintained by the inclusive fitness benefits of helping (Fig. 2.1). Furthermore, it is unlikely that all kin-based helping in extant species was acquired via a stepwise process involving an intermediate that exhibited delayed dispersal but not helping. Just as some singular breeders are derived from cooperatively breeding ancestors, some species with kin-based social behavior, but no helping, are derived from ancestral species that exhibit the complete set of behaviors that characterize kin-based cooperative breeders (Edwards and Naeem 1993; Peterson and Burt 1992; Cockburn 1996). In cases where

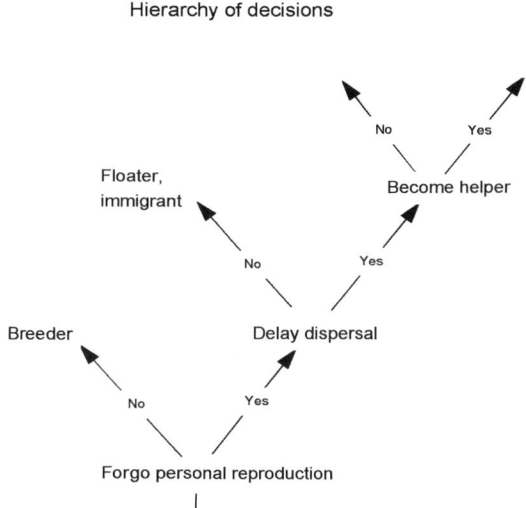

Figure 2.1. The hierarchical structure of decisions leading to cooperative breeding.

cooperative breeding is an historic legacy, the evolutionary stability of delayed dispersal may depend in part on inclusive fitness benefits of helping, but fitness data suggest that these benefits are rarely large enough to maintain delayed dispersal on their own (Stacey and Koenig 1990a).

On the other hand, the existence of species that exhibit delayed dispersal without helping demonstrates that delayed dispersal can be maintained in the absence of inclusive fitness gains from helping (Gayou 1986; Veltman 1989; Birkhead 1991; Ekman et al. 1994; Frith et al. 1997; Urban et al. 1997; Walls and Kenward 1998; Jansen 1999; Robinson 2000; Green and Cockburn 2001; Nakamura et al. 2001). Examination of the functional utility of delayed dispersal in such species will provide key insights into the evolution and maintenance of the patterns of philopatry that characterize many cooperative breeders.

Species with delayed dispersal of offspring that do not help have attracted little attention until recently. In part, this is because their young do not exhibit the apparent altruism that has excited the interest of sociobiologists for the past 40 years. Because there is cogent evidence demonstrating that delayed dispersal can be maintained in the absence of inclusive fitness gains of helping at the nest, however, we regard delayed dispersal as an independent decision that does not require inclusive fitness benefits via helping behavior. Any inclusive fitness benefits of helping would certainly augment benefits of delayed dispersal, but they appear to be neither necessary nor, in most cases, sufficient to explain why dispersal is delayed.

Delayed dispersal: where to wait

Delayed dispersal is merely one of several alternatives available to individuals that are unable to find a suitable reproductive vacancy (Koenig et al. 1992). Such individuals may wait on the natal territory for a vacancy to emerge, disperse and attempt to settle elsewhere, or roam between territories as a "floater."

We can best understand what is unique about staying home by considering the full gamut of dispersal options available to young birds. Young may disperse alone or in coalitions of relatives. After dispersal they may settle and become site-faithful or roam widely. Socially, they may remain solitary, form pairs, or join groups. Groups that form after dispersal will typically be comprised of non-relatives, except when individuals disperse in coalitions, as occurs in acorn woodpeckers, in which single-sex breeding units are often comprised of same-sex relatives (Koenig and Mumme 1987). The decision to stay home means that an individual will remain sedentary at its natal site, and will usually live in a social group with relatives, often including its parents.

Although Brown (1987) and Koenig et al. (1992) suggested that the potential fitness of floaters is a critical factor in why young birds stay home, the option of waiting outside the natal territory, once breeding is delayed, has received little empirical attention. If young birds face constraints on independent breeding, the ultimate cause of delayed dispersal can be best understood by determining why it is better to wait on the natal territory than elsewhere.

In part, the answer to this question lies in an investigation of the survival and reproductive advantages, current and future, of waiting on the natal territory, which should be independent of any fitness gained through helping. Survival advantages may include gains due to nepotism and gains arising from the quality of or familiarity with the natal site relative to other locales. The natal site may also yield reproductive advantages if staying on a high-quality territory means increased access to potential breeding partners that are attracted to the natal site (Kraaijeveld and Dickinson 2001), access to a portion of the natal site (Woolfenden and Fitzpatrick 1984), or nepotistic assistance in competing for reproductive vacancies (Brown 1987). Such benefits must be weighed against the costs of staying. For example, offspring remaining in the natal territory may experience limited access to vacant habitat, while floaters may search larger areas, thereby increasing the frequency with which they locate reproductive vacancies (Koenig et al. 1992; Kokko and Ekman 2002). Limited access to habitat that is "suitable" for reproduction has been a main theme in discussions of delayed dispersal (Brown 1969; Koenig and Pitelka 1981; Stacey and Ligon 1987). However, while this argument applies to the decision to postpone independent reproduction it cannot be extrapolated to the decision to delay dispersal. While offspring may forgo independent reproduction for a lack of high-quality habitat or mates (Komdeur 1992; Pruett-Jones and Lewis 1990) they do not necessarily have to wait for a breeding vacancy in the natal territory. Therefore postponed reproduction and delayed dispersal do not necessarily have the same cause, although constraints arguments are often invoked to explain both phenomena.

Constraints on successful reproduction are ubiquitous and it is difficult to identify a particular intensity of constraint that results in either deferred reproduction or delayed dispersal (Hatchwell and Komdeur 2000). There are a number of species that do not delay dispersal although they experience breeding constraints that appear to be as pronounced as those exemplified by cooperative breeders (Smith 1978; Carmen 2004). These examples show that variation in habitat quality and a shortage of breeding vacancies (territories or mates) are not sufficient to account for delayed dispersal.

Dispersal decisions prior to the age of first reproduction

The distribution of resources, particularly food, may explain variation in the time of dispersal. Approximately 3% of avian species worldwide are classified as cooperative breeders (Russell 1989; Clarke 1995; Arnold and Owens 1998; Chapter 1) and the majority exhibit delayed dispersal (Stacey and Koenig 1990a). In addition, there are a number of species that exhibit delayed dispersal but do not exhibit helping, and the frequency of these species is probably underestimated. Much of the extraordinary variation in winter social systems and space use by birds appears to be tied to food supply, so resource distribution during the non-breeding season may play a critical role in allowing young to delay dispersal and remain on their natal territories during their first breeding season.

An example of this is the western bluebird, in which sons commonly remain on their natal territories for winter, but only occasionally stay through the first breeding season and help (Kraaijeveld and Dickinson 2001). The basis of winter territoriality in western bluebirds is mistletoe (*Phoradendron villosum*), a highly clumped berry resource. The hypothesis that winter food accounts for the retention of sons and their subsequent localized dispersal can be tested by comparing dispersal and sociality of western bluebirds breeding in habitats with and without mistletoe, a study that has yet to be performed. In other cases, food supplies may not be sufficient to allow birds to stay on their breeding grounds for winter, or parents may be able to stay, but do not retain offspring due to a reduction in survival with increasing group size.

Environmental and life-history correlates of cooperative breeding

The tight connection between year-round residency and cooperative breeding (Arnold and Owens 1998) may be further refined by considering whether or not the social system is influenced by kinship outside the breeding season. For example, it may be that selection for retention of families on their territories for winter is a key predictor of kin-based helping. Of course, any investigator using phylogenetic comparative methods must also recognize the circularity of correlative analyses: year-round residency and retention of offspring through the winter could be either a predisposing factor or an inevitable consequence of selection on families to live in groups during the subsequent breeding season.

Several authors have suggested a link between a high adult survival rate and cooperative breeding. (Rowley 1965; Fry 1977; Brown 1987; Arnold and Owens 1998). This association could arise simply because life-history traits and, more specifically, high survival will directly influence the relative values of current and future reproduction, favoring postponement of reproduction (Stearns 1992). Species with high survival place a greater value on future reproduction and should therefore be more willing to forgo inferior breeding opportunities in the current year to increase opportunities for future reproduction. The longevity argument is that there is a bias favoring cooperative breeding in long-lived species because the option of forgoing reproduction is available to species with a longer life expectancy but not to shorter-lived species.

A weakness in this argument is that postponement of reproduction does not necessarily coincide with delayed dispersal. While a longer life gives individuals more leeway as to the breeding opportunities they will accept, it has nothing to say about whether the offspring should wait on the natal site or elsewhere. Alternatively, low adult mortality could influence dispersal behavior by reducing the rate at which vacant territories become available (Russell and Rowley 1993b; Arnold and Owens 1998; Hatchwell and Komdeur 2000). In other words, the trade-off is mediated through the effect of survival on the turnover of breeding vacancies. Again, it is important to bear in mind that any correlation between delayed dispersal and adult survival may simply mean that increased survival is a consequence, rather than a cause, of delayed dispersal.

Cooperative breeding is most frequent in the southern hemisphere, particularly in Australia. Cockburn (1996) suggested that phylogenetic considerations may be a key factor in this distribution, but here we consider the possibility that climate and winter food supply play a significant role. The retention of young will ultimately be a balance between the costs of retention for parents and the benefits that accrue to parents through increased survival and reproductive success of their offspring. The costs of retention may very well be higher in colder climates where energetic needs are increased and winter food supplies are diminished (Ekman and Rosander 1992).

A role of climate is implied by the rarity of delayed dispersal among northern-hemisphere birds (Russell 2000). This could be linked to a seasonal environment with harsh winters. In the temperate and sub-boreal climates of the northern-hemisphere land masses, pronounced winter conditions may restrict sharing of food in territorial or colonial species, even though they are site-faithful. In contrast, the landmasses of the southern-hemisphere are largely tropical and subtropical so the costs of retaining offspring could be lower.

The comparative analysis by Russell (2000) shows that seasonal variation in access to energy can explain large-scale patterns in the seasonal timing of dispersal. She demonstrated that the rarity of species with delayed dispersal in the northern hemisphere is associated with a tendency for offspring to disperse prior to winter. This analysis provides compelling evidence for a role of winter energy resources in delayed dispersal. In contrast, she found no seasonal peak in the southern hemisphere.

Together, these results indicate that less seasonal habitats provide conditions that favor offspring staying with their parents. This key finding supports the contention of Rowley (1968) and Ford et al. (1988) that aseasonality promotes cooperative breeding, but it contrasts with the conclusion of Du Plessis et al. (1995) that cooperation is associated with seasonal environments in South African birds.

BENEFITS OF DELAYED DISPERSAL

There are two issues that must be taken into account when considering dispersal decisions, and especially the benefits of delayed dispersal. First, offspring are not making a unilateral decision. In general, it is likely that parents can exercise a degree of control over offspring dispersal, although they cannot force offspring to stay. Therefore, for dispersal to be delayed, the interests of the parents and offspring must broadly coincide. Second, to understand the evolution of delayed dispersal it is important that its benefits are not simply the benefits of group living. In other words, there must be a benefit that is explicitly related to retention on the natal territory and/or association with kin, or, put differently, "a special value to home." If this is not the case, there is no particular reason to expect family formation rather than simply group formation.

Delayed dispersal may provide benefits through prolonged association with the natal site or prolonged association with kin, and with parents in particular. These alternatives are not mutually exclusive. The delayed independence observed in species like ducks, swans, geese, and cranes, which live in non-sedentary groups with long-lasting family associations, points to the importance of benefits of prolonged contact between parents and offspring, independent of association with the natal site (Black and Owen 1987, 1989; Alonso and Alonso 1993; van der Jeugd 1999). In these non-sedentary species, the dominants are the last to become independent, suggesting that young compete to stay with their parents (Black and Owen 1987).

While benefits of remaining with relatives in non-sedentary species suggest that delayed independence can be favored by the benefits of nepotism alone, they do not preclude additional benefits of associating with the natal site. It is important to note that association with parents and the natal site co-occur in sedentary species and that both should be considered of potential importance in determining the benefits of delayed dispersal.

All that is required for maintenance of delayed dispersal within a population is that delayers do better by delaying than by taking advantage of alternative options. Species with flexible natal dispersal strategies such as Siberian jays (Ekman *et al.* 2002), carrion crows (Baglione *et al.* 2002a, 2002b), and brown thornbills (Green and Cockburn 2001) are particularly interesting in this context, because the consequences of pursuing alternative options can be observed. However, delayers as individuals do not have to outperform other birds in their cohort that disperse and breed in their first season. It is entirely possible that delayers are individuals unsuccessful or unlucky in competition for breeding space and mates. If the option is to breed on a low-quality territory and incur breeding and survival costs of early independence with little chance of producing offspring, it does not take much of a survival or future reproductive benefit to favor remaining on the natal territory. Delayed dispersal in this case can be best understood by considering the ways in which enhanced survival or enhanced opportunities for reproduction may lead to the minimal fitness increases needed to improve the lot of an individual that already has reduced fitness due to reduced competitive success at the onset of breeding.

The empirical evidence is, however, equivocal regarding which offspring postpone dispersal. A number of studies have demonstrated sibling rivalry in which the stronger offspring stay with parents (Black and Owen 1987; Strickland 1991; Ekman *et al.* 2002). These studies indicate that staying is a preferred option and that there are benefits to be gained from staying. Other studies suggest that the opportunity to stay is assumed by poor-quality phenotypes with a low potential to compete for available vacancies (Richner 1990).

Increased access to high-quality territories or mates

In species without helpers, staying home will be favored if it enhances an individual's opportunity to breed or fill a vacancy on a high-quality territory. This acquisition of a breeding territory could also operate through a "budding" process (Woolfenden and Fitzpatrick 1984), which might be favored by nepotism or by a correlation between natal-site quality and priority of access to high-quality sites or mates.

At first glance, inheritance appears to be a likely outcome of staying on the natal territory, but empirical data have shown that retained offspring only sometimes inherit the natal territory in cooperative breeders, an observation that is linked to incest avoidance (Koenig *et al.* 1998; Komdeur and Edelaar 2001a; Chapter 9). Young birds that assume vacancies in their natal groups would risk mating with a parent or sibling unless both their mother and father were dead. As incestuous mating is rare, the filling of a vacancy is often followed by dispersal of helpers of the deceased sex (Koenig *et al.* 1998) or dispersal of breeders of the surviving sex (Piper and Slater 1993).

In the fairy-wrens (*Malurus* spp.) the problem of incest avoidance has an unusual resolution. While young males that lose their fathers remain in their natal groups and help rear their mothers' young, the young are usually sired by unrelated males from outside the group through extra-group mating and fertilization (Brooker *et al.* 1990; Dunn *et al.* 1995; Double and Cockburn 2000). The non-territorial long-tailed tit also appears to have arrived at an unusual solution to the risk of inbreeding between mothers and their strongly philopatric sons. Adults have relatively high mortality and so females may run a risk of pairing with sons that have recruited into the local breeding population. Therefore, pairs that have bred successfully usually divorce before the following breeding season, the female moving to pair with a male from a different family (Hatchwell *et al.* 2000). Territorial inheritance is also rare in species without helpers and in those with infrequent helping, so inheritance is unlikely to provide a general explanation for delayed dispersal (Ekman *et al.* 2001b; Kraaijeveld and Dickinson 2001).

Even though inheritance is rare, the quality of the natal site may still be important if it is spatially autocorrelated with the quality of nearby sites. If individuals elect to stay only on high-quality territories, and if the ability to detect and fill vacancies diminishes with distance, then offspring remaining on high-quality territories would have increased access to high-quality vacancies. In this scenario individuals are queuing for good territories, a hypothesis first put forth by Zack (1990). Thus, delayed dispersal offers potential fitness benefits, not only while an offspring is waiting to become a breeder, but also after it has dispersed and begun to breed independently (Ekman et al. 1999).

High-quality territories may also serve as attractants for mates, a benefit that may be particularly important in cases where reproduction is delayed because one sex is in short supply (Kraaijeveld and Dickinson 2001). In the western bluebird, for example, females may be attracted to territories with abundant mistletoe. Sons that stay home often mate in spring with yearling females that were attracted to and joined their winter group in fall.

Finally, variation in habitat quality may be important in selecting for delayed dispersal if the quality of the natal site influences an individual's condition or competitive ability (Kraaijeveld and Dickinson 2001). An offspring's decision to stay should depend on site quality if staying on a high-quality site means being in better condition and having greater energy reserves when it comes time to compete for a vacancy. Decision-making of this sort would be irrelevant if all sites were equivalent, suggesting that variation in territory quality is important to the outcome of competition for vacancies. An example is the Seychelles warbler, where there is good empirical evidence that individual dispersal strategies are influenced by habitat quality (Komdeur and Edelaar 2001b).

Site quality and individual quality are likely to be correlated, hence if high-quality individuals compete more effectively for superior reproductive vacancies, their mates will also tend to be in superior condition. These correlations should augment the benefits of staying home where high-quality territories are clumped.

The importance of variation in habitat quality for the evolution and maintenance of delayed dispersal has recently been challenged by Kokko and Lundberg (2001), who modeled dispersal as a trade-off between habitat quality and degree of crowding. In a game-theoretical approach, where individuals were assumed identical in their ability to search out and compete for high-quality territories, simulations indicated that delayed dispersal is unlikely to be maintained by variation in habitat quality alone. The model allows individuals to make dispersal decisions based on trade-offs between habitat quality and degree of crowding, which is in turn determined by life-history traits.

Although Kokko and Lundberg (2001) incorporated the valuable idea that the physical properties of the habitat may be compromised by the degree of crowding, they assumed that all individuals have equal access to high-quality sites. As such, their model did not incorporate the mechanisms that we outline above to explain delayed dispersal as a product of individual reproductive decisions.

The assumption of equal access to vacancies was relaxed in a model by Kokko and Ekman (2002) that incorporated a dominance structure within broods, resulting in siblings queueing for territorial vacancies. More consistent with empirical data on species with delayed dispersal (Black and Owen 1987, 1989; Strickland 1991; Ekman et al. 2002), their model shows that offspring may prefer to delay dispersal for benefits gained in the natal territory (the "safe haven" effect) even when they cannot inherit and when they suffer reduced ability to search for vacancies.

The potential "safe haven" effect of habitat quality can be augmented by nepotism. If survival is food-limited, parents should be more willing to concede food to retained offspring in high-quality habitat, leading to a correlation between territory quality and degree of nepotism, and thus increasing the benefits of staying home on high-quality territories (Ekman and Rosander 1992).

In addition to nepotistic sharing of food, a high-quality natal territory, and access to high quality neighboring territories, offspring that stay home may receive assistance from relatives when competing for breeding vacancies, as occurs in the Florida scrub-jay (Woolfenden and Fitzpatrick 1984). This form of nepotism combines parental assistance with sharing of space via territorial budding, where the young are granted a portion of the parents' territory on which to breed. As a second example, the likelihood of attaining a reproductive vacancy in acorn woodpeckers depends in part on the size of the coalition of same-sex family members seeking to disperse together, indicating a group-size effect on breeding access that is a direct outgrowth of staying home (Hannon *et al.* 1985). The same mechanism could apply when young assume vacancies individually as long as parents and siblings provide aid during the competition phase and, if boundary disputes arise, perhaps afterward.

Increased survival: the value of the natal site

When considering the survival benefits of delayed dispersal there is an implicit assumption that "home" has a unique value and is a superior place to wait. There are two obvious properties of "home" that distinguish it from elsewhere and that potentially contribute to its superiority: site familiarity and relatedness to other residents. We first address the implications of site familiarity.

When offspring disperse to find a vacancy, they are likely to compete with local birds. This may be difficult if retained offspring have a competitive edge in disputes over territories, as is generally the case for residents compared to intruders in territorial species. This outcome can be explained by an arbitrary rule, an asymmetry in resource-holding potential, or an asymmetry in the value of the territory to the competitors (Parker 1974; Maynard Smith and Parker 1976; Maynard Smith 1979). Empirical tests in territorial birds have generally supported the value-asymmetry hypothesis (Krebs 1982; Jakobsson 1988; Beletsky and Orians 1996; Hatchwell and Davies 1992), with the suggestion that territory familiarity plays a key role. Therefore, assuming that territory ownership is beneficial, remaining on the natal territory may yield a substantial benefit in competitive interactions compared to the alternative option of early dispersal to find a new territory. Of course, even individuals that delay dispersal will have to compete for a breeding position eventually, but by deferring the contest the number of competitors may be reduced by winter mortality.

Increased survival: nepotism

The natal territory is unique in that it is the only place where the offspring can associate with their parents. When young stay home, the relationship with their parents is likely to contain elements of parental care, mutualism, and competition. Where more than one offspring stays, dominance interactions and competition among siblings may be pronounced. For example, in acorn woodpeckers competition is expressed at an early age as dominance interactions among fledglings (Stanback 1994), and later on among joint-nesting sisters that compete by destroying each other's eggs (Koenig *et al.* 1995).

Defensive behaviors, interactions over food or at roosts, and access to mates should therefore be complex products of competition, dominance, cooperation, and kinship, the general expectation being that kinship should mitigate competitive interactions (Hamilton 1964). More importantly, parents will have an incentive to give preferential benefits to their offspring and such preferential treatment should in turn provide an incentive for offspring to stay.

Nepotistic behavior may also be extended to relatives that have dispersed to neighboring territories. For example, cooperatively breeding long-tailed tits (Hatchwell *et al.* 2001a) and non-cooperative red grouse (Watson *et al.* 1994) exhibit differential treatment of neighbors in relation to kinship during the winter. Many of the nepotistic interactions that result from delayed dispersal may also apply to species like these with at least some limited dispersal.

Starvation and predation are the two main threats to offspring survival that can be modulated by remaining with parents. If survival is food-limited, parents may gain by allowing offspring preferential access. Defense against predators is generally thought to be costly behavior that should be directed preferentially toward close kin, although costs of defensive behavior have rarely been measured (Sherman 1977). When nepotism is dependent on site quality, it may be difficult to separate habitat quality issues from benefits of nepotism.

Nepotistic concession of food

Parental tolerance in which retained offspring are allowed preferential access to food has been found in a number of bird species (Scott 1980; Barkan *et al.* 1986; Ekman *et al.* 1994; Pravosudova 1999) and may be more general than currently suspected. Parental concession of food through restraint on aggression is a "non-behavior" and can easily escape notice. However, its importance may be profound in family groups that form through retention of young.

Social conflict at foraging sites can also have a strong impact on priority of access to food in avian flocks (Ficken *et al.* 1990; Ekman and Lilliendahl 1993).

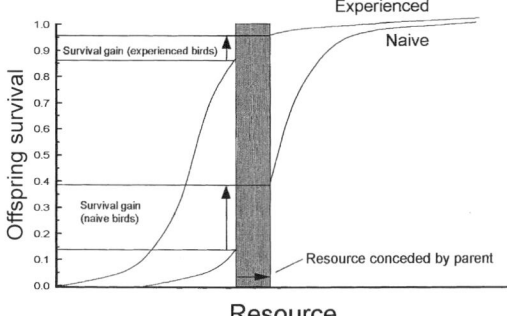

Figure 2.2. The role of offspring experience in the payoffs of parental nepotism. The payoffs are influenced by the impact of food concession on parent and offspring survival with the underlying assumption that the impact will be lower on experienced than on naive offspring. The figure shows a lower survival payoff from the same amount of food conceded to experienced compared to inexperienced offspring.

Expression of nepotism may depend upon social context if its main function is to protect offspring from interference competition with non-kin (Scott 1980; Black and Owen 1987). Offspring need not consume more food as a consequence of nepotism; rather, they may benefit because access to food is more constant and predictable. Such predictability may be as important to starvation risk as is direct consumption rate (McNamara and Houston 1990). When access to energy is predictable, individuals can afford to store less energy in body fat to buffer the risk of starvation (Ekman and Hake 1990). Such a reduction in the size of fat reserves can be beneficial if fat loads lead to increased risk of predation (Witter and Cuthill 1993).

The incentive for parents to concede resources depends on their own costs and on the impact of preferential access to food on offspring survival (Ekman and Rosander 1992; Fig. 2.2). Risk of starvation has a non-linear relationship with resource abundance, so the quality of the natal site will influence the costs and benefits of food sharing (Ekman and Rosander 1992). This trade-off can be measured in terms of increased risk of starvation for the parents and reduced risk for their offspring. The relationship between starvation risk and access to food is what could drive parents to concede food to their offspring. As their personal survival prospects increase, parents should be more willing to share resources. Similarly, as the impact of food sharing on offspring survival increases, parents should be increasingly willing to incur food-sharing costs. A general conclusion is that evolution will promote concession of food by parents when they are more competent foragers and therefore face a lower risk of starvation (and thus higher survival) than their inexperienced offspring. Conversely, the incentive for preferential treatment of offspring will be altogether absent if parental survival prospects are poor.

The costs to parents of conceding food are low when their survival is high. However, it is not the level of survival that is important, but the rate of change in survival, which depends on the overall survival prospects. Thus the rate is low when survival is high. Simultaneously the direct fitness gains from parental concession of food will be influenced by kinship in accordance with Hamilton's rule. To calculate the direct fitness gain from conceding food the benefit to the offspring has to be devalued by relatedness. Once the parents are below the region with high survival, concession costs accelerate, while simultaneously the benefit to an offspring is devalued by half (relatedness $r = 0.5$, Fig. 2.2).

Parental concession of food in this scenario is coupled with habitat quality. This once again emphasizes that nepotism and variation in habitat quality are not mutually exclusive explanations for delayed dispersal. Rather, the influence of habitat quality is filtered through the behavior of the parents in their propensity to concede food. It is this filtering that leads to nepotism, allowing offspring preferential access to resources.

Nepotism in defense

The effects of group size on defensive behavior and risk of predation have been addressed extensively in birds, but the importance of group composition, and kinship in particular, has received little attention. In groups, risk of predation is diluted through selfish herd effects (Hamilton 1971) and reduced by increased vigilance (Pulliam 1973) and alarm calling. The first mechanism is indiscriminate in benefiting kin and non-kin alike, but vigilance and alarm calling are behaviors that can be used preferentially to benefit kin (Sherman 1977).

Alarm calls are often assumed to carry costs that are specifically borne by the caller and so are often assumed to be altruistic. However, such signals may also be interpreted as selfish behavior (Sherman 1985; Clutton-Brock *et al.* 1999b). Therefore, it is important to demonstrate that alarm calling varies in relation to the social environment.

For example, parents may favor their own offspring, and in some cases even more distant relatives, by being more vigilant to attacks when their young are present and by calling more frequently when their own offspring are present. Such nepotistic alarm calling is known for several species of mammals (Sherman 1977, 1985; Hoogland 1983; Cheney and Seyfarth 1985). In birds, Siberian jay parents provide protection by being more vigilant in the company of their offspring (Griesser 2003), and via nepotistic alarm calling in winter flocks (Griesser and Ekman 2004).

Nepotism at roosts

Kinship may also mitigate competitive interactions in communal roosts of winter groups. Close huddles of roosting birds during the non-breeding season have been recorded in many bird species (Beauchamp 1999; Chapter 7). These may be temporary associations among individuals that are usually solitary or live in pairs, such as treecreepers (Cramp and Perrins 1993) and the wren (Armstrong and Whitehouse 1977). Among group-living species these huddles have a stable composition and often occur every night during the non-breeding season. For example, green woodhoopoes (Williams *et al.* 1991) and acorn woodpeckers (Du Plessis *et al.* 1994) roost together in cavities, while babblers (*Turdoides* spp.) (Bishop and Groves 1991; Gaston 1977) and long-tailed tits (Cramp and Perrins 1993) roost in linear arrays.

Individuals in communal roosts of bushtits benefit through reduced thermoregulatory costs (Chaplin 1982), but unlike nepotistic concession of food and nepotistic defense, these benefits are mutual in that both parents and offspring reduce overnight energy expenditure. However, for communal roosting to be a benefit of delayed dispersal rather than simply a benefit of group living, it must be shown that behavior at roosts is nepotistic and/or that there are advantages to roosting with kin rather than non-kin.

There has been very little research on these issues, but some suggestive evidence exists. For example, in jungle babblers (Gaston 1977) and Arabian babblers (Bishop and Groves 1991), where groups are usually nuclear families, the end positions of the linear huddles are always taken by dominants. By contrast, winter groups of long-tailed tits include a substantial portion of non-relatives (Hatchwell *et al.* 2001a) and dominants always occupy central roost positions following a short period of jockeying for position during roost formation (A. McGowan and B. Hatchwell, unpublished data). These examples suggest that there may be differential treatment of kin in communal roosts, although this has yet to be demonstrated.

The relative importance of nepotism versus territory quality

Because parents are generally nepotistic, it may appear trivial to invoke preferential treatment as a key benefit of delayed dispersal. However, the crucial point is not nepotism *per se*, but for how long parents provide such favors to their offspring. Time is the key feature of delayed dispersal, and it is the latency to departure from the natal territory that begs explanation.

While all avian parents may be nepotistic to some degree, selection acts on the point at which parental care is terminated. Modeling the trade-offs involved both demonstrates the logic of parental concession of resources, and helps to delineate the optimal duration of parental care. The inclusive fitness gain for nepotistic parents depends on offspring skill level and survival probability. As the offspring gain experience, parents gain less by conceding food (Fig. 2.2). The potential for direct fitness gains from nepotism eventually declines to zero.

There is now substantial evidence for "extended parental investment" in species with delayed dispersal. Evidence of nepotism, where parents provide retained offspring access to food and protection against predators suggests that staying with parents may increase survival. Long-term studies in a variety of taxa have shown that longevity accounts for most of the variation in lifetime reproductive success (Clutton-Brock 1988, Newton 1989), so the impact of delayed dispersal on survival may be more important than any gains due to increased access to high-quality breeding sites and mates. Indeed, it is theoretically possible that delayed dispersal is maintained through enhanced survival alone (Kokko and Johnstone 1999).

If nepotism is to promote delayed dispersal, favoring offspring that wait for a vacancy with kin rather than elsewhere, it must result in a demonstrated increase in offspring fitness. A number of studies suggest that winter survival is enhanced for offspring postponing dispersal in the company of their parents (Black and Owen 1987; Ekman et al. 2000; Green and Cockburn 2001; Kraijeveld and Dickinson 2001). However, of these examples, only the Siberian jay study (Ekman et al. 2000) effectively separated survival effects from effects of territory quality. On territories that had a mix of philopatric offspring and young birds that dispersed into the group, survival was enhanced only for the offspring of the local pair. Hence, in this species, enhanced survival of delayed dispersers appears to be a function of nepotism, not of territory quality.

The generality of nepotism as a factor favoring delayed dispersal is currently unknown. The behavioral dynamics of families interacting in winter, when resources are scarcest, and the impact of kin-based interactions on subsequent survival and reproductive success of group members, are likely to offer novel insights into the costs and benefits of prolonged association with kin. The extent of between-group and within-group discriminatory behavior is poorly understood, perhaps because of the relatively limited research effort invested in studying birds during the non-breeding season. Such discrimination clearly occurs, however, as evidenced by experimental removal of fathers in Siberian jay family groups. Only retained offspring, and not non-family group members, left when removed fathers were replaced by unrelated immigrants, providing support for the hypothesis that presence of parents is essential to the decision to be philopatric (Ekman and Griesser 2002).

COSTS OF DELAYED DISPERSAL FOR OFFSPRING

In this chapter we assume that constraints on independent breeding are usually a cause rather than a consequence of delayed dispersal. However, once dispersal is delayed, both cooperation and competition may influence an individual's prospects of becoming a breeder. Because dispersal is usually female-biased in birds, the most obvious of these costs is competition with same-sex relatives for opportunities to breed. This represents a cost for parents to the extent that competitive interactions of one offspring with another reduce the average reproductive success of offspring. This argument is analogous to Clark's (1978) model of local mate competition in the somewhat different context of facultative sex-ratio manipulation. If offspring compete directly with the same-sex parent, the parents' fitness will be even more strongly affected.

Such interactions are well known to have reproductive consequences in winter flocks of unrelated birds (Otter et al. 1999). Within-sex dominance hierarchies may also be important in interactions involving kin (Wiley and Rabenold 1984). Many studies of cooperative breeders have documented intragroup conflict, particularly over reproduction (Cockburn 1998), and identical mechanisms should govern transactions within groups that are comprised of parents and retained young.

Whether staying home has costs in terms of the ability to search for and locate breeding vacancies is not clear. Apart from the limited option of territorial inheritance, offspring can either wait in the natal territory until a vacancy emerges nearby or they can disperse before one becomes available. In choosing between these two options, behavioral plasticity may provide subordinate group members with the best of both worlds: prolonged association with parents and the natal site during the non-breeding season and the ability to choose between dispersing and staying as the breeding season approaches.

In western bluebirds, for example, most females disperse far from home in fall to join new groups of breeders and their philopatric sons from the prior spring (Kraaijeveld and Dickinson 2001). A small proportion of females delay dispersal to stay the winter in their natal group, and then usually disperse to breed in the spring. Their behavior is in sharp contrast with that of philopatric males, which disperse just a few hundred meters from their natal groups in spring, usually with an immigrant female that joined their winter group. This suggests that prolonged attachment to the natal territory and parents does not necessarily lead to natal philopatry and that the costs and benefits of breeding close to home may not be the same for males as for females. It thus seems reasonable to expect that philopatric males might also vary in their post-winter dispersal behavior.

Dispersal can take many forms. Young birds may become "floaters" while searching for vacancies. However, it is not clear that floaters have a better chance of detecting a vacancy than do territorial residents that make regular forays. Birds that delay dispersal may be better able to recognize vacancies, however, and may detect vacancies more quickly by recognizing individuals' location and songs. The selective basis of floating versus settling is likely to be a complex outcome of the costs of acquiring and defending space and the association between site fidelity and the probability of breeding.

Dispersal strategies are likely to be profoundly influenced by incest avoidance (Johnson and Gaines 1990; Weatherhead and Forbes 1994). If pairs form in winter flocks, dispersers will have a higher probability of finding a breeding vacancy or available mate within a flock of non-relatives than within a flock of relatives. Therefore, although retained offspring may have increased access to high-quality territories and to the benefits of prolonged parental investment, these advantages must offset the costs of competition with relatives and reduced availability of unrelated mates (Zack and Ligon 1985; Walters 1990; Ekman *et al.* 2001b). The selective pressure that incest avoidance exerts on offspring dispersal strategies will depend critically on the costs of inbreeding and as yet this cost has been measured in few bird species (Keller and Arcese 1998, Koenig *et al.* 1999; Chapter 9).

COSTS OF DELAYED DISPERSAL FOR PARENTS

Parental concession of food to retained offspring has been confirmed for several species (Scott 1980; Barkan *et al.* 1986; Black and Owen 1987; Ekman *et al.* 1994; Pravosudova 1999). Such concessions should be considered extended parental care as they are identical to investment in offspring earlier in life. As such it has an implicit cost to the parents, the magnitude of which should be a key determinant of the decisions of parents to retain young and concede food. There is currently no compelling evidence as to how large these costs might be. In part, this may be because teasing apart parental effects from other confounding factors is an empirically difficult task.

An experimental field study of the tufted titmouse explicitly tested the hypothesis that retention of young reduces parental survival (Pravosudova and Grubb 2001). Overwinter survival was higher for parents with retained offspring than for pairs whose offspring had been experimentally removed. While this result appears to contradict Ekman and Rosander (1992), who modeled retention of offspring as a trade-off between the benefits of increased offspring survival and the cost to parents of food sharing, the apparent positive effect of offspring retention on the survival of tufted titmouse parents is confounded with group-size effects as the size of parent–offspring groups had not been reduced. Consequently, despite being experimental, this study does not falsify the hypothesized costs of retention for parents, because it is still possible that the general group benefit outweighed the cost to parents of conceding food.

A number of processes can cause a reduction in personal fitness directly attributable to associating with offspring and absent when parents associate with non-relatives. Examples include reduced feeding due to concession of food to offspring and increased conspicuousness due to nepotistic vigilance. Simple observation can detect these differences in behavior of breeding-aged adults interacting with kin versus non-kin. However, tests of parental costs can be achieved only by controlling for flock size and group composition, which may be a far less tractable proposition.

CONCLUSION

Delayed dispersal involves a complex interplay between the costs and benefits for parents of retaining young and the costs and benefits for young of remaining on their natal territories and prolonging their period of interaction with kin. Nepotism is the only benefit unique to remaining with parents, and the potential fitness gain from what is effectively prolonged parental care is a viable explanation for delayed dispersal by offspring. However, because territory or site quality influences the costs of food sharing for nepotistic parents and the benefits of remaining at home for delayers, spatiotemporal variation in resources remains a critical component of any analysis of delayed dispersal. This argues for experiments addressing offspring retention and the behavior of parents toward offspring in those systems where resources can be easily manipulated.

Measurement of the fitness consequences of different dispersal strategies, through either observation or experiment, represents a formidable challenge. In particular, the potential for confounding effects of group size, habitat quality and individual quality requires cautious interpretation of both demographic and experimental data. Behavioral analysis of interactions within and between non-breeding groups is, perhaps, a more promising area for research. Such studies have certain advantages because, in many respects, interactions with kin and non-kin are relatively straightforward to study empirically, although it must be recognized that the mechanism of kin recognition may play a crucial role in determining the degree of kin discrimination possible by nepotistic individuals.

In general, we believe that these aspects of the behavioral ecology of cooperative breeders have been largely neglected, despite the crucial importance of delayed dispersal in "setting the scene" for helping behavior. It is likely that many species besides those discussed here exhibit variable dispersal strategies that will prove tractable for investigating the role of ecology and behavior in individual dispersal decisions.

3 · Fitness consequences of helping

JANIS L. DICKINSON
University of California, Berkeley

BEN J. HATCHWELL
University of Sheffield

Cooperatively or communally breeding birds are species in which individuals live in groups of three or more breeding-aged adults, all of which care for young at a single nest (Brown 1987; Stacey and Koenig 1990a). Most cooperative breeders retain young that delay breeding and help their parents raise siblings. Additional forms of cooperative breeding include polygamous groups with multiple cobreeders of one or both sexes and, more rarely, groups with unrelated helpers. Cooperative breeding is rare, occurring in only about 3% of avian species worldwide, and is particularly common in Australian birds (Brown 1987; Russell 1989; Arnold and Owens 1998). Its prevalence in Australasia can be accounted for phylogenetically due to a particularly high frequency in the Corvida (23%) (Russell 1989; Edwards and Naeem 1993; Clarke 1995).

Theoretical and comparative treatments of avian cooperative breeding have usually dealt with the full range of avian social systems (Brown 1987; Koenig *et al.* 1992; Hartley and Davies 1994; Arnold and Owens 1998). This practice has demonstrated that cooperative breeders share many important characteristics, such as year-round residency, high survivorship, small clutch sizes, and, in many cases, constraints on independent breeding (Brown 1987; Stacey and Koenig 1990a; Arnold and Owens 1998, 1999). Specific limitations on independent breeding vary from one species to the next, and involve a variety of resources, including food, territories, suitable nest or roosting sites, and a lack of skill or mates (Smith 1990).

The primary focus of this chapter is helping at the nest by retained offspring. Although we consider both direct and indirect benefits of helping, we do not attempt to provide a full review of more generalized group benefits such as shared vigilance, the selfish-herd effect (Hamilton 1971), sharing of information (Brown 1988), and cooperative defense. Neither do we attempt a comprehensive review of the diversity of cooperative systems or reproductive conflicts within groups (Emlen 1991; Cockburn 1998). Instead, we adopt a perspective that uses intraspecific variation in social strategies as a means for exploring the evolutionary ecology of cooperative breeding with the aim of guiding future experimental field studies and comparative analyses.

We start with the assumption that patterns of non-breeding sociality and natal dispersal canalize the opportunities individuals have to interact with and help close relatives. These patterns include retention of young in their natal group following their nutritional independence, localized dispersal, continuous association with parents and other relatives after dispersing to breed, and behavioral preferences for interacting with kin. As in the previous chapter, we view helping as a stepwise process in ecological time. Because most helping at the nest is kin-directed, the propensity to help can be viewed as one possible outcome of a series of decisions, beginning with the decision to remain in proximity to the natal group. This can also be looked at from the parents' standpoint as the decision to allow young to stay home after the breeding season. As clarified in Chapter 2, this does not necessarily mean that the stepwise progression reflects the order of evolutionary events, because many species without helpers are derived from cooperatively breeding ancestors (see Chapter 1). Neither do we argue that helping itself plays no selective role favoring delayed dispersal. Rather, the heuristic value of viewing the process of helping as a series of ordered events is that it permits us to explore pathways to kin-directed helping as a decision-making process for individuals. This makes the behaviors associated with helping empirically tractable by clarifying ways in which experiments and observational studies can be structured to address the

---

Ecology and Evolution of Cooperative Breeding in Birds, ed. W. D. Koenig and J. L. Dickinson. Published by Cambridge University Press.
© Cambridge University Press 2004.

fitness consequences of each set of options along the way. In systems where young stay home and help, the process begins with prolonged contact with parents and other close kin.

Advances in molecular genetics have shown that helping and breeding are not mutually exclusive options. On the other hand, the selective factors favoring cooperative polygamy often appear to be different from those favoring helping (Brown 1987). The differences are straightforward in species like dunnocks in which multiple breeders are unrelated (Burke et al. 1989). When group members are related, cooperative polygamy typically does not involve incest. Instead, co-breeders tend to be related within a sex and unrelated to the breeders of the opposite sex (see Chapters 9–11). This situation arises either when outsiders join a group to fill a reproductive vacancy, at which time same-sex offspring of the surviving breeder ascend to cobreeding status within the group, or when same-sex relatives disperse together in coalitions and gain access to a reproductive vacancy in another group (Koenig 1981; Hannon et al. 1985; Piper and Slater 1993; Magrath and Whittingham 1997).

The distinction between communal breeding based on shared parentage and that based on collateral kinship is not as clear as once envisaged (Hartley and Davies 1994, Cockburn 1998), but the great diversity of social organization and mating systems among communal breeders suggests that there is no single evolutionary route to cooperation. Intragroup cooperation and conflict among multiple breeders of either sex are considered elsewhere (Chapters 10 and 11). Here we focus on helping behavior, particularly within kin groups, because this is the context within which most helping at the nest occurs. We do not address cases of cooperative polygamy that are thought to have evolved via sexual conflict among non-relatives (Davies 1992), although we discuss how cooperative polygamy may arise when there is elevated competition for breeding vacancies as a result of natal philopatry and kin-directed helping.

We also do not discuss at length the direct reproduction that may be gained by helpers either within their own social group or in neighboring groups. The existence of covert reproduction by helpers has been recognized only recently (Cockburn 1998), and where it occurs it can clearly affect estimates of the fitness payoffs of delayed dispersal and deferred independent breeding (Richardson et al. 2002). However, our focus is on the fitness consequences of helping behavior and the factors that increase the potential for kin-directed helping; it is not clear that direct reproduction by helpers is a benefit of helping *per se*. Instead, it may be viewed as one of a suite of benefits young birds gain by staying at home and for which they are selected to pay by helping to raise non-descendant kin (Gaston 1978, Kokko et al. 2002).

Our underlying argument is that the current functional utility, measured in terms of the fitness benefits and costs in ecological time, accounts for the maintenance of a costly trait like helping. We view examination of the evolutionary origins of helping (Chapter 1) as a distinct approach requiring different logic and non-mutually exclusive tests (Reeve and Sherman 1993). The challenge, when addressing current function, is to analyze current selection and current fitness benefits while avoiding misinterpretation due to confounding variables.

The past decade has seen the progression from long-term demographic studies (Stacey and Koenig 1990a) to experimental tests of important hypotheses for the current functional utility of helping. Here we aim to describe the theoretical and empirical ontogeny of these tests by giving an historical summary and providing a critical analysis of empirical findings. Much of this discussion focuses on the difficulties of separating the benefits of helping from the benefits of staying with relatives and the complex relationship between philopatric helping, demography, and constraints on independent breeding.

HISTORICAL PERSPECTIVE ON KIN-DIRECTED HELPING AT THE NEST

Helpers usually accrue lower mean fitness returns by helping than they do by breeding independently (Stacey and Koenig 1990a; Emlen 1991). Obligate cooperative breeding, as occurs in white-winged choughs, is rare (Heinsohn 1991b), and cases in which the average helper derives benefits from helping that fully compensate for failing to breed are also rare (Rabenold 1984; Bednarz 1988; Heinsohn 1991a). More often, when the inclusive fitness returns of helping and breeding have been compared, helpers are making the best of a bad job (Reyer 1984; Woolfenden and Fitzpatrick 1984; Koenig and Mumme 1987; Emlen and Wrege 1989; Dickinson et al. 1996; MacColl and Hatchwell 2002). Therefore, the obvious question to ask is why offspring remain on their natal territory and help instead of dispersing to breed.

Why stay?

Most cooperative breeders live on all-purpose territories, but some are colonial nesters that do not defend breeding territories, necessitating a general theory to account for delayed dispersal in species that vary in nest dispersion and spacing behavior (Emlen 1982a). Early on, the question of why helpers help was divided into two questions: why young birds remain in their natal groups (why stay?) and why they help feed young (why help?) (Emlen 1982a). Brown (1987) later split the question of why stay into two, asking why young birds delay breeding and, once they delay, why they remain on their natal territories.

An important source of controversy in the field of cooperative breeding has been the ecological basis of delayed breeding and retention of breeding-aged offspring in their natal groups. Offspring are expected to stay home if the benefits they receive due to increased survival or increased probability of current or future reproduction exceed the benefits they would receive if they were to float or attempt to disperse to another site. By staying home, offspring may incur costs due to increased competition with neighbors and relatives, but these may be counterbalanced by special properties of home that are not available elsewhere (Ekman *et al*. 2001a; Clutton-Brock 2002).

Selander (1964) first proposed habitat saturation as an explanation for delayed breeding, suggesting that young birds stay on their natal territories due to a shortage of adequate breeding territories and in order to benefit from the experience gained by helping. Brown (1969) added significantly to this idea by introducing the concept that association with the natal territory would be favored in saturated habitats both because competition for breeding vacancies is intensified and because floating is difficult. This "habitat saturation" hypothesis was further developed by Verbeek (1973) and Brown (1974), who suggested that a lack of available habitat would make staying home a better option than floating. Koenig and Pitelka (1981) later formalized the "marginal habitat" hypothesis, suggesting that the key feature distinguishing territorial cooperative breeders from non-cooperative species is a steep gradient in quality of available territories and a paucity of habitat intermediate in quality. These early ecologically based models were the starting point for explicit hypotheses regarding what distinguishes cooperative from non-cooperative breeders, with a distinct focus on species with all-purpose territories. Subsequently, the hypothesis of habitat saturation was generalized by Emlen (1982a), who proposed that other ecological constraints, such as availability of food or seasonal constraints on breeding, could explain helping in non-territorial cooperative breeders.

Emlen's (1982a) hypothesis invoked three classes of constraints on independent breeding to explain why young birds delay and stay, suggesting that staying should be favored not only when females and territories are in short supply, but also when dispersal costs are high or when available breeding opportunities are relatively poor, in terms of either the likelihood of fledging young or the number of young parents can fledge. Most empirical studies have treated constraints on independent reproduction as constant within a season, but Emlen (1982a) proposed that within-year variation in reproductive constraints may influence the decision to delay dispersal and help.

Indeed, there are several cooperatively breeding species in which failed breeders choose to become helpers, termed "redirected helping" by Emlen (1982a). In the long-tailed tit, all of whose helpers are failed breeders, MacColl and Hatchwell (2002) have shown that the probability of breeding successfully declines as the season progresses. The switch from breeding to helping occurs when the expected fitness payoff of breeding falls below that of kin-directed helping. The argument that redirected helping is a "best of a bad job" strategy employed at the end of a temporally constrained breeding season has been suggested for several other cooperative breeders and is discussed further below (Emlen 1982a; Lessells 1991; Dickinson *et al*. 1996). After Emlen (1982a) published his general theory of constraints on independent breeding, attempts to identify the precise ecological conditions leading to retention of young proliferated (Hatchwell and Komdeur 2000).

A critical prediction of the constraints hypothesis is that helpers are making the best of a bad job and would become breeders if given the chance. In the acorn woodpecker, experimental removal of the sole breeder of one sex created a "power struggle" over the resulting reproductive vacancy involving a large number of helpers from other territories (Hannon *et al*. 1985). The contests often lasted for several days and the vacancies were usually won by coalitions of same-sex siblings dispersing together.

Evidence for a shortage of breeding opportunities was also provided by Pruett-Jones and Lewis (1990) in an experiment on superb fairy-wrens. Helper males dispersed to fill vacant territories created by experimental male removal, but only when a female breeder was also present on the territory, indicating that female mates were limiting. Females are not the only limiting factor in superb fairy-wrens, however, because removal of the breeder female in a different population, with predominantly male helpers, resulted in relatively rapid female replacement (Ligon et al. 1991).

There have been surprisingly few studies of mate limitation as a route to helping, but a recent study provides further support for this concept. In western bluebirds, males whose mates were removed had only a 15% chance of renesting with a new female compared to an 83% chance for intact pairs whose nest and eggs were removed (Dickinson 2004). If they did not get a new mate, experimentally widowed males became lone territory holders, helpers, or non-infanticidal replacement males on territories of actively nesting widowed females. Because some males held territories alone after removal, a local shortage of females appears to explain why males adopt the occasional strategies of helping and replacement.

Stacey and Ligon (1987, 1991) added a new perspective on the potential importance of variation in habitat quality for the evolution of cooperative breeding, proposing the "benefits of philopatry" hypothesis as an alternative to the marginal-habitat hypothesis of Koenig and Pitelka (1981). The new approach focused not on a shortage of marginal habitat, but on the quality of available territories relative to the quality of territories typically exporting young. This hypothesis predicts that young birds will stay on high-quality territories because the direct benefits of increased survivorship and access to high-quality territories, combined with indirect benefits due to helping, exceed the fitness expectations for individuals dispersing to breed independently on available, low-quality territories. The idea was important because it focused on individual assessment and demonstrated that the decision to delay dispersal and help should be based not on the average fitness of helpers versus breeders, nor on the absolute availability of breeding habitat, but on the relative fitness consequences of the helping and breeding options available to an individual at any given point in time. In the benefits-of-philopatry hypothesis, the benefits of helping and staying are no longer viewed separately. An individual that gains inclusive fitness benefits by helping on a high-quality territory should not move to a low-quality territory where its inclusive fitness benefits, through direct reproduction, will be comparatively low.

The early 1990s saw the first experimental evidence for the simultaneous importance of both habitat saturation and variation in habitat quality with work by Komdeur (1992) on the Seychelles warbler. Transplanting warblers to previously unoccupied islands resulted in independent breeding until territories began to fill up. After territories filled, individuals chose helping (or cobreeding, Richardson et al. 2001) on high-quality, insect-rich territories, over independent breeding on lower-quality territories. They remained with parents on low-quality territories only after territory vacancies in low-quality habitat were filled (Komdeur 1992; Komdeur et al. 1995). Furthermore, breeding vacancies created by breeder removals were filled only by helpers from territories of equivalent or poorer quality and never by helpers from superior territories, for which helping remained a better option than breeding.

These experiments provided an independent assessment of territory quality, based on extensive sampling of insects, and demonstrated that individuals could make appropriate fitness-maximizing decisions when choosing among a complex set of reproductive options. While the conclusions of this study will surely be reinterpreted based on molecular inference of parentage, analyzing the fitness consequences of helping as a composite of individual fitness-based decisions raised the bar for empirical studies, leading to empirical tests of the fitness consequences of helping as a function of individual quality and individual opportunity.

Territory inheritance is often considered of potential importance in the evolution of group living (Wiley and Rabenold 1984; Lindström 1986; Blackwell and Bacon 1993). However, inheritance is not a common outcome of staying home in cooperatively breeding birds (Koenig et al. 1999; Komdeur and Edelaar 2001a). In acorn woodpeckers, offspring typically do not ascend to breeding status unless unrelated breeders of the opposite sex have filled a reproductive vacancy in their group (Koenig et al. 1999). Occasionally, when a vacancy arises, other birds are prevented from filling it by helpers that have stayed on the territory and are of the same sex as the deceased parent. In such cases the group may forgo breeding for up to two years, an observation that is theoretically consistent with the idea that incest is costly. Under incest avoidance, one of two things must happen for an offspring to breed in its natal group. Either both parents must die or the helper must become a cobreeder with its same-sex parent after its opposite-sex parent dies and is replaced by a new breeder from outside the group. Simultaneous death of both parents is improbable and sharing parentage with offspring is potentially costly for parents. Hence, inheritance tends to be rare in avian cooperative breeders and instead, the breeding territory is more commonly shared with offspring by "budding" off a portion of the parents' breeding territory (Woolfenden and Fitzpatrick 1984; Komdeur and Edelaar 2001b).

Zack (1990) suggested that philopatric offspring of cooperative breeders (usually males) tend to breed closer to their parents than do offspring of closely related species that are also year-round residents, but breed as pairs. On the basis of this difference, he proposed that delayed breeding and local dispersal were causally related and hypothesized that non-breeders are waiting for nearby high-quality territories to open up, essentially queuing for breeding positions.

Like Stacey and Ligon (1987, 1991), Zack (1990) focused on variation in habitat quality rather than degree of saturation, because habitat saturation and cooperative breeding are not always linked, even in territorial species. Zack argued that the costs of dispersal may not differ between cooperative and noncooperative breeders, but that the differences lie in the steepness of the decline in the quality of breeding opportunities with distance from the natal site and the potential direct fitness benefits of staying. In more contemporary terms, Zack's (1990) argument predicts low spatial variance (high spatial autocorrelation) in productivity of territories whereas neither Stacey and Ligon (1987) nor Koenig et al. (1992) made predictions about the spatial component of variance in territory quality. There is some evidence in support of queuing in the cooperatively breeding *Campylorhynchus* wrens of Venezuela (Zack and Rabenold 1989) and in the Siberian jay, a species with delayed dispersal but no helping (Ekman et al. 2001b).

In the 1990s, there was a shift toward a more inclusive approach to the evolution of delayed dispersal and helping, with increasing recognition that habitat saturation and the benefits of philopatry are complementary theories (Emlen 1991). The marginal-habitat and benefits-of-philopatry hypotheses were augmented by Walters et al. (1992a), who proposed that shortage of a single critical resource could explain cooperative breeding in red-cockaded woodpeckers. Walters et al. (1992a) provided experimental support for the critical importance of cavity clusters, demonstrating that artificially created clusters result in dispersal of helpers to breed on previously unoccupied territories. In a detailed analysis of dispersal patterns and reproductive success, however, Walters et al. (1992b) found that associating with the natal territory increases reproductive success, because the stay-and-foray strategy allows individuals (primarily males) to compete effectively for nearby vacancies. The two results together appear to provide answers to the questions, "why delay?" and "why stay?", as proposed by Brown (1987). Individuals, usually males, delay because of a shortage of cavity clusters, and stay in part due to direct fitness benefits of associating with the natal territory.

Concurrently, Koenig et al. (1992) developed a model that combines ideas on habitat saturation into a single predictive framework, which included variation in the fitness of individuals staying at home, dispersing, and floating. They suggested that the key difference between constraints arguments and variance arguments like benefits of philopatry is whether the focus is on extrinsic constraints on independent breeding or on intrinsic benefits of delayed dispersal. Their review provided a summary of the suite of ecological conditions favoring young remaining on the natal territory and was conciliatory in its inclusion of most preceding ecological models for territorial species. We believe, however, that the key contribution of the approach used by Stacey and Ligon (1987) was not its focus on intrinsic benefits, but its emphasis on individual fitness-based decisions, which changed the way in which empirical researchers partitioned their data. This empirical focus on individual reproductive strategies accounts for most of the progress toward understanding avian cooperative breeding in the last decade.

Why help?

A handful of researchers has examined the direct and indirect fitness consequences of helping as defined by Brown (1980). Potential costs of helping include reduced survival due to increased risk of predation or increased energetic expenditure (Heinsohn and Legge 1999).

Other direct costs include reduced probability of breeding later in the season, due to energy expended helping earlier on, or reduced opportunities for extra-pair fertilizations. On the other hand, helping may yield direct benefits with increases in the helper's future breeding success due to learning that takes place on the natal territory. Indirect benefits resulting from increased production of non-descendant kin are additive with these direct benefits, and arise from increased productivity of the parents' current nest, increased survival of recipient young, and increased parental survival or future breeding success. It is also possible that helpers gain nothing from helping *per se*, but simply help in exchange for the direct benefits they gain from being allowed to stay on the natal territory, an hypothesis that has been termed "payment of rent" (Gaston 1978) or "pay to stay" (Mulder and Langmore 1993; Kokko *et al.* 2002).

The common practice of comparing the inclusive fitness of helpers with that of independent breeders indicates the potential importance of extrinsic constraints on independent breeding (Emlen and Wrege 1989; Dickinson *et al.* 1996), but is of limited value in addressing the benefits of helping. In order to determine the costs and benefits of helping, helper effects must be extricated from the benefits of delaying and staying, an endeavor that is possible only in limited circumstances, for example when philopatric delayers vary in whether or not they help (Magrath and Whittingham 1997).

Direct fitness benefits
Most hypothesized direct benefits of helping remain difficult to test even in species in which helpers can be compared with birds that delay breeding, but do not help (Table 3.1). First, comparison of survival of helpers and non-helping delayers is interpretable only if we can rule out the possibility of systematic bias in the tendency to disperse off the study area or systematic differences in individual quality. Consider the results of Rabenold (1990), who reported for stripe-backed wrens that more industrious helpers had lower survivorship than their less industrious counterparts when matched by sex, group, and year. This result is compelling because it was statistically significant even when the analysis was restricted to males, most of which dispersed within two territories of their natal sites. While it is still possible that the more industrious helpers tended to disperse off the study area, the study suggests significant direct fitness costs of helping. Further analysis indicated that the estimated costs of helping were more than compensated by indirect fitness gains (Rabenold 1990).

Helping may also be costly in colonial pied kingfishers in which primary helpers contribute more to provisioning and have reduced survival compared to secondary (unrelated) helpers and non-helping "delayers" (Reyer 1984). In this case, primary helpers had biannual inclusive fitness equivalent to that of secondary helpers, because, like the stripe-backed wrens, they gained indirect benefits that compensated for reduced survival.

Another important source of direct fitness is proposed in the "skills" hypothesis, which was originally put forth by Skutch (1961) to explain why young delay and aid parents rather than breeding on their own. More recently, this hypothesis has been renamed the "experience" hypothesis, referring to the idea that young birds that delay dispersal and become helpers gain direct fitness benefits through experience that enables them to become more productive when they are finally able to breed. Tests have therefore focused on whether the experience gained from helping allows helpers to perform better as breeders than if they did not attend a nest at all.

The results of these tests are equivocal. In group-territorial red-cockaded woodpeckers, two-year-old novice breeders that helped as yearlings did not outperform two-year-old novice breeders with no helping experience (Khan and Walters 1997). These authors avoided confounding age with experience, but birds with helping experience were philopatric, while inexperienced delayers were not, raising the possibility of a confound with individual quality. If high-quality individuals tend to disperse and become delayers, then this would work against finding a difference even if there were experience-derived benefits.

Two researchers have attempted to test the skills hypothesis by comparing helpers with delayers that remain on the natal territory without helping. In white-fronted bee-eaters, novice breeders that helped as yearlings were not more successful than novice breeders that delayed and remained in their natal groups without helping (Emlen and Wrege 1988). In contrast, Seychelles warbler females that helped as yearlings had much higher success as novice breeders than did inexperienced delayer females (Komdeur 1996). Delayers failed to place their nests in stable tree forks and spent less time incubating than did females with helping experience. This is a dramatic result, and the magnitude of the effect raises

Table 3.1. *Classification of hypotheses for the current functions of helping and cooperative polygamy, excluding numerical benefits of group size that accrue regardless of breeding or social system*

Category	Applicability	Explanation of benefit	Reference
(a) Helping at the nest			
Why delay?			
Shortage of marginal habitat (territory shortage)	All systems	Ecological constraint	Koenig and Pitelka 1981
Shortage of mates combined with incest avoidance	All systems	Ecological constraint	Emlen 1982a
High costs of dispersal	All systems	Ecological constraint	Emlen 1982a
Shortage of high quality breeding opportunities	All systems	Survival and reproductive benefit	Emlen 1982a
Inexperience; lack of foraging skill	All systems	Delayed maturation	Heinsohn 1991a
Why stay?			
Queue for high-quality territory	Kin-based systems	Territory quality spatially correlated	Zack 1990
Benefits of philopatry[a]	Kin-based systems	Fitness of helper on high-quality territory > fitness of breeder on low-quality territory	Stacey and Ligon 1987
Survival benefits of remaining with parents	Kin-based systems	Prolonged brood care	Ekman and Rossander 1992
Survival benefits of remaining on natal territory	All systems	Territory quality, ecological constraints	Lindström 1986
Territory budding	Kin-based systems	Territory quality, ecological constraints	Woolfenden and Fitzpatrick 1984
High cost of floating	All systems	Greater survival in natal group	Brown 1987; Koenig et al. 1992
Why help? Indirect fitness benefits			
Current indirect fitness benefits	Kin-based systems	Increased productivity of parents' nest[b]	Brown 1980
Future indirect fitness benefits	Kin-based systems	Increased parental survival	Mumme et al. 1989
Why help? Direct fitness benefits			
Reciprocal help from recipient broods	All systems	Helping is reciprocated by fledged chick	Ligon and Ligon 1978a
Reciprocity	All systems	Helping is reciprocated by parent	Skutch 1961
Experience	All systems	Helping increases later breeding success	Mulder and Langmore 1993
Pay to stay[c]	All systems	Breeders tolerate helpers, not delayers; survival of stayers must be increased	
Future direct fitness benefits	All systems	Increased access to mates via increased social status or prestige	Zahavi 1995
(b) Cooperative polygamy			
Why cobreed?			
Benefits of philopatry[a]	Kin-based systems	Fitness of joiner on high-quality territory	Stacey and Ligon 1987
Benefits of coalition formation	All systems	Groups size effect on access to vacancies	Hannon et al. 1985
Sexual conflict	All systems	Benefits of polygamy	Davies 1989

[a] Benefit crosses over from staying to helping (or cobreeding) category, because hypothesis is based on comparison of inclusive fitness benefits of helping (or cobreeding) at home versus direct fitness benefits of breeding on available territories.

[b] Includes increased growth rate, reduced predation, increased fledging/nesting success, increased fledging condition, increased post-fledging survival

[c] Benefit crosses over from helping to staying category, because there must be benefits of staying that account for why helpers pay to stay

the question of why young birds have not been selected to recognize suitable nesting sites based on experiencing successful nests as juveniles or as non-helping delayers.

One possibility is that delayers are birds of poor quality and that the observed pattern is due to individual quality differences rather than experience. Two-year-old females that helped as yearlings had breeding success comparable to that of two-year-old females that bred as yearlings, indicating that the experience of helping in the first breeding season is superior to doing nothing, but is equivalent to the experience gained from independent breeding.

A further direct benefit of helping that has yet to be tested empirically is the "group augmentation" hypothesis (Brown 1987; Kokko et al. 2001). This benefit is based on the idea of "delayed reciprocity" (Ligon and Ligon 1978a; Wiley and Rabenold 1984), which envisages offspring repaying the care provided by helpers at some point in the future when helpers become breeders. Group augmentation describes a situation where individuals survive or reproduce better in large groups so that it pays to recruit new members by increasing group productivity, or even by "kidnapping" the members of other groups (Heinsohn 1991a). The evolutionary stability of reciprocal helping has been questioned, but Kokko et al. (2001) have shown that, in theory, the more generalized benefits proposed by the group-augmentation and delayed-reciprocity hypotheses can be evolutionarily stable, at least under certain conditions. The effects of group augmentation may, in practice, be difficult to distinguish from those of kin-selected helping, but group augmentation does not require kinship within cooperative groups (Clutton-Brock 2002).

Indirect fitness benefits

Most evidence that indirect fitness benefits are important arises from two key sources. First is the finding that helpers are more likely to help rear close than distant relatives, and second is demographic data indicating that nests with helpers fledge more young than nests of similarly-aged breeders without helpers. Natal philopatry, a common pattern in species with helpers, increases the likelihood of kin-biased helping, even in the absence of behavioral preferences to interact with close kin. In white-fronted bee-eaters (Emlen and Wrege 1988), Galápagos mockingbirds (Curry 1988a), Seychelles warblers (Komdeur 1994a), and western bluebirds (Dickinson et al. 1996) individuals are more likely to help raise close kin than distant kin. In these studies, the identities of potential helpers, potential recipients, and their proximity to each other were not manipulated experimentally. Therefore, the options available to potential helpers varied and it was not always clear that offspring discriminated and helped close kin as opposed to using a mechanism based on spatial proximity.

In pied kingfishers, males unable to breed on their own usually do not become unrelated (secondary) helpers as long as they have at least one parent still alive and can thus help related individuals (become a primary helper) (Reyer 1984, 1990). Because pied kingfishers are colonial, this pattern is unlikely to be explained by proximity. However, failure to help less-related pairs such as a parent and step-parent or an unrelated pair may result from reproductive competition and eviction from the territory or nest area, rather than from the helper's preference for rearing close kin (Shields 1987). For example, a son may be evicted from his natal group when his father dies simply because his mother's new mate does not have a genetic interest in providing him access to the territory or nesting area and regards him as a competitor. Additional behavioral studies are required to examine the relative importance of kinship, dominance, and aggression for group composition in cooperative breeders.

The best evidence of a preference for rearing kin over non-kin comes from long-tailed tits, in which helpers are failed breeders that must decide whom to help after they have already dispersed to breed on their own. Winter flocks, consisting of both close relatives and unrelated immigrants, may provide information on kinship that allows philopatric individuals, usually males, to recognize relatives and direct their helping efforts at kin. The ranges of neighboring flocks overlap extensively, and when two flocks share relatives, their ranges overlap more extensively than when they do not (Hatchwell et al. 2001a). This suggests that winter sociality and winter space-use patterns are determined in part by kinship.

Although birds that experience breeding failure do not always help, when they do, they help at nests of close relatives and do not simply select the closest nest (Russell and Hatchwell 2001). When nests failed, naturally or by experimental chick removal, the failed breeders helped at nests of relatives over equidistant nests of non-relatives, effectively demonstrating that aid is governed by a preference to help close kin rather than by spatial proximity (Russell and Hatchwell 2001). Furthermore, if close relatives with active nests are not available, failed breeders do not become helpers. In this case, cross-fostering experiments showed that discrimination is achieved by individuals recognizing broodmates as relatives and not using spatial cues (Hatchwell *et al.* 2001b).

In most demographic analyses of the indirect benefits of helping, the mean inclusive fitness of helpers is compared with the mean for independent breeders. A first-line approach is to ask whether mean fledging success is relatively high at nests with helpers. The mechanism for increased productivity of nests with helpers may be reduced predation (Rabenold 1990) or reduced risk of starvation (Curry and Grant 1990). The two are interrelated because increased food delivery can reduce predation by increasing nestling growth rates and shortening the time young are in the nest.

Although it is difficult to measure the association between help and survival of independent young, a few researchers have been able to follow juveniles after independence (Curry and Grant 1990; Rabenold 1990). In most open populations, low return rates and the possibility that dispersal and nestling condition covary make it difficult to test for an effect of help on recruitment.

On the other hand, the effect of help on fledging success may underestimate indirect fitness benefits if a helper's provisioning enhances the condition of fledglings and their survival to breeding age (Waser *et al.* 1994). For example, brood size at fledging is unrelated to the number of helpers in long-tailed tits because nestling starvation is infrequent. Nevertheless, there is a strong positive association between the recruitment of offspring as breeders and the number of helpers (Hatchwell *et al.* 2003). Effects such as these, if masked by biases in dispersal, may explain the occurrence of multiple helpers even though fledging success rarely increases with addition of helpers beyond the usual one.

Demographic (non-experimental) measures of the indirect benefits of helping may be problematic if they confound effects of help with the quality of breeders and territories producing excess young (Brown *et al.* 1982). This confound arises because the more productive a pair or the better the territory, the more likely it is that young will survive and act as helpers. Three approaches have been implemented to circumvent the correlation between the presence of a helper and the parents' productivity: paired comparisons, helper-removal experiments, and helper-addition experiments.

The paired-comparisons approach involves examining the success of pairs in sequential years with and without helpers. This approach is seriously flawed for the following reason. A pair must be successful to go from having no helpers one year to having helpers the next. In contrast, unsuccessful pairs with helpers will tend to go from having helpers one year to having none the next. In the first case, the method has selected unaided pairs with above average reproductive success, while in the latter case it has selected helped pairs with below average reproductive success. The method is consequently biased against finding an effect of help, and negative results based on self-paired comparisons, such as those reported for pinyon jays (Marzluff and Balda 1990), American crows (Caffrey 2000), and laughing kookaburras (Legge 2000b) must be viewed with caution, while the magnitude of positive results may be underestimated (Woolfenden and Fitzpatrick 1984; Walters *et al.* 1992b). Unfortunately, this confound counters Cockburn's (1998) argument that such comparisons are better than experiments.

A more convincing approach can be employed when non-helpers and helpers coexist within the same social group, as occurs in white-browed scrubwrens (Magrath and Yezerinac 1997; Magrath 2001). Neither helpers nor non-helpers affected group reproductive success in a four-year study (Magrath and Yezerinac 1997). In a later analysis based on seven years of data, group size increased group productivity but only for yearling females. Controlling for territory quality, yearling females in groups of three or more had higher seasonal reproductive success than yearling females breeding in pairs (Magrath 2001). As the later analysis did not distinguish groups with subordinate helpers from groups with subordinates that did not help, it is not yet clear whether the group-size effect with yearling females is a direct effect of help. However, in the white-browed scrubwren system and others like it, there is considerable potential for addressing the benefits of helping while avoiding confounds with group size and territory quality.

Helper-removal experiments have been few and are also not problem-free. First, experimental helper removal influences both helping and group size, raising the possibility that failure to observe a helper effect is due to the fact that there is a benefit, but it is not high enough to counter-balance the cost of living in a larger group (Koenig and Mumme 1990). Second, experiments may erroneously support a helper effect if helper removal disrupts the experimental group, reducing the success of unhelped pairs. Mumme (1992b) removed helpers from Florida scrub-jay nests and found a significant reduction in productivity relative to controls. Experiments with Seychelles warblers (Komdeur 1994b) and grey-crowned babblers (Brown et al. 1982) also demonstrated significant effects of helping. Interestingly, Seychelles warbler helpers hindered on medium-quality territories with more than one helper, indicating that reproductive competition or sharing of resources can be costly and supporting the idea that potential confounds should be carefully considered.

Although they are subject to the same group-size confound as helper-removal experiments, self-paired helper-addition experiments do not involve social disruption and thus may provide superior tests of the effect of help. By preventing predation, Haydock (1993) experimentally increased bicolored wren group size on territories that were historically held by pairs. The next season, fledging success at nests with experimentally "created" helpers was greater than at control nests without helpers. In this case, it is still possible that fledging success improved due to increased group size rather than help per se; the general group benefit of shared vigilance may have led to greater feeding rates or reduced nest predation.

Although comparative data are currently not sufficient to identify patterns of variation in helper effects, there appear to be fundamental differences between case studies providing experimental support for an effect of help and existing counter-examples. We suggest two main ways in which helper effects should vary. First, helping should be less effective in precocial than altricial species. Second, juvenile helpers, which are not of breeding age, may have less impact on parental productivity than adult helpers. In contrast with experimental studies of altricial species, experimental removal of juvenile helpers in the common moorhen had no significant effect on the survival of young to independence (Leonard et al. 1989). Common moorhens differ from most cooperative breeders, however, in having precocial young and in having juvenile helpers, whose potential sacrifice is minimal compared to that of yearlings with breeding potential. In western bluebirds, which have both juvenile and adult helpers, it is only the adult helpers that increase overall rates of food delivery to nestlings (Dickinson et al. 1996). Classes of helpers may vary considerably in the help they provide and additional data are needed before we can make broad generalizations regarding the effects of help in cooperative breeders.

Helpers may also derive indirect fitness benefits if the aid they give increases their parents' survival and future reproductive success. These benefits are comprehensively, but not easily, measured as future indirect fitness benefits (Mumme et al. 1989). When helpers feed chicks, parents often exhibit a reduction in their provisioning rates (Hatchwell 1999; Chapter 4), but this does not always lead to an increase in survival (Kahn and Walters 2002). In 73% of 22 species with helpers, helping reduced the feeding rates of one or both parents (Hatchwell 1999). Of eight cases where breeder workload was reduced and survival estimated, five (62.5%) showed an increase in breeder survival. Interestingly, helping increased breeder male survival in acorn woodpeckers even though it did not affect parental feeding rates (Mumme and de Queiroz 1985). Currently, empirical data suggest that future indirect fitness benefits are potentially important, with the caveat that non-experimental measures of effects of help on breeder survival are confounded with breeder and territory quality, both of which may be higher for groups producing helpers (Magrath 2001).

In spite of these problems, non-experimental tests of an effect of help, based on multiple lines of evidence, suggest that increased production of non-descendant kin is a primary benefit of helping in cooperative breeders (Emlen 1991). These multiple lines of evidence include increased fledging success, food delivery rates, and nestling growth rates at nests with helpers. We suggest that recent emphasis on direct benefits has tended to minimize the quality and magnitude of evidence in support of indirect benefits (Cockburn 1998; Clutton-Brock 2002). This is an understandable reaction to an historical tendency to interpret kin-directed cooperation as evidence for kin selection, and we agree that more rigorous tests of the kin-selection hypothesis are required. On the other hand, the mean fitness benefit due to help is rarely enough to compensate individuals for failing to breed independently and so is only part of the story for why helpers help. Measures of indirect fitness benefits of helping must be combined with information on the availability and fitness expectations of other options. Further examination of direct benefits of helping will certainly help to complete the picture.

CAN HELPING AND STAYING BE UNCOUPLED?

One of the inherent difficulties in studies of the "why stay?" and "why help?" questions is that the two behaviors are inevitably closely coupled. Many authors have explicitly recognized year-round residency as a shared characteristic of species with kin-directed helping behavior (Russell and Rowley 1993b; Clarke 1995; Arnold and Owens 1999), and most previous investigations of the fitness consequences of helping have considered the decision to remain on the natal territory into the breeding season as the switchpoint. Brown (1987) suggested that the greatest insights into cooperative breeding would come from comparisons of species in which delaying, breeding, and helping are uncoupled. This approach has been put to effective use in a few intraspecific comparisons where such uncoupling is feasible (Reyer 1984; Emlen and Wrege 1988; Komdeur 1996; Khan and Walters 1997).

Studies of western bluebirds (Dickinson et al. 1996), Galápagos mockingbirds (Curry and Grant 1990), long-tailed tits (Hatchwell et al. 2002), Siberian jays (Ekman et al. 1994), and white-browed scrubwrens (Magrath and Whittingham 1997) also allow for empirical separation of the act of helping from the act of remaining on the natal territory. In the first three cases, helpers help while they have nests of their own nearby or following failure of their own breeding attempt, effectively helping without delaying dispersal or reproduction. In the Siberian jay they delay dispersal for up to three years without ever helping at all. In white-browed scrubwrens, males that remain on their natal territories vary in whether they help or not, and this variation is tied to the potential for direct fitness benefits via paternity in the current nest (Magrath and Whittingham 1997). Such breeding systems permit tests of hypotheses for the evolution and maintenance of staying or helping, and highlight characteristics that distinguish species with helpers from non-cooperative species (Chapter 2).

Redirected helping, where birds help following a failed breeding attempt or loss of a mate, also provides an interesting perspective on helping because it demonstrates that young birds do not necessarily give up the opportunity to help by dispersing locally and attempting to breed. Redirected helping is the sole source of helpers in long-tailed tits (Gaston 1973; MacColl and Hatchwell 2002). It is also one route to helping in a variety of other cooperative breeders, including pinyon jays (Balda and Bateman 1971), green woodhoopoes (Ligon and Ligon 1990b), white-fronted bee-eaters (Emlen and Wrege 1988), Mexican jays (Brown 1987), European bee-eaters (Lessells 1990), and western bluebirds (Dickinson et al. 1996; Dickinson and Akre 1998).

In western bluebirds, males help only rarely (7% of pairs have helpers, range 3–16%), but exhibit extraordinary plasticity in being able to switch from breeding to helping throughout their lives (Dickinson et al. 1996). Redirected helping appears to be a consequence of mate loss, rather than simple nest failure. In long-tailed tits, the frequency of helping is much higher (54% of broods have helpers) and helpers are males that disperse locally, attempt to breed independently, and become helpers at the nests of relatives when their own nests fail (Gaston 1973; MacColl and Hatchwell 2002). These observations indicate that local dispersal and the presence of relatives provide the permissive conditions for kin-directed helping.

Helping is also uncoupled from staying in species with simultaneous breeder-helpers, which occur in western bluebirds (Dickinson et al. 1996) and Galápagos mockingbirds (Curry and Grant 1990). In both these species, sons that have nests next door to their parents sometimes feed at both their parents' and their own nests. Genetic information is lacking for the mockingbirds, but in western bluebirds fitness estimates indicate that annual inclusive fitness of simultaneous breeder–helpers is high relative to non-breeding or redirected helpers and may even be higher than that of same-aged breeders that do not help at all (Dickinson and Akre 1998). Redirected and simultaneous helping are important because they suggest a simple route to facultative helping whereby a tendency to disperse to breed near kin sets the stage for helping that is expressed when territories are adjacent or when the option to breed is unavailable.

The premier example of young staying on the natal territory without helping is provided by the Siberian jay, in which young of both sexes are retained for up to two breeding seasons, but are not permitted within 25 m of the nest (Ekman *et al.* 1994). A similar phenomenon has been reported in green jays (Gayou 1986) and Australian magpies (Veltman 1989). Examples of retention of young in species without helping are important because they demonstrate that offspring can benefit by remaining on the natal territory even if there are no indirect benefits of helping at the nest. Such species provide opportunities to investigate the causes and consequences of delayed dispersal without the confounding effect of help, and are discussed at greater length by Ekman *et al.* (Chapter 2).

In contrast, gray jays are prevented from feeding at the nest by their parents, but later feed fledglings (Strickland and Waite 2001). A meta-analysis performed by Strickland and Waite (2001) suggests that parents prevent helping at the nest in some species of jays in order to reduce nest predation. This interpretation is supported by reduced parental feeding rates in non-helping species as well as reduced clutch sizes, smaller group sizes, and relaxation of parental aggression toward retained breeding-aged offspring after the chicks have fledged. Cases in which helping occurs only after young have fledged reinforce the message that helping is extremely plastic in its expression and can be broken down into components, each of which may be addressed separately in phylogenetically controlled comparative studies.

STEPWISE REPRODUCTIVE DECISION-MAKING AND KIN-DIRECTED HELPING

Twenty years of dialogue on ecological factors leading to helping at the nest has led to the current emphasis on opportunistic and adaptive decision-making by individuals within cooperatively breeding populations. Assessment is explicit in the benefits-of-philopatry hypothesis (Stacey and Ligon 1987) and Zack's (1990) hypothesis, but it is also explicit in theoretical treatments of within-group dynamics, beginning with Vehrencamp (1983a, 1983b) and culminating in the more recent revival of reproductive-skew theory (Reeve *et al.* 1998; see Chapter 10). Here we expand the "why stay – why help" framework to explore the utility of viewing helping as a series of sequential and sometimes reversible decisions. Unlike previous treatments, we include the period from the time young fledge until the start of their first potential breeding season and take from Emlen (1982b),

Stacey and Ligon (1987), and Zack (1990) the idea that continuous assessment of alternative options is a critical component of kin-directed helping. This approach provides a new basis for comparison of cooperative with non-cooperative breeders.

Resident species are much more likely to be cooperative than are migratory species (Brown 1987; Arnold and Owens 1999), but here we are more interested in the dispersal strategies adopted by individuals within species and populations. In particular, we focus on variation in the timing of dispersal by offspring and consider the implications of that variation for the emergence of kin-directed cooperative breeding. Brown (1987) suggested that the benefits of the "stay and foray" strategy may explain why young birds remain on their natal territories after the first year of life. As we discuss below, however, it is possible to take this idea one step backwards and suggest that the benefits of kin-based sociality outside the breeding season are predisposing factors with respect to helping.

One prevailing idea is that dispersal strategies are driven, at least in part, by incest avoidance (Johnson and Gaines 1990; Weatherhead and Forbes 1994). The key argument is that sex-biased dispersal should evolve as a mechanism for avoiding incest, providing that the costs of dispersal do not exceed inbreeding costs. Greenwood (1980) proposed that in systems where males gain access to mates by controlling resources, as in most birds, females gain less by staying and so should be the ones to disperse. Thus, female-biased dispersal is the typical pattern in passerine birds and philopatric recruitment of female fledglings tends to be low, except in island populations (Arcese 1989).

The connection between philopatry and helping is explicit in all models for the evolution of helping that involve kin selection, and there is good empirical support for this link. For example, long-tailed tits exhibited the typical pattern of female-biased dispersal in "mainland" sites, but in an isolated site both sexes exhibited philopatry, suggesting that as the costs of dispersal increased, females were more likely to stay close to home (Russell 2001). This difference in dispersal strategies across populations resulted in a significantly higher proportion of female helpers in the isolated population than in mainland populations.

The selective context of delayed dispersal is discussed by Ekman *et al.* (Chapter 2) who point out that prolonged interactions with offspring occur both through retention of offspring on the natal territory and, more rarely, through retention of offspring in mobile or even migratory family groups. Here we are interested in the opportunities that prolonged association with families provide, the population consequences of these associations, and the ways in which viewing helping as a series of decision points can lead to more informative empirical and comparative studies.

First, we explore the consequences of remaining with family outside the breeding season. How do life-history traits and ecological factors compare among species in which young spend the winter with relatives and those in which young birds leave their natal group early to join non-relatives? There is a need for further study of ecological and behavioral factors favoring retention of young in their natal groups during the non-breeding season; potential avenues for research are again discussed in Chapter 2.

Second, we can analyze the breeding consequences of remaining in the neighborhood with parents into the next breeding season, answering the question, "why stay in the neighborhood?" Neighborhood effects can be examined using intraspecific comparisons of fitness of immigrants (dispersive young) and residents (philopatric young) that breed (Bensch et al. 1998). Such analyses are difficult in open populations due to an inability to distinguish immigrants from just off the study area from those that have dispersed into the population from a long distance away.

A more refined analysis would involve examining the effect of distance from the natal site on the breeding success of males, controlling for other factors that may correlate with distance, such as natal condition, breeder age and seasonal timing. For example, familiar neighbors enhance breeding success in red-winged blackbirds (Beletsky and Orians 1989). Similar analyses could be used to investigate whether proximity to parents influences survival or success of yearling breeders in species with low levels of kin-directed helping. In western bluebirds, males return home to winter with their parents even after they have bred successfully (Kraaijeveld and Dickinson 2001). Although there are currently no data indicating that wintering on the natal site enhances survival, philopatric sons wintering with both parents were twice as likely to breed on the study area as were philopatric sons with just one living parent. This suggests that prolonged interaction with kin may have long-term benefits, such as increased winter survival and the opportunity to help in the face of mate loss or seasonal decline in breeding opportunities.

Third, we can investigate why individuals are constrained from breeding independently once they have remained in the neighborhood. This is the stage at which questions regarding breeding constraints should be focused in both single-species and comparative studies. If mates are the constraint, then mate removal can potentially force individuals into other options like helping, floating, holding a territory as a lone individual, or joining non-relatives, including helping at nests of unrelated widowed females (Dickinson and Weathers 1999). By experimentally forcing males into these options, it is possible to estimate how the options rank in terms of inclusive fitness.

The fourth step, "why help?" tests for increases in inclusive fitness that result from helping and that cannot be achieved by simply remaining on the natal territory. There are three routes to kin-directed helping: remaining on the natal territory, helping while simultaneously breeding on a nearby territory, or returning to the natal group after breeding failure or mate loss (Fig. 3.1). Effects of help on helper inclusive fitness are best addressed by comparing helpers with individuals that are essentially doing everything the helper does without helping. This involves controlling for variables such as age, experience, territory quality, parental quality, and individual quality, a goal that can be achieved with multivariate statistics given sufficient long-term data or through careful partitioning of data into comparison groups that differ by just one explanatory variable. In some systems, helpers can be compared with individuals that stay home without helping, controlling for inherent phenotypic differences between helping and non-helping birds. In other systems, simultaneous breeder–helpers can be compared with breeders that are also in close proximity to their parents, but that do not help. This sort of fine-tuning of the questions can help to elucidate the selective and phenotypic determinants of helping.

Breaking the fitness consequences down in this way provides a productive framework for single-species field studies and comparative studies to elucidate ecological correlates of cooperative breeding. While these decision points do not necessarily reflect the order of evolutionary events (see Chapters 1 and 2), they do reflect the series of decisions made by individuals, and thus provide important information regarding current selective pressures.

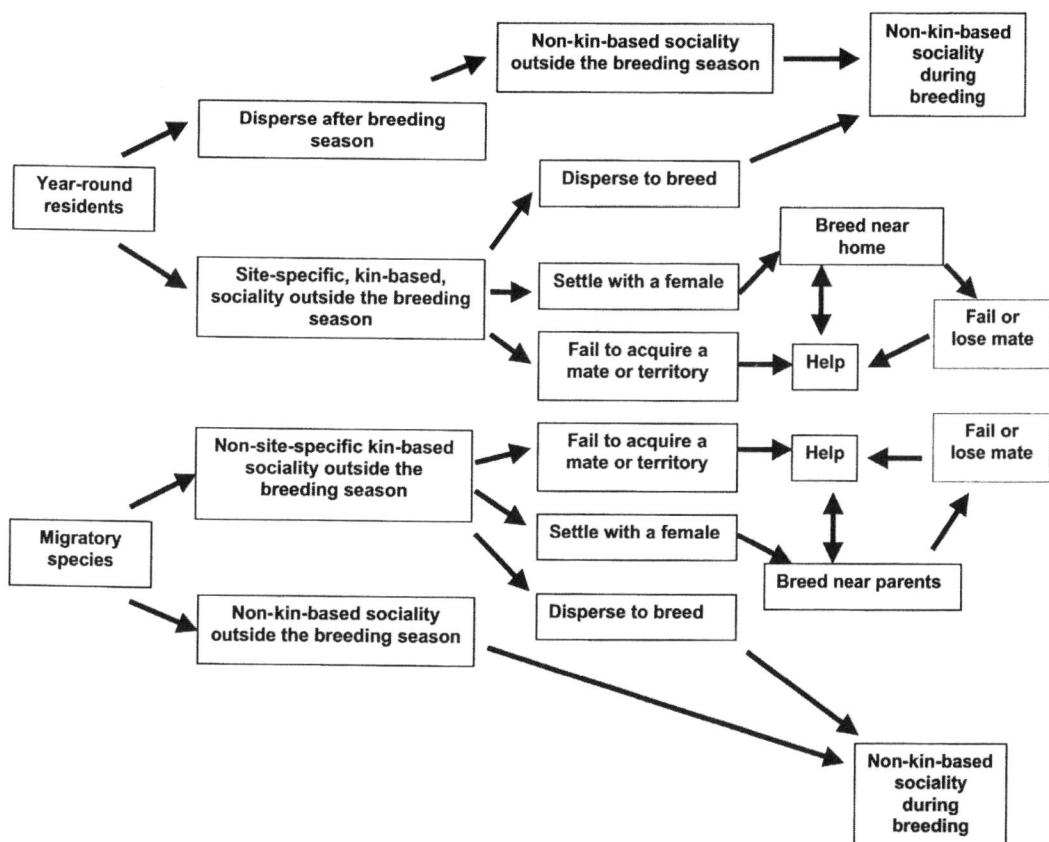

Figure 3.1. Routes to kin-directed helping behavior as a function of patterns of migration and social behavior outside the breeding season.

THE IMPORTANCE OF BEHAVIORAL PLASTICITY: REDIRECTED HELPING AS A ROUTE TO KIN-BASED COOPERATIVE BREEDING

Behavioral plasticity has the potential to allow individuals to make the transition from singular breeding to cooperative breeding and back again by allowing for opportunistic choices about whether or when to help. The first permissive condition, extended contact between parents and their offspring, increases the likelihood that offspring will remain nearby and provides young birds the opportunity to learn the identities of relatives (Komdeur and Hatchwell 1999). Extended contact with kin may be a consequence of the advantages of prolonged brood care or other benefits of kin-directed social behavior outside the breeding season (Ekman et al. 1994), high costs of dispersal (Greenwood 1980), or increased access to nearby territories or mates for young remaining on high-quality territories (Kraaijeveld and Dickinson 2001).

These factors need not lead to retention of young on the natal territory after the non-breeding season; rather, they may simply result in prolonged contact between parents and offspring and continued familiarity after young disperse to breed locally. In the colonial white-fronted bee-eater, neighborhoods or "clans" form within colonies (Emlen and Wrege1989); the "exploded clans" found in some noncolonial, territorial, cooperative breeders may be very similar in function. Examples include the extended family groups of western bluebirds (Dickinson et al. 1996) and long-tailed tits (Russell 2001; Hatchwell et al. 2001a, 2001b) and the coteries of bell miners (Clarke and Fitz-Gerald 1994). Here, we refer to these extended networks of relatives as "kin neighborhoods" (after Ligon and Ligon 1990b) and suggest that the benefits of living in kin neighborhoods are underestimated by the majority of studies, which focus only on cooperatively breeding groups.

The usual pattern among territorial cooperative breeders is for young to remain on the territory and help, a behavior that has lower fitness payoffs than the alternative of independent breeding (Brown 1987). If offspring breed near their parents and the costs of feeding young are low enough, little benefit may be required to offset the costs of redirecting care to the parental nest. Theoretically, helping need not be kin-directed if helpers accrue direct fitness benefits from joining a breeding group, but settling close to parents also provides opportunities for indirect benefits. Hence, it is more probable that helping will be a beneficial strategy if relatives are nearby and available to be helped. Redirected helping should reinforce behavioral plasticity and assessment, permitting individuals to adjust their probability and intensity of helping in response to current ecological circumstances (MacColl and Hatchwell 2002).

Comparative studies have revealed that adult survivorship is higher in cooperative breeders than in noncooperative species (Arnold and Owens 1998). High survival could be a conserved life-history trait that predisposes certain lineages to be cooperative. For example, if kinship plays a role in helping, adult offspring in lineages with high survival are more likely to have living parents than are offspring in lineages with low survival. Within such predisposed lineages, species may exhibit cooperation when exposed to the appropriate ecological conditions, while in lineages without this predisposition, cooperation would not be predicted even under conditions expected to select for such behavior (Owens and Bennett 1995; Arnold and Owens 1998, 1999). Therefore, the extent of behavioral plasticity in helping and the facility with which it can be expressed in particular phylogenetic lineages has become an important issue in assessing the role of life-history traits in the evolution of cooperative breeding (Hatchwell and Komdeur 2000; Chapter 1).

Although Arnold and Owens (1998, 1999) considered year-round residency, they did not investigate the association between cooperative breeding and year-round, family-based, territoriality. Paired comparisons that ask whether survivorship of species that stay in family groups outside the breeding season is higher than that of closely related species without family flocks are required to elucidate the non-breeding benefits of living with kin. Similarly, among species with family-based winter territoriality, we can investigate whether survivorship, and thus the potential for parent–offspring overlap, is higher in species that retain young on the family territory during the breeding season than in species whose young stay over the winter, but disperse to breed as yearlings. This framework can potentially lead to more explicit tests of hypotheses for differences between cooperative and noncooperative breeders.

LONG-TAILED TITS: A CASE STUDY OF REDIRECTED HELPING

Long-tailed tits are atypical cooperative breeders because helping is uncoupled from delayed reproduction. In a given season, all adults in a population attempt to breed in monogamous pairs, but failed breeders may become helpers at the nests of close relatives living nearby. Males are the philopatric sex, and most helpers are brothers or sons of one member of the helped pair. Russell (1999) compared breeding constraints, dispersal, and demography of long-tailed tits with those of four non-cooperative but ecologically similar species occupying the same habitat: great tit, blue tit, wren and treecreeper. The aim was to determine the key differences among these species that might have led to the evolution of cooperation in long-tailed tits but not the other species.

First, the fact that all long-tailed tits attempt to breed each year suggests that constraints on independent breeding are weak. Moreover, Russell (1999) found that constraints on independent breeding, including a shortage of nest sites and breeding vacancies, were no higher in long-tailed tits than in the non-cooperative species, although the former were much less likely to be successful in their breeding attempts because of higher rates of nest predation.

Second, a capture–recapture analysis of marked juveniles using data from a large-scale, controlled banding database revealed that long-tailed tits did not exhibit greater local recruitment of juveniles as breeders relative to the non-cooperative species. This result appears to contradict Zack's (1990) hypothesis that cooperative breeders have lower dispersal than non-cooperative species. However, it is possible that although local recruitment is similar across species, there may be fewer juveniles dispersing long distances in long-tailed tits. That is, they may have a shorter "tail" in the dispersal distribution. Even if this is the case, it appears that a similar proportion of juveniles in each of the five species recruits close to the natal site.

Third, although local recruitment appears to be similar in this cross-species comparison, Russell (1999) was able to demonstrate dramatic differences in the demography of the five species, and this has profound implications for a kin-based cooperative system. Adult survival varied little among species (Siriwardena et al. 1998), so the probability of offspring having surviving parents did not differ. The striking difference was that the recruiting offspring of the non-cooperative species were survivors from the broods of the 65–80% of pairs that were successful, while in long-tailed tits local recruits were the product of just the 25–30% of pairs that were successful (Hatchwell et al. 1999; Russell 2001). As a consequence, the probability of long-tailed tits having a close relative, such as a brother, in the neighborhood in following years is two to three times higher than in the otherwise similar non-cooperative species.

The timing of dispersal also differs markedly: long-tailed tits remain in kin-based groups from fledging through to the start of the following breeding season, whereas in the non-cooperative breeders, dispersal from the natal group follows shortly after nutritional independence of juveniles. Flocks of adults and juveniles may form subsequently in these species, but they are not composed of kin. The long kin association in long-tailed tits is probably important in enabling the discrimination of kin from non-kin (Komdeur and Hatchwell 1999; Russell and Hatchwell 2001; Hatchwell et al. 2001b).

This interspecific comparison suggests that the formation of kin flocks outside the breeding season is important in providing the permissive conditions for kin-based cooperative behavior in the following year. However, this must also be coupled with a pattern of juvenile mortality that ensures the existence of "kin neighborhoods." Therefore, in answer to our earlier question, we suggest that there may be key differences in life-history parameters other than dispersal and that these dictate whether family groups or flocks of non-relatives form outside the breeding season. The significance of the pattern of juvenile mortality has not been explored in any systematic analysis of cooperative and non-cooperative species. Finally, this comparison raises the familiar problem of whether the demographic differences identified are the cause or an effect of the cooperative breeding system of long-tailed tits.

POPULATION CONSEQUENCES OF REDIRECTED HELPING: THE SUPERSATURATION HYPOTHESIS FOR THE EVOLUTION OF KIN-BASED COOPERATIVE POLYGAMY

The steps we have identified as leading to kin-directed helping (Fig. 3.1) may result in increased constraints on independent breeding and alter life-history parameters, which will result in a feedback influencing the fitness consequences of helping. First, if staying home reduces mortality, then as more individuals of the philopatric sex stay or return home to help, we expect an increase in the disparity in survival of the helping sex relative to the non-helping sex. The breeding sex ratio will become increasingly biased in favor of the sex that stays home, leading to further constraints on independent breeding. Indeed, a shortage of mates is one of the key constraints invoked to explain kin-directed helping (Emlen 1982a; Pruett-Jones and Lewis 1990). Although the adult sex ratio has been correlated with the frequency of groups containing male helpers (Rowley 1965; Emlen 1984), we are aware of no comparison of adult sex ratio biases in cooperative and non-cooperative species. Comparative studies are required to determine if sex-ratio biases are associated with kin-directed cooperative breeding.

Brown (1987) acknowledged that retention of young could increase the constraints on independent breeding due to increased annual survival of breeders and retained offspring in groups, an idea also proposed by Russell and Rowley (1993a, 1993b), who emphasized the low turnover in cooperative breeders, and Walters et al. (1992b), who emphasized the demographic consequences of the stay-and-foray strategy. Although it is difficult to distinguish cause from effect, we propose that opportunistic, kin-biased helping will act as a positive feedback loop, potentially resulting in habitat "supersaturation" (Fig. 3.2), defined here as an excess of individuals beyond the number that would be supported if young were unable to remain in or return to their natal groups.

Supersaturation is distinct from habitat saturation in referring to an actual increase in carrying capacity that is a direct consequence of a relatively simple change in social behavior. It is also distinct from the "Allee effect" in that there is a sudden change in the maximum population size caused by increased carrying capacity due to greater tolerance of conspecifics, whereas

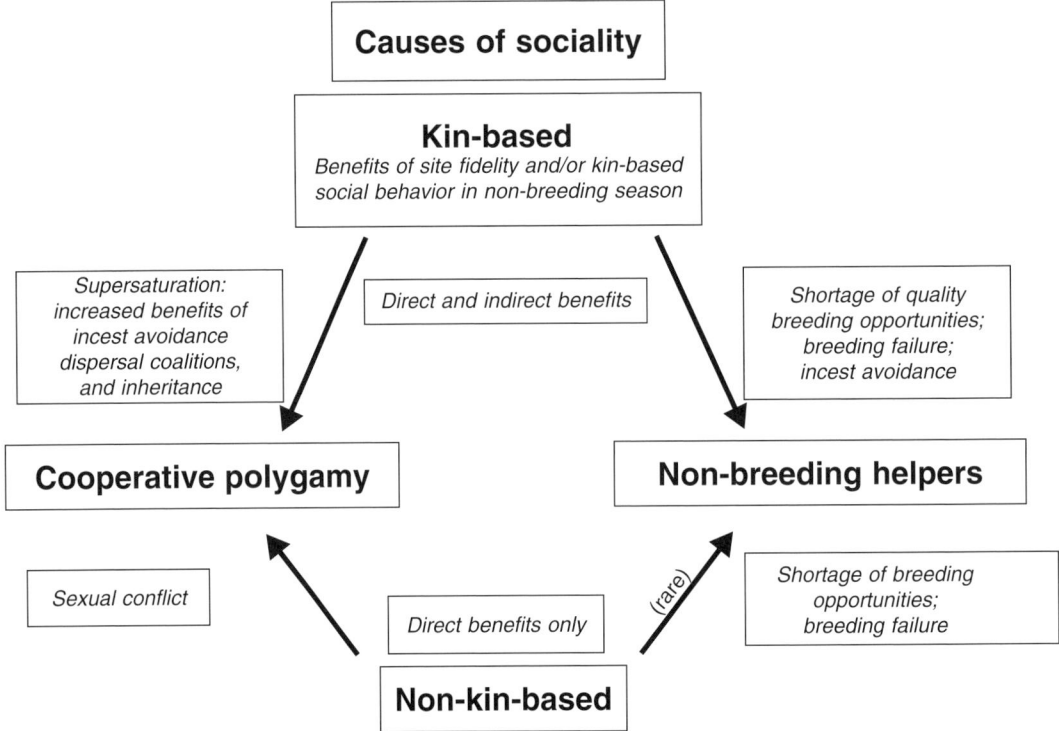

Figure 3.2. Population consequences of retention of offspring and kin-directed helping behavior.

in the Allee effect, density-dependent potential for social interaction influences population growth rate (Allee 1931; Stevens and Sutherland 1999). Once competition for space is reduced through young birds associating with their natal groups, the effective carrying capacity of a population will increase. This is because, while larger groups tend to occupy more space, the increase in space use rarely keeps up with the number of individuals in the group, so per capita use of monopolizable space declines.

Ultimately, the costs of increased local competition for food and breeding opportunities on the natal territory will increase to the point that reproduction and offspring retention are limited. Nevertheless, the expected outcomes are that more individuals will be supported within a given amount of space and that competition for independent breeding opportunities will intensify.

Carrying this scenario to its logical conclusion, increased parental survival and nesting productivity due to help will ultimately result in increased competition for breeding opportunities outside the group, not only for the philopatric sex, but for the dispersive sex as well. Supersaturation may lead to retention of offspring of both sexes and to retention of offspring beyond the number that effectively help, even to a point where "helpers" hinder. Negative impacts of helpers on parental reproductive success will result in a potential conflict between parents and offspring over whether offspring should be allowed to stay. The situation is further complicated because the benefits of ascendance to breeding status within groups and the benefits of dispersal in coalitions should increase as breeding competition intensifies, an idea supported by comparison of two populations of acorn woodpeckers in which group size is linked to degree of habitat saturation (Stacey 1979a).

If supersaturation is a root cause of kin-based cooperative polygamy, this may explain why species with cobreeding relatives also tend to have non-breeding helpers, whereas species with unrelated cobreeders do not (Davies 1992; Faaborg et al. 1995; Briskie et al. 1998).

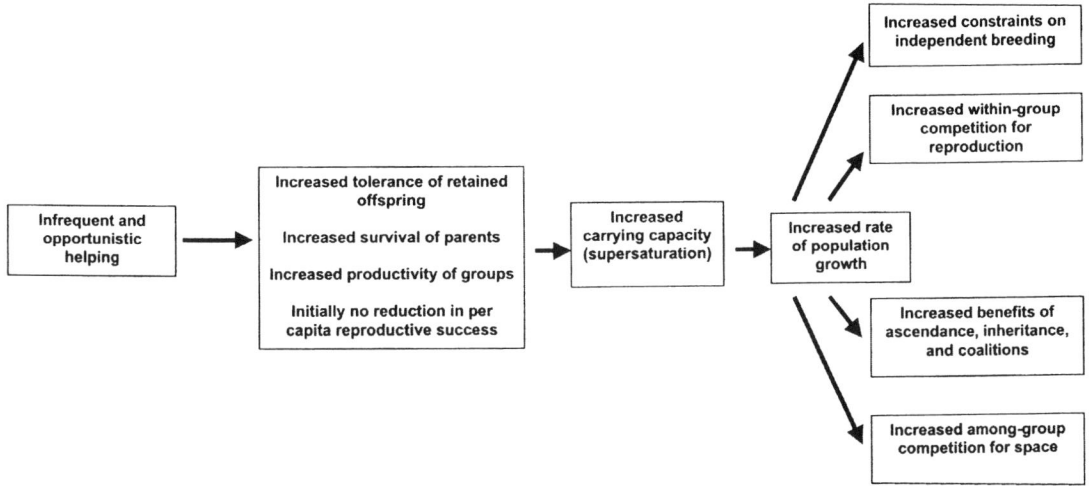

Figure 3.3. Comparison of selective factors and consequences of kin-based and non-kin-based helping behavior.

This supports two routes to cooperative polygamy, one via selection for retention of offspring and another via sexual conflict (Davies 1992; Fig. 3.3). Over time, emergent properties of group living, such as the storage granaries of acorn woodpeckers (Koenig and Mumme 1987), the cavity clusters of red-cockaded woodpeckers (Copeyon et al. 1991), and the communal roosting behavior of green woodhoopoes (Du Plessis 1992) will increase the value of natal philopatry, resulting in the suite of cooperative and competitive behaviors apparent in the most complex of avian societies (Koenig and Mumme 1987).

One possible outcome of intensification of local competition for space due to extreme philopatry is a socio-genetic structuring of populations that leads to a hierarchy of social organization. For example, bell miners have three tiers of groups: colonies, coteries, and nest contingents, or breeding groups with helpers (Clarke and Fitz-Gerald 1994). Microsatellite markers have revealed a genetic structure within each of these tiers (Painter et al. 2000). Colony similarity declines with distance and genetic similarity is greater within than among colonies. Coteries differ within high-density colonies and nest-tending contingents are more similar than are coteries. Although we cannot reconstruct the ontogeny of these populations, the pattern of socio-genetic structuring is the expected outcome of intense kin-assisted competition for breeding vacancies and space.

CONCLUSION

We have avoided presenting a comprehensive summary of the evidence for and against particular hypothesized benefits of helping, because this valuable function has been fulfilled by other recent reviews (Cockburn 1998; Hatchwell and Komdeur 2000). Instead, we have presented a limited review focused on questions, approaches, and study systems that offer the greatest promise for addressing what we regard as the fundamental, but as yet unanswered, question in the field: what selection pressures cause one species or population to exhibit cooperative behavior while other (often closely related) species do not?

By focusing on behavioral plasticity and assessment, we have attempted to elucidate ways in which information on mechanisms, fitness consequences, and behavioral choice can be combined to increase our understanding of the current functional utility of helping at the nest and its associated behaviors. In contrast with Cockburn (1998), we view kin-directed helping as a syndrome that involves a series of behavioral steps, each of which has its own impact on individual fitness. These steps include patterns of dispersal that change the demographics, life history parameters, density, and kin structure of populations, resulting in both increased competition for breeding vacancies and increased opportunities to associate with close kin. In the latter case, these opportunities are enhanced not only by postponing dispersal, but also by dispersing to breed close by and retaining associations with kin after dispersal. Our hope is that this dissection of helping at the nest will lead to novel approaches to addressing the ecological and life-history factors that have proven intractable in the past, stimulating a new generation of experimental field studies of single species and targeted, phylogenetically controlled, comparative studies aimed at understanding the evolution and maintenance of helping at the nest and the emergent properties of group living in birds.

ACKNOWLEDGMENTS

We thank Walt Koenig, Rob Magrath, and Jan Ekman for comments on earlier versions of this chapter.

4 · Parental care, load-lightening, and costs

ROBERT G. HEINSOHN

Australian National University

Helping behavior is enigmatic as it appears to entail an individual sacrificing personal reproduction while assisting others in their breeding attempts. Over the past 40 years, the field of cooperative breeding has developed a rich body of theory to explain helping behavior, and enough cooperative species have been studied in detail to establish common ground and test theory. Indeed Emlen (1997a) stated that the original paradox of cooperative breeding had largely been resolved with the widespread confirmation that (1) helpers are often individuals that are constrained from breeding due to a shortage of quality breeding opportunities, (2) helpers unable to obtain breeding positions in the current year frequently improve their chances of becoming breeders in the future, and (3) helpers frequently obtain large indirect benefits by helping to rear collateral kin. With identification of these direct and indirect benefits to helpers, Emlen suggested that the original questions asked by researchers in this field would appear to be "largely answered."

In contrast, Cockburn (1998) concluded that "we are still some way from understanding the adaptive significance of helping behavior although we are poised for a reinvigoration of the study of cooperation through a number of conceptual, empirical, and technical advances." Clearly, conceptual breakthroughs have been made, but many important questions also remain unanswered. In particular, our understanding of the varying level of helper contributions within and between species and how these contributions benefit breeders and helpers remains poor.

The approach to cooperative breeding has often been to compare the fitness benefits of philopatry and help with the alternative options of dispersing to float or dispersing to breed (Emlen 1982a; Reyer 1990; Walters *et al.* 1992b). Evaluation of the fitness rewards for each strategy presumably leads to an understanding of the adaptive consequences of a particular decision. Implicit in this approach is that the outcome reflects all the costs and benefits of dispersal versus non-dispersal, and helping versus non-helping. Although this may be true, it unfortunately does not lead to an appreciation of the nature of each cost and benefit. In fact, the above approach has tended to treat helping behavior as a discrete strategy with two levels (dispersal or philopatry plus help), when in reality the extent of help varies greatly and philopatry may even occur without help (Chapter 2). Helping behavior is thus a continuous variable bounded only by zero at the lower end.

My goal here is to evaluate how the costs and benefits of care limit both helpers and breeders, especially whether and how much an auxiliary individual should contribute, and how breeders should seek or respond to contributions from helpers. I argue that these variables are inextricably linked, and advocate a comprehensive life-history approach to understanding the behavioral decisions of whether, and by how much, individuals should help others to breed.

THE NATURE OF THE PROBLEM

Investment in one's own offspring, or the offspring of others, should reflect the trade-off between the costs and benefits of such behavior, weighted by the probable relatedness to those individuals (Hamilton 1964). Consider the Seychelles warbler. Komdeur (1994a) showed that helpers much prefer to feed nestlings that are more closely related to themselves, an important result that emphasized the lability and adaptive nature of helping behavior in this species. Intriguingly, however, the figures presented in his article show that helpers raising apparently full sibs do not feed as much as the

---

Ecology and Evolution of Cooperative Breeding in Birds, ed. W. D. Koenig and J. L. Dickinson. Published by Cambridge University Press.
© Cambridge University Press 2004.

parents, even though both parents and helpers would presumably gain the same fitness reward. Relatedness was assumed to be 0.5 in both cases, but a recent molecular study has shown that mean relatedness of nestlings to both parents and helpers is considerably lower than 0.5 due to a high level of contribution to clutches from subordinate females (Richardson et al. 2002). The Seychelles warbler is one of the few species in which an experimental approach has shown a clear effect of group size on productivity (Komdeur 1994b), leading to the question of why group members do not contribute more. Indeed, it could also be asked why helpers do not work as hard or harder to raise less-related individuals, because any additional increment in reproductive success could compensate for the lower relatedness.

In the same study population, Komdeur (1994b) showed that the overall provisioning rate to nestlings went up with the first helper but leveled out with additional helpers. Whereas the female parent maintained her delivery rate, the male reduced his contribution in response to the presence of helpers, in spite of evidence that extra food translated into more and heavier fledglings. Taken together, Komdeur's work shows that the dual questions of whether to help, and how that help is utilized by parents, are complex and driven by multiple variables.

In some species, younger individuals are not as good at providing parental care as older individuals (Boland et al. 1997a). However, such age-specific ability is not a universal explanation for patterns of help, because helpers can work as hard as, or harder than, the breeders (Reyer and Westerterp 1985). Other species have philopatric individuals that fail to help at all, or that only help if they have the incentive of direct paternity (Veltman 1989; Davies 1992; Magrath and Yezerinac 1997). Coercion from parents might also be important (Mulder and Langmore 1993). Some helpers regularly aid non-relatives (Dunn et al. 1995), whereas others forgo the opportunity to raise close kin (Boland et al. 1997b; Magrath and Whittingham 1997). Together, these observations suggest a large range of costs and benefits to helping that combine in different measure to determine whether, and by how much, helping should occur. Alongside the decisions made by potential helpers, breeders must decide whether to accept their help and whether to use it for production of extra or higher-quality offspring or to reduce their own parental expenditure.

THE BENEFITS OF ALLOPARENTAL CARE

The adaptive benefits of helping, as distinct from philopatry, have been reviewed thoroughly (Brown 1987; Koenig et al. 1992; Emlen 1997a; Cockburn 1998). Here I note that the hypothesized benefits fall into two major categories: the enhanced production of non-descendant kin (indirect benefits), and benefits that increase the chance of survival or direct reproduction, either immediately or in the future (direct benefits). The second category includes enhanced social prestige (Zahavi 1990), the payment of "rent" in return for enjoying the benefits of philopatry (Mulder and Langmore 1993), parentage itself (Davies 1992), enhancement of territorial or group quality by increased production of group members (Woolfenden and Fitzpatrick 1984; Kokko and Johnstone 1999), formation of alliances to aid in competitive situations (Ligon and Ligon 1990b), and enhancement of skills for later reproduction (Heinsohn et al. 1988; Komdeur 1996).

Kin selection has often been interpreted as providing the primary fitness benefit of helping behavior (Brown 1987; Emlen 1997a). However, many studies do not show a positive relationship between the number of helpers and production of young (Magrath and Yezerinac 1997; Magrath 2001). Even when the relationship does exist, it is often difficult to establish whether it is driven by helper contributions rather than territory or breeder quality. Only a few experimental studies provide convincing evidence for the former (Brown et al. 1982; Komdeur 1994b; see Chapter 3). An alternative kin-selected benefit of help is increased survival of the breeder through reduced parental effort (Crick 1992). The benefits of such "load-lightening" have also been difficult to assess due to problems similar to those encountered in the case of group size (Cockburn 1998). For example, survival might covary with number of helpers, parental quality, or territory quality.

Parents can thus benefit from help in several ways, either as additional to their own (referred to as "additive") or by reducing their own workload ("compensation"), or as some combination of the two (Hatchwell 1999). However, given that help is sometimes withheld from relatives or directed at non-relatives, combined with the lack of a helper effect in other species, there would appear to be two challenging possibilities for at least some cooperative breeders: either helping might be selected against even when kin benefit in the short term, or kin selection might not be the driving force for some cases of helping.

In order to separate indirect and direct benefits of helping behavior, it is necessary to understand when, and by how much, helping should occur purely for kin-selected reasons. Hamilton's rule states that a helper should only help when $rB > C$. Any unit of care from a helper has two effects: it incurs a cost (C) and it produces a benefit (B) that is weighted by r, its relatedness to the breeder. However, Hamilton's rule can equally be used in reverse to ensure that the breeder gains a net benefit from the helper's contribution. That is, helping is only beneficial to the breeder when $B > rC$. If a breeder and helper are closely related and incur similar costs in caring for the young, it is not immediately clear how much help should be sought or given.

As Hamilton's rule indicates, it is important to focus not only on the benefits of helping, but also on the costs. Although his model is genetic and applies to a fixed level of helping, costs and benefits should also be considered in any model predicting how much aid helpers should give. In general, analyses of helping behavior in birds have been strongly biased towards the benefits while neglecting the cost component (Heinsohn and Legge 1999).

HOW COSTLY IS CARE?

The costs of providing care to offspring have been investigated primarily in biparental systems, and include reduced body condition, reduced survival, and reduced future fecundity (Clutton-Brock 1991; Ketterson and Nolan 1994). In comparison, the costs of providing alloparental care have been relatively neglected. However, if parents are limited by the costs of care, and can adjust their level of investment to mitigate these costs, then helpers should be restricted in the same way, and show similar flexibility.

Although philopatry carries obvious costs, such as competition for breeding opportunities (Koenig et al. 1995) and risk of mortality while waiting for reproductive opportunities, the costs of helping *per se* are not well documented. A physiological cost of helping was first demonstrated in a cooperatively breeding cichlid fish *Lamprologus brichardi* by Taborsky (1984) who showed that helpers grow more slowly than non-territorial fish. The benefits they receive to offset this cost are the protection from predators afforded by a safe territory and an increase in the size of the clutch raised by the related individuals they help.

In birds, Reyer and his colleagues demonstrated a physiological cost of helping in the pied kingfisher. Helpers in this species are always male and come in two forms. "Primary" helpers are offspring that remain with the breeding pair throughout the nesting period in non-breeding condition, during which time they expend as much energy as the breeders to provision young at the nest (Reyer and Westerterp 1985). In contrast, "secondary" helpers, which are not related to the breeders, are recruited after the eggs have hatched, and only if food is in short supply. Secondary helpers do not work nearly as hard as the breeders, and are in reproductive condition (Reyer et al. 1986). Although they do not appear to gain direct reproduction when recruited, they may enhance their probability of breeding with the female in future years (Reyer 1990). This elegant contrast between the two types of strategy suggests that only those helpers seeking inclusive fitness will bear both "psychological castration" (Reyer et al. 1986) and the physiological costs associated with high levels of alloparental care.

Helping is also costly in white-winged choughs. Cooperative breeding in this Australian passerine is enforced by a difficult foraging niche that requires large amounts of time to dig for invertebrates in soil and leaf litter. Choughs have an extended four-year period of skill development before reaching sexual maturity, but even fully mature breeders must have at least two helpers to breed successfully (Heinsohn et al. 1988; Heinsohn 1991c). Each additional helper, up to group sizes of 14, means additional food brought to the nest and increased productivity through reduced nestling starvation. However, one- and two-year-old helpers, which are most limited by inferior foraging ability, contribute the least, and even withhold food deliveries (Heinsohn et al. 1988; Boland et al. 1997b). When supplementary food is experimentally provided at the nest, small groups supply as much food to nestlings and produce as many fledglings as large groups, and young birds contribute as much as older individuals (Boland et al. 1997b). Thus, it is the inability or unwillingness to provision at higher rates that normally limits young choughs from helping as much, and small groups from producing as many young as larger groups.

The cost of helping in white-winged choughs is only detected when helpers contribute excessively. For example, one-year-old helpers contribute to incubation only when group size is under seven individuals, and lose weight in proportion to the amount of time they spend on the nest (Heinsohn and Cockburn 1994). Incubation occurs during the cool months of early spring, and time out from foraging appears to entail energetic costs. In the absence of data indicating that they are somehow "forced" to incubate, it seems likely that these young birds help in this fashion only when their contribution is essential because of a lack of older helpers (Heinsohn and Cockburn 1994).

A similarly revealing example of an energetic cost to helping comes from a mammal, the meerkat (Clutton-Brock et al. 1998). In these cooperatively breeding mongooses, non-breeding adults commonly babysit young pups at the burrow and have to forgo foraging for long periods. The energetic costs of this activity are high: over the 24-hour shift, the average babysitter loses 1.3% of its body weight compared with other group members that continue foraging and gain 1.9% of their body weight. Over the entire reproductive effort, top babysitters lost on average 3.8% of their body weight and some lost as much as 11%. Babysitting young at the burrow is an essential activity that serves to guard pups from avian and terrestrial predators, but interestingly is never performed by the breeding pair themselves. Like choughs, meerkats are sensitive to group size and modify the extent of their help accordingly. Non-breeders in smaller groups perform a larger share of the babysitting and bear greater costs to achieve the required corporate effort. For further discussion, see Chapter 13.

These studies have three important implications. First, becoming a helper can have profound implications for an individual's life-history, including suppressing, or at least delaying, sexual maturation (Taborsky 1984; Reyer et al. 1986).

Second, helping is not necessarily automatic and, within species, is a flexible response set by the needs of the breeders and the costs to the helper (Reyer and Westerterp 1985; Heinsohn and Cockburn 1994; Clutton-Brock et al. 1998). Such flexibility is important for demonstrating that helping behavior is adaptive (Jamieson and Craig 1987a; Heinsohn and Cockburn 1994; Komdeur 1994a; Clutton-Brock et al. 1998). Among species, helpers can contribute as much alloparental care as if they were breeding themselves (Reyer and Westerterp 1985), less than they would as parents but still enough to increase productivity (Komdeur 1994b; Dickinson et al. 1996), or they may remain philopatric without contributing any help at all (Veltman 1989; Magrath and Whittingham 1997; Chapter 2).

Third, although attempts to measure the costs of helping have been few, in many the costs may be difficult to detect. This is chiefly because of the natural tendency of helpers to limit such alloparental investment according to their ability (Pettifor et al. 1988; Komdeur 1996; Boland et al. 1997b). Because recruitment of helpers and expression of help might be based on a combination of the needs of breeders and helper ability, measurements of costs and how the upper limit to helping behavior is set may only be possible through experiments or carefully controlled comparisons. Further, care of young can also occur in more than one form, leading to a potential division of labor between the sexes and various age classes (Clutton-Brock et al. 1998). The costs and benefits of helping in such complex societies may be difficult to compare using a single currency.

SHORT- VERSUS LONG-TERM COSTS OF CARE

Logically, the long-term costs of helping can be analyzed in the same fashion as ordinary parental care. Potential costs include reduced body condition, future survival, and fecundity (Clutton-Brock 1991). Good data are available from stripe-backed wrens, where Rabenold (1990) showed that helpers provisioning at high rates have lower survival. This shows that the decision to help may have cascading effects throughout the individual's lifetime, not just within one breeding season. To the extent that this is true, all apparent benefits of helping must be discounted by reductions in future survival or fecundity, with the implication that helping might not always be the best strategy while waiting for a breeding position. Conversely, helpers in stripe-backed wrens and some other species stand a high chance of never gaining a breeding position, in which case the probability of eventual success could determine the value of working for immediate inclusive fitness.

A confounding explanation for differences in future survival and fecundity is that helpers vary in quality. Those with low chances of independent reproduction might even devote more time or effort to helping. For example, some individuals that help for long periods before breeding themselves have lower success than those that reproduce sooner, and are usually interpreted as being of lower quality (Dickinson et al. 1996; Marzluff et al. 1996). Helper quality and costs incurred through helping are difficult to distinguish as the cause of decreased future fecundity. One way would be to reduce the costs to some helpers experimentally (Boland et al. 1997a) and then monitor their future breeding success compared with same-aged individuals who begin breeding without a helping period.

A GENERAL MODEL OF ALLOPARENTAL CARE

Fig. 4.1 is a graphical representation of the costs and benefits to breeders and helpers of providing care to a current brood. It utilizes two general cost curves and two general benefit curves that are likely to be common in nature. A general linear measure of parental care is on the x-axis for all three graphs. The benefit curves (Fig. 4.1a) depict the number of offspring produced (y-axis) for any given level of care. I have chosen a general function in which offspring number (a proportion of the maximum clutch size) increases with parental care such that:

$$\text{Offspring} = 1 - e^{-ap} \quad (1)$$

where $p = $ care and a is a constant. I have chosen this function as it embodies the following characteristics: (1) monotonic increase in offspring with increasing parental care, (2) rapid increases early, but progressive slowing, and (3) potential for differing rates of increase. Importantly, it allows approximation of two entirely different types of functions without losing its mathematical generality. Curve B_1 (high a) approaches a step function in which initial investment is extremely rewarding but additional units of parental care have little effect. Curve B_2 (low a) approaches linearity and embodies situations in which additional care continues to translate into further offspring. Note that offspring number could be explicitly included by using the function, Offspring = $k(1 - e^{-ap})$ where $k = $ the maximum number of offspring (Cant and Field 2001), but this would not change any of the conclusions drawn here.

On the x-axis are three levels of care given to the current brood of offspring. P denotes the care that is given by a breeder in the absence of helpers, and H is the level when a helper also provides care. P' is the level a breeder may reduce its own care to in response to having help. For simplicity, the helper's contribution ($H - P$) is fixed at an arbitrary level between zero and that contributed by the parent ($P \leq H \leq 2P$). The breeder can either maintain its level of care such that

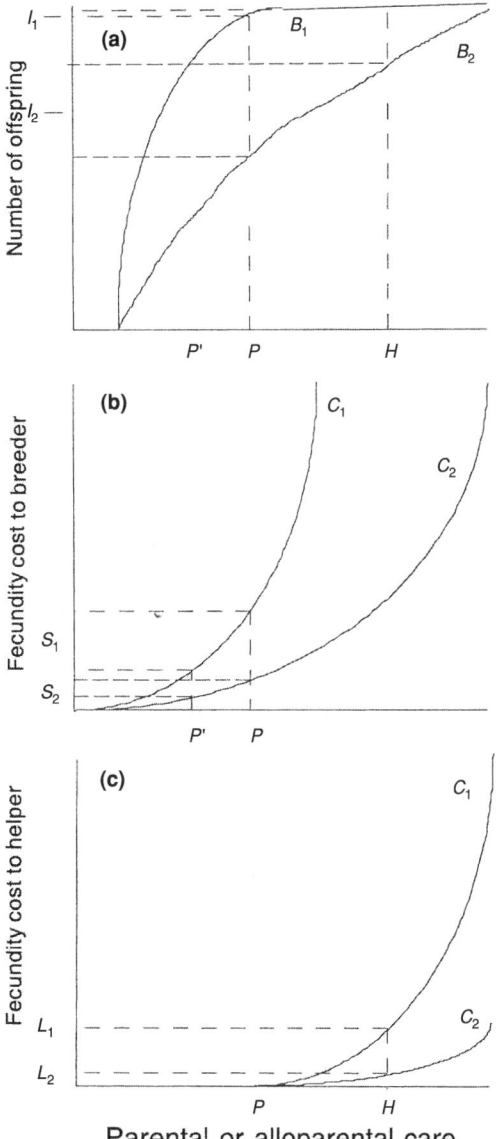

Figure 4.1. Cost and benefit curves for breeders and helpers when they contribute parental and alloparental care to the current brood of young. Benefit curves depict the number of offspring produced whereas cost curves show the loss of future fecundity. (a) Two contrasting benefit curves (B_1 and B_2), together with the increments in offspring production (I_1 and I_2) brought about by an additive helper contribution of $H - P$. (b) Two contrasting cost curves (C_1 and C_2), together with the fecundity savings (S_1 and S_2) that the breeder would make if it reduced its care to P'. (c) The same cost curves displaced to the right to show the fecundity cost to the helper (L_1 and L_2) of contributing $H - P$ alloparental care.

the helper's contribution is additive, or it can reduce its care to compensate (either completely or partially) for that provided by the helper (Hatchwell 1999). The increment in production of offspring when care is additive is denoted by I_H, or I_1 and I_2 respectively for curves B_1 and B_2, such that $I_1 < I_2$.

Figure 4.1b shows two possible cost curves, C_1 and C_2, for the breeder for any given value of parental care it gives. Costs are measured as losses in future fecundity and can be subtracted directly from curves B_1 and B_2. The general form is:

$$\text{Cost} = e^{bp} - 1 \quad (2)$$

where b is a constant. This function was chosen for its acceleration of costs at higher levels of care. Costs of care accelerate more quickly in C_1 (high b) than in C_2 (low b). The cost savings to the breeder associated with reducing its level of care P to P' is S_B, or S_1 and S_2 respectively. In Fig. 4.1c, the same curves C_1 and C_2 have been shifted to the right and are used to show the costs of care to helpers when they contribute to rearing the brood. The costs of contributing $H - P$ care are denoted by L_H (loss to helper), or L_1 and L_2 respectively.

These graphs allow the visualization of the circumstances when breeders and helpers are best served by contributions from helpers, and whether this help should be additive or accompanied by a compensatory reduction in care by breeders. All fecundity costs and benefits can be translated into "offspring equivalents" by weighting outcomes according to the relatedness between breeders and helpers. For simplicity I assume initially that only one of the breeders will respond to helper contributions, although the fundamentals of the model do not change if both breeders respond to the available help.

SIMPLE MODEL

In the first model, the breeder has one of two choices: it can either maintain its level of care such that the helper's contribution is additive, or it can reduce its care by $H - P$ to compensate completely for that provided by the helper. If the level of help available from the ith helper is fixed, three general rules concerning use of that help apply from a breeder's perspective:

Rule 1. Additive helping is beneficial when $I_H > rL_H$, where r is the relatedness between the breeder and the helper and rL_H = loss of future offspring equivalents through fecundity costs to the helper. In other words, when the increase in offspring outweighs the indirect fitness lost through costs to the helper.

Rule 2. Compensation is beneficial when $S_B > rL_H$, (r as above), that is, when cost savings attributable to reducing parental care outweigh the indirect fitness lost through costs to the helper. This rule always holds when breeder and helper have the same cost curve. To demonstrate, the inequality becomes $e^{ap} - e^{a(2P-H)} > r(e^{a(H-P)} - 1)$, which is true for all likely values of r ($r \leq 0.5$) and assuming $H > P$.

Rule 3. Assuming full paternity or maternity for the breeder, compensation is better than additive help when $S_B > I_H$: that is, when the cost savings attributable to reducing parental effort are greater than the reproductive benefits of additional effort beyond that of a simple pair.

Three rules also apply from the helper's perspective:

Rule 4. Additive helping is beneficial when $(r_1 + r_2)I_H > L_H$, where r_1 and r_2 are the helper's relatedness to the breeders. That is, when the increase in indirect fitness benefits of helping outweighs the long-term fecundity costs to the helper.

Rule 5. Compensation is beneficial when $r'S_B > L_H$, where r' is the mean of ($r_{1,n}$ and $r_{2,n}$) in all n future breeding attempts accounting for S_B. That is, when the fitness gains attributable to reducing the breeder's parental care outweigh the long-term fecundity costs to the helper.

Rule 6. Compensation is better than additive help when $r'S_B > (r_1 + r_2)I_H$, that is, when the fitness gains attributable to reducing the breeder's parental care outweigh the fitness benefits through increasing the number of offspring produced.

It is interesting to note that $r' \leq (r_1 + r_2)$. This is because future offspring of the breeder might be less related to the helper than those in the current brood. For example, although the helper might assist both its parents initially to raise full sibs ($r_1 + r_2 = 0.5 + 0.5 = 1.0$), potentially leading to I_H full offspring equivalents, only one parent might survive and retain a breeding position in later years ($0.5 \leq r' \leq 1.0$) potentially leading to $r'S_B$ offspring equivalents.

Table 4.1 evaluates the six general scenarios concerning payoffs I_H, S_B, and L_H that may arise when

Table 4.1. *The conditions defining when additive help only or compensation only are beneficial for both helpers and breeders*

Inequality	$L_H < S_B < I_H$	$L_H < I_H < S_B$	$I_H < L_H < S_B$	$I_H < S_B < L_H$	$S_B < I_H < L_H$	$S_B < L_H < I_H$
When is additive help beneficial?						
Breeder	Always	Always	Low r	Low r	Low r	Always
Helper	High r	High r	Never	Never	Never	High r
When is compensation beneficial?						
Breeder	Always	Always	Always	Low r	Low r	Low r
Helper	High r'	High r'	High r'	Never	Never	Never
Best strategy						
Breeder	Additive	Compensation	Compensation	Compensation	Additive	Additive
Helper	Either	Either	Either	Either	Additive	Additive

breeders and helpers are faced with the possible cost/benefit combinations defined by the families of curves in Fig. 4.1. As $S_B \geq L_H$ whenever the breeder and helper have the same cost curve, and help is either additive or completely compensatory in this model, the three inequality scenarios for these situations are presented first. As an example, the likely inequality that results when costs and benefits to both breeder and helper are defined by B_1 and C_1 (Fig. 4.1) is $I_H < L_H < S_B$. For the breeder, rule 1 only applies when r is low. Rule 2 states that compensation is always beneficial under these circumstances, and Rule 3 shows that compensation is the better of the two strategies. By comparison, additive help is not beneficial for the helper (Rule 4), and compensation is only beneficial to the helper when r' is high (Rule 5). Finally, Rule 6 shows that either compensation or additive help may provide the better outcome depending on the relative values of r' versus $(r_1 + r_2)$. If $r' = (r_1 + r_2)$, then compensation is always the best strategy.

The remaining possible outcomes in Table 4.1 refer to when helpers have steeper cost curves than breeders, such that $L_H > S_B$ is possible. For example, when B_1 applies to both helper and breeder, but the helper has cost curve C_1 and the breeder has C_2, the inequality $I_H < S_B < L_H$ is likely. Alternatively, when B_2, C_1, and C_2 are in force, then inequality $S_B < L_H < I_H$ is likely.

PARTIAL-COMPENSATION MODEL

I next ask when partial compensation is in the best interests of either the breeder or the helper. In this model, the helper's contribution remains fixed according to its ability, whereas the breeder can reduce P to P' such that $(2P - H) \leq P' \leq P$.

Breeder's perspective

From the breeder's point of view, if $S_B > I_H$, then full compensation is always the best outcome. However, a mix of partial savings (S'_B) and partial additive help (I'_H) is more beneficial than complete additive help when the following conditions are met:

$$S'_B + I'_H > I_H \quad \text{(Condition 1)}$$
$$S'_B + I'_H > r L_H \quad \text{(Condition 2)}$$
$$S_B < I_H \quad \text{(Condition 3)}$$

These conditions can be evaluated in two scenarios, when the breeder and helper have either the same or different cost functions.

(a) Breeder and helper have same cost function
Substituting for Condition 1, we get:

$$(e^{bP} - e^{bP'}) + \left((1 - e^{-a(H-(P-P'))}) - (1 - e^{-aP})\right)$$
$$> (1 - e^{-aH}) - (1 - e^{-aP})$$
$$e^{bP} - e^{bP'} + e^{-aP} - e^{-a(H-(P-P'))} > e^{-aP} - e^{-aH}$$
$$e^{bP} - e^{bP'} - e^{-a(H-(P-P'))} > -e^{-aH}$$

(Inequality 1)

P' clearly must fall below the threshold defined by Inequality 1. The plausibility of Condition 1 can be examined by evaluating Inequality 1 at the extreme value of

P', that is, when $P' = 2P - H$, and $H = 2P$. This gives:

$$e^{bP} - 1 - e^{-aP} > -e^{-2aP}$$
$$e^{bP} - 1 > e^{-aP} - e^{-2aP} \quad \text{(Inequality 2)}$$

Inequality 2 demonstrates that Condition 1 remains plausible even when the breeder reduces its care by the maximum amount, which should not happen since $S_B < I_H$ (that is, full compensation is not beneficial).

Substituting for Condition 2, we get:

$$(e^{bP} - e^{bP'}) + ((1 - e^{-a(H-(P-P'))}) - (1 - e^{-aP}))$$
$$> r(e^{b(H-P)} - 1)$$
$$(e^{bP} - e^{bP'}) + e^{-aP} - e^{-a(H-(P-P'))} > r(e^{b(H-P)} - 1)$$
$$\text{(Inequality 3)}$$

P' must again fall below some threshold value to satisfy the condition. The plausibility of Condition 2 can be examined by evaluating Inequality 3 at the extreme value of P'. That is, when $P' = 2P - H$, and $H = 2P$. This gives:

$$e^{bP} - 1 - e^{-aP} + e^{-aP} > r(e^{bP} - 1)$$
$$e^{bP} - 1 > r(e^{bP} - 1)$$
$$r < 1 \quad \text{(Inequality 4)}$$

Inequality 4 demonstrates that Condition 2 remains plausible even when the breeder reduces its care by the maximum amount.

Finally, substituting for Condition 3, we get:

$$e^{bP} - e^{bP'} < e^{-aP} - e^{-aH} \quad \text{(Inequality 5)}$$

P' must fall above some threshold to satisfy this condition.

Inequalities 1, 3, and 5 combine to give:

Rule 7. When breeder and helper have the same cost function, savings from partial compensation, S'_B, can be beneficial to the breeder for some window of values P' defined at the upper end by some value of P' satisfying both

$$e^{bP} - e^{bP'} - e^{-a(H-(P-P'))} > -e^{-aH} \quad \text{and}$$
$$e^{bP} - e^{bP'} - e^{-aP} - e^{-a(H-(P-P'))} > r(e^{b(H-P)} - 1)$$

The lower end of the window is defined by the value of P' that satisfies Inequality 5:

$$e^{bP} - e^{bP'} < e^{-aP} - e^{-aH}$$

In summary, given that full compensation is always the best outcome when $S_B > I_H$, Rule 7 defines the precise window of conditions for which partial compensation is beneficial to the breeder.

(b) Breeder and helper have differing cost functions

When breeder and helper have differing cost curves $C = e^{bx} - 1$ and $C = e^{dx} - 1$, such that $d > b$, Condition 2 becomes:

$$e^{bP} - e^{bP'} + e^{-aP} - e^{-a(H-(P-P'))} > r(e^{d(H-P)} - 1)$$
$$\text{(Inequality 6)}$$

Inequality 6 leads to a lower threshold of P', and gives:

Rule 8. When breeder and helper have different cost curves, savings from partial compensation, S'_B, can be beneficial to the breeder for some window of values P' defined at the upper end by some value of P' satisfying both

$$e^{bP} - e^{bP'} - e^{-a(H-(P-P'))} > -e^{-aH} \quad \text{and}$$
$$e^{bP} - e^{bP'} + e^{-aP} - e^{-a(H-(P-P'))} > r(e^{d(H-P)} - 1)$$

The lower end of the window is again defined by the value of P' that satisfies Inequality 5:

$$e^{bP} - e^{bP'} < e^{-aP} - e^{-aH} \quad \text{(Inequality 7)}$$

The window of values for P' is narrower for Rule 8 than for Rule 7, and the breeder should generally favor more compensation when the helper has a steeper cost curve (Table 4.2).

Helper's perspective

Using analogous logic to that for conditions 1 to 3 above, the conditions for when partial savings to the breeder are beneficial from the helper's perspective can be constructed as follows:

$$r'S'_B + (r_1 + r_2)I'_H > (r_2 + r_2)I_H \quad \text{(Condition 4)}$$
$$r'S'_B + (r_1 + r_2)I'_H > L_H \quad \text{(Condition 5)}$$
$$r'S'_B < (r_1 + r_2)I_H \quad \text{(Condition 6)}$$

Thresholds can be constructed in the same manner as for rules 7 and 8. Qualitative predictions for helpers (with respect to breeders) can be made using simple logic. In all of Conditions 4, 5, and 6, the left-hand side of the inequalities are relatively smaller than in Conditions 1, 2, and 3, chiefly because $r' \leq (r_1 + r_2)$ assuming there is no extra-pair paternity in the current brood. It follows that the window of values for which compensation is beneficial for helpers is shifted to the left. That is, P' is smaller, and there is more compensation. This leads to two additional rules:

Rule 9. When breeders and helpers have the same cost function, savings to the breeders from partial compensation, S'_B, can be beneficial to the helpers for some window of values for P' such that all values are smaller (that is, higher compensation) compared to Rule 7.

Rule 10. When breeders and helpers have different cost curves, savings from partial compensation, S'_B, are beneficial for the helpers for some window of values of P' shifted to the left (that is, increasing compensation) compared to Rule 8.

IMPLICATIONS OF THE MODEL

Whereas most previous analyses distinguish between dispersing to breed and remaining as a helper, the present model isolates the costs and benefits both to the parent and to the auxiliary individual if the latter provides alloparental care. In at least some cooperative breeders there is no apparent effect of helpers on production of young (Cockburn 1998; Hatchwell 1999). This has led to the alternative hypothesis that parents benefit from helpers through reducing their own levels of costly care (Crick 1992). A benefit for the helper in this scenario may be enhanced production of relatives in the future. The chief value of the models presented here is that they define, for both breeder and auxiliary, all of the possible kin-selected benefits in a single set of trade-offs, in particular establishing whether an individual that is already philopatric should help, and whether that help should be used by breeders in an additive or compensatory fashion. Importantly, it also recognizes the kin-selected cost to the breeder caused by the workload of its related helper.

The simple model developed first allows identification of the circumstances when breeders and helpers can potentially benefit from help. In this model, the breeder can choose either to accept the help offered as additional to its own, or it can reduce its own care by the same amount (compensate fully) and thus make savings in future fecundity. In addition to the six formal rules, four important generalizations can be drawn from Table 4.1:

(1) When L_H (loss to helper) is greater than S_B (savings to breeder) and I_H (increase in offspring when help is additive), it is never beneficial for an individual to help, although the breeder can benefit if its relatedness to the helper is low.

(2) A breeder can potentially benefit (depending on its relatedness to the helper) from both additive and compensatory help in every possible inequality scenario, whereas a helper can benefit by each type of help in only three out of six scenarios.

(3) Even when they both potentially benefit, differences in the required relatedness suggest that there is a window of conflict between the breeder and helper in every inequality scenario.

(4) There are potential scenarios, all involving high relatedness between breeder and helper, when it is not in the best interests of the former to either accept or demand help from the latter. In these cases, the loss of eventual offspring for the helper incurs too great an inclusive fitness cost for the breeder.

The extended form of the model then predicts when partial compensation is beneficial from both the breeder's and the helper's perspective. The model identifies windows of variable width in which the breeder and helper can benefit by the breeder reducing its parental care. The windows for both breeder and helper overlap, with the helper's shifted to the left (favoring higher compensation, Fig. 4.2). Both breeder and helper will have optimal points within their window in which the difference between cost savings and loss of direct benefit is maximized. Four general predictions follow (Table 4.2):

(1) The helper will prefer a higher level of compensation than the breeder (Rule 9). There may be a conflict over the extent of compensation, as the breeder could maximize its fitness at the expense of the helper. However, there also exists common ground under which both can benefit from partial compensation.

(2) The windows for helper and breeder will be smaller when they have different cost functions, in particular, when the helper has lesser ability than the breeder (Rules 8 and 10). In this case, both parties will prefer a higher level of compensation.

(3) The breeder will prefer to compensate more when it is more closely related to the helper (Rules 7 and 8).

Table 4.2. *Predictions from the partial compensation model*

Variable	Effect	Breeder		Helper	
		Rule	Prediction	Rule	Prediction
Same cost function for breeder and helper	Breeder and helper have same ability	7	Largest P' and least compensation	9	Smaller P' and more compensation than 7.
Different cost function for breeder and helper	Breeder has greater ability than helper	8	Smaller P' and more compensation than 7	10	Smaller P' and more compensation than 7, 9.
Increase r	Closer breeder relatedness to helper	7, 8	Smaller P' and more compensation	—	None
Increase r_1, r_2, or r'	Closer present/future helper relatedness to breeders	—	None	9, 10	Larger P' and less compensation

(4) The helper will prefer less compensation when the present and expected future relatedness between helper and both breeders is high (Rules 9 and 10). The area of conflict between breeder and helper is likely to decrease as r (relatedness of breeder to helper) and r' (mean relatedness of helper to both breeders in future breeding attempts) increase.

These four predictions are illustrated in Fig. 4.2.

The model also defines the boundary, from both the breeder's and the helper's perspective, of when help ceases to have a kin-selected benefit. This is a useful distinction because it allows identification of when other direct benefits of helping are important. For example, one hypothesis explaining the provisioning of help is that helpers are in effect "paying rent" to breeders for the advantages of remaining on a territory (Mulder and Langmore 1993; Cockburn 1998). The cost of this rent is hypothesized to be made up for by the eventual direct benefits, for example, inheriting the territory or mating access to the female. Thus "rent," a form of short-term forfeiture of fitness, could be defined as

Rent paid by helper = total cost from helping − kin-selected benefit

On average, rent is predicted to be less than the total direct benefits eventually gained by the helper, since otherwise it would not be worth paying. However, such a scenario could occur through manipulation or deceit (Connor 1995; Heinsohn 1991b). A negative value would imply that help is more than compensated for by the immediate kin-selected benefits, and thus that no rent is being paid.

The ten rules generated by the model demonstrate that there are likely to be conflict zones under which help is beneficial to the breeder but not to the helper. This is complementary to Emlen's (1982b) early theoretical demonstration of conflict over whether an auxiliary should remain and help, and whether the breeder or helper should forfeit fitness to maintain a mutually beneficial relationship. My examination of every combination of cost–benefit scenario in Table 4.1 shows that alongside situations when it is not in the breeder's or auxiliary's interests for the latter to provide care, the breeder generally has more opportunities to benefit. For example, if $L_H > S_B$ and I_H, the auxiliary cannot benefit from help but the breeder might if their relatedness is sufficiently low. The entire component of help in this situation, if it occurs at all, must be considered "rent."

Similarly, a breeder can always potentially benefit from compensation, whereas it is only beneficial to the helper when $S_B > L_H$, and r' is sufficiently high. When the breeder can adjust its care to partially compensate for the help it receives, different rules determine whether it is beneficial for both helper and breeder (see Rules 7 to 10).

Clearly, breeders are in a better position to evaluate the kin-selected benefit of obtaining active help as it only depends on their relatedness to one individual

Parental care, load-lightening, and costs   77

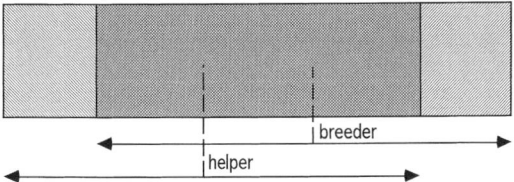

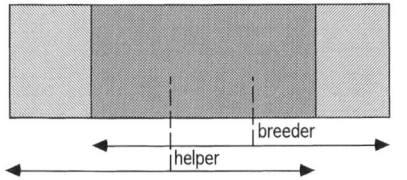

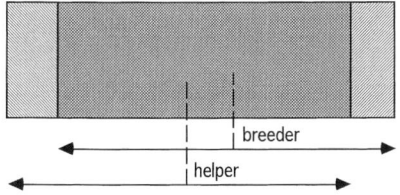

Figure 4.2. Illustration of the four predictions outlined in Table 4.2. Figures denote "windows" of values when compensation is beneficial to breeders and helpers. Hatched regions show areas of overlapping benefit. (a) Predictions 1 and 2, showing smaller window for helper, and smaller window for both when cost curves for breeders and helpers differ. (b) Predictions 3 and 4, showing that breeders prefer more compensation as r increases, and helpers prefer less compensation as r' increases. This leads to a smaller area of conflict between breeder and helper.

(the helper) in one reproductive event. Helpers, by comparison, must assess their likely relatedness to two breeders in all future reproductive events. A likely outcome is greater precision and flexibility in adaptive behavior from breeders, depending on their relatedness to the helper, compared to behavior by helpers that may, by necessity, be based on species-specific "rules of thumb" that estimate probable relatedness in the future.

SUPPORT FOR THE MODEL

Cost curves affect help and survival

Reproductive effort is predicted to be inversely related to its impact on residual reproductive value, leading to trade-offs between age-specific mortality and fecundity (Williams 1966). Although the precise shape of the cost curve for reproduction has rarely been measured, most workers agree that it is most likely approximated by a concave-up function (Fig. 4.1) such that costs accelerate with increasing energetic expenditure (Crick 1992; Hatchwell 1999). Although some studies suggest a linear relationship, this may be because the costs have only been measured across small ranges of proximate parental investment. This underscores our limited understanding of the costs of care, because few long-lived animals are likely to invest at levels that fall in the steepest part of the cost curve (Partridge 1989).

Pied kingfishers provide an excellent example of a cooperative breeder in which help is known to be costly and to vary with relatedness. Primary helpers provide more care and survive less well than secondary helpers (Reyer et al. 1986). Since primary helpers are closely related to the young, this is exactly the difference predicted by the model. Helpers in stripe-backed wrens that help more also survive less well; help in this analysis was always directed at close relatives (Rabenold 1990).

More examples exist of helpers apparently increasing the survival of breeders, including pied kingfishers (Reyer and Westerterp 1985), splendid fairy-wrens (Russell and Rowley 1988), Florida scrub-jays (Woolfenden 1975), bicolored wrens (Rabenold 1990), and white-browed sparrow-weavers (Lewis 1982), but some studies have showed little or no effect, including acorn woodpeckers (Koenig and Mumme 1987), western bluebirds (Dickinson et al. 1996), and common moorhen (Eden 1987). That the factors affecting survival are many and complex, and therefore difficult to analyze, has been suggested by a number of workers (Koenig and Mumme 1987; Crick 1992; Cockburn 1998). Again pied kingfishers provide good evidence that breeders respond to the cost of reproduction, as these birds actively recruit helpers in poorer years and when their brood size has been artificially increased (Reyer and Westerterp 1985).

In his discussion of load-lightening, Crick (1992) suggested that breeders and mature helpers on the same territory are likely to have similar cost functions. However, one general difference between breeders and helpers is their age. This difference has several consequences, one of which is the lesser experience and skill of helpers.

Indeed, one of the earliest hypotheses for the adaptive advantage of helping is that it allows younger individuals to practice parenting skills (Skutch 1961), an hypothesis supported by Seychelles warblers in which birds with helping experience have been shown to be better breeders (Komdeur 1996). In extreme cases, obligate coexistence between parents and juveniles may result when juveniles develop their skills very slowly. In white-winged choughs, juveniles do not mature sexually until they are four years old, during which time they rely on their parents for extended support (Rowley 1978; Heinsohn et al. 1988). The documented costs to individuals when they contribute high levels of alloparental care confirm that they operate from a steeper cost curve than older birds (Heinsohn and Cockburn 1994). Thus skill level may determine the level at which a helper contributes and, from Table 4.1 and Rule 8, helpers may require a higher level of relatedness (r') to the breeders when their care is more costly than that of breeders. Since chough groups consist of very close, and sometimes even inbred, kin (Heinsohn et al. 2000); such close relatedness may be required for poorly skilled helpers even to attempt such costly help.

Benefit curves affect help and how it is used

Good evidence that the shape of the benefit curve affects how breeders use help comes from Hatchwell (1999) and Legge (2000b), who showed that provisioning at the nest is overwhelmingly additive in two species that experience high rates of nestling starvation. Apparently help can easily be converted into additional young in these species. Excellent examples of strong effects of additional provisioning come from white-fronted and red-throated bee-eaters (Crick and Fry 1986; Emlen and Wrege 1991), pied kingfishers (Reyer and Westerterp 1985), and white-winged choughs (Heinsohn 1992). Help can also be additive for the first x helpers, but followed by compensation for additional helpers (Seychelles warblers; Komdeur 1994b). In extreme cases help can be additive over several helpers (up to 12), as in white-winged choughs (Heinsohn 1991c, 1992). Clearly, both the shape of the benefit curve in Fig. 4.1, and where the additional care from the ith helper falls on that curve, determines whether help should be additive, compensatory, or a compromise between the two.

Conversely, in species that do not experience nestling starvation, and therefore where additional help has a smaller effect on productivity, load-lightening is common (Crick 1992; Cockburn 1998; Hatchwell 1999). Help in these situations is best visualized as falling on the benefit curve after the point at which it flattens out (Fig. 4.1), and the diminishing returns of increased offspring are small compared to the benefits of reducing care.

Combined effects of costs and benefits

An important hypothesis for the lack of a helper effect on production of young seen in many studies has been proposed by Magrath (2001). He found that production of young in white-browed scrubwren groups was only boosted by helper contributions when the breeding female was an inexperienced yearling. A meta-analysis of a large sample of cooperatively breeding species showed that group size has a larger effect on reproductive success under relatively poor conditions, including both adverse environmental conditions and when breeder quality or experience is poor. Thus increased group productivity may in fact be available in many species for which research, perhaps carried out in good conditions, showed a negative result.

The model presented here would predict that, when conditions are poor, breeders face steeper cost curves and are thus more likely to compensate for any help they receive in the form of cost savings. However, the benefit curve would also steepen under poor conditions, and a likely outcome would be either complete or partial compensation, with the contribution from helpers leading to a strong helper effect as suggested by Magrath's (2001) analysis.

Legge (2000b) discussed three examples in which nestlings starved but helpers did not increase the overall rate of food delivery to the nest. Both laughing kookaburras and white-throated magpie-jays (Innes and Johnston 1996) lose young through nestling starvation and competition, but Legge (2000b) interpreted the lack of additive care in kookaburras as most likely due to the high cost of care. In this situation a very steep cost curve overrides high potential benefits from producing more young.

In white-winged choughs, food from additional helpers continues to ameliorate starvation up to group sizes of 14 (Heinsohn 1991c, 1992). However, choughs are one of very few cooperative species in which help has been shown to be energetically expensive if performed at high levels (Heinsohn and Cockburn 1994). Three different studies of choughs at different sites demonstrate how costs and benefits from providing care can vary. In one study where foraging conditions were poor, group size had no effect on overall feeding rate (Rowley 1978). In another where choughs had access to watered fields and gardens, group size had a marked positive effect (Heinsohn 1995). In a third experimental study where the control foraging conditions were again poor, feeding rates increased only when large amounts of food were supplied (Boland et al. 1997a). Taken together, these studies demonstrate that when foraging is difficult and help is costly, group members may prefer to make energetic savings rather than obtain the higher breeding success possible from higher rates of provisioning, presumably because this latter option will prove more costly in the long term.

A LIFE-HISTORY APPROACH TO THE PROVISION OF HELP

Males and females may frequently pursue different life-history strategies with respect to parental and alloparental care in cooperatively breeding species. Earlier I noted that male breeders in Seychelles warblers reduced their care in response to increased alloparental care, whereas females were more likely to maintain their care at the same level (Komdeur 1994b). Hatchwell (1999) found this pattern to be general across a large number of species such that male breeders are more likely to reduce their level of care in response to helpers whereas female breeders only do so when starvation is uncommon. Interestingly, Hatchwell also found that female breeders exhibit either compensation or additive care, but that males often partially compensate for the help they receive. One parsimonious explanation for this pattern is that males in larger groups are less confident of their paternity and opt for at least some enhancement in survival when help is available (Hatchwell 1999). In contrast, females, being more confident of maternity of the offspring in the nest, opt for increased output of offspring when this benefit is available.

Analogous patterns have also been detected for male and female helpers. Helping behavior is more common in males, but a general explanation for this sex bias has been lacking. Cockburn (1998) reviewed six hypotheses, but perhaps the most striking aspect of his compilation is a previously unnoticed pattern concerning female help. By using only non-correlational studies involving comparisons of the same breeders with and without help, or experimental manipulations of helper numbers or critical resources, improvements in the productivity of young were present in only two out of six studies in which helping was male-biased, and in those two the male helpers were not related to the breeders. However, when helpers were either predominantly female or of both sexes, a positive effect was found in seven of seven studies.

Cockburn's analysis was based on a small number of studies, and paired analyses such as those used in some of the studies may produce an inherent bias away from detecting a helper effect (see Chapter 3). However, if confirmed, the trend may have major implications for the field of cooperative breeding. Most important is the possibility that males and females help for different reasons. Specifically, females may be more likely to help for the inclusive fitness benefits of increasing production of young, whereas males may help for more direct fitness benefits. In particular, the higher frequency of male philopatry and inheritance of the natal territory suggests that they stand to gain more from direct benefits such as social prestige, enhancing the local territory, or gaining direct parentage. Because male helpers are more common than female helpers, this raises the exciting possibility that cooperative breeding in birds is driven by direct benefits to helpers.

If some helpers are primarily motivated by the immediate returns of inclusive fitness, whereas others bank more on future reproduction of their own, we would also predict that the former will bear greater energetic costs than the latter. This is particularly true of species in which a large proportion of helpers fail to secure a breeding position (Rabenold 1990) and thus immediate inclusive fitness may be the major source of fitness they achieve in their lifetime. The patterns also suggest a general sensitivity from both helpers and breeders to their degree of confidence in their relatedness to offspring.

That helping behavior, like parental care, must be considered a life-history trait with consequences throughout an individual's lifetime received an important boost in a comparative analysis of Arnold and Owens (1998), who extended earlier analyses indicating strong taxonomic biases in the distribution of cooperative breeding (Russell 1989; Cockburn 1996; Chapter 1) to show that the trait is strongly associated with, and might even result from, high longevity.

This controversial result (see Chapter 2) has the important implication that although habitat saturation may be the proximate cause of philopatry in many species, it is probably low annual mortality that leads to such crowding and lack of space. If a life-history trait such as longevity increases the likelihood of cooperative breeding, then patterns of help must be as likely a target for selection over an individual's lifetime as patterns of direct reproduction (Saether 1990). In particular, some attributes of the costs and benefits analyzed in this paper suggest that helpers, but not breeders, must make their decisions based on likely projections of their relatedness to the breeder's offspring in future breeding attempts, whereas breeders are more driven by their absolute relatedness to the helper. Thus, the extent of alloparental care given by the average helper could be as species-specific as parental care itself, and might help explain interspecific differences in this behavior.

In conclusion, whether an auxiliary individual should help or not, and how much help is given, are complex questions with differing answers from both the breeder's and the auxiliary's viewpoints. Although the extent of helping behavior is known in some species to relate to ability, kinship, or paternity, in most cases the costs and benefits that lead to any particular pattern of helping behavior remain unclear. The huge variation in the extent of help can be interpreted only by combining knowledge of the type of benefits sought with how the costs of helping limit individuals in the present and future. The analysis presented in this paper goes some way towards clarifying the kin-selected costs and benefits of helping behavior in this decision-making process.

ACKNOWLEDGMENTS

I thank Chris Boland, Janis Dickinson, Walt Koenig, Sarah Legge, Rob Magrath and an anonymous reviewer for helpful comments, and the Australian Research Council for funding.

5 · Mating systems and sexual conflict

ANDREW COCKBURN
Australian National University

Cooperative breeding occurs where more than two individuals contribute to a single brood of young, so it is certain that some individuals cannot be parents of some of the young they are rearing. By universal acclaim, studies of this phenomenon represent some of the most heroic field studies ever initiated, with several researchers following the fate of individuals within groups for decades (Stacey and Koenig 1990a).

Despite the admirable and detailed data achieved in these studies, there has been remarkably little crossing over from the study of avian cooperation into the burgeoning literature concerned with sexual selection. Indeed, a lot of research on cooperative breeding birds totally ignores the sex of the "helpers" that have prompted so much interest (Wright *et al.* 1999). No doubt part of the reason for ignoring gender arises because some cooperatively breeding birds are sexually monomorphic even as adults. Until recently, such monomorphism precluded identification of the sex of helpers without difficult and intrusive surgery. In addition, lack of obvious dimorphism or dichromatism might have been taken to indicate that sexual selection was of restricted importance in these species. However, as pointed out by Burley (1981), sexual conflict can lead to selection for concealing sexual identity. Such selection for indistinguishability may be particularly important in group-living species.

This chapter looks anew at the role of sex differences in the evolution of cooperative breeding, prompted by several considerations. First, most practical problems caused by monomorphism have been removed by the availability of cheap and easy molecular methods for determining sex (Griffiths *et al.* 1998). Second, it is now clear that drabness or monomorphism is no impediment to sexual intrigue, as evidenced by classic studies of the dunnock (Davies 1992). Third, the advent of molecular methods for resolving parentage have shattered any illusions that all supernumerary birds can be viewed as faithful eunuchs or that intense sexual selection is not prevalent in others. It is also unlikely that it is possible to compress the diversity of cooperatively breeding birds into a single model (Cockburn 1998), despite attempts to do so (Emlen 1995, 1997a).

Most important and worrying, I will show that the traditional conceptual focus of the study of cooperative breeding on the issues of philopatry and help in family groups has led us to ignore important conclusions that have emerged from the study of sexual conflict and selection in birds. The most important of these is the near universal acceptance that female birds have considerable ability to regulate paternity in their own interests, even though male behavior may evolve in retaliation to female control of reproduction (Gowaty 1996a; Magrath and Heinsohn 2000; Putland 2001).

I approach this problem by first attempting to classify the mating patterns that have thus far been described in cooperative breeders. As a platform to understanding this diversity I then reiterate some of the main hypotheses used to explain sexual conflict over reproduction in birds. I then review what is known of each of the mating patterns among cooperative breeders, attempting an explanation in terms of the benefits that females obtain from mate choice and/or multiple mating, and how males can influence the expression of that choice. I argue that the negotiations between males proposed by reproductive-skew theory (Chapter 10) represent a small and largely unsubstantiated subset of these models. Throughout, I speculate on possible forms of sexual conflict and selection that might emerge among the complex societies exhibited by some cooperative breeders, and that are unlikely to be seen in simpler societies.

Ecology and Evolution of Cooperative Breeding in Birds, ed. W. D. Koenig and J. L. Dickinson. Published by Cambridge University Press.
© Cambridge University Press 2004.

TOWARDS A CLASSIFICATION OF MATING SYSTEMS IN COOPERATIVE BREEDERS

What is a cooperative breeder?

I maintain the convention that cooperative breeding occurs where more than two individuals combine to rear a brood of young. This excludes some social systems that resemble those found among cooperative breeders very closely, such as the polygynandry exhibited by penduline tits (Schleicher *et al.* 1997), where acquisition of extra mates occurs through desertion of the original mates, so only one bird provisions the brood, and Siberian jays, where philopatric young remain in their natal group but provide no alloparental care when their parents breed (Kokko and Ekman 2002). I exclude such systems because much of the interest in cooperative breeders lies in the paradox provided by alloparental care. However, it is important to bear in mind that one resolution of the conflicts that I will describe is for birds to decline to provide alloparental care.

Who participates in the mating system?

Conventional classifications of mating systems depend on how many individuals of each sex participate in mating. Where mating associations are reasonably stable, we are able to define monogamy (1 male, 1 female in an exclusive relationship), polygyny (1 male, >1 female), polyandry (>1 male, 1 female) and polygynandry (>1 male, >1 female). A number of problems bedevil the application of this classification to cooperative breeders:

(1) A clear lesson from modern molecular ecology is that we cannot infer mating relationships by counting the number of individuals. For example, nests of the bell miner can be provisioned by as many as 29 males (Painter *et al.* 2000), yet the mating relationships among the social pair that form within colonies are clearly monogamous (Conrad *et al.* 1998). Similarly, superb fairy-wrens live in territorial groups containing as many as five sexually active males, yet the majority of paternity is obtained by extra-group males during predawn forays that would never be detected during conventional daytime censuses (Mulder *et al.* 1994; Double and Cockburn 2000).

(2) Even with molecular data, how do we classify the mating system of birds with small broods? Five male eclectus parrots can attend a nest containing just one or two young (Heinsohn and Legge 2003), so it is obvious that most males have not sired an offspring. However, eclectus parrots can produce more than one brood each season and are likely to be very long-lived, so all males might share paternity over time. Even where more young are produced in a brood, the young in a brood cannot be treated as statistically independent, and reproductive dominance of a brood by one male might be reversed in successive broods, as seen in acorn woodpeckers (Haydock and Koenig 2002). Males may therefore share paternity over time, but may need to accept lack of paternity in some of the broods they help rear.

(3) There are numerous well-studied cases where social partnerships are ephemeral, as individuals change their social bonds with high frequency. For example, in the *Manorina* miners described above, not only do many birds feed at a single nest, but many males, including breeders, feed at as many as 11 nests at the same time (Clarke 1989; Dow and Whitmore 1990).

(4) Some individuals live year-round in cohesive groups that split into pairs or subgroups to breed. For example, in the Mexican jay, large groups break up into pairs that take exclusive responsibility for nest construction. However, males within the group that lack partners or active nests closely follow the female. These males will provision her brood if she allows them extra-pair matings (Li and Brown 2000, 2002). In this case the female mates within the group but outside the pair bond, leading Brown and Brown (1990) to dub the society "uncooperative."

(5) Most troublesome, only a subset of the members of the group that associate to rear young show any interest in mating. This can either be because of the extended period needed for gonadal development, which can be delayed for several years (Rowley 1978), or because of incest taboos (Chapter 9). For example, mortality of breeders can lead to circumstances where the senior male and female living in a group are nuclear family relatives. This can lead to abandonment of the territory by the breeding female in red-cockaded woodpeckers (Daniels and Walters 2000b), usurpation of the territory by immigrant male Florida scrub-jays (Woolfenden and Fitzpatrick 1984) and refusal to breed despite strong fitness penalties in acorn woodpeckers (Koenig *et al.* 1999). For the purpose of describing mating systems

it is important to consider whether nuclear family relatives should be treated as potential mating partners. For example, in superb fairy-wrens, initial evidence suggested that helper males gained negligible within-group paternity (Mulder et al. 1994). However, when only helpers that are not sons of the breeding female are considered, their share of within-group paternity is more than 20% (Cockburn et al. 2003).

Brown (1978, 1987) provided the most recent attempt to document the diversity of cooperative breeding systems, but his analysis occurred before the advent of data from molecular ecology, so a new synthesis is needed. Here I approach the problem by describing mating systems from first principles, following the advocacy of Vehrencamp and Bradbury (1984) that we should first identify patterns, and then try to explain them. This is a completely different approach from attempting to fit variation in social organization into a theoretical straightjacket. Such conceptual restrictions not only have the potential to cause neglect of variants that do not conform to prior expectation (Cockburn 1998), but can also cause us to prejudge adaptive explanations, for example by implying that monopolization of reproduction is under male rather than female control.

In constructing a classification, I follow numerous authors who have advocated that we should distinguish *mating groups* from *social groups* (Brown 1980; Haydock et al. 2001). The mating group is defined to include all the birds within the social group that are likely to participate in reproduction. For example, neither young white-winged chough helpers nor sons of Florida scrub-jay females would be considered part of the mating group. However, they would both be part of the social group, defined as the group of individuals that live together on a territory or, in the absence of territoriality, the individuals that forage and coexist as a cohesive group during the breeding season. Members of the social group may or may not provide help in rearing the brood.

A literature survey revealed 30 species that are both cooperative breeders and have been subjected to reasonably detailed molecular analysis of parentage (Table 5.1). Some molecular data were available for a few additional species where the data were either too few or insufficiently detailed to answer critical questions about the mating system. However, I have included four species (eclectus parrot, Galápagos mockingbird, rifleman, and moustached warbler) in which molecular data are unavailable but where breeding biology is sufficiently characterized to presume that the mating system is distinctive.

I then used the molecular data and observations of copulation behavior to assess the number of birds in social groups that participate in matings. Among females, I follow Brown (1987) in considering four main patterns:

(1) *Joint nesting*, where more than one female contributes eggs to a nest.
(2) *Coloniality*, where females can nest in extremely close proximity, such as the same tree or the same complex nest.
(3) *Plural breeding*, where two or more females in the social group build separate nests that, while built within a single territory, are at least moderately dispersed.
(4) *Singular nesting*, where usually only one female breeds on the territory or within the mating group at any one time.

Among joint-nesting and colonial nesters, I distinguish the case where a single male is involved ("polygynous joint nesting") from the case where more than one male participates in mating. For the latter, I distinguish "egalitarian polygynandry," where mating is shared equally or where the most successful male is difficult to predict from behavioral cues, from "flexible polygynandry," where within-group mating access is predictably associated with dominance or some other behavior of the males. Where singular nesters associate and mate with more than one male, I use similar criteria to distinguish between "egalitarian" and "contextual" polyandry. Most peculiar, in fairy-wrens, helpers are common but mating is dominated by extra-group fertilization, which takes place at "hidden leks" where females visit males advertising from song perches before dawn. In some group-living species, just one male and female from the mating group dominate reproduction ("true monogamy with helpers"). In other cases, supernumerary individuals show little fidelity to any group ("unattached helpers").

There are two caveats necessary in interpreting this classification. First, I ignore comparatively rare behaviors unless they help illuminate questions of interest. For example, splendid fairy-wrens occasionally nest plurally (Rowley et al. 1989), but are generally singular nesters and are classified as such. Similarly, female

Table 5.1. *A descriptive classification of the mating systems of cooperatively breeding birds*

Type	Name	Group composition		Female nesting	Relatedness			Sex of helpers/ coalition partners	Stability of supernumerary associations	Description and additional notes	Examples	References
		Males	Females		Male breeders	Female breeders						
1	Joint-nesting polygyny	1	>1	Joint	na	High		Female	High	Combined with high extra-group mating	Seychelles warbler	Richardson *et al.* 2001, 2002
2a	Egalitarian polygynandry	>1	>1	Colonial	Low	Low		Both	Low	Intense coloniality	Red-browed buffalo-weaver	Winterbottom *et al.* 2001
2b				Joint	High	High		Both	High	Female relationships are stabilized by egg-destruction to coordinate reproduction	Acorn woodpecker	Haydock *et al.* 2001; Haydock and Koenig 2002
											Guira cuckoo	Quinn *et al.* 1994
2c				Joint	Variable	Variable		Both	High	May frequently mate incestuously	Pukeko	Lambert *et al.* 1994; Jamieson *et al.* 1994; Jamieson 1997
											Tasmanian native hen	Gibbs *et al.* 1994; Goldizen *et al.* 2000; J. Buchan and A. Goldizen personal communication
											Common moorhen	McRae 1996b; McRae and Burke 1996
3a	Flexible polygynandry	>1	>1	Plural	Low	Low		Male	Low	Females compete for male helpers	Alpine accentor	Hartley *et al.* 1995; Davies *et al.* 1995, 1996; Heer 1996
											Dunnock	Burke *et al.* 1989
											Smith's longspur	Briskie *et al.* 1998

3b				Plural, sometimes colonial	Low	Low	Male	Low	Males compete for high-quality females; females depend on males because of predation/infanticide	Eclectus parrot	Heinsohn and Legge 2003
4a	Plural breeding	>1	>1	Plural	High	High	Male	High	Extra-pair matings with dominant males elsewhere inferred from copulations	Galápagos mockingbird	Curry 1988b
4b				Plural	High	High	Male	Low	Supernumerary males trade paternity for care	Mexican jay	Li and Brown 2000, 2002
4c				Plural?	Low	Low	Male	Low	Males including extra-group males chase females and force copulations	Stitchbird	Castro et al. 1996; Ewen et al. 1999; Ewen and Armstrong 2000
4d				Plural	High	High	Male	Low	Complete fidelity within pairs. Supernumerary males trade paternity for future pairing	Noisy miner Bell miner	Poldmaa et al. 1995 Conrad et al. 1998
5	Egalitarian polyandry	>1	1	Singular	High	na	Male	High		Galápagos hawk Brown skua	Faaborg et al. 1995; DeLay et al. 1996 Miller et al. 1994; Young 1999
6a	Contextual polyandry	>1	1	Singular	High	na	Male	High	Unrelated males gain paternity; sons also provide help	Stripe-backed wren Bicolored wren	Rabenold et al. 1990; Piper and Slater 1993; Piper 1994 Haydock et al. 1996

(cont.)

Table 5.1. (cont.)

Type	Name	Group composition		Female nesting	Relatedness		Sex of helpers/ coalition partners	Stability of supernumerary associations	Description and additional notes	Examples	References
		Males	Females		Male breeders	Female breeders					
6b				Singular	Variable	na	Male	High	Unrelated subordinates gain paternity; sons rarely help	White-browed scrubwren	Magrath and Whittingham 1997; Whittingam et al. 1997
6c				Singular	Low	na	Both	High	Paternity shared when newly formed coalitions have merged	Arabian babbler White-winged chough	Lundy et al. 1998 Heinsohn et al. 2000
7a	Hidden leks	>1	>1	Singular, occasionally plural	High	High	Both	High	Mating largely extra-group	Splendid fairy-wren	Brooker et al. 1990; Rowley and Russell 1997
7b				Singular	Moderate	na	Male	High	Mating largely extra-group when more than one male is present	Superb fairy-wren	Mulder et al. 1994; Dunn and Cockburn 1999; Double and Cockburn 2000; Green et al. 2000; Cockburn et al. 2003
8a	True monogamy with helpers	1	1	Singular	na	na	Male	Low	Failed breeders return to help relatives; extra-pair mating rare	European bee-eater	Jones et al. 1991
							Mostly male			Long-tailed tit	Russell and Hatchwell 2001; Hatchwell et al. 2002

8b		Singular	na	na	Male	Moderate	Failed breeders return to help relatives; extra-pair mating common	Western bluebird	Dickinson and Akre 1998; Dickinson 2001
8c		Singular	na	na	Both	High	Exclusive mating despite unrelated subordinates	Florida scrub-jay	Quinn et al. 1999
								Laughing kookaburra	Legge and Cockburn 2000
8d		Singular	na	na	Male	High	Extra-pair mating rare	Red-cockaded woodpecker	Haig et al. 1994
9a	Unattached-helper systems	Singular	Low?	na	Male	Low	Permanent males provide help and pair with young they raise	Rifleman	Sherley 1989, 1990
9b		Singular	Low?	na	Male	Low	Helpers move between territories and usurp or succeed the dominant	Moustached warbler	Fessl et al. 1996
9c		Singular	Low?	na	Male	Low	Females often copulate with "helpers" and switch nests within seasons	Hoopoe	Martin-Vivaldi et al. 2002

white-winged choughs sometimes nest jointly, but brood reduction is rapid so only the progeny of a single female survive. Second, some societies show bewildering diversity at the level of the individual territory, with polygyny, polyandry, monogamy, and polygynandry co-occurring frequently within a single population. I classify such mixtures as polygynandrous.

With the exception of egalitarian polyandry, all of the major groupings have distinct variants. These arise for several reasons:

(1) There are differences in the composition of the social group such as whether one or more sex provides help, and the stability of the associations between supernumeraries and the dominant breeders.
(2) The coalitions of same-sexed individuals that participate in mating can be unrelated, or closely related because of philopatry or because of dispersal in same-sex groups of relatives.
(3) The prevalence of extra-group mating varies.
(4) Incestuous matings apparently occur in some species, despite associative cues that should enable incest avoidance.

Including these variants, I recognize 22 distinct social/mating systems among the 34 species (Table 5.1). Most of the rare systems that enjoy representation by more than one species gain this predominance only because close relatives have been sampled (*Manorina* miners, *Campylorhynchus* wrens, gallinules, and *Prunella* accentors), eroding the possibility of using phylogenetically based comparative methods to discern patterns. In addition, some unlisted species of cooperative breeders are sufficiently distinctive that they cannot yet be accommodated within the major categories without detailed molecular studies, and may in the future warrant the erection of additional categories. For example, pale-winged trumpeter social groups are unusual because they comprise unrelated males and unrelated females, yet female supernumeraries appear not to be allowed to participate in reproduction (Sherman 1995a, 1995b). Similarly, in the white-throated magpie-jay, groups of females defend territories and apparently pair with a single male, while males roam between several territories (Langen 1996a). The reproductive tactics of the females in both these species appear to be complicated and the relative success of females will only be determined with molecular analysis.

Such diversity may initially seem to indicate that I am an incorrigible splitter. In my defense, I have direct field experience with 21 of the 34 species, and with close relatives of two more. I suspect that even switches between minor categories, such as from eclectus parrot to dunnock, are sufficiently great to be comparable to traditional major dichotomies that have interested students of avian mating systems, such as the transition from monogamy to polygyny.

I first briefly review how tensions can arise between males and females in mating systems. I then use the perspective of sexual-conflict theory to identify common themes that influence the diversity of mating systems in cooperatively breeding birds.

SEXUAL-CONFLICT THEORY

Female choice

The role of female choice in mating systems and sexual selection has been one of the most active areas in behavioral ecology (Andersson 1994). Females are often more selective about their sexual partners than males. This selectivity may occur because females have greater constraints on their ability to increase their fecundity through promiscuity, unless they can ensure rearing of their young by another individual or individuals. There are three important questions that underlie this theory. When should a female form an exclusive relationship with a single male? Why do females of some species all prefer the same small subset of males (the lek paradox)? Last, and of particular importance to cooperative breeders, when should a female seek copulations with more than one male (Jennions and Petrie 2000; Tregenza and Wedell 2000)?

Benefits from female choice fall into four broad classes:

(1) *Direct material benefits* – Females can extract or obtain direct benefits from the males that court or mate with them. For example, females can obtain nuptial gifts from the male or access to high-quality territories or nesting sites. In altricial birds, where young have high metabolic demands and are exposed to predators in a nest, a female often depends on male care in order to rear her young and to improve her own ability to survive to produce another brood. It is likely that males will only be willing to provide care if they sire some of the young directly,

or if their relatedness to the young means that any costs they suffer are compensated by the indirect fitness associated with enhancing the fitness of related individuals. We therefore expect direct benefits such as nest construction, protection and defense, or provisioning and defense of young to be an important contributor to female decisions over mating. In complex societies, females have the opportunity to gain these benefits from more than one male, promoting polyandry. In addition, females living in groups may face particularly intense harassment from males unless they copulate with them, so they could provide matings to allay the risk of harassment or infanticide.

(2) *Genetic compatibility* – Females need to mate with a fertile male whose genotype is compatible with their own. Fertility could be compromised because of basic deficiencies in the male or because his sperm has been depleted in earlier matings. Depletion may be a particular problem if all females prefer the same male. Genetic incompatibility arises for a variety of reasons. First, nuclear-family incest can cause inbreeding depression through the exposure of deleterious recessives or by the loss of heterosis. Second, it may also be advantageous to avoid mating with individuals that are too distantly related in order to minimize penalties associated with disrupting coadapted gene complexes. This makes obvious sense in the extreme case where a female avoids mating with a heterospecific male in order to avoid producing infertile offspring. Last, the genotypes of some individuals can prove incompatible or deficient. Among birds, the best evidence comes from mating patterns associated with chromosomal inversions in the white-throated sparrow (Thorneycroft 1976) and patterns of infertility and extra-pair mating in tree swallows (Kempenaers *et al.* 1999).

(3) *Improving the quality of young* – Females may be able to obtain benefits for their offspring by selecting aspects of male phenotype that indicate the genotype of the male would enhance the viability or attractiveness of their offspring. Females could achieve similar effects by mating with several males if the male with superior sperm usually fertilized her eggs. Female discrimination is particularly evident in lekking species, where males display at arenas that are visited for the purpose of copulation. Cooperative breeders often have limited choice over their initial settlement, as many live year-round on territories and cannot compete for vacancies during an annual settlement phase, and they can be long-lived, so vacancies become available only rarely (Arnold and Owens 1998). Møller (1992) suggested that in these circumstances birds should be particularly likely to prospect for extrapair matings as a way of capturing good genes from males.

(4) *Increased diversity of young* – Females might also benefit from increased diversity of their brood. By sampling many males they might increase the likelihood of producing superior offspring, or they might reduce the extent of competition between sibs because the genotypes of those young are dissimilar.

By contrast, males will typically benefit from gaining exclusive access to as many females as possible. They could do this by brokering access to a resource that is important to the female, persuading the female to exchange fidelity for access to the resource (Gowaty 1996a). However, resource defense becomes problematic in cooperative groups when several males share a territory. Even when females share a territory with a single male, they can continue to derive genetic benefits covertly by mating with extrapair males. Male defenses against extra-pair mating include guarding the female during her fertile period, and the various mechanisms of sperm competition, including production of large, sperm-rich ejaculates, copulating frequently, displacing sperm from previous matings, and timing copulations to maximize the probability of fertilization.

MATING SYSTEMS

With these considerations in mind, I now return to the mating systems known from cooperatively breeding birds. I sketch the chief features of each system in order to determine whether intrasexual and intersexual conflict may have shaped the evolution of the system.

Type 1: joint-nesting polygyny

The social system of the Seychelles warbler was originally viewed as a rare variant of cooperative breeding based on philopatry of female young and kin-associated altruism. Predominant female help is extremely rare among cooperative breeders (Brown 1987), for reasons that remain poorly understood and controversial (Cockburn 1998; Haig 2000). In an extraordinary study involving the manipulation of the entire world population of the species, Komdeur and his collaborators have shown that groups form when available territories are fully exploited (Komdeur 1992; Komdeur et al. 1995). Extreme habitat gradients exacerbate the advantages of philopatry for females fledged in high quality habitat. Junior females that remain on their territory contribute care as helpers (Komdeur 1994a). By contrast, male young are much less likely to help rear offspring, and instead bud off small micro-territories that they use as a platform to gain a territory suitable for breeding (Komdeur and Edelaar 2001a, 2001b).

However, molecular analysis reveals that female "helpers" frequently lay eggs (Richardson et al. 2001, 2002), so direct as well as indirect kinship effects are important in this unusual case of female-biased cooperation. The probability that helpers gain maternity is not precisely known, and because the modal brood size is one, it may require data collected over a considerable period to sort this out. However, the proportion of nests in which "helpers" lay an egg is likely to be considerably greater than 50%, so the mating system must be viewed as joint-nesting polygyny.

An additional nuance arises in this species because some 40% of young are sired by extra-group males (Richardson et al. 2001, 2002). Although currently unclear (Chapter 9), I believe that the available data are compatible with a primary role for incest avoidance. The population has probably always occurred in small numbers and in addition has recently been through a severe bottleneck of only 29 individuals. Genetic variation may therefore be limited, which suggests limited incentives for extra-pair mating for good genes (Petrie and Lipsitch 1994; Petrie et al. 1998; Griffith 2000). However, the small population size may increase local relatedness. Local clustering of male relatives could be exacerbated by the male tactic of forming new territories by budding off from their natal territory (Komdeur and Edelaar 2001a, 2001b). In addition, females do not show consistent choice of male genotypes. Only one of 20 successful extra-pair sires was chosen by two females, and four of 20 males that were cuckolded were also successful in obtaining extra-pair mating. One female even mated with two different extra-pair sires. These results suggest a refined inbreeding avoidance system. Inbreeding could be avoided at the time of pairing. Examples of such a system are those of small island populations of Darwin's finches, where females avoid pairing with males that sing the same song as their father (Grant and Grant 1996). Such a solution is unlikely in Seychelles warblers because females in good habitats are under selection to remain philopatric.

The Seychelles warbler is the only species that can currently be classified as a polygynous joint-nester. The strongest evidence for a comparable system in other birds comes from the magpie goose. Early observations suggested that this species frequently bred in stable trios comprising two females and a male (Frith and Davies 1961). A more recent molecular analysis using rapid amplified length polymorphism analysis suggested that birds in trios could be close relatives (Horn et al. 1996), but the analysis was unable to allow firm conclusions. Recent observations have documented substantial levels of brood parasitism (Whitehead and Tschirner 1991) and shown that more than one male may attend and defend the nest (Whitehead 1999), so further studies are desperately needed.

Type 2: egalitarian polygynandry

Egalitarian polygynandry occurs where it is difficult to predict which of a group of males will gain fertilizations among a group of females, even if dominance is present among the males. Instead, paternity is generally shared, particularly when success is examined over several nests (Haydock and Koenig 2002). Egalitarian polygynandry appears inevitable when several males form coalitions and where females engage in joint nesting or in colonial nests. The evolution of joint nesting is discussed elsewhere (Chapter 11) and is dealt with only briefly here. The ability of two females to contribute to the nest may be associated with male incubation, which facilitates the ability of a second female to approach the nest. Joint nesting may also initially be closely linked to brood parasitism. Brood-parasitic goldeneyes are often close relatives of the females that they parasitize, suggesting that their breeding system is mutually beneficial rather than parasitic (Andersson and Åhlund 2000; Andersson 2001).

Incubating males benefit from egg-dumping if they can mate polygynously with the second female. However, his control is easily subverted if one or more females mate with additional males. This association between equal distribution of mating among males and joint nesting in females has not been previously recognized. Indeed, reliable molecular data have only just become available, no doubt because of the exceptional difficulty that arises in reconstructing parentage where both maternity and paternity are uncertain. However, the strength of the association leads me to suspect that the ability to exploit several males may be a driving force stabilizing the evolution of joint nesting. Males may be unable to resist sharing paternity because of the difficulty of simultaneously guarding more than one female, particularly where the males are committed to incubation.

This stabilizing role may be critical because joint nesting poses some severe problems for females. It may be difficult for two females to coordinate egg-laying and incubation. In addition, the optimum clutch size for incubation or brood size for provisioning may be exceeded (Chao 1997).

This group of species is also united by the frequency with which copulation-like behavior occurs in unusual contexts. Acorn woodpeckers, for example, engage in "pre-roost mounting" in which any individual can mount any other group member. The behavior peaks during the breeding season, but can occur any time of the year. (MacRoberts and MacRoberts 1976). In captivity, male red-browed buffalo-weavers often force copulations on other males (Winterbottom et al. 2001). Homosexual matings also occur frequently in gallinules. In pukeko, female–female copulations are more common than male–male copulations (Jamieson and Craig 1987b). Males copulate with each other in large groups, where dominants may divert subordinates from seeking heterosexual copulations by allowing them to copulate with them (Jamieson and Craig 1987c). By contrast, the primary contexts for female–female copulations is where senior females mate with junior females just prior to or during egg-laying, in contrast to heterosexual copulations that start months before egg-laying. Males always share paternity, but in populations comprising related females, one female can dominate egg-laying (Jamieson 1997). Jamieson and Craig (1987c) speculate that homosexual matings could regulate the number of eggs laid in the communal nest, but this is unresolved. In groove-billed anis, the majority of "copulations" occur when a female mounts a male, though many of these reverse copulations occur outside the breeding season (Bowen et al. 1991). While all of these behaviors have been noted occasionally in other birds, their frequency in birds facing conflict over reproduction suggest that further investigation is warranted to determine whether false copulations provide a means of resolving (or exacerbating) conflict over parentage.

(a) Nests with many brood chambers
Several bird species build nests containing several nesting chambers, reaching a pinnacle in the remarkable nests of the sociable weaver. Cooperative breeding occurs in several of these species, but the mating system has been resolved only in the red-billed buffalo-weaver. Interest was stimulated in this species because males have a conspicuous phallus-like structure that is unique among birds (Winterbottom et al. 1999). Although some nests were attended by just a single male, most nests (80%) have coalitions of two to four usually unrelated males (Winterbottom et al. 2001). Males share paternity, but also lose 19% of fertilizations to extra-group males. The phallus is not intromittent during copulation, but is stimulated by rubbing, leading to "orgasm" and ejaculation after more than 10 minutes of mounting. Unfortunately, there is no molecular data on the mating system in the majority of group-nesting weavers that lack elongated stimulatory organs. This group certainly warrants comparative analysis.

(b) Egg-tossing and clutch coordination
Female crotophagine cuckoos (anis and guira cuckoos) and acorn woodpeckers resolve conflicts over clutch size and timing of initiation of the clutch by tossing or otherwise destroying each other's eggs from the nest until all participants are ready to proceed with the brood, limiting the ability of any one female to dominate reproduction (Vehrencamp 1977; Mumme et al. 1983a; Koenig et al. 1995; Macedo and Bianchi 1997).

Acorn woodpeckers and crotophagine cuckoos differ in that the latter form pairs within groups, whereas in acorn woodpeckers females do not form pair bonds with individual males. Observational research on groove-billed anis, which have not been subjected to molecular analysis, suggest that one pair can monopolize reproduction at the nest (Vehrencamp et al. 1986). However, that view has not been sustained by molecular analysis of relatedness in the guira cuckoo, which is a close relative and has a similar social system (Quinn et al. 1994).

In both groups, sharing of reproduction among males is not harmonious, as males can be infanticidal or destroy the nest if unconvinced they have obtained paternity (Koenig 1990; Macedo et al. 2001). Nest destruction induces the female to renest, presumably allowing the cuckolded male to gain fertilizations. It is therefore probably important for females to copulate with as many males as possible, promoting egalitarian sharing.

(c) Incestuous gallinules

The third variant occurs in cooperatively breeding gallinules. These birds show considerable similarities in sociality despite interspecific and intraspecific variation in the degree of relatedness among group members. However, where relatedness is high, incestuous matings appear to be common. Such incest is surprising, given the strong incest taboos observed in other cooperatively breeding birds (Chapter 9). Although repeated inbreeding could purge populations of the deleterious recessives that lead to inbreeding depression, data from the common moorhen suggest that these birds are likely to suffer from inbreeding depression (McRae 1996b). I doubt, however, that the incidence of incest is closely related to cooperative breeding. Habitat use and dispersal behavior of rails and gallinules may promote inbreeding quite generally. These birds can often disperse successfully across considerable barriers, which enables them to exploit the fragmented nature of their wetland habitats (Taylor and van Perlo 1998). Indeed, what is arguably the greatest radiation of bird species took place when rails and gallinules dispersed throughout the Pacific islands, subsequently evolving into numerous flightless forms (Steadman 1995), with species on individual islands evolving via repeated colonization rather than sympatric speciation (Trewick 1997; Coyne and Price 2000). Palaeontological estimates suggest that this radiation may have accounted for more than 20% of all birds before colonization by humans led to a catastrophic mass extinction (Steadman 1995). The efficacy of this repeated colonization suggests that this group may be particularly good at colonizing from very small founder populations. Such colonization would be facilitated if there were no restraints on incest, or colonization might have selected for incest. Examination of mating patterns in species that are not cooperative would be illuminating.

Jamieson (1997) suggests that polygynandrous gallinules have weaponry that gives them the capacity to inflict serious injuries on conspecifics. This may increase the cost to males of seeking to monopolize fertilizations, further encouraging male egalitarianism.

Type 3: flexible polygynandry

The mating options of males that live in groups containing more than one breeding male and female become more complicated when females can breed simultaneously in different nests. Two outcomes appear possible. In plural breeding the senior birds break up into pairs that nest separately within the territory, while in flexible polygynandry stable associations between females and males develop only during the fertile period and competition erupts between females for access to males.

(a) Competition for male provisioning

In the latter outcome male dominance typically occurs, but secondary females can undermine male control by copulating with additional males, inducing them to provision their offspring. The outcomes from this can be very unstable. For example, in a small population of the dunnock at the Cambridge Botanic Garden, individual territories have variable numbers of males and females, so that polyandrous, polygynandrous, monogamous, and polygynous associations co-occur (Davies 1985). Such diversity results in part because males and females follow separate rules in defending space. Males defend as large an area as possible regardless of food availability, while experimental food supplementation causes female ranges to decline in size, and pushes the modal mating system from polyandry to polygyny (Davies and Lundberg 1984). While this result is superficially consistent with mating outcomes being determined by male brokering of resources, females gain benefits from having several males to provision their young and use a variety of tactics to subvert male dominance (Davies 1989).

Females may use high copulation rates as a form of reproductive competition over male access, as males generally exclusively feed the brood where they have had greatest copulatory access (Davies et al. 1992, 1995, 1996; Hartley et al. 1995; Briskie et al. 1998). Females sometimes solicit at such extreme rates that copulations are declined. Males use counter-tactics such as production of massive quantities of sperm (Birkhead et al. 1991; Nakamura 1990), and pecking the cloaca of the female until she ejects a droplet of sperm from the previous male with which she has copulated (Davies 1983). However, multiple paternity is common, and females frequently have access to more than one male.

The best long-term data on male tactics in these species come from the alpine accentor (Nakamura 1998a,1998b). Male dominance is pronounced, and males queue for many years to attain the alpha position. Dominance conveys an overall advantage in access to paternity, although dominant males can be cuckolded completely in some broods (Burke *et al.* 1989; Hartley *et al.* 1995; Briskie *et al.* 1998). However, the size of dominance queues may be constrained because of dilution of the advantages of dominance if the size of male coalitions becomes too great (Hartley and Davies 1994; Soltis and McElreath 2001).

(b) Nest-defense polygynandry
The eclectus parrot has long puzzled biologists because of its striking reverse dichromatism (Heinsohn *et al.* 1997; Heinsohn and Legge 2003). Females occupy a nest cavity located high in emergent rainforest trees continuously for as much as nine months of the year, and are fed by males throughout this period. Almost one-quarter of nest trees have more than one active nest, so it is the nest rather than the tree or territory that is defended by the female. All food provided to nestlings is gathered by males, but is transferred via the female. After fledging, females remain at the nest cavity and males provide care directly to chicks. Female reproductive success is highly variable, with success contingent on ownership of a cavity that does not flood during the wet season. Females may compete violently, and sometimes fatally, for the best cavities. A female probably copulates with all the males that attend her, and these males are unlikely to be related to her or to each other. Males visit more than one tree, and may provision at nests where they are likely to gain paternity. As for accentors and longspurs, females may benefit from polyandry by improved provisioning, and males may provision wherever they are allowed access to paternity. However, provisioning the female may be as important as provisioning the young because of her need to occupy the nest site continuously.

Type 4: plural breeding

In plural-breeding species, a group that is cohesive outside the breeding system splinters into stable social pairs that build individual nests. There can be considerable interference between the breeding pairs. In addition, there can be additional unpaired supernumeraries that interfere with the breeding pairs. The failure of pairs within plural breeding groups to form year-round territories suggests either that there are general benefits to group living (Brown 1987) or that it is impossibly costly for dominant birds to attempt to exclude other birds from reproduction, so that they restrict defense to a nest within a territory. The relationships between pairs and between pairs and helpers are highly variable, leading to distinctive outcomes in the four groups for which data are available.

(a) Extra-pair mating with superior males
Evidence for selection of high-quality males comes from the Galápagos mockingbird. Although molecular analysis is unavailable, copulations are prolonged and occur in such predictable circumstances that inferences about their function are possible (Curry 1988b). Females mate most frequently with their partner, particularly if he is dominant within the larger social group. However, they will also accept copulations from group males that are dominant to their own partner, and from extra-group males. Although extra-pair mating with within-group dominants could be construed as courtship designed to improve social position, extra-group mating indicates that females may be using dominance as a cue for male quality.

(b) Extra-pair mating with potential provisioners
The situation in Mexican jays is quite different. Nests are built cooperatively by pairs of males and females that form within the group. Mating by the female outside the pair but within the group is extremely frequent, and fertilizations from this source involve 63% of broods and 40% of offspring (Li and Brown 2000). Males that gain extra-pair matings are usually unpaired or lack an active nest, and hence are likely to be of lower quality. These supernumerary males pursue females throughout the fertilization period. Males are more likely to help at the nest if they have sired young in the brood, suggesting that the primary benefit of extra-pair mating is additional provisioning that enables the female to reduce her own workload (Li and Brown 2002).

(c) Forced copulations by supernumerary birds
The stitchbird (or hihi) exhibits an unusual form of copulation where males force females to the ground in a front-to-front position (Castro *et al.* 1996). Females resist copulation and give distress calls during the

grappling. Males gaining fertilizations from forced copulations are usually unpaired and sire 35% of nestlings spread across 80% of broods (Ewen and Armstrong 2000). Males that are pair-bonded to females are sensitive to extra-pair mating and reduce their feeding rate in proportion to extra-pair copulation attempts (Ewen *et al.* 1999). Cooperative breeding involving multiple males and females occurred in a population studied by Castro *et al.* (1996), but unfortunately there was no joint nesting by females and feeding by supernumerary males was extremely rare in the population from which molecular data are available (Ewen *et al.* 1999). Homosexual copulations are common (Ewen and Armstrong 2002). The population studied by Castro *et al.* (1996) used nestboxes, and both populations stem from recent translocations to small islands. Thus the breeding situation may be atypical. However, the limited data from unmanipulated populations suggests a male-biased adult sex ratio, and the remarkable copulation behavior and sperm storage organs indicate that the mating system in natural habitats also involves intense sperm competition.

How has male ability to force copulations on females evolved in this species? One possibility arises from the observation by Castro *et al.* (1996) that copulations were particularly prevalent at a feeder visited by several birds, a situation resembling natural aggregations at a flowering tree. The feeding biology of these birds may therefore represent a case of males being able to broker a critical resource, as occurs in some lekking hummingbirds (Stiles and Wolf 1979), leading to a mixed mating system of defense of a nest site, or attempts at extra-pair mating at a site where females congregate. Another explanation is suggested by the unusual nesting behavior of this species, which is unique among the Meliphagidae, to which the stitchbird belongs. While all other meliphagids build cup nests concealed in thick vegetation or suspended from vegetation extremities, stitchbirds build large platform nests in tree cavities (Higgins *et al.* 2001), possibly facilitating male ambush at or near the nest. Ordinarily, grappling on the ground should be a costly affair, increasing susceptibility of both males and females to predators, but these birds are found in New Zealand, islands lacking ground predators, and it is possible that this absence has reduced the cost of grappling on the ground. We are unlikely to make progress with this system until it is understood what prompts some males to pair and supernumerary males to provision young. Thus, the assignment of this species within Table 5.1 should be regarded as tentative.

(d) Helping for future vacancies

The last variant of plural breeding occurs in *Manorina* miners, where females build separate nests within large groups. Extremely large numbers of males can attend each nest, yet each female mates monogamously with a single male (Pöldmaa *et al.* 1995; Conrad *et al.* 1998), who becomes the primary provisioner at the nest. Supernumerary males appear to provision for two reasons. First, close relatives of the female from within the social group are likely to provision the nestlings (Clarke 1989), but these birds should not be considered members of the mating group. Unrelated males also contribute a large proportion of care. Such males may be competing for future mating opportunities, because when the dominant dies, the female pairs with the unrelated male that has fed her most in previous attempts (Clarke 1989). Unrelated feeding can therefore be viewed as courtship, allowing the female to assess likely direct benefits and broker future mating opportunities accordingly.

Type 5: egalitarian polyandry

In Galápagos hawks and some populations of brown skuas, groups of males form associations with a single female and paternity is distributed randomly among males. Group size is typically small, but as many as eight Galápagos hawks can cooperate with a single female to rear young. The biology of these egalitarian polyandrous species is extraordinarily similar, providing the best evidence for convergent evolution of mating systems in cooperative breeders. This syndrome, drawn from accounts in Young (1999) and Faaborg and Bednarz (1990), includes the following features:

(1) Cooperative breeding is rare in close relatives of these species.
(2) The relationships between birds are long-lived and very stable. The commonest changes that occur within social groups result from the death of a male, which usually reduces group size rather than allowing opportunities for replacement. Hence, unassisted pairs may have their origins in cooperative groups.

(3) The annual productivity of all birds is very low, and there is no evidence that groups have enhanced reproductive success relative to pairs.
(4) They live and forage on islands.
(5) The male coalitions are established during pre-breeding aggregation in non-territorial flocks that occupy a central part of each island before breaking into groups which move on to one of the permanent year-round territories.
(6) They are raptorial birds with reversed size dimorphism.
(7) Reproduction is shared among males peacefully, without the overt competition that often characterizes polygynandrous systems.

Confinement to islands need not be a predisposing factor. Although molecular data are not available, Malan et al. (1997) provide evidence consistent with egalitarian polyandry in the pale chanting goshawk, a raptor widely distributed in southern Africa. In this species, polyandry only emerges in superior habitats.

I have already introduced Jamieson's (1997) hypothesis that egalitarian relationships in cooperative breeders may be associated with possession of weapons such as claws and spurs that increase the cost of fighting. Both skuas and Galápagos hawks are large predators. Formation of a coalition may increase the probability of carving out a new territory or usurping residents. This resembles the pattern found in African lions, an egalitarian cooperative breeder (Packer et al. 1991, 2001). However, unlike in lions, the females are larger than the males, so female cooperation may be unnecessary to defend against infanticide by invading male coalitions. In addition, males have a reasonable probability of becoming the sole breeder on a territory once other members of the initial coalition die.

Type 6: contextual polyandry

The systems I have classified under this heading are those where the distinction between the mating and social group takes on greatest significance. In all these societies, there are many groups where the mating group comprises only a single male and female, even though many individuals related to the dominants may provision the young. However, polyandry emerges when females find themselves in a group with more than one male to which they are unrelated, such as when a new female fills a vacancy in a group containing a previously fledged male helper.

(a) Sons of the females are philopatric and are likely to help

In *Campylorhynchus* wrens polyandry is rare, but occurs predictably where the breeding female shares a territory with more than one unrelated male. Comparable evidence of an advantage comes from unrelated pied kingfisher helpers (Reyer 1990). While related helpers are generally accepted, only groups struggling to provide adequate resources for the brood will accept unrelated helpers. The extent of reproductive access by these males has not been studied with molecular techniques, but unrelated helpers are likely to gain some direct benefits and also to enhance their probability of taking over the territory when the dominant male dies.

(b) Sons of the females are often philopatric but are unlikely to help

As in the *Campylorhynchus* wrens, male white-browed scrubwren groups form via two paths. Some male offspring of the female are philopatric, but there is also dispersal of males into new groups. There are strong dominance hierarchies based on age regardless of relatedness. Females do not mate with their sons, but allow unrelated subordinates considerable paternity (Whittingham et al. 1997). Unrelated subordinates are much more likely to provision young than are sons of the female (Magrath and Whittingham 1997). The difference may arise because unlike in *Campylorhynchus*, where big groups fledge more young (Rabenold 1984), male help in scrubwrens has little impact on offspring fitness (Magrath and Yezerinac 1997) unless the female is breeding for the first time (Magrath 2001). Because sons cannot be living with females breeding for the first time, there can be no inclusive fitness benefits other than deferred effects resulting from load-lightening in their parents. There is considerable extra-pair paternity in scrubwrens (10% of nestlings attended by coalitions of males unrelated to the female), but sires have not been identified, so we do not yet know its adaptive significance.

(c) New coalitions of males initially share paternity

In a more complex variant of the *Campylorhynchus* system, most groups of white-winged choughs and Arabian babblers are formed via recruitment of young of a monogamously breeding pair (Rowley 1978;

Zahavi 1990). Offspring of both sexes provide important assistance to the dominant pair in rearing young. However, occasionally new groups form as the result of factions from different groups combining, and under those circumstances males from different factions share reproduction (Lundy et al. 1998; Heinsohn et al. 2000). Unfortunately, paternity analysis has not determined whether one faction eventually gains predominance because the other factions die out or through the assertion of dominance.

In these species it has been argued that helping is a signal of prowess as a coalition partner (Wright 1997). Arabian babblers compete to help or provide nest defense even to the extent of preventing other individuals doing so (Zahavi 1990). In white-winged choughs, young birds pretend to help even if they are incapable of foraging at a rate that leaves them with surplus resources for provisioning (Boland et al. 1997b). These species therefore provide behavioral evidence of reproductive "transactions" among males (Chapter 10). The principal benefit of polyandry to females may be to stabilize the factions into a coherent group, which is important where help is critical for effective reproduction (Heinsohn 1992).

Type 7: hidden leks

The most surprising revelation from the application of molecular techniques has been the discovery that although fairy-wrens (*Malurus* spp.) live in social groups where supernumeraries are recruited through natal dispersal, the majority of fertilizations (>65%) are obtained from males living outside the social group (Brooker et al. 1990; Mulder et al. 1994). Such extreme infidelity is associated with massive cloacal protuberances that allow sperm storage in males (Mulder and Cockburn 1993; Tuttle et al. 1996), and conspicuous extra-group courtship displays (Brooker et al. 1990; Mulder 1997). However, in contrast to other species with these adaptations, copulations are rarely observed and are generally within-pair, and courtship displays do not lead to fertilizations (Green et al. 2000). Despite similarities in courtship, dichromatism, and extra-group infidelity, there is considerable variation in social organization within fairy-wren species (Rowley and Russell 1997). The most important dichotomy is between societies where both males and females are philopatric (7a) and those where only males are philopatric (7b).

We understand the evolution of this mating system best in the superb fairy-wren. Brooker et al. (1990) originally proposed that incest avoidance was the ultimate evolutionary cause of this remarkable dependence on extra-group mating, as about 20% of social pairings in the splendid fairy-wren are between nuclear family members. However, in the superb fairy-wren all females seek extra-pair copulations, yet the incidence of incestuous pairings is much lower, and is virtually confined to mothers and their sons. Dunn and Cockburn (1999) showed that females always cuckolded younger males regardless of relatedness, and suggested that the availability of extra-group mating allowed incestuous pairing, rather than the presence of incestuous mating necessitating extra-group mating.

Double and Cockburn (2000) used radiotelemetry to determine why copulations are rarely observed. Females initiated fertilizations before dawn by flying directly to the male's singing post on his own territory. Females gain no direct benefits from the preferred male, as extra-group males neither provision the young nor desist from courtship. Females not only control fertilizations, but most females prefer the same male phenotype. Preferred males molt into epigamic plumage months before the breeding season, after which they immediately commence extra-group courtship (Dunn and Cockburn 1999; Green et al. 2000). Early molt is costly and can only be performed by older, high-quality males (Peters 2000; Peters et al. 2000).

Female choice for a limited group of males is inconsistent with inbreeding avoidance or other compatibility models. Rather, it suggests that females are choosing good genes by assessing male phenotypes. Wagner (1998) has coined the well-chosen term "hidden leks" to describe such reproductive behavior in socially monogamous species.

How can such a system evolve and be maintained without male defection, which is predicted in all theory pertaining to this question (Kokko 1999; Shellman-Reeve and Reeve 2000)? Three factors may help stabilize the system. First, where females initiate extra-pair fertilizations on the territory of the male, males are tied to a base where they can be located and may be forced to remain in socially monogamous relationships (Gowaty 1996b).

Second, dominant males attack helpers that are experimentally removed during parental care, consistent with punishment (Mulder and Langmore 1993). Males substantially reduce care when they have helpers (Dunn and Cockburn 1996), allowing them to devote more effort to extra-pair courtship (Green et al. 1995). The greater level of courtship may have allowed females to assess more males, and the enforced help provided by helpers liberates them to choose mainly for genetic rather than direct benefits (Mulder et al. 1994). However, females without helpers may allow their partners some paternity to ensure continued care.

Third, in a system where only a few males gain much extra-group success and males have to wait for several years before they become competitive as extra-group sires (Dunn and Cockburn 1999), other males may have to make the best of a bad job by seeking within-group fertilizations. They are therefore compelled to provide some care to avoid jeopardizing this avenue to reproductive success. Females whose helpers are removed during their fertile period are reluctant to continue with reproduction, suggesting that males and females negotiate an adequate level of care with their partner (Dunn and Cockburn 1996). Males may generally regulate their provisioning to a level well below their maximum capacity (Macgregor and Cockburn 2002). Any incentive for helper males to defect is reduced by direct benefits. Helpers unrelated to the female gain about 20% of within-group fertilizations (Cockburn et al. 2003), and helpers of attractive males gain substantial extra-group success, apparently by acting as satellites and parasitizing their dominants during the predawn forays of the females (Double and Cockburn 2000, 2003).

Type 8: true monogamy with helpers

I define true monogamy as occurring when a single male and female gain within-group parentage in more than 90% of broods where multiple within-group parentage is possible. Such fidelity appears to be limited to only a few cases.

(a) Failed breeders help close relatives and no extra-pair mating

There appears to be little competition over paternity in European bee-eaters or long-tailed tits, where failed breeders return and redirect help to their close relatives (Lessells 1990; Jones et al. 1991; Lessells et al. 1994; Russell and Hatchwell 2001; Hatchwell et al. 2002). Returning to help is easily understood in terms of the classic arguments of kin altruism. In this case, the breeding group is the monogamous pair and the failed breeders are making the best of a bad job by helping rear their relatives. This variant is probably common among birds that only occasionally breed cooperatively.

(b) Low within-group sharing but high extra-pair paternity

Provisioning in male western bluebirds is confined almost exclusively to the case where young males are present in close proximity to both their social parents, though they sometimes provision a same-aged brother (Dickinson et al. 1996). These helpers sometimes feed simultaneously at their own nest and at the nest of a relative, but are more likely to be failed breeders returning home. Extra-pair mating is also common in this species (20% of fertilizations), but helpers are not usually the beneficiaries, and beyond any incest taboos, females prefer to mate with males older than their social partner (Dickinson 2001). Thus, although helpers prefer to provision offspring of the birds that rear them, some of those offspring will be only partly related to the helper. Dickinson and Akre (1998) have shown that this lack of relatedness has negligible effects on likely inclusive fitness benefits.

(c) and (d) Exclusive monogamy despite unrelated subordinates

In the remaining variants, females neither mate polyandrously despite on occasion having more than one unrelated male in the group, nor commonly seek extra-pair copulations. In laughing kookaburras and Florida scrub-jays supernumerary unrelated birds of both sexes are occasionally present within the group (8c), while in red-cockaded woodpeckers only males remain philopatric and provide care (8d).

Explaining the absence of a behavior is more difficult than explaining its occurrence, as occurrence may be associated with conspicuous behaviors where contextual analysis is possible. Indeed, progress with explaining the absence of extra-pair mating in birds that do not breed cooperatively is at best modest (Petrie and Kempenaers 1998; Hasselquist and Sherman 2001). However, the absence of polyandry is particularly surprising in cooperatively breeding birds since females have easier access to extra-pair mating than is true for birds where only one male lives on the territory. Subordinates in these cooperatively breeding species are not simply reproductively suppressed, as their titers of the major reproductive hormones are similar to those of dominant breeders (Chapter 8).

Explanations for the absence of subordinate reproduction and extra-pair fertilizations vary. In Florida scrub-jays, Quinn *et al.* (1999) explained the absence of extra-pair matings by suggesting that only high-quality mates attain territories, so the variance in male quality is low, making it unlikely that females could easily find a better-quality male through extra-group mating. This explanation is hardly consistent with persistence of mate choice in lekking species. Legge and Cockburn (2000) suggested that because dominant male kookaburras are the main provisioners, the relationship between the female and her partner may be too important to jeopardize. A comparable argument has been suggested for raptors where females are heavily dependent on males during incubation and nestling care (Warkentin *et al.* 1994; Negro *et al.* 1996). This explanation could have some generality. Adult males also dominate provisioning in Florida scrub-jays (Stallcup and Woolfenden 1978). However, predominant male provisioning is true of many other species, so further comparative data are required before any conclusions can be reached. Exploration of patterns of paternity in cooperatively breeding hornbills should prove particularly interesting, as females in this group are completely dependent on males, being sealed inside the nest cavity for the duration of parental care. The one study of a (non-cooperatively breeding) hornbill thus far revealed complete monogamy (Stanback *et al.* 2002).

Type 9: unattached helpers

In some territorial species helpers move regularly between territories instead of forming attachments to a territory or group of breeders. Future mating opportunities are apparently the primary motive for this behavior.

(a) Helping to rear future mates
One of the most remarkable motivations for cooperative behavior comes from riflemen, a small New Zealand passerine. Riflemen have not been subjected to genetic analysis, but behavioral data suggest that helpers in this society are of two sorts (Sherley 1989, 1990). "Casual" helpers move between territories and contribute little care in each. These helpers compete well for any vacancies created by the death of breeding males. "Regular" helpers form a more stable association with the breeding pair and provision at a rate comparable to the parents. These helpers have a high probability of subsequently mating with the young they provision. While this mating system is highly idiosyncratic, it has special interest with the new realization that the small group containing the rifleman forms the outgroup to all other passerines (Barker *et al.* 2002; Ericson *et al.* 2002).

(b) and (c) Help as courtship
In moustached warblers, unrelated floaters assist the female in incubation, feeding of nestlings, and defense of chicks (Fessl *et al.* 1996). Females sometimes switch mates between successive broods, but only if they have received previous assistance. Similarly, unmated male hoopoes frequently visit the nests of pairs, and will both provision young and copulate with the female. These copulations rarely lead to paternity, but females often switch mates between successive broods, suggesting that courtship is the primary motivation for help (Martín-Vivaldi *et al.* 2002).

CONCLUSIONS

What are the main benefits females derive from mate choice?

Females in many cooperatively breeding species seek direct benefits by trading paternity for provisioning by males. Benefits are of two sorts: enhanced reproductive success and load-lightening (Hatchwell 1999; Legge 2000b). Females also use provisioning to assess the suitability of future mates, particularly in the case of unattached helpers and *Manorina* miners. These indirect benefits are likely to promote polyandry.

Comparing the frequency of within-group polyandry is difficult because of the haphazard reporting of data in many molecular studies. Some authors do not even report the number of broods sampled, let alone disaggregate data in a way that enables determination of the incidence of multiple paternity in the crucial case where breeding females are living with unrelated males. A good example of how such data should be reported is provided by Whittingham *et al.* (1997).

In addition, life-history features constrain comparisons. The incidence of multiple paternity is obviously irrelevant for broods containing a single young, and if paternity is allocated randomly among two males, we

Table 5.2. *Incidence of within-group polyandrous broods where females are living with more than one male to which they are unrelated*[a]

Mating system	Species	Modal brood size	% polyandrous broods (N broods)
Egalitarian polygynandry	Red-billed buffalo-weaver	3	46 (13)
	Acorn woodpecker	3	28 (65)
	Pukeko[b]	5	100 (13)
	Common moorhen	9	100 (1)
Flexible polygynandry	Alpine accentor	3	50 (38)
	Dunnock	3	46 (26)
	Smith's longspur	4	77 (31)
Plural breeding[c]	Mexican jay	3	63 (51)
	Stitchbird	3	35 (34)
	Noisy miner	2	0 (35)
	Bell miner	2	0 (13)
Egalitarian polyandry	Galápagos hawk	2[d]	100 (6)
	Brown skua	2[d]	38 (21)
Contextual polyandry	Stripe-backed wren	2	31 (13)[e]
	White-browed scrubwren	3	53 (19)
	White-winged chough	3	56 (9)
Hidden leks[f]	Superb fairy-wren	3	23 (86)
True monogamy	Western bluebird	5	7 (28)
	Laughing kookaburra	3	8 (12)

[a] Where it can be discerned, only broods with more than one young are included. Except where indicated, sources are as for Table 5.1.
[b] Data from the largely unrelated groups studied by Jamieson *et al.* (1994).
[c] In the case of plural breeding, all sires in the larger group or neighborhood are considered potential participants in within-group polyandry.
[d] Broods with only one young are common, but only broods of two are included in the summary.
[e] Number of broods may include broods of a single young.
[f] Broods with two or more within-pair offspring are rare because of high extra-group paternity when females live in groups (M. Double and A. Cockburn, unpublished data).

expect 50% of broods of two and 25% of broods of three to be sired by just one of them. Nonetheless, the quantitative data suggest that within-group polyandry occurs commonly in species where males are unrelated to the female (Table 5.2). Such data probably underestimate polyandry, as paternity access may change between successive broods (Haydock and Koenig 2002). The failure of polyandry to occur occasionally in *Manorina* miners is a surprising result, as the complex nature of social groups in these species seem likely to facilitate covert mating.

What is the evidence that males negotiate mating access?

Male–male competition has clearly influenced the evolution of cooperative breeding, as evidenced by morphological and behavioral adaptations that reflect intense sperm competition in many species. Mate-guarding is also common, but not inevitable. Indeed, mate-guarding is potentially a significant cost for cooperative breeders (Komdeur 2001). However, there is scant evidence that males negotiate levels of paternity among themselves.

Strong evidence for the occurrence of "transactions" between males comes from three different mating systems. Egalitarian polyandry provides the best evidence, as agreement to form a coalition appears inextricably linked to peaceful sharing of paternity. Less well understood is the polyandry that emerges when coalitions in white-winged choughs and Arabian babblers merge to form new groups. Last, there is evidence from fairy-wrens that dominant males may coerce subordinates to provide help.

None of these cases bears much resemblance to the reproductive transactions envisaged in models of reproductive skew. Indeed, formation of coalitions in cooperative polyandry and in white-winged choughs and babblers may have more to do with gaining initial access to a territory or female than with subsequent apportionment of mating. More dramatically, not only is the situation in fairy-wrens best viewed in terms of pay-to-stay models (Gaston 1978; Kokko *et al.* 2002), but coercion of helpers by dominant male fairy-wrens appears to deliver unprecedented reproductive control to females. In its simplest but most pervasive form, skew theory argues that the allocation of paternity to subordinate males reflects concessions negotiated between the dominant and subordinate male in order to retain the services of the subordinate as a helper, and hence ignore the interests of the female (Reeve 2000). Magrath *et al.* (Chapter 10) argue that little progress will be made in development of reproductive-skew models without addressing the role of females in regulating mating opportunities. I suspect that the situation is more serious, and that discussions of skew need to be subsumed into discussions of sexual conflict (see also Chapter 11).

What is the evidence that males and females negotiate mating access to resolve sexual conflict?

The advantages that accrue to females and attendant disadvantages to males that accrue from polyandry mean that sexual conflict between male and female interests is rife among cooperatively breeding birds. Opposite extremes in the expression of conflict are represented by the success of male stitchbirds in gaining paternity from forced copulations, and the rampant infidelity of female fairy-wrens. Evidence for subtle transactions over paternity access is almost ubiquitous. Some instances, such as the paternity bartering in flexible polygynandry, have been subject to sophisticated experimental analysis (Davies *et al.* 1992, 1996). Others, such as the greater level of paternity achieved by male fairy-wrens in groups without helpers, have proved difficult to manipulate because of the extreme sensitivity of the birds to even minor manipulations of their social circumstances (Dunn and Cockburn 1996; Macgregor and Cockburn 2002). Understanding of most other systems is in its infancy, and it may remain so while theory concentrates primarily on interaction between males.

Do females choose genetic benefits?

Although there is overwhelming evidence that females mate polyandrously within their group in order to obtain direct fitness benefits, some of the genetic benefits of multiple mating might also be achieved through within-group polyandry. However, the only evidence pertaining to this point comes from observations of copulations in Galápagos mockingbirds, a species desperately requiring molecular analysis. Discriminating whether fitness benefits are derived from genetic advantages of multiple mating or from additional provisioning or defense in multi-male groups will be difficult, though the issue could potentially be resolved by cross-fostering experiments.

Stronger evidence for female choice for genetic benefits comes from extra-group mating. In many species of cooperative birds, extra-group mating occurs at negligible levels. However, extra-group fertilizations comprise 10% or more of all paternity in species exhibiting joint-nesting polygyny (Seychelles warbler), egalitarian polygynandry (red-browed buffalo-weaver), contextual polyandry (white-browed scrubwren) and true monogamy (western bluebird), and they are the defining feature of the hidden leks of fairy-wrens. Patterns of mate choice in superb fairy-wrens and western bluebirds are compatible with selection of good genes from males. All the evidence that genetic compatibility is important is currently associated with incest avoidance. While this primarily acts by precluding the formation of pair bonds, it may influence extra-pair mating in Seychelles warblers.

Where now?

The diversity of social and mating systems among cooperative breeders is poorly understood, but the diversity represents one of the most exciting, if bewildering, strands in the study of social evolution. In particular, we are currently unable to predict where new species will lie within the framework I have presented, as the membership of most groups is confined to a single species or clusters of close relatives. Even the rare cases of convergence, such as between Galápagos hawks and brown skuas, and between white-winged choughs and Arabian babblers, do not lead to clear predictions of where else such systems might be found. Nonetheless, the diversity of mating relations suggests that, with further work, cooperative breeders offer unique paths to improve our understanding of mating-system theory in general. It is clearly time to shed our obsession with philopatry and helping behavior, and to move on to explore the rich pickings offered by these remarkable creatures.

ACKNOWLEDGMENTS

The Australian Research Council has consistently supported my research on cooperative breeding. Walt Koenig made useful comments on the manuscript, and Jan Ekman, Joey Haydock, Rufus Johnstone, Walt Koenig, Hanna Kokko, Annie Kazem, and Jon Wright allowed me access to unpublished data or manuscripts. I am indebted to the ornithologists who have shown me cooperatively breeding species in the field, and owe special thanks to the spekboomvelders.

6 · Sex-ratio manipulation

JAN KOMDEUR
University of Groningen

Modern evolutionary theory is based on the idea that individuals are selected for their ability to efficiently translate resources into genetic contributions to future generations. Fisher's (1930) theorem states that in sexually reproducing organisms, frequency-dependent selection should lead to an evolutionarily stable strategy of equal expenditure by parents on offspring of the two sexes. Thus, where costs of producing males and females differ, parents may be selected to invest more heavily in the cheaper sex to equalize investment ratios within a population.

Fisher's theorem assumes that the fitness effects of producing sons and daughters are the same for each parent, resulting in all parents producing the same ratio of sons and daughters. Trivers and Willard (1973) argued that where selective pressures on the two sexes vary, the reproductive value of male and female offspring may also differ. This favors individual parents that bias their broods toward the more "valuable" sex, specifically the sex that contributes more to parental fitness relative to its production cost (Trivers and Willard 1973; Charnov 1982). Such facultative biasing by individual parents can occur despite strong selection for equal investment in daughters and sons within the population. This theory contradicts Fisher's theorem because it predicts unequal allocation of resources in sons and daughters at the level of the population (Frank 1990; Pen and Weissing 2000).

Charnov (1982) discussed various theoretical reasons why individuals should vary their investment in male and female offspring. Strong empirical evidence supporting many of these ideas has come from several taxa with well-understood mechanisms for the adjustment of offspring sex ratios, particularly haplodiploid insects, and sex-allocation theory is often cited as one of the best developed in evolutionary ecology (Trivers and Willard 1973; Charnov 1982; Frank 1990). However, the success of applying sex-allocation theory to vertebrate taxa with chromosomal sex determination, particularly birds, has been less successful (Sheldon 1998; Yezerinac 1999; Pen 2000; Hasselquist and Kempenaers 2002; Komdeur and Pen 2002).

Several obstacles have been recognized. First, up until recently, reviews were unanimous in the belief that facultative adjustment of offspring sex ratio in vertebrates was rare, of minor magnitude, and of little or no adaptive significance (Williams 1979; Charnov 1982; Clutton-Brock 1986; Bull and Charnov 1988). Sex determination is almost ubiquitously associated with chromosome heterogamety, constraining the physiological or genetic mechanisms for skewing the sex ratio at birth (Williams 1979, 1992; Krackow 1995). Thus, the very possibility of adaptive sex-ratio manipulation at laying in birds has been questioned. Second, the results of empirical sex-ratio studies in vertebrates are often interpreted within the framework of classic sex-allocation theory based on invertebrate species, and fail to take into account the complexities of vertebrate sex determination and life-histories (Pen and Weissing 2002). Third, to test models of sex-allocation strategies, detailed knowledge of the fitness functions for parents and offspring of both sexes is required (Leimar 1996; Koenig and Walters 1999; Pen and Weissing 2000). For the majority of populations these data are not available and they are not easily obtained (Lessells *et al.* 1996).

The key to testing sex-allocation theory in birds is identifying the specific circumstances operating on species that affect either the relative cost of producing each sex or the reproductive potential of the sexes. A good example of this is the study of red deer, in which birth weight affects subsequent male fitness, but not the fitness of females. Differences in dominance rank

---

Ecology and Evolution of Cooperative Breeding in Birds, ed. W. D. Koenig and J. L. Dickinson. Published by Cambridge University Press.
© Cambridge University Press 2004.

between hinds affect reproductive success of their sons more than that of their daughters. High-ranking females are in better condition and can therefore invest more resources in their offspring than subordinate females. Sons produced by high-ranking females are stronger and have a higher reproductive success than sons produced by low-ranking females. High-ranking females consistently bias their sex ratio toward male calves while subordinates produce an excess of daughters (Clutton-Brock et al. 1984). However, red deer hinds that produced male calves in one year calved later the following year, and were almost twice as likely to be barren as hinds that reared female calves (Clutton-Brock et al. 1981).

Until recently, another problem was the lack of good data on avian brood sex ratios. Earlier studies relied on sex determination based on size or plumage differences between the sexes in nestlings, restricting studies to sexually dimorphic species. Most of these studies measured the sex ratio about the time of fledging (the "secondary" brood sex-ratio). However, a secondary sex-ratio bias does not necessarily mean that the sex ratio at hatching (the "primary" sex ratio) is biased. A biased secondary sex ratio might be the by-product of differential mortality due to sexual size dimorphism, different requirements of male and female nestlings, brood reduction, or differential allocation of parental care to chicks of different sexes.

The problems of establishing sex ratios of young birds before and at hatching have now been solved with the development of simple molecular sexing techniques using DNA (Griffiths et al. 1996, 1998; Lessells and Mateman 1996, 1998). As such, the number of studies investigating nestling sex ratios in birds has increased sharply and sex-ratio adjustment has now been demonstrated in several bird species. Because of potential biases, such studies have generated considerable controversy (Festa-Bianchet 1996; Bensch 1999; Krackow 1999; Lessells and Quinn 1999; Palmer 2000; Hasselquist and Kempenaers 2002). However, despite these difficulties, there remains broad interest in questions of adaptive sex ratios in birds.

Helper systems are particularly good models for testing the occurrence of adaptive sex allocation for two reasons. First, the most common form of cooperative breeding involves a breeding pair being assisted by offspring from previous broods, and juveniles of one sex are often more likely to stay and assist with parental care than the other. The value of sons and daughters therefore depends on the costs and benefits for parents of receiving help (Pen and Weissing 2000). Second, many studies on cooperatively breeding birds are sufficiently long-term that they can provide extensive demographic data and knowledge of the key life-history parameters and the fitness functions for parents and offspring of both sexes (Stacey and Koenig 1990a).

Differences in the propensity of young to disperse from the natal territory or help at the nest were originally used to predict individual-based patterns in sex-ratio bias (Clark 1978; Gowaty 1993). For example, a biased offspring sex ratio toward the dispersing sex could be the result of selection to avoid competition between philopatric siblings. Similar to an argument proposed by Hamilton (1967), a bias toward the dispersing sex could then reduce the cost generated by increased competition directly associated with the number of philopatric offspring. The converse to this situation is when helping by philopatric individuals reduces the cost because they increase the reproductive success of their parents in the future.

Here I begin with an outline of basic sex-allocation models and discuss the difficulties in their application to birds. I then consider a variety of specific social and ecological circumstances that could drive variation in adaptive sex allocation in birds exhibiting a cooperative social system. I review empirical studies of sex-ratio variation in cooperatively breeding birds both at the population level and at the individual level, and determine how well the observed sex ratio can be explained by traditional sex-allocation models. Lastly, I outline some of the unresolved issues in sex-ratio studies and suggest future research objectives.

Sex allocation is the quantity on which selection acts, whereas the sex ratio merely describes the relative numbers of sons and daughters. The two need not be equivalent. Unfortunately, sex allocation is more difficult to measure than the sex ratio and most studies have only addressed the latter. Therefore I assume that sex ratio reflects sex allocation, a practice that in consistent with the majority of studies to date.

CLASSIC SEX-ALLOCATION THEORY
The major ideas

Classic theory of sex allocation is founded on four major ideas (Charnov 1982). The first and foremost is that populations are always pulled by frequency-dependent selection toward an equilibrium in which the total investment in the sexes is equal (Darwin 1871; Fisher 1930). In sexually reproducing species, this idea is in turn based on two assumptions. First, the total reproductive value of males and females in a population is equal because every offspring has one mother and one father. Second, parents should divide their resources between the production of sons and daughters so as to maximize their genetic contribution to future generations. It follows that individuals of the minority sex have a greater per capita share of their genes in future generations, putting a premium on the production of that sex. This holds true regardless of which sex is in the minority, and hence an equal sex ratio is the unique evolutionarily stable strategy. This has become one of the most widely cited theories in evolutionary biology (Frank 1990) and has commonly been cited as the explanation for why sex ratios at the population level are generally constrained to parity.

The second idea on which classic sex-allocation theory is based is that of sex-specific kin competition and group structure. Hamilton (1967) was the first to point out that the assumptions underlying Fisher's (1930) equal-allocation hypothesis are violated if individuals living in groups interact more with each other than with other members of the population and if the effects of these interactions are focused disproportionately on one sex. If this is true, selection should favor a sex ratio biased toward the sex experiencing less kin competition. In many cases, this is likely to be the dispersing sex (Clark 1978; Bulmer and Taylor 1980). Conversely, selection may also favor an overproduction of the sex that improves conditions for kin. This might be the helping sex in cooperatively breeding birds (Emlen *et al.* 1986; Lessells and Avery 1987).

The third idea is that relative fitness costs and benefits of producing sons or daughters are not identical, as assumed by Fisher (1930), but may vary according to parental condition, and that selection favors parents that "individually optimize" the sex ratio accordingly (Trivers and Willard 1973). Females in good condition are predicted to produce the sex with a higher variance in reproductive success whereas females in poor condition should produce the sex with lower variance in reproductive success. These systematic deviations in sex ratio around some mean condition are expected largely to cancel out in the local breeding population, thus maintaining population-wide sex ratios at parity (Trivers and Willard 1973).

The fourth idea is that of genetic conflict over the sex ratio. As an example, a disparity between the cost of producing one sex over the other is predicted to result in a sex ratio skewed toward the cheaper sex at independence (Fisher 1930). As a result, at the Fisherian equilibrium, the more expensive sex has a higher individual reproductive value, owing to its relative scarcity. An offspring's gene that increases its chances of being the expensive sex might, therefore, be favored by selection, even if the gene's action compromises the total number of offspring produced by the parents (Trivers 1974; Trivers and Hare 1976).

Difficulties applying the theory to birds

The results of empirical sex-ratio studies in birds have often been interpreted within the framework of classic sex-allocation theory, even though the complexities of sex determination and life-histories of birds clearly violate a number of the assumptions of the standard models. These assumptions include:

(1) *Sex-ratio manipulation is without cost to the individual in control* – Whether this is true depends on the mechanism of sex-ratio manipulation. To adjust the sex ratio, females, the heterogametic (WZ) sex, must exercise either pre- or post-ovulation control (Emlen 1997b; Hardy 1997; Oddie 1998). Pre-ovulation control could occur through the regulated production or release of W and Z gametes (Krackow 1999), whereas post-ovulation control could occur through sex-selective reabsorption of the ova in the oviduct or dump-laying of eggs of the "unwanted" sex (Krackow 1995; Emlen 1997b; Sheldon 1998). If sex-ratio manipulation requires selective killing of offspring at some point during development, this is likely to result in a loss of invested resources or a reduction in lifetime reproductive success compared with controlling the sex ratio at conception (Myers 1978; Cockburn 1990). Modeling has suggested that even small costs of sex-ratio control may overcome the adaptive value of adjusting the primary sex ratio (Pen *et al.* 1999). The inclusion of the costs of sex-ratio control typically leads to less-biased sex ratios than predicted by standard models. However, it has recently been demonstrated

that sex-ratio control in the Seychelles warbler arises through a pre-ovulation control mechanism with virtually no costs (Komdeur et al. 2002; see below). Thus, this assumption may be valid for birds.

(2) *Non-overlapping generations* – The life-cycle of birds involves complicated interactions between overlapping generations to an extent that makes theory much more difficult (Cockburn et al. 2002). Birds therefore face a fundamental decision about how much to invest in a particular reproductive episode, complicating predictions for the adaptive sex ratio (Zhang et al. 1996).

(3) *Fixed total amount of parental resources for reproduction* – Parental resources in this context are synonymous with parental investment as defined by Trivers (1972). Parental investment generally includes more than one resource (such as time and energy), and any single resource can often be invested in different ways. In addition, parental investment can be considered to include any resource, such as a sexually selected trait, that differentially affects the fitness returns of sons and daughters and that is variable in its availability to breeding females. Thus, measuring total investment is impractical.

(4) *Uniparental control and a single short period of investment* – Birds usually have extended parental care, often by both parents, which makes it very difficult to estimate relative investment in sons and daughters, especially if differential mortality takes place during the period of parental care (Komdeur and Pen 2002).

(5) *Random mating* – Birds often have complicated, highly structured societies, negating the assumption of random mating within a population, which forms the basis of models predicting equal investment in sons and daughters (Fisher 1930).

(6) *Absence of sibling competition* – In birds where nestlings hatch at different times or aggressively interact with each other and sometimes commit siblicide, opportunities arise for competitive effects that alter the relative value of each sex (Bortolotti 1986; Stamps 1990; Legge et al. 2001). If certain combinations of offspring sexes exacerbate conflict they should be selected against because they may increase the probability of brood reduction (Bortolotti 1986; Bednarz and Hayden 1991; Legge et al. 2001). On the other hand, these same combinations may be favored under particular circumstances, for example when food is limited and brood reduction becomes advantageous.

SEX-RATIO BIAS AT THE POPULATION AND INDIVIDUAL LEVEL

Of 36 empirical studies on nestling sex ratios in socially monogamous bird species reviewed by Ewen (2001), 19 reported nestling sex ratios at the population level and only two (11%) found significant deviations from parity, both toward males (zebra finch: Clotfelter 1996; Cooper's hawk: Rosenfield et al. 1996). This is in accord with Fisher (1930), Trivers and Willard (1973), and the previously noted observation that population-wide sex ratios that deviate from parity are rare in birds (Williams 1979; Charnov 1982; Clutton-Brock 1986; Bull and Charnov 1988).

How much weight can be given to studies finding a significant bias compared to those finding no such bias? Both studies providing evidence for a significant male bias either have been challenged (Kilner 1998) or are inconsistent with the results of prior studies (Meng 1951; Rosenfield et al. 1985). Unfortunately, these latter studies lacked sufficient power to be confident of their negative results (Ewen 2001). Altogether, of the 17 studies reporting non-significant population sex ratios, 10 (59%) lacked such power.

Other problems compound this statistical issue. For example, studies reporting equal primary sex ratios are difficult to assess due to the difficulties in distinguishing among the numerous forces predicting parity. As a consequence, the best tests of population sex-ratio models thus far involve assessing species with sexual size dimorphism and species that breed cooperatively.

In sexually size-dimorphic species, sex ratio is predicted to be biased at termination of parental care toward the smaller, cheaper sex (Fisher 1930). A comparative analysis of all published studies showed that population-level sex ratios at fledging, but not at hatching, are on average biased toward the smaller sex (Pen and Weissing 2002), contrary to previous analyses based on fewer species (Clutton-Brock 1986). Other studies on species with sexual size dimorphism have also failed to detect consistent primary sex-ratio biases at the population level (Hartley et al. 1999; Radford and Blakey 2000), with the sole exception of a recent study on the highly size-dimorphic blue-footed booby (Torres and

Drummond 1999). Thus, it appears that differential mortality of the larger sex is a general phenomenon (Dijkstra et al. 1998). Since mortality of the larger sex reduces the cost differential between the sexes, the adaptive value of sex ratios biased toward the smaller sex may be much smaller than suggested by size differences alone.

In contrast to primary sex-ratio control at the population level, there is good evidence for facultative variation of sex ratios in birds. Of 33 empirical studies on primary sex ratios at the facultative level in social monogamous bird species, 15 (46%) found significant relationships with either habitat quality or some characteristic of the parents such as paternal attractiveness, maternal condition or sexual size dimorphism (Ewen 2001). Furthermore, of the 18 studies reporting non-significant facultative sex ratios, 12 (67%) lacked power and four were inconsistent with other studies on similar species that reported significant sex ratio biases (Ewen 2001). Unfortunately, few studies measured the accrued fitness benefits to the parents of sex-ratio adjustment in relation to potentially explanatory variables (Komdeur and Pen 2002).

SEX-ALLOCATION THEORY AND COOPERATIVE BREEDING

Bird species with sex-specific helper systems are excellent models for testing allocation theory for the reasons discussed above. Many studies have reported that the helpers appear to increase the reproductive success or survival of their breeding parents (Brown 1987; Stacey and Koenig 1990a; Emlen 1991; Cockburn 1998). Helping by philopatric individuals can be thought of as reducing the overall cost of their production because they "repay" their parents. Such repayment has been argued to constitute a form of "local resource enhancement" favoring an overproduction of the philopatric, helping sex (Emlen et al. 1986; Lessells and Avery 1987; Koenig and Walters 1999; Pen and Weissing 2000).

Unfortunately, there are other potential costs and benefits of producing helpers, most of which have been ignored in the calculation of the fitness consequences of helping (Cockburn 1998; Heinshohn and Legge 1999; Cockburn et al. 2002). For example, the direct benefits of help for the breeding pair may be offset later by competition for food or reproductive conflicts between the breeding pair and additional helpers. Furthermore, an overproduction of the philopatric helping sex may lead to intensified interactions between same-sex siblings, bringing them into competition for access to resources, including food or vacant territories, and to mates.

Some studies have reported that the presence of one or more philopatric offspring has a negative effect on parental fitness (Komdeur 1994b; Legge 2000a; Ewen et al. 2001; Koenig et al. 2001). Under these circumstances breeding pairs should invest more heavily in the more dispersive sex, assuming the genetic return from parental investment in the helping sex is devalued by competition over resources ("local resource competition": Clark 1978) or by competition for mates ("local mate competition": Hamilton 1967). For example, Clark (1978) noted a male-biased offspring sex ratio in bush babies and argued that this was the result of selection favoring a bias toward the dispersing sex in order to avoid competition between philopatric siblings (females). Clark (1978) also suggested that if the presence of relatives enhanced reproductive success, the converse would be true, and parents should then overproduce the philopatric sex ("local resource enhancement"). Each of the above hypotheses can be considered a special case of either negative (mate or resource competition) or positive (resource enhancement) frequency-dependent selection that lead to biased sex ratios at the population level.

In cooperatively breeding species, where helping may be sex-specific, models for mate competition, resource competition and repayment were originally generated to predict population-wide patterns of primary sex-ratio bias. Nonetheless, these models may also be applied at the level of the subpopulation, the individual family, or within a brood (Koenig and Walters 1999). This is because both enhancing and competitive effects probably operate to varying extents among families within populations for at least two reasons. First, each breeding pair may have different optima depending upon whether they already have some helping offspring, and the fitness effects of helpers is likely to be a function with diminishing returns. And second, the territory of the breeding pair may not be of a quality that can support extra helpers. Helpers may even experience a net loss of inclusive fitness if they use scarce resources on a territory. In both situations the presence of a single additional helper in a social group may be sufficient to cause females to increase their production of the more dispersive sex.

SEX-RATIO BIAS IN COOPERATIVE BREEDERS

The population level

Published data on sex-ratio bias at the population level in cooperatively breeding species is summarized in Table 6.1. The 13 species included constitute 32% of all published studies on avian primary sex ratios, considerably higher than one would expect given that only about 3% of avian species are cooperative breeders (Arnold and Owens 1998). Of the 17 studies conducted on these species, six (35%) found significant sex-ratio biases and attempted to explain them with one or more explanatory variables.

In twelve of the studies, helping is sex-biased and there were apparent helping benefits to the breeding pair, thereby meeting the requirements for testing the local-resource-enhancement model (Table 6.1). However, evidence to support this model is weak. Only four studies (33%) found a primary sex-ratio bias toward the helping sex: Seychelles warbler on Aride island (Komdeur et al. 2002); bell miner (Clarke et al. 2002), red-cockaded woodpecker (Gowaty and Lennartz 1985), and Harris's hawk (Bednarz and Hayden 1991). Eight other studies found no support for this model, including those on the pied kingfisher at Lake Victoria, green woodhoopoe, western bluebird, sociable weaver, noisy miner, black-eared miner (which found a primary sex-ratio biased toward the non-helping, dispering sex), red-cockaded woodpecker (using much larger sample sizes than the earlier Gowaty and Lennartz study; Walters 1990, Koenig and Walters 1999), and acorn woodpecker.

Only one study, that of the laughing kookaburra (Legge 2000a, 2000b) met the requirements for testing the local-resource-competition model, namely sex-specific helping and helping that is apparently disadvantageous to the breeding pair (Table 6.1). However, in contrast to the predictions of this model, broods were not male-biased at the population level (Legge et al. 2001).

Four studies reported on sex-specific philopatry and no helping benefits to the breeding pair, a requirement for testing the local-mate-competition hypothesis: Seychelles warbler on Cousin Island, pied kingfisher at Lake Naivasha, laughing kookaburra, and bell miner (Table 6.1). None of these studies found a primary sex-ratio bias toward the non-philopatric sex. In fact, in the bell miner, sex ratio at the population level was biased toward the philopatric sex. It was not possible to test the models for mate competition, resource competition, and enhancement in the yellow-faced honeyeater, the crescent honeyeater and the eclectus parrot. In the first two species it is unknown whether there is sex-specific helping and dispersal, and in all three species it is unknown whether and how helpers affect reproductive success of the breeding pair.

This summary suggests that population-level primary sex ratios deviating from parity are rare, and that evidence to support the resource-enhancement, resource-competition, and mate-competition hypotheses for biased sex ratio is lacking. All of the species reported in Table 6.1 (including three studied at two locations) met the requirements to test at least one of the three hypotheses to assess the potential for population-level sex-ratio control, and only three found evidence to support one of the hypotheses.

An additional problem is that multiple selective pressures may act in opposition, thereby obscuring patterns (Koenig and Dickinson 1996; Grindstaff et al. 2001). For example, the absence of a sex-ratio bias toward males, the helping sex, in sexually size-dimorphic western bluebirds and acorn woodpeckers may be due to larger males being more expensive to produce. Alternatively, because males tend to disperse more locally than females, they are more likely to compete with same-sex parents and siblings, and thus the benefits of help may be offset by the costs of local competition.

A further limitation among many published sex-ratio studies is inappropriate analyses. Analysis of sex-ratio data should use broods as their units of replication. This is because the sexes of nestlings within a brood are not necessarily independent of one another (Lessells et al. 1996; Questiau et al. 2000). Eight of the 17 studies in Table 6.1 reported sex ratios based on the total pool of nestlings, rather than broods, leading to pseudoreplication. Furthermore, five of the eight studies analyzing brood sex ratios lacked sufficient power to allow much confidence to be placed in the observed negative result, because they fell below the minimum sample size of 88 broods, which is required before any confidence can be placed in a negative statistical result (Ewen 2001).

The facultative level

There have been 11 studies on eight cooperatively breeding species investigating facultative control of offspring

Table 6.1. Studies investigating sex ratio at the population level in cooperatively breeding birds[a]

Taxon	Smaller sex[b]	Dispersing sex[b]	Helping sex[b]	Effect of helpers on reproductive success[b]	N broods[c]	N nestlings[c]	Hatching sex ratio (% M)[c]	Fledging sex ratio (% M)[c]	Reference
Eclectus parrot	—	F	M	Unknown	—	209	—	46.4	Heinsohn et al. 1997
Red-cockaded woodpecker	—	F	M	Increase	85	168	58.9*	58.9*	Gowaty and Lennartz 1985
	—	F	M	Increase	—	984	49.6	—	Walters 1990; Koenig and Walters 1999
Acorn woodpecker	F	F	M	Increase	—	392/837[d]	47.2	50.2	Koenig et al. 2001
Green woodhoopoe	F	—	F	Increase	—	233	45.5	45.5	Ligon and Ligon 1990a
Pied kingfisher									
Lake Victoria	—	—	M	Increase	—	38	—	47.4	Reyer 1990
Lake Naivasha	—	—	M	None	—	104	—	52.9	Reyer 1990
Laughing kookaburra	M	M, F	M, F	M: none; F: decrease	66/88[d]	189/175[d]	47.1	48.6	Legge et al. 2001
Harris's hawk	M	F	M	Increase	—	262	57.0*	—	Bednarz and Hayden 1991
Seychelles warbler									
Cousin Island	—	M	F	None	177	177	53.7	53.7	Komdeur et al. 1997
Aride Island	—	M	F	Increase	86	142	13.4*	13.4*	Komdeur et al. 2002
Western bluebird	M	—	M	Increase	393[e]	1691	51.3	51.3	Koenig and Dickinson 1996
	M	—	M	Increase	549	2187	—	51.9	Koenig and Dickinson 1996
Sociable weaver	—	F	M	Increase	—	148/124[d]	56.8	54.0	C. Doutrelant, unpublished data
Noisy miner	—	—	M	Increase	43	114	50.0	50.0	Arnold et al. 2001
Bell miner	—	F	M	None	—	230/89[d]	60.0*	48.3	Jones 1998
	—	F	M	Increase	243	437	55.6*	—	Clarke et al. 2002
Black-eared miner	—	F	M	Increase	34	86	37.2*	—	Ewen et al. 2001

[a] * = $P < 0.05$.
[b] M = male, F = Female. Dash indicates absence.
[c] Dash indicates not sampled.
[d] First number: nestlings used to determine hatching sex ratio; second number: nestlings used to determine fledging sex ratio.
[e] Biased sample: sex ratio based on broods without brood reduction.

sex ratios (Table 6.2), starting with the red-cockaded woodpecker. In this monogamous species males often assist the breeding pair with rearing young while females disperse (Gowaty and Lennartz 1985) and the presence of male helpers increases the productivity of the breeding pair (Heppell *et al.* 1994). Females were described as having never been on the study site before (without tenure) or as having been observed on the study site before (with tenure). Females without tenure produced significantly more sons than tenured females (Table 6.2). Gowaty and Lennartz (1985) suggested that this was adaptive because females with tenure were already likely to have produced sons present in the study site, and thus such females were able to reduce competition for mates among their offspring by overproducing daughters.

A better test would have been to include both female tenure and previous reproductive success on the study site, since females breeding unsuccessfully on the study site would not be predicted to overproduce daughters. Such a scenario has recently been reported for the western bluebird, where the presence of a helping son increases the productivity of the breeding pair (Dickinson 2004). Sons often remain with their parents on the natal territory either as non-breeding helpers or as breeding helpers. In the latter case, sons breed independently with immigrant females adjacent to their parents' nest, but feed at both their parents' and their own nests simultaneously (Dickinson and Akre 1998). Breeding females responded facultatively to the presence of non-breeding male helpers in their group (indicating a competition for female partners) by producing more daughters, and to the presence of breeding male helpers in their group (indicating no competition for female partners) by producing equal numbers of sons and daughters, consistent with the local mate-competition hypothesis (Table 6.2).

Additional support for facultative sex-ratio adjustment due to local resource enhancement or local resource competition comes from two species, the Seychelles warbler and the bell miner. The Seychelles warbler is a rare island endemic, and until 1988 occurred only on Cousin Island in the Seychelles. On this island, the warbler population has reached carrying capacity, and many breeding pairs are aided by helpers, which are usually daughters from previous broods. Having "helpers" around is costly for parents inhabiting poor territories that have less insect food, because such birds deplete insect prey. On high-quality territories, the presence of one or two helpers increases the reproductive success of breeding pairs, but the presence of more helpers decreases the reproductive success (Komdeur 1994b). As predicted by the local-resource-competition hypothesis, unassisted breeding pairs maximize their inclusive fitness by modifying the sex of the single-egg clutch toward sons, the dispersing sex, when breeding on poor territories. Breeding pairs produced sex ratios toward parity when breeding on medium-quality territories. As predicted by the local-resource-enhancement model, breeding pairs on high-quality territories without helpers or with one helper biased the sex ratio toward daughters, whereas females with two helpers already present produced mainly sons (Table 6.2).

This explanation is also supported by experimental work. First, helper-removal experiments confirmed that sex-ratio bias yields more helpers. Specifically, when females on high-quality territories had one of their two helpers removed, they switched from producing all sons to producing 83% daughters (Komdeur *et al.* 1997). Second, in efforts to conserve this species, an additional population was established on nearby Aride Island in 1988. Experiments in which the same parents were transferred between islands confirmed that the sex-ratio differences were related to territory quality. Breeding pairs transferred from low- to high-quality territories switched from producing 90% sons to producing 85% daughters. Pairs switched between high-quality territories showed no change in sex ratios, producing 80% daughters before and after the switch (Komdeur *et al.* 1997).

All birds transferred to Aride formed pairs and established territories in the high-quality habitat, and during the three years following the translocation female warblers skewed their clutch sex ratio strongly toward daughters (Table 6.2). However, this pattern could not be explained by the local-resource-enhancement model. Due to the absence of habitat saturation, daughters dispersed rather than remaining as helpers on their natal territories. They remained unpaired for some time because breeding pairs were producing so few sons. Given the absence of habitat saturation on Aride, it would have been a better strategy for the artificially translocated breeding pairs to produce equal number of sons and daughters. Given that warblers rarely colonize new

Table 6.2. Studies investigating facultative control of offspring sex ratios in cooperatively breeding birds based on some ecological or social variable[a]

Species	Dispersing sex[b]	Helping sex[b]	Effect of helpers on reproductive success	N broods[c]	N nestlings[c]	Hatching sex ratio (% M)[c]	Fledging sex ratio (% M)[c]	Reference
Red-cockaded woodpecker								Gowaty and Lennartz 1985
Without tenure	F	M	Increase	—	19	—	84.2	
With tenure	F	M	Increase	—	63	—	60.3	
Acorn woodpecker								Koenig et al. 2001
Low territory quality	F	M	Decrease	—	200	54.5	—	
High territory quality	F	M	Increase	—	639	49.0	—	
No helpers	F	M	—	—	373	52.0	—	
With helpers	F	M	—	—	467	48.9	—	
Green woodhoopoe								Ligon and Ligon 1990a
0–2 helpers	—	F	Increase	—	94	54.3	—	
3 helpers	—	F	Increase	—	74	64.9	—	
Laughing kookaburra								Legge et al. 2001
No helpers	M, F	M, F	M: none; F: decrease	19/22[d]	56/44[d]	33.9	32.5	
All female helpers	M, F	M, F	M: none; F: decrease	6/7[d]	18/9[d]	66.7	85.7	
All male helpers	M, F	M, F	M: none; F: decrease	27/29[d]	80/66[d]	33.8	43.8	
Seychelles warbler (Cousin Island)								Komdeur et al. 1997
Unassisted breeding pairs								
Low-quality territory	M	F	Decrease	57	57	77.2**	—	
Medium-quality territory	M	F	None	27	27	51.9	—	
High-quality territory	M	F	Increase	32	32	12.5**	—	

Category								Reference
Breeding pairs, high-quality territory								
1 helper	M	F	Increase	15	15	6.7	—	
≥2 helpers	M	F	Decrease	13	13	84.6**	—	Komdeur et al. 2002
Seychelles warbler (Aride Island)								
No helpers, high-quality territory	M, F	None	—	86	142	13.4	13.4	
Western bluebird								
No helpers	F	M	Increase	549	2187	—	51.9	Koenig and Dickinson 1996
Non-breeding helper at home when mother laid eggs	F	M	Increase	21	—	—	34.3*	Dickinson 2004
Helper breeding when mother laid eggs	F	M	Increase	18	—	—	45.5	
Sociable weaver								
No helper	F	M	Increase	17/20[d]	53/52[d]	45.6	37.7	C. Doutrelant, unpublished data
≥1 helper	F	M	increase	30/26[d]	96/72[d]	62.2	62.5**	
Bell miner								
Low territory quality	F	M	Decrease	41	78	35.9*	—	Ewen et al. 2003
High territory quality	F	M	Increase	201	357	58.7	—	Ewen et al. 2003
No helpers	F	M	Increase	7	10	92.9	—	Clarke et al. 2002; Ewen et al. 2003
With helpers	F	M	Decrease	52	95	52.6	—	Clarke et al. 2002; Ewen et al. 2003

[a] * = $P < 0.05$; ** $P < 0.01$
[b] M = male, F = Female. Dash indicates absence.
[c] Dash indicates not sampled.
[d] First number: nestlings used to determine hatching sex ratio; second number: nestlings used to determine fledging sex ratio.

islands and have lived in a saturated environment for at least the last 100 years (Komdeur *et al.* 2004), it is perhaps unsurprising that the birds were apparently unable to optimize the sex ratio of their offspring under these novel conditions.

Two other studies provide evidence consistent with the local-resource-competition model but not the resource-enhancement model. In the bell miner, breeding pairs are monogamous (Conrad *et al.* 1998) and helping is primarily by sons of the breeders (Painter *et al.* 2000). Daughters disperse before reaching maturity. The presence of helpers decreased the productivity of breeding pairs on low-quality territories and viceversa on high-quality territories (Clarke 1989; Clarke *et al.* 2002). Breeding pairs with low food availability produced female-biased sex ratios, whereas breeding pairs with high food availability had male-biased primary sex ratios (Table 6.2), the difference being significant (Ewen *et al.* 2003). There was no indication, however, that females without helpers produced more sons than those with helpers. Although more males were produced when food resources were abundant, there was no evidence for there being benefits associated with the presence of more sons. This suggests that factors other than resource enhancement associated with helpers influence a female's preferential allocation to sons.

The laughing kookaburra is an excellent example of how populations may experience more than one selective pressure on the sex ratio. In this species both male and female helpers assist socially monogamous pairs and both sexes disperse. However, daughters disperse at a younger age than sons (Legge and Cockburn 2000). The pattern of helping is complicated, because the presence of male helpers apparently has no effect on fledging success while female helpers depress productivity (Legge 2000a). Thus, there is evidence of both local-enhancement and competition effects at the level of individual families. Breeding females respond facultatively to increases in the number of female helpers by producing more male eggs and fledglings (Legge *et al.* 2001). This effect is a slight variant from the normal interpretation of an enhancement effect, because kookaburras seem to be avoiding the "anti-enhancement" effect of having too many detrimental females. However, countering this hypothesis is the finding that unassisted females produced mainly female eggs and fledglings, even though males are apparently the more "helpful" sex.

The female-biased sex ratio of unassisted pairs could also be interpreted as a resource-competition effect, because these pairs were on very small territories that probably could not support philopatric sons. Other support for the resource-competition effect comes from breeding pairs assisted by male helpers which produced female-biased clutches, that is, produced more daughters which are likely to disperse.

A study of green woodhoopoes also failed to provide strong support for the local-resource-enhancement model. In this species, most breeding pairs were assisted in rearing offspring by up to three helpers (Ligon and Ligon 1990b). Offspring of both sexes are philopatric, though daughters are more likely than sons to become helpers, and helping daughters provide more assistance than helping sons. The number of helpers present was positively associated with reproductive success. Females with two or fewer helpers produced slightly more females than females with many helpers (Table 6.2), as predicted, but the difference was not statistically significant.

A detailed study on facultative control of primary sex ratio in the acorn woodpecker found no support for either the local-resource-enhancement model or the local-resource-competition model (Koenig *et al.* 2001). Unlike many cooperative breeders, both male and female acorn woodpeckers frequently remain as non-breeding helpers in their natal groups. Koenig *et al.* (2001) calculated that the average male was 6% more helpful than the average female. Local resource enhancement should select for a female-biased sex ratio on high-quality territories with large facilities for storing acorns. However, the sex ratio at hatching was not biased in relation to territory quality (Table 6.2). In this species, cobreeding coalitions are a combination of siblings or of parents and their offspring. Consequently, the potential for local resource competition among males exceeds that for females, and local resource competition should select for a female-biased sex ratio. However, primary sex ratios produced by breeding pairs with varying numbers of male helpers present remained at parity (Table 6.2).

Taken together, several studies show evidence for facultative variation in primary sex ratio in cooperative breeding birds, but few are able to provide clear explanations for such control. In future studies, care should be taken to ensure that analyses are appropriate.

Three of the 11 studies in Table 6.2 reported sex ratios based on the total pool of nestlings, rather than broods, and seven of the eight studies analyzing brood sex ratios lacked sufficient power to be confident of their results.

EXPERIMENTAL APPROACHES TO ADAPTIVE SEX ALLOCATION

In order to demonstrate causal relationships between sex-ratio variation and properties of organisms or their environment it is necessary to carry out experimental manipulations, a common practice in the study of other life-history traits (Lessells 1991). Experiments are also necessary to demonstrate trade-offs between alternative sex-allocation decisions and to study their fitness consequences. Such experiments have only recently begun in the study of sex allocation in birds, and only one of these involved a cooperative-breeding species, proof that this field is still far from maturation. Experimental manipulations causing Seychelles warbler pairs to change territory quality or number of helpers present resulted in corresponding changes in the sex of their single-egg clutches (Komdeur et al. 1997, 2002). An additional experiment involved selecting unassisted breeding pairs on low- and high-quality territories that were feeding a nestling of the putatively adaptive sex. By swapping nestlings immediately after hatching, some breeding pairs were forced to raise either a foster son or a foster daughter, allowing comparison of the subsequent inclusive fitness gains for pairs raising the (putatively) less and more adaptive sex. Inclusive fitness was estimated as the sum of estimated fitness obtained through the breeding offspring (grandchildren) and the fitness obtained through the helping offspring (in the form of extra offspring produced by the breeding pair through help). On low-quality territories breeding pairs raising foster sons gained significantly higher inclusive fitness benefits than by raising foster daughters, and vice-versa on high-quality territories with breeding pairs raising foster daughters (Komdeur 1998). This provides good evidence that sex allocation in the Seychelles warbler is adaptive for the breeding pair. However, given the recently discovered high rate of extra-pair paternity (40%: Richardson et al. 2001) and complex mating system in which many female "helpers" in fact lay eggs (Richardson et al. 2002), the long-term inclusive fitness functions for the breeding pair of producing sons and daughters is currently uncertain and is in need of reassessment.

MECHANISMS AND COSTS OF SEX-RATIO CONTROL

In order to quantify the adaptive benefits of sex-ratio control, understanding of its mechanisms and costs is essential (James 1993; Krackow 1995, 1999). If certain combinations of same-sexed or different-sexed offspring exacerbate conflict, they could be selected against because they lead to brood reduction, or selected for because they achieve brood reduction efficiently in times of low food resources (Cockburn et al. 2002).

One mechanism for adaptive brood sex-ratio manipulation is biasing the sex of the offspring with laying order (Krackow 1999). In sexually size-dimorphic species chicks of the smaller sex may be at a considerable disadvantage when competing for food with faster-growing opposite-sex siblings (Yom-Tov and Ollason 1976; Bortolotti 1986; Stamps 1990; Oddie 2000). Mothers may seek to temporarily offset this disadvantage by promoting the smaller, cheaper sex up to or early after hatching. Of the four cooperatively breeding species with sexual size dimorphism in which sex ratio has been studied, only two have been found to show biases toward the cheaper sex early after hatching. In the Harris's hawk, female nestlings are 43% heavier, but the first to hatch is a male in 69% of 95 broods (Bednarz and Hayden 1991). In the laughing kookaburra, where female nestlings are 7% heavier, sibling competition is aggressive and sometimes fatal, and nest productivity is determined by competitive interactions between the oldest two nestlings (Legge et al. 2001). Overall first-hatched nestlings were predominantly male (63%, $N = 92$), and second-hatched nestlings predominantly female (32%, $N = 82$), whereas the sex of the third-hatched nestling was unbiased. Hatching a fast-growing female after a male potentially achieves at least two goals, including destabilizing the age-based dominance hierarchy and functioning as a bulwark against siblicide leading to the loss of the second and third chicks (Legge et al. 2001).

In the noisy miner, on the other hand, male nestlings are 9% heavier and hatch first at 95% of nests (18 nests; Arnold et al. 2001). These authors argue that through their helping behavior, large healthy sons enhance the future reproductive success of their parents to a greater extent than daughters. Finally, no sex-biased hatching sequence effects were observed in the bell miner, where male nestlings are 4% heavier than females (Clarke *et al.* 2002). Clearly the situation with respect to the empirical relationship between sex allocation and hatching sequence remains ambiguous.

Although some of the above studies imply sex-ratio control at hatching depending on the ecological circumstances, it is more difficult to determine whether this control occurs before egg-laying. A study on captive eclectus parrots has provided such evidence (Heinsohn *et al.* 1997). Individual females can produce extremely long runs of chicks of one sex. For example, one female produced 30 sons before producing a single daughter. Another produced 20 sons before fledging 13 daughters in a row. Because females are the heterogametic sex in birds, adjustment of the clutch sex ratio could arise either by pre- or post-ovulation control mechanisms. Pre-ovulation control could occur through segregation distortion at the first meiotic division or through differential provisioning of ova of different sexes to influence the order in which they are released from the ovary (Ankney 1982; Krackow 1995; Oddie 1998).

Post-ovulation control could operate through sex-selective reabsorption of the ova in the oviduct (pre- or post-fertilization) or dump-laying of eggs of the "unwanted" sex (Emlen 1997b). A key difference between pre- and post-ovulation mechanisms of adjustment is that post-ovulation control presumably requires skipping a day when an egg could have been laid and, in the case of dump-laying, wasting the resources that were provisioned to that egg. Skipping day(s) at the start or during the ovulation sequence would result in either delayed clutch completion or a smaller clutch.

The Seychelles warbler exhibits pre-ovulation control of hatchling sex ratio (Komdeur *et al.* 2002). Typically, warblers produce only single-egg clutches, but by translocating pairs to vacant habitat of high quality, most females were induced to produce two-egg clutches. Overall, females skewed clutch sex ratios strongly toward daughters. This bias was evident not only in the first egg, but also in the second egg laid one day later. Although a sex bias in the first egg may arise through either pre- or post-ovulation mechanisms, the bias observed in second eggs could only arise through pre-ovulation control. The determination of the actual pre-ovulation mechanisms, however, requires further investigation.

LIMITATIONS OF PAST APPROACHES

This review suggests that population-level primary sex ratios deviating from parity in cooperative-breeding birds are rare, and evidence in support of a general hypothesis for biased sex ratio in these species is lacking. Investigation into facultative adjustment of primary sex ratio by breeding females has been more successful, but results are still quite variable. Furthermore, at either the population level or the individual family level there has been a general lack of consistent support for sex-allocation hypotheses either within or between species. Currently there would appear to be no single framework within which to predict sex-ratio bias in cooperative breeders.

As previously mentioned, there are several difficulties with the interpretation of population and individual sex-allocation patterns from empirical studies. First, because sex of nestlings within a brood may not be independent, analysis of sex-ratio data should use broods as their units of replication (Lessells *et al.* 1996; Questiau *et al.* 2000). Some studies still present tests on the total pool of nestlings, resulting in pseudoreplication. Care should also be taken to confirm independence of multiple broods from the same female, which can be readily achieved by randomly selecting subsets of the data containing only one brood per female.

Second, samples are frequently too small to detect any but the largest deviations from equality with reasonable statistical power. Negative results based on small sample sizes can result in false negative evidence (Type II error) for a sex ratio bias. On the other hand, inappropriate analyses can also result in false positive evidence (Type I error) of sex-ratio bias (Wilson and Hardy 2002). It is not always apparent that authors interpret their findings with this in mind: few studies to date have tested the power of the results (Koenig and Dickinson 1996; Sheldon 1998; Ewen 2001), although it is sometimes acknowledged that sample sizes are small (Arnold *et al.* 2001).

Third, many studies lack the rigorous quantification of expected bias before support or rejection of any hypothesis is made. We need to know whether an influential parameter, such as habitat quality and helper benefits for each member of the breeding pair, has been measured accurately (Westerdahl et al. 1997; Koenig and Walters 1999; Pagliani et al. 1999) both at the population and at the individual level. For example, to test the local-resource-enhancement hypothesis adequately, there is a need for an experimental demonstration that helpers have a positive effect on the reproductive success of the breeding pair, and for an accurate knowledge of the inclusive fitness gains of breeding sons versus daughters (grandchildren). Because the reproductive success of a breeding pair is also affected by other variables, such as brood size, age, and experience of the breeders and their helpers, efforts to gain congruence between expected and observed biases often fail (Koenig and Walters 1999). One should keep in mind that estimates of lifetime reproductive success should include the reproductive success of all sons and all daughters produced over the breeding female's and male's lifetime (Komdeur 1998; Koenig and Walters 1999).

There is the further complication that in cooperatively breeding species, parents (including the social pair and cobreeders, if present) and non-parents (usually offspring) all contribute to the care of broods. Because different individuals may have different optimal sex allocations, there can be conflicts of interest over the allocation of care to offspring. For example, if female helpers compensate for reduced parental effort by the breeding male, they might allow him to invest more effort in seeking extra-pair fertilizations and thus be to his benefit. On the other hand, the presence of several female helpers in the group may be a disadvantage to the dominant female because of conflicts between the breeding female and additional females over who should reproduce and who should not. Additional females may sneak in one of their own eggs (McRae 1996a; Jamieson 1999) or even destroy eggs already present that are not theirs (Mumme et al. 1983a; Koenig et al. 1995). Such differing benefits may result in conflict over sex allocation between the male and female parent. Even though females determine the sex of eggs, males at least potentially have a means to influence the secondary sex ratio, by feeding sons and daughters differentially (Dhondt and Hochachka 2001).

Fourth, there is the question of how generalizable are reported sex-ratio patterns. This is exemplified in recent studies of primary sex ratio in great tits. One study reported a significant relationship of primary sex ratio with laying date but not with paternal size (Lessells et al. 1996). A later study found the opposite pattern, with no relationship between primary sex ratio and laying date but a significant relationship of primary sex ratio with paternal size (Kölliker et al. 1999). A third study reported no relationship with either laying date or paternal size (Radford and Blakey 2000).

A fifth problem involves the questionable extrapolation of sex-allocation patterns from the level of the population to the level of the individual. The results of facultative control have largely been interpreted within the population-wide models for local resource enhancement and competition. Although it seems that individual variation can be in accordance with predictions of population models (Emlen 1997b; Koenig et al. 2001), there has been debate over the applicability of such an interpretation (Frank 1987, 1990; Koenig and Walters 1999). For example, it would be a logical error to use the resource-enhancement hypothesis to explain the average level of sex allocation in the population, and to explain the variation in sex allocation between individuals in the population. For example, the model of Pen and Weissing (2002) suggests that if offspring of one sex become helpers at the nest and parents adjust the sex ratio to varying benefits of help, then at the population level the sex ratio may be unbiased or even biased toward the non-helping sex.

Finally, it is difficult to differentiate between alternative hypotheses that predict qualitatively similar biases in sex ratio. One key example is distinguishing between Fisher's (1930) equal-allocation hypothesis, predicting parity in population sex ratios, and population parity under a Trivers and Willard (1973) model. Just as problematically, a lack of support for a given hypothesis may result from the predicted bias being offset by additional and inverse forces (Grindstaff et al. 2001). For example, in most species with helpers, and even within individual families of birds, both the local-resource-enhancement and local-resource-competition hypotheses can be operating simultaneously because different groups are exposed differently to factors such as local competition and number of helpers present in the group.

CONCLUSION

The advent of molecular sexing techniques has meant that the study of sex allocation in birds is enjoying a welcome renaissance. Biased sex-ratios show that sex ratio modification is occurring, probably even before ovulation. However, the current data on sex allocation in cooperative breeders are variable and the results are often ambiguous. Too few species have been studied sufficiently well for any patterns of sex allocation to emerge. However, some key empirical patterns tell us almost nothing about the adaptive cause and do not fit comfortably within the framework provided by available theory. It is clear that we need a better understanding of when sex ratio manipulation should be expected (theory), and when it occurs (data).

Studies on cooperatively breeding birds are suitable for testing sex-allocation theory because they are among the longest-running studies available and often provide comprehensive data on the life-history and fitness of individuals with which to interpret the observed patterns. However, results need to be repeatable and preferably experimental if we are to achieve a unified theoretical framework with which to understand the evolutionary template for this fundamental attribute of a species' life-history.

ACKNOWLEDGMENTS

I am grateful to Walt Koenig and Janis Dickinson for the invitation to write this review, and to Janis Dickinson, Walt Koenig, Mathew Berg, Ido Pen, John Ewen, and Sarah Legge for providing valuable comments.

7 · Physiological ecology

MORNÉ A. DU PLESSIS
University of Cape Town

Despite elegant work by Reyer and his colleagues on the proximate endocrinological and physiological mechanisms of cooperative breeding (Reyer and Westerterp 1985; Reyer *et al.* 1986), no more than peripheral mention of such factors was made in Brown's (1987) comprehensive review of cooperative breeding. Indeed, most research on cooperatively breeding birds has until recently focused on its functional consequences rather than the mechanisms that underpin it. However, evolutionary and mechanistic approaches complement each other and much can be learned by considering both (Sherman 1988; Mumme 1997; Creel and Waser 1997).

This situation contrasts from that of mammals, where physiological and other proximate causes of cooperative breeding have featured prominently (Solomon and French 1997). The degree to which this is beginning to change is highlighted here and in the following chapter.

Here I provide a general review of the physiological mechanisms and behaviors used by cooperatively breeding birds to survive and reproduce in their environment. As a simplifying framework, I consider the breeding and non-breeding periods separately. These equate roughly to the heuristic dichotomy of, first, factors that determine why cooperative breeders live in groups and, second, why non-breeders help during the breeding season ("group-living" versus "alloparental" effects as defined by Koenig and Mumme 1990).

PHYSIOLOGICAL FACTORS: THE NON-BREEDING SEASON

Food and foraging

Diet
Different workers have made conflicting predictions about the nature of constraints on cooperative breeding.

For example, Emlen (1982a), no doubt thinking of bee-eaters, suggested that cooperative breeders tend to be diet specialists, whereas Brown (1987), thinking primarily of corvids, concluded that cooperative breeders are most likely to be omnivores. In fact, there remains no demonstrated link between diet and the incidence of cooperative breeding (Ford *et al.* 1988; Du Plessis *et al.* 1995; Arnold and Owens 1999; Langen 2000).

Despite the lack of any overall relationship between cooperative breeding and diet, one might predict that the need for defending food resources as a group might arise under two broadly different scenarios. First are cases where the nutritional quality of the diet is so low that the contribution of non-breeding helpers is required for reproductive success. Species fitting into this category include several cooperative breeding herbivorous and folivorous species, including ostriches (Bertram 1978), pukeko (Craig and Jamieson 1990), hoatzin (Strahl 1988) and mousebirds (*Colius* spp.) (Prinzinger 1988). In these species, the intrinsic benefits related to food provisioning by non-breeding helpers are likely to play a significant role. Second are cases where the spatial or temporal distribution of a critical food resource make the costs of resource defence prohibitive for a pair of birds. Examples may include nectarivores, such as several Australian honey-eaters, that breed cooperatively and rely on a resource that is notoriously variable both spatially and temporally.

In few cooperatively breeding birds does diet play as important a role in the evolution of sociality as in the hoatzin. This folivorous bird, at 750 g, is well below the theoretical minimum body mass predicted for homeotherms with ruminant-like digestive systems (Dement and van Soest 1983). The physiological constraints that folivory and foregut fermentation place on this species are reflected in their slow growth

---

Ecology and Evolution of Cooperative Breeding in Birds, ed. W. D. Koenig and J. L. Dickinson. Published by Cambridge University Press.
© Cambridge University Press 2004.

rate, behavioral thermoregulation, generally sedentary habits, and delayed maturation (Strahl 1985, 1988). Other morphological adaptations related to their folivorous habits include an oversized, heavy crop, a long flightless period after leaving the nest, and well-developed predator escape behavior of the young. The primary effect of having helpers is enhanced growth rate and survival of young. The secondary effect is the reduction of energetic costs of reproduction of the female breeder, thus allowing her to renest more swiftly (Strahl and Schmitz 1990).

Group hunting

Social hunting is uncommon among birds and described for only a few group-living raptors. In the two species in which this phenomenon has been documented, two quite different situations exist, neither of which has been studied with respect to their energetic consequences.

The first, the Harris's hawk, is highly social in that the majority of individuals live in groups. Groups are generally composed of subgroups of up to three birds that keep constant visual contact with one another and with other subgroups. When hunting, these subgroups "leap-frog" throughout the home range in search of prey (Bednarz 1988; Faaborg and Bednarz 1990). When a prey item is discovered, several hawks may swoop on it simultaneously. If the potential prey item finds cover, several hawks will surround it, while one or two individuals will walk or fly into the vegetation where the prey item hides. Once the prey is killed, all group members congregate at it and feed.

Faaborg and Bednarz (1990) suggested that the importance of group foraging may be limited to temporal food availability of large prey items to foraging birds. Groups may be able to procure more prey on a more consistent basis than pairs, which could lead to increased survival and fitness benefits. These benefits may be particularly important to young group members that are less proficient foragers than are adults. This explanation thus revolves around the energetic benefits obtained by inexperienced individuals, and might suggest that cooperative breeding in this species is a direct consequence of cooperative hunting behavior. Alternatively, Koenig *et al.* (1992) suggested that social activities such as group hunting may be secondarily derived after families formed as a result of ecological constraints, rather than the driving force behind cooperative breeding itself.

In the second species, the pale chanting goshawk of South Africa, hunting success of individuals is low as a result of dense cover. Consequently, there are occasions when a failed solitary hunting effort results in cornered, yet unthreatened, prey. In these instances, family members join the original hunter in flushing out the prey from its cover by trampling into the shrubs. Only large rodents are usually caught during social hunts, and the prey item is not shared between cooperators. However, juveniles that are relatively incompetent solitary hunters stand an equal chance of capturing prey during social hunts and may be able to balance their energy budgets only because of the returns gained from social hunting (Malan 1998).

Food storage

A number of bird species in north temperate areas store food that they relocate and eat months, and sometimes even years, later (Sherry 1985). Such food storage is generally thought to be an adaptation to enable individuals to cope with unpredictability and variability in food supply and appears to be related to group living and cooperative breeding in several species.

In the acorn woodpecker, for example, winter is an especially challenging time because of their high dependence on stored acorns that often contain appreciable amounts of tannins (Koenig and Heck 1988). Koenig (1991) demonstrated that high tannin levels reduce protein availability to woodpeckers and thus diminish the nutritional value of their primary winter food. Nevertheless, acorn stores appear to be critically important to winter survival and reproductive success in this species throughout its range (Stacey and Koenig 1984; Koenig and Mumme 1987), and are postulated to be an important constraint leading to the evolution of cooperative breeding behavior. The coincidence of this species' subsistence on a largely nutrient-poor diet with inclement winter conditions is likely to result in a high premium for adaptations enabling acorn woodpeckers to increase what is otherwise likely to be a negative energy budget.

Physiological constraints appear to alter costs and benefits in ways that influence the functional consequences of behaviors (Weathers *et al.* 1990). In the case of the acorn woodpecker, protein limitation is consistent with many of the unusual physiological characteristics of this species. Protein limitation may also be an important factor influencing the costs and benefits of cooperative breeding. The additional brooding and feeding capabilities provided by group members beyond a pair in the form of both additional breeders and non-breeding helpers may be particularly valuable as a consequence of the slow development of nestlings and their extended vulnerability to external conditions. Thus, Weathers *et al.* (1990) suggested that acorn woodpeckers' unusual social behavior might be related, at least in part, to their peculiar physiology. Unfortunately, this hypothesis is countered by two lines of evidence. First, acorn woodpeckers living in Colombia breed cooperatively, yet store no acorns (Kattan 1988). Second, the congeneric Lewis' woodpecker, despite often living in the same oak savanna habitats as acorn woodpeckers and also depending substantially on stored acorns, does not live or breed in groups (Bock 1970).

A second example is the pinyon jay. In autumn, these birds converge on seeding pinyon pine trees in relatively large flocks and spend several hours extracting seeds from pitch-laden cones (Ligon 1978). Up to 57 seeds may be held by a single bird in the distended esophagus (Vander Wall and Balda 1981). The flock then takes off and can fly as far as 10 km back to their home range, where the birds alight on the ground. They then proceed to store seeds below ground. Ligon (1978) calculated that a single flock of about 300 pinyon jays in central New Mexico cached up to 4.5 million seeds in a mast year. These seeds then act as a larder of stored food for future use when other foods are scarce or absent. Thus, the birds receive nutrients, energy, nest and roost sites, and stimuli to breed from the pinyon pine trees, and in return the tree relies on the bird for safe seed dispersal (Marzluff and Balda 1990).

A third example is the Florida scrub-jay, which has served as a model of cooperative breeding for several decades (Woolfenden and Fitzpatrick 1986). DeGange *et al.* (1989) reported that despite living in a relatively stable environment, Florida scrub-jays store acorns during half the year, retrieving them during the other half. Apparently only about one-third of stored acorns are retrieved each year. DeGange *et al.* (1989) proposed that acorn stores allow Florida scrub-jays to survive periods of low arthropod availability, and that this may facilitate delayed dispersal by juveniles, thereby contributing to the maintenance of permanent group territoriality and cooperative breeding in this species.

Again, however, caution is suggested by the failure of a closely related form to exhibit similarly social tendencies. The western scrub-jay also stores food where it occurs in the oak savannas of the western USA, yet has only been recorded to breed cooperatively (albeit infrequently) in a single population located in the extreme southern end of its range in Oaxaca (Burt and Peterson 1993).

Solar-enhanced digestion
Group-living mousebirds, like other members of the order Coliiformes, sun-bask frequently throughout the day (Rowan 1967; Decoux 1988a). This behavior typically involves perching on bare branches and facing into the sun in an upright position, with legs apart and the short hair-like feathers on their bellies raised to expose skin to the sun. The behavior is not limited to the winter months and suggests that for some unknown reason mousebirds require or benefit from additional heat beyond that produced by normal metabolism. Dean and Williams (1999) suggested that sunning behavior in mousebirds speeds up the enzymatic reaction in their digestive systems in order to hydrolyze starches, thereby serving an important digestive function. The behavior may thus represent a remarkable evolutionary solution to the problem of extracting the most easily digested fraction from large amounts of vegetable matter by a small flying bird.

Development of foraging skills
Brown (1985) postulated that non-breeding members of cooperatively breeding species, which are typically younger individuals, might forego breeding because they have yet to develop adequate foraging skills, and are consequently in relatively poor condition.

One species where this has been suggested to be true is the white-winged chough. Apart from occasional occurrences of intergroup dispersal (Heinsohn 1991a), philopatry is strong and group members tend to be close relatives. Birds apparently are incapable of independent reproduction until they are four years old (Rowley 1978). However, individuals of all ages contribute to reproduction including nest building and incubation. The white-winged chough appears to be unable to raise young without helpers and can thus be regarded as one of very few birds that is a truly obligatory cooperative breeder (Rowley 1978; Heinsohn 1991c, 1992). Heinsohn and Cockburn (1994) showed that although adult birds generally do not lose body mass during incubation, young helpers lose mass in proportion to the amount of incubation that they perform, independent of any effect of group size.

Thermoregulation

Communal roosting

Communal roosting in birds may yield several adaptive benefits including thermoregulatory benefits, protection from predators, and increased foraging efficiency (Beauchamp 1999). Here we are primarily concerned with the first of these.

The presence of nearby companions is thought to reduce the energetic demands for thermoregulation through mechanisms such as huddling and wind reduction. With communal cavity roosting, temperature within the cavity may be increased, further decreasing energetic demands during cold nights (Du Plessis and Williams 1994).

Three species or groups of species have been studied with regard to these factors. First is the green woodhoopoe. The avoidance by this species of roosting in ostensibly safe open sites, such as thorn-covered branches of acacias, coupled with the high risk of predation at the roost, prompted Ligon and Ligon (1978b) to propose that they are obligated to roost in tree cavities because of their inability to maintain normothermic body temperatures when exposed to cold conditions. This hypothesis was further supported by the finding that rates of oxygen consumption are relatively constant in the thermal neutral zone, but woodhoopoes become hypothermic when exposed to 19 °C, even though they increase their oxygen consumption (Ligon et al. 1988). This led to a proposed scenario for the evolution of group living and cooperative breeding in this species based on their thermoregulatory insufficiency (Stacey and Ligon 1987; Ligon et al. 1988; Ligon and Ligon 1988). The theory posits, first, that the thermoregulatory insufficiency of green woodhoopoes has led to the evolution of cavity roosting, second, that their dependence on tree hollows, which are often in short supply, mandate extreme philopatry, and third, that offspring retention on the parental territory led to the evolution of cooperation, including helping to rear young and territorial defense.

This scenario was subsequently questioned by Williams et al. (1991), who used wild-caught green woodhoopoes and found that birds in good condition were easily able to maintain their body temperatures at temperatures as low as $-10\,°C$, almost certainly lower than those encountered by birds in Africa. However, birds that fell below two standard deviations below the mean body condition of adults displayed the same inability to sustain body temperature as the birds studied by Ligon et al. (1988). Williams et al. (1991) concluded that body condition is fundamentally important in maintaining body temperature, particularly during inclement weather. Subsequently, Du Plessis and Williams (1994) found that a bird roosting with four conspecifics was able to reduce its night-time energy expenditure by at least 30% at ambient temperatures of around 5 °C. Similarly, Boix-Hinzen and Lovegrove (1998) showed that woodhoopoes huddling in groups were able to conserve 12–29% of their daily energy expenditure when compared to non-huddling woodhoopoes at ambient temperatures of 20 °C and below.

Acorn woodpeckers are a second species in which the energetic consequences of communal roosting have been studied. This species mitigates the effects of subsisting on a nutrient-limited diet during the non-breeding season both behaviorally and physiologically (Du Plessis et al. 1994). First, large numbers of acorns are stored during fall specifically for use during periods when other food sources are either scarce or unavailable (MacRoberts and MacRoberts 1976). Second, adult field metabolic rates are 30% lower than predicted from adult body mass (Weathers et al. 1990). Third, roosting in cavities provides an improved thermal environment, the benefits of which vary with weather conditions and cavity features. Fourth, roosting with conspecifics further ameliorates the effects of low temperatures when multiple bodies warm the cavity micro-environment. For example, an 80 g woodpecker roosting in a cavity with three conspecifics when the ambient temperature outside the cavity is 0 °C would potentially expend 4.44 kJ h^{-1}, a saving of 17% compared to the estimated expenditure of 5.35 kJ h^{-1} if it sleeps alone outside. Savings are further augmented if there is wind or if more individuals share the roost cavity.

The last group studied with respect to communal roosting is the mousebirds (order Coliiformes), an order restricted to sub-Saharan African and comprising six species, all of which breed cooperatively (Decoux 1988b). In at least four of these, torpor occurs in response to low body mass (Bartholomew and Trost 1970; Prinzinger et al. 1981; Prinzinger 1988). Mousebirds are also know for their huddling, or clustering, behavior, which has been shown to be important for reducing energy expenditure (Brown and Foster 1992; Prinzinger et al. 1981). In white-backed mousebirds, for example, clustering behavior is considered essential for effective thermoregulation and the avoidance of pathological hypothermia at low ambient temperatures (McKechnie and Lovegrove 2001a).

McKechnie and Lovegrove (2001b) found that, unlike most other birds, speckled mousebirds do not maintain body temperature with respect to a constant set point. Instead, body temperature decreased during the course of the rest phase, with the highest cooling rates observed at moderate ambient temperatures. Restricted food was associated with significant reductions in rest-phase body temperature and metabolic rate. Thus, metabolic suppression normally associated with entry into torpor and the defense of a torpor set point were largely absent.

Key to understanding the unusual thermoregulatory patterns shown by mousebirds is their clustering behavior (Fry et al. 1988). Clustering behavior is an important component of thermoregulation and is necessary for the defense of a constant rest-phase body temperature in both the white-backed (McKenchnie and Lovegrove 2001a) and speckled mousebirds (McKechnie and Lovegrove 2001b). Clustering behavior also appears to be important in the avoidance of pathological hypothermia at low ambient temperatures in white-backed mousebirds (McKechnie and Lovegrove 2001a) and is an important mechanism for reducing rest-phase energy expenditure. Brown and Foster (1992) reported savings of 31% in a group of four speckled mousebirds at 16 °C, whereas Prinzinger et al. (1981) recorded a reduction in energy expenditure of 45.1% in a group of three red-backed mousebirds at 8 °C. McKechnie and Lovegrove (2001a) recorded energy savings of 50% in white-backed mousebirds in a cluster of six birds at 15 °C. The energy savings that speckled mousebirds are able to make by means of clustering behavior are hence similar to, or greater than, the energy savings associated with other kinds of presumably adaptive hypothermic responses.

Communal nest structure
Sociable weavers in southern Africa construct a single compound nest that they utilize for roosting throughout the year. Even when not rearing offspring, sociable weavers return to their nest chamber during the middle part of the day, apparently to escape from solar radiation in summer and possibly to reduce energy demands in winter. The daily process of building and maintaining the nest requires an appreciable investment of both time and energy; thus, the adaptive benefits are probably considerable. Air temperatures within occupied chambers in winter can be elevated by the metabolic heat of birds by as much as 23 °C above external temperature (White et al. 1975). During most of the night the air temperature within nest chambers remains near levels that are believed to be within their thermoneutral zone. Williams and Du Plessis (1996) measured field metabolic rate (FMR) of free-living sociable weavers during winter, and found that their energy expenditure, averaging 48.7 kJ day^{-1}, was much lower than that expected for birds of similar body size living in more mesic environments. They concluded that this was due to a combination of the savings achieved by communal roosting and possibly reduced basal metabolic rate.

Vigilance

Sentinel behavior is a cooperative system of vigilance found in stable social groups of many birds and mammals living in relatively open habitats (Bednekoff 1997). Group members take turns being vigilant and sounding the alarm when danger is sighted. This allows individuals to forage in relative safety and presumably the whole group becomes more efficient in both foraging and predator avoidance (Clutton-Brock et al. 1999b; Wright et al. 2000, 2001).

A number of studies have determined that total group sentinel effort increases with group size, while effort per individual decreases with group size. Using food supplementation, Wright et al. (2000) demonstrated that in Arabian babblers there was a detectable cost to an individual's contribution to sentinel behavior as evidenced by a reduction in body mass. Thus, an individual that was provided additional food was able to spend longer periods on sentinel duty without a loss in body mass. Unsupplemented group members incompletely compensated for these increases by reducing their own sentinel effort. Differences in individual body mass within groups reflected natural and experimental variation in sentinel effort. Thus, the indirect physiological consequences, as manifested in daily energy expenditure, may have been important in the evolution of group territoriality at least in those species where sentinel behavior is important for survival, but relatively costly.

PHYSIOLOGICAL FACTORS: THE BREEDING SEASON

Although energy was proposed early on as one of the most important factors in shaping cooperative breeding behavior in birds (Brown 1982, 1986), relatively little empirical work has followed. Quantitative energy budgets are usually only measured in relation to an individual's size, sex, habitat, and diet (Nagy *et al.* 1999), and not in relation to social context. This mismatch between the proposed importance of energetic factors and social behavior has largely to do with the difficulties of gathering these types of data both under free-living and captive conditions. However, the doubly labeled water (DLW) technique (Lifson and McClintock 1966) has opened up the possibility of obtaining such data, even within complex social contexts such as that presented by cooperative breeders. Despite these opportunities, surprisingly few physiological ecologists have risen to the challenge. Here I briefly summarize some of the evidence regarding the energetic costs of helping; more detailed information is presented in Chapter 4.

Indirect evidence

There is good evidence that in some cooperative breeding systems, individual compensation occurs in the rates at which nestlings are provisioned (Hatchwell and Russell 1996). This happens through the activity of additional group members and is often associated with decreased individual provisioning rates by each member of the group (Hatchwell 1999), which logically translates into reduced energy expenditure. Although few studies have gone beyond this, Heinsohn and Cockburn (1994) demonstrated that in white-winged choughs, young helpers lose mass in proportion to the amount of incubation they perform independent of any group-size effect. Such evidence indirectly suggests that there is a burden to food provisioning and that individuals will generally reduce their own contributions when given the chance (Hatchwell 1999).

Direct evidence

Although qualitative evidence that nestling growth improves with the presence of non-breeding helpers has been available for some time (Ligon 1970; Woolfenden 1978; Reyer 1980), Dyer (1983), studying red-throated bee-eaters, was the first to conclusively demonstrate this effect. Similarly, Taborsky (1984) demonstrated a physiological cost of helping in the cooperatively breeding cichlid fish *Lamprologus brichardi*, in which helpers grew more slowly than non-territorial fish. The benefits offsetting this apparent cost were the protection from predators afforded by a safe territory (without having to bear the cost of territory defense) and an increase in the size of the clutch raised by the related individuals that they helped. More recently, various costs of helping have also been documented in the cooperatively breeding meerkat (Chapter 13).

Pied kingfisher

The physiological consequences of helping in birds were first quantified by Reyer and his colleagues in the pied kingfisher using the DLW technique (Reyer and Westerterp 1985; Reyer *et al.* 1986; Reyer 1990). In this species there are two types of helpers. Primary helpers, usually offspring, are related to the breeders and are with the breeders throughout the nesting period (Reyer 1990). Although primary helpers remain in non-breeding condition, they work as hard as the breeders in provisioning the young at the nest (Reyer and Westerterp 1985; Reyer *et al.* 1986). In contrast, secondary helpers are unrelated to the breeders and are recruited only after the eggs have hatched and only when food is in short supply. They supply food mostly to the female breeder, presumably to enhance their own future chances of mating with her. They also do not work nearly as hard as breeders and, unlike primary helpers, remain in reproductive condition throughout the breeding season. Thus, secondary helpers incur a slight increase in their energy expenditure in exchange for building a potential future relationship with the breeding female. This trade-off is apparently based on an adult sex ratio strongly biased toward males, leaving females in short supply. By contrast, primary helpers assume the energy expenditure of breeders, but are "psychologically castrated" and offset these costs through indirect fitness benefits (Reyer *et al.* 1986; Chapter 8).

As the daily energy expenditure of adults increased, the amount of food delivered to nestlings rose linearly at both Reyer's study areas (Lake Naivasha and Lake Victoria, Kenya), but with significantly different slopes. Thus a Lake Victoria bird achieved a lower feeding contribution than one at Lake Naivasha for the same amount of energy expended, due to poorer hunting conditions

Physiological ecology 123

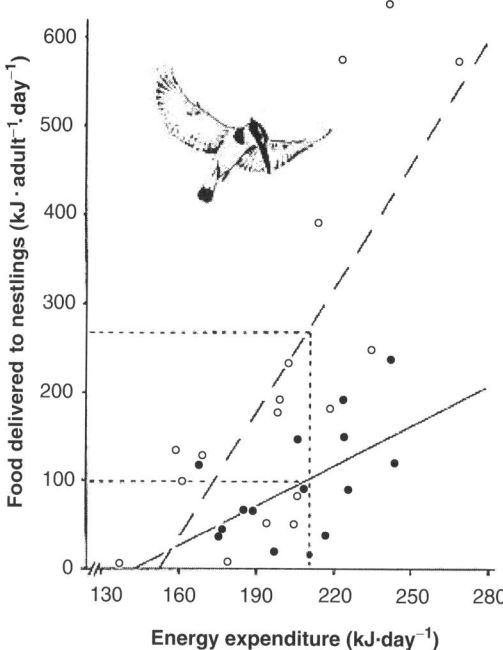

Figure 7.1. Amount of food delivered to pied kingfisher nestlings versus daily energy expenditure of feeding adults. Solid circles: Lake Victoria; open circles: Lake Naivasha. Dotted lines are the upper limit of energy expenditure and resulting feeding capacities for the two sites. Used with permission from Reyer (1990).

at Lake Victoria. Overall, pied kingfishers expending less than 210 kJ day^{-1} maintained or increased their body weight, while those that spent more than 210 kJ day^{-1} lost an average of 3 g day^{-1} (3.8% of mean body mass) (Fig. 7.1). Thus, 210 kJ day^{-1}, about 4 × basal metabolic rate (BMR), appears to represent a physiologically determined energy threshold that can only be exceeded for a short period without a dangerous decline in body condition. Because of the differences in foraging conditions, a parent at Lake Victoria was able to bring a maximum of 102 kJ day^{-1}, while at Lake Naivasha it could deliver as much as 267 kJ day^{-1}, or 2.6 times more.

A pied kingfisher nestling requires about 90 kJ day^{-1} to maintain body mass (Reyer 1990). At Lake Victoria it was not possible for parents to raise a brood of four successfully without the help of non-breeders, while at Lake Naivasha an unassisted breeding pair was able to provide enough energy to raise four young without the assistance of helpers (Fig. 7.2).

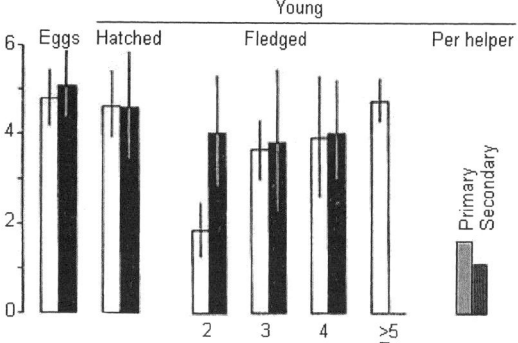

Figure 7.2. Mean clutch size, number of young hatched, and number of young fledged (± SD) for different group sizes of pied kingfishers at Lake Victoria (open blocks) and Lake Naivasha (filled blocks). Also shown are the additional young surviving per primary and per secondary helper at Lake Victoria. Used with permission from Reyer (1990).

Reyer and Westerterp (1985) conducted a series of experiments testing the hypothesis that the different demands of nestlings, together with the varying energetic stress on parents, provides the proximate mechanism for their treating secondary helpers differently at the two study sites. While Lake Naivasha breeders were able to raise two to three young per parent without the assistance of helpers, they accepted secondary helpers only when experimentally challenged to raise four to five young per parent. The reverse experiment at Lake Victoria was equally consistent with the hypothesis that secondary helpers are recruited in order to reduce the energetic stress of reproduction.

Acorn woodpecker
The eggs of acorn woodpeckers are small, even by the standards of a taxon that lays small eggs. They also have one of the shortest incubation periods of any bird, only 11 days (Weathers *et al.* 1990). Despite, or perhaps because of, their extreme altriciality at hatching, nestling acorn woodpeckers grow relatively slowly. Based on Ricklefs' (1968) equation, this species' *k*-value is only 69% of the expected value and is close to that predicted for tropical species (Ricklefs 1976), which are generally regarded as slow-growing.

Weathers *et al.* (1990) attributed this slow growth to diet, which consists of both insects and pieces of acorns. The latter contain relatively little protein (3.9 to 7.1% of dry weight) and significant amounts of tannins (Koenig and Heck 1988). Tannins bind proteins and are known to depress the growth of domestic chickens when present in even smaller quantities (Helsper et al. 1996). Even so, acorn woodpeckers apparently do not maximize the energetic content of their stores by choosing to cache either the largest acorns or those containing the most energy (Koenig and Benedict 2002). Further, based on the fact that several other species of *Melanerpes* woodpeckers display similar ecology and slow nestling growth, and that at least a third of the 21 species within this genus are known to breed cooperatively, Weathers et al. (1990) suggested that low-protein diets, slow growth, and cooperative breeding might be interrelated.

Nest-cavity temperatures during the breeding season are often so low that nestlings less than three weeks old rapidly become hypothermic unless they are brooded continuously. Weathers et al. (1990) also found that the amount of time adults can devote to brooding is constrained by the adults' foraging and other activity requirements. Thus, one important contribution of nest helpers is to aid in brooding nestlings.

The resting metabolic rate (RMR) of acorn woodpeckers is unusual in two respects. First, it apparently has no thermoneutral zone and progressive hypothermia appears to take place at ambient temperatures below 20 °C, reducing RMR through the Q_{10} effect. Second, BMR is almost 40% higher than expected for a bird of its size (Weathers et al. 1990).

Florida scrub-jay

Schoech (1996) found that food-supplemented groups of Florida scrub-jays bred on average 16 days earlier than did non-supplemented controls. Non-breeders of both sexes had lower body mass and lipid content than breeders, but food-supplemented and control non-breeders were equally likely to become breeders. Thus, non-breeders do not appear to forgo breeding as a result of being food-limited nor is it likely that non-breeders' inefficiency as foragers results in their remaining reproductively inactive (see Chapter 8).

Arabian babbler

Anava et al. (2000, 2001a, 2001b, 2001c) have performed an intensive study of the energetic aspects of cooperative breeding among Arabian babblers. Using a logistic growth curve, they determined a growth rate constant (k) for nestlings of 0.45, 18% higher than that predicted for a passerine of its body mass. Asymptotic body mass of fledglings was 46 g, only 63% of adult body mass and low compared to other passerines. Energy intake retained and accumulated in tissue decreased with nestling age and amounted to only 29% of the total metabolizable energy intake over the nestling period. However, energy content per gram of body mass increased with age and averaged 4.48 kJ g^{-1} body mass. Thus, the growth rate of babbler nestlings is relatively fast compared to other passerine species, but fledgling mass is relatively low.

Anava et al. (2001b) measured FMR of 10-day-old nestlings from small (two to three individuals), medium (four and five individuals), and large (six or more individuals) groups. There was an increase in body mass and FMR from small- to medium-sized groups, but a leveling off or decrease in large groups. This suggests that there is an optimum group number for provisioning nestlings, above which there may be a negative effect.

FMR for babblers provisioning nestlings was 2.01 kJ g^{-1} day^{-1} (Anava et al. 2001b) compared to 1.65 kJ g^{-1} day^{-1} for non-breeding babblers not feeding nestlings (Anava 1998). Thus, helping increased FMR by 22%. The FMR for breeding babblers is 1.36 kJ g^{-1} day^{-1} higher than BMR, and that of non-breeding, non-helping babblers is 1.00 kJ g^{-1} day^{-1} higher than BMR. This suggests that 73.5% of the increase in energy expenditure above BMR is used for maintenance and 26.5% for the provisioning of nestlings (Anava et al. 2001b).

Contrary to expectation, Arabian babblers did not show any difference in energy expenditure between seasons. This was due to their consuming relatively energy-rich diets in summer and water-rich diets in winter (Anava et al. 2000). Most of the metabolizable energy was provided by invertebrates in both seasons.

All group members provisioned nestlings at similar rates, and individual visitation rates declined with group size (Anava et al. 2001b). FMR of adult females, but not of other group members, decreased linearly with group size. This energy saving could allow primary females in larger groups to start a new nest more quickly than those in smaller groups. FMR for all babblers was 61–66% of the value predicted for a passerine of its body mass provisioning nestlings, and was 3.11 × BMR, similar to the mean value of 3.13 × BMR reported for a number of terrestrial species (Tatner 1990).

Table 7.1. *Summary of studies providing evidence for a link between physiological factors and the fitness consequences of living or breeding cooperatively*

Species	Method[a]	Proposed mechanism	Fitness effect	Reference
Pied kingfisher	DLW	Helper contributions reduce energy expenditure of breeders	Increased survivorship of nestlings and breeding adults	Reyer and Westerterp 1985
Green woodhoopoe	BMR	Communal roosting reduces individual energy expenditure in relation to group size	Increased winter survival of adults living in large groups	Du Plessis and Williams 1994
Acorn woodpecker	BMR	Communal roosting reduces individual energy expenditure in relation to group size	Increased winter survival of adult males	Du Plessis *et al.* 1994
Meerkat	DLW	Helper contributions reduce energy expenditure of lactating female	Increased production of surviving pups	Scantlebury *et al.* 2002

[a] DLW = doubly labeled water study; BMR = basal metabolic rate measurements.

DRAWING THE LINK BETWEEN ENERGY EXPENDITURE AND FITNESS PARAMETERS

The ultimate test of whether physiological factors have been significant in the evolution of cooperative breeding lies in drawing a link between energy expenditure, degree of sociality, and fitness. This requires not only intensive physiological study, but also long-term population data with which to examine the selective advantages and disadvantages that are associated with various strategies. Not surprisingly, few studies have as yet attempted to make such a link (Table 7.1).

First, Reyer and Westerterp (1985) demonstrated for pied kingfishers that energy expenditure is indeed a currency of fitness, specifically in situations where foraging conditions are challenging. Groups with helpers are able to raise significantly more young than those without helpers as a result of adult group members cumulatively meeting the energetic requirements of young. Further, adult survivorship is enhanced by the presence of helpers, who "lessen the energetic load" of breeders. To this end, Reyer (1984) demonstrated that the contribution of one primary helper improves the survival of breeding males by over 3% and that of breeding females by almost 25%.

Second, in areas where nocturnal temperatures sometimes drop below freezing, green woodhoopoes that roost with conspecifics in cavities can conserve 30% or more of their night-time energy expenditure, even at relatively mild ambient temperatures (Du Plessis and Williams 1994). These authors proposed that the proportionately higher survival of individuals living in large groups is related to the energetic benefits obtained by way of roosting communally in cavities during the winter months. Further, the difference in the seasonal pattern of mortality between individuals living in groups of different sizes was only significant at the climatically harsher of their two study sites, where nocturnal temperatures occasionally dropped below freezing. Du Plessis and Williams (1994) proposed that the potential fitness benefits gained may vary between populations, but that it probably plays an important role in balancing the cost–benefit equation of sociality in this species.

Third, Du Plessis *et al.* (1994) showed that the benefits of communal cavity roosting by acorn woodpeckers in California was correlated to an increase in the survival of adult males over the winter period, largely due to the apparently poor survival of adult males living in groups without helpers. This was not the case for adult females, however. They concluded that although other factors are certainly important, this result broadly supports the hypothesis that the thermal benefits of roosting in groups may contribute to the higher survivorship in adult male acorn woodpeckers.

Finally, Scantlebury et al. (2002) showed that among cooperatively breeding meerkats, helpers reduce the energy expenditure of reproductive females. They estimated that 10 non-breeding helpers were able to reduce the breeding female's workload by the equivalent of reducing the litter by one pup. Further, they demonstrated that lactating breeders living in large groups were able to produce more milk than those with fewer helpers. Such females were also able to come into estrus sooner than their small-group counterparts. Together these two lines of evidence translate into increased reproductive success as a result of the breeding female's release from her full workload (Clutton-Brock et al. 2001c). More recently, based on the same study of meerkats, Russell et al. (2003b; see also Chapter 13) demonstrated that there are several reasons why short-term energetic costs may not lead to substantial long-term fitness costs for helpers in cooperative vertebrates. First, there is evidence that the body condition of helpers influences their tendency to help (Clutton-Brock et al. 2002). Second, helpers can compensate for their increased energetic investment during the non-breeding season (Russell et al. 2003a). Third, when helpers invest heavily in one breeding event, they may reduce their subsequent helper inputs depending on their own condition (Clutton-Brock et al. 2002). It remains to be seen what effect helper contribution (as a result of the reduction in workload) has on the long-term survival of meerkat breeders.

CONCLUSION

Can we make any general predictions about the role that physiological factors may play in the occurrence of cooperative breeding in relation to factors such as body mass or the environments in which the birds live? If energetic factors related to body size were an important driving force in the evolution of cooperative breeding behavior, one might predict that this should be most prevalent among smaller birds. In support of this, Du Plessis et al. (1995) found that there was a disproportionately high incidence of cooperative breeding among South African birds smaller than 100 g; conversely, birds larger than 100 g seldom breed cooperatively. Unfortunately, phylogeny is likely to be confounding this relationship between body size and the incidence of cooperative breeding (Bennett and Owens 2002; Chapter 1).

Similarly, if energetic factors have been a key proximate mechanism underpinning the evolution of cooperative breeding, one might predict that it would occur disproportionately more frequently in colder or more unpredictable environments. It is generally thought that the incidence of cooperative breeding is highest in Australia (22% of 258 species in the family Corvidae: Russell 1989) and South Africa (20% of 217 well-studied species: Du Plessis et al. 1995) compared to a global figure of about 3% of all species (Brown 1987). Unfortunately, little is known about the relative incidence of cooperative breeding in South and Central America, tropical Africa, and Indo-Malaysia. But with the apparently high proportion of cooperative breeding among Australian and southern African birds, the unpredictable effects of the El Niño Southern Oscillation phenomenon might provide a golden thread worth pursuing. Recently, Bennett and Owens (2002) performed an analysis using species representing 139 families and concluded that phylogenetic hotspots of cooperative breeding are not restricted to the continents derived from Gondwanaland, but are also over-represented in one or two Eurasian and North American families, thus throwing open this question even wider.

In an analysis of 16 carnivore species, Creel and Creel (1991) suggested that cooperative breeding was related to energetic costs of gestation and lactation. They showed that litter mass, litter growth rate, and total energetic investment in the offspring were highest in communally breeding carnivores. Similarly, in canids, Moehlman and Hofer (1997) found a correlation between energetic costs of reproduction and the incidence of alloparental behavior. Despite these results, a strong relationship between the energy needed for reproduction and cooperative breeding clearly does not apply to mammals in general nor to other groups, such as raptors, the avian group most ecologically equivalent to carnivores. Thus, physiological constraints would appear to be at best a predisposing factor leading to the evolution of alloparental behavior, even in mammals. Although such factors may be important, they invariably fail to resolve the more difficult challenge of explaining why such cooperative behaviors often fail to be expressed in closely related species that have largely similar physiological capabilities and constraints, matching demographic profiles, and virtually identical environments.

Finally, I return to the two divergent pathways by which physiology may influence a species' tendency to breed cooperatively. First, for species that live in closely knit groups throughout the year, I propose that most, if not all, the fitness benefits of breeding cooperatively are likely to stem from group living per se rather than from alloparental effects (Koenig and Mumme 1990) associated with helping behavior. Thus, the type of physiological factors that, in combination with ecological and demographic factors, might predispose such species to cooperative breeding can take several forms that have little to do with the breeding season itself. Examples include situations in which there are benefits to roosting communally (mousebirds, acorn woodpeckers, and green woodhoopoes), hunting in groups (Galápagos hawks, pale chanting goshawks, and African wild dogs), and reductions in foraging time lost as a result of having to be vigilant (Arabian babblers and meerkats).

Second are cases in which the energetic effects of assistance provided by non-breeding helpers during the breeding season itself are important by reducing the breeders' workload leading to increased survival (pied kingfishers), or by allowing females to invest more in breeding than they would be able to otherwise.

This puts a somewhat counterintuitive spin on our understanding of cooperative breeding. Specifically, the alloparental contributions of non-breeding helpers may often have relatively little effect on the evolution of cooperative breeding, at least in species where helping is regular (Du Plessis et al. 1995). On the other hand, opportunistic cooperative breeders probably derive fitness benefits exclusively during the reproductive period. Ultimately, the distinction between regular and opportunistic cooperative breeding is not absolute, and there will therefore be species that have physiological and other adaptations that confer a fitness advantage to individuals in both breeding and non-breeding season. Further, many of the physiological peculiarities that have thus far been described in cooperatively breeding species are likely to be secondarily derived, and thus have played little if any role in the origin of group living or helping behavior.

In summary, we remain a long way from being able to perform broad comparative analyses that will allow us to unravel the role that physiological constraints have played in the evolution of cooperative breeding. Such constraints are, however, likely to have played a more important role than has generally been acknowledged.

ACKNOWLEDGMENTS

Walt Koenig and Janis Dickinson have demonstrated exceptional understanding and good humor with the incremental completion of this manuscript. Joe Williams deserves credit as the primary catalyst of my interest in physiological ecology. Hilary Buchanan and Danelle du Toit assisted in bringing the bibliography up to speed. Walt Koenig and David Ligon provided constructive suggestions for improving an earlier version of the manuscript. Finally, Andrew Russell kindly allowed me access to unpublished material.

8 · Endocrinology

STEPHAN J. SCHOECH, S. JAMES REYNOLDS[1], RAOUL K. BOUGHTON

University of Memphis, [1]*Present address: University of Birmingham*

In the nearly 80 years since Skutch (1935) coined the term "helper-at-the-nest," cooperative breeding has attracted considerable interest, to no small extent because helping to raise non-descendant young violates a primary tenet of Darwinian theory. This "paradox" of how cooperative breeding could have evolved and subsequently have been maintained was partially resolved first by Hamilton (1963), who introduced the concept of kin-selected benefits by individuals that assist in rearing related individuals other than their own offspring, and later by Brown (1978), Koenig and Pitelka (1981), and Emlen (1982a), who developed the hypothesis that cooperatively breeding species were constrained by specific habitat requirements that induced philopatry, thus setting the stage for helping behavior.

Here we focus on the contributions of field endocrinology to our proximate-level understanding of cooperative breeding. Given that hormones are involved in mediating virtually all aspects of an organism's life and affect functions as diverse as gut absorption, blood production, and reproductive and agonistic behaviors, we can expect that they will also play an important role in the various kinds of cooperative and competitive interactions characteristic of cooperative breeders.

BACKGROUND

Reproductive hormones

Two endocrine axes are of primary interest here: the hypothalamo–pituitary–gonadal (HPG) axis and the hypothalamo–pituitary–adrenal (HPA) axis. The HPG axis consists of a region of the forebrain known as the hypothalamus, the pituitary that lies immediately below, and the gonads (Fig. 8.1). In response to stimulatory environmental or endogenous cues, the hypothalamus secretes gonadotropin-releasing hormone (GnRH). GnRH travels to the anterior pituitary via a blood portal system where it stimulates the release of luteinizing and follicle-stimulating hormones (LH and FSH, respectively). These blood-borne gonadotrophic hormones induce the seasonal recrudescence of the gonads and accessory structures of the reproductive tract in a seasonal breeder or gonadal maturation during puberty in a non-seasonal breeder. In the mature gonad, LH and FSH orchestrate gonadal function.

In addition to producing gametes, mature testes and ovaries also produce and secrete sex steroid hormones. Although the two best-known sex steroid hormones, testosterone (T) and 17β-estradiol (E_2), are primarily thought of as male and female hormones, respectively,

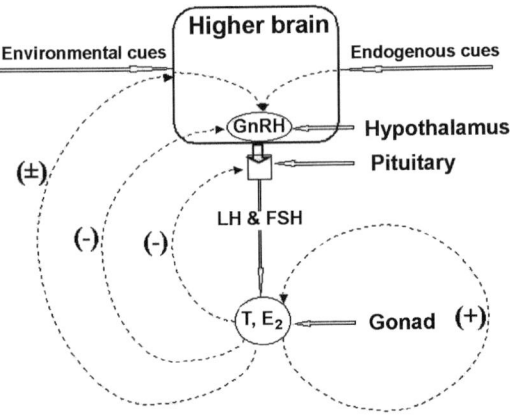

Figure 8.1. A schematic of the hypothalamo–pituitary–gonadal (HPG) axis. Both environmental and endogenous cues can influence the axis, both positively and negatively. Gonadal steroids, in addition to inhibitory upstream effects via negative feedback, also exert positive effects at the gonad and other tissues, and can influence behaviors at the level of the brain.

Ecology and Evolution of Cooperative Breeding in Birds, ed. W. D. Koenig and J. L. Dickinson. Published by Cambridge University Press.
© Cambridge University Press 2004.

this is an oversimplification since they are found in both males and females. Sex steroid hormones have local effects and are essential for the maturation of ova and sperm. They also travel via the bloodstream to exert effects upon multiple target tissues, often in synergy with other hormones. Some of these effects are clearly visible as those secondary sexual characteristics that readily allow us to assign an individual to a given gender. For example, facial hair and breasts in humans and courtship plumage, spurs, and wattles of male birds are, to various degrees, under the control of sex steroid hormones. Additionally, specific receptors in several areas of the brain bind sex steroids to induce behaviors that are associated with reproduction, such as territoriality, courtship, nest building, and care of eggs or young (Ball 1991, 1993; Buntin 1996).

Stress hormones

The hypothalamo–pituitary–adrenal (HPA) axis is also of import to studies of cooperative breeding. Similar to the HPG axis, an endocrine cascade is initiated in response to endogenous or environmental cues. The hypothalamus responds to neural inputs from the "higher brain" by secreting corticotropin-releasing hormone (CRH) that acts upon the pituitary to induce secretion of adrenocorticotropin hormone (ACTH) into the peripheral bloodstream. The primary target of ACTH is the adrenal cortex that, in turn, responds by producing and releasing glucocorticoids. The avian glucocorticoid is corticosterone (CORT). Glucocorticoids are secreted in response to stressful stimuli and are an essential part of an individual's response to stressors encountered in its environment (Wingfield 1988; Moore et al. 1991). A primary action of CORT is to increase blood glucose levels to fuel an individual's response to the stressor, be it "fight" or "flight." While CORT is essential in facilitating responses to environmental challenges, it induces breakdown of muscle tissue and thus chronically elevated levels can severely compromise an animal and, if unchecked, may result in death (Selye 1978; Sapolsky 1992). In addition to these catabolic effects that could cause an animal to shift from a reproductive to a survival mode, glucocorticoids have been shown to directly inhibit reproductive behavior in numerous taxa (Wilson and Follett 1975; Moore and Zoeller 1985). Elevated CORT levels in birds can cause low levels of sex steroid and luteinizing hormones, and can result in incomplete gonadal development or compromised gonadal function.

Within the context of cooperative breeding, CORT levels have been examined with respect to their role in keeping non-breeding helpers reproductively inactive. Several researchers have proposed that helpers might be "psychologically castrated" (Rowley 1965; Carrick 1972; Brown 1978), presumably as a result of their relatively low social status within their social group. Later researchers proposed that one mechanism whereby this scenario could take effect was via CORT, that could be elevated in response to dominant–subordinate interactions (Reyer et al. 1986; Schoech et al. 1991, 1997; Wingfield et al. 1991).

Parental hormones

Another hormone that has drawn interest in the context of cooperative breeding is prolactin (PRL), a protein hormone produced in and secreted by the anterior pituitary. In some avian species, including at least two cooperative breeders (Florida scrub-jay and Mexican jay), the secretion of PRL is induced by vasoactive intestinal polypeptide (VIP), a hypothalamic neuropeptide (Macnamee et al. 1986; Maney et al. 1999; Vleck and Patrick 1999). However, having been examined in only a handful of species, whether VIP is a universal avian prolactin-releasing factor remains to be determined. Regardless of the upstream inputs that result in the release of PRL into the peripheral blood system, its involvement in mediating parental behaviors, including nest building, incubating, and feeding of young, is well documented in many avian taxa (Buntin 1996). Although PRL has broad-ranging effects in animals, it is its association with parental care that makes it a viable candidate to mediate helping behaviors that are expressed in cooperative breeders.

Hormones and behavior: fundamental questions

Are non-breeding helpers physiologically capable of breeding? By assessing circulating levels of reproductive hormones in breeders and non-breeding helpers, one can gain insight into the status of the reproductive axes of helpers as compared with breeders. If non-breeding helper males are found to have low levels of T compared to breeder males, one might infer, first, that their gonads are not fully developed and they are probably reproductively compromised and, second, that areas of the brain that bind T will remain inactive due to the relatively low T signal and this might result in the failure of a bird to express sexual behaviors. In addition, given the clear role of T in mediating aggression and the relatively low levels of aggression that characterize many cooperative groups, such a finding might help play a role in the maintenance of group harmony.

Does the HPA axis play a role in mediating cooperative breeding? Given that corticosterone can negatively affect the reproductive axis at multiple levels (see above) and that many cooperative groups are characterized by dominance hierarchies, one might hypothesize that subordinate helpers are stressed as a result of their status. Examination of CORT levels might, therefore, reveal a mechanism whereby reproductive quiescence is enforced upon non-breeders.

Are helping behaviors mediated by the parental hormone prolactin? Given the role of PRL in multiple aspects of parental behavior, it is not a great leap to hypothesize a function for PRL in mediating alloparental, or helping, behaviors.

Here we first briefly introduce the discipline of endocrinology and how it can give insight into mechanisms underlying cooperative behaviors. Second, we utilize case studies to present a review of endocrine research on avian cooperative breeders. Our coverage is selective, and only studies that consider fundamental questions about possible endocrine mediation of cooperative behaviors are included. Finally, we conclude with a section that synthesizes our knowledge by examining both commonalities and paradoxes from the studies considered. We point out the difficulties of drawing definitive conclusions about the role of hormones in cooperative breeding, mostly because only a small number of studies, few of which have been experimental, have been conducted on cooperative breeders. Additionally, studies to date have considered only circulating hormone levels and there are other factors that can regulate hormone function independent of plasma levels. These include plasma-binding proteins that can affect the delivery of lipophilic steroid hormones to specific target tissues and can also increase their longevity in circulation (Breuner and Orchinik 2002), intracellular enzymes that either convert hormone precursors to the active form of the hormone or, alternatively, deactivate a hormone by converting it to an inactive form (Schlinger et al. 1999; Soma et al. 2000), and hormone receptors (or variation in numbers) at target tissues that are essential for an endocrine signal to be received and subsequently to induce a behavioral or physiological response (Balthazart 1983).

CASE STUDIES

Pied kingfisher

The first study of a cooperatively breeding species that incorporated endocrine techniques was conducted by Reyer and his colleagues on pied kingfishers in Africa. Pied kingfishers have two types of helpers, "primary" and "secondary." Primary helpers are the sons of at least one member of the breeding pair that they assist. They generally return to the breeding sites with the breeders and help them defend the nest site against conspecifics during the pre-laying stage of nesting. After the clutch is complete, primary helpers continue to defend the nest site against predators and conspecifics and later help feed the young. In contrast, secondary helpers, which are also males, assist breeders to which they are unrelated, and are accepted as helpers apparently only when ecological conditions make it difficult for the breeding pair to provision their young adequately, and only after the breeding female has completed her clutch.

Because secondary helpers are usually excluded from the nest site by breeder males, Reyer et al. (1986) hypothesized that they might pose a significant threat to the breeder male's assurance of paternity and might have different reproductive capabilities than primary helpers. They examined this by comparing LH and T titers and found no differences in plasma LH levels between breeders and either category of helper (Fig. 8.2). In contrast, plasma T levels in breeders and secondary helpers were similar and roughly twice those of primary helpers.

The fact that secondary helpers have T levels equivalent to those of breeders whereas those of primary helpers are markedly lower suggests that primary helpers may be reproductively compromised. Although sample sizes were small, this conclusion was supported by laparotomy and sperm-milking data that found primary helpers had testes less than half the size of those of breeders and that ejaculates of such helpers contained no active sperm.

What causes the differences between the two types of helpers? Reyer et al. (1986) concluded that low T levels in primary helpers were not attributable to either differential handling time (i.e., CORT effects upon T)

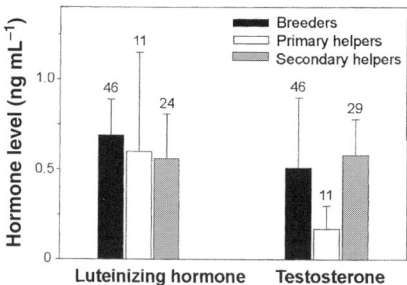

Figure 8.2. Plasma levels of LH and T in breeder and helper male pied kingfishers. Data are presented as means and standard errors. Samples sizes are shown above the error bars. Adapted from Reyer et al. (1986); used with permission from Blackwell Verlag.

or age. Instead, foreshadowing the "challenge hypothesis" of Wingfield et al. (1990), they suggested that the challenges secondary helpers experience as they aggressively interact with male breeders for acceptance into the group and as competitors for females lead to elevated T levels, in marked contrast to primary helpers that reside with the breeders in a stable group prior to and throughout the breeding season. Reyer et al. (1986) also postulate that maintaining low T levels might be selectively advantageous, given that primary helpers gain a significant amount of inclusive fitness by assisting to rear related young and that elevated T usually reduces paternal care (Schoech et al. 1998). Additional reasons underlying low T levels in primary helpers that one might invoke are incest avoidance (see chapter 9) and a social situation devoid of the intra-pair stimulation required for breeding (Schoech et al. 1996a).

Florida scrub-jay

Florida scrub-jays are non-migratory and live in a year-round territory that all group members help to defend against conspecifics. Groups consist of a breeding pair and, in the extreme, up to eight non-breeding helpers, although mean group size is three and approximately half of territorial groups consist of only a breeding pair (Woolfenden and Fitzpatrick 1984). Most helpers are the offspring of the breeding pair of a group. When this is not the case, it is invariably because one of the breeders has died and the surviving member of the pair has re-paired, or the helper's parents have divorced, or the unrelated helper has moved into the territory from

elsewhere (Woolfenden and Fitzpatrick 1984, 1990, 1996). Florida scrub-jays are socially and genetically monogamous (Quinn et al. 1999) and pairs generally remain together until one member dies.

Plasma T levels in breeders and helpers have been compared in two studies conducted in four different years (Schoech et al. 1991; 1996a). T levels in non-breeding helpers during 1989 were uniformly low enough to suggest that these individuals were reproductively incapable. This finding prompted a follow-up study to assess at what level the reproductive axes of non-breeding helpers were down-regulated. Schoech et al. (1996a) reasoned that by measuring baseline levels of LH and the capability of the anterior pituitary to respond to a GnRH challenge, a better understanding of the function of various components of the HPG axis would be gained. For example, if baseline LH levels were comparable to those of breeders while T levels remained low, this would suggest that, while both the hypothalamus and pituitary of helpers were fully functional, the testis was incapable of responding to the upstream signals. Similarly, LH response to injections of GnRH would allow evaluation of pituitary function (Fig. 8.1).

No differences in baseline LH levels between breeders and non-breeding helpers were found. The responses to GnRH injections of breeders and non-breeding helpers revealed a rapid increase in LH levels when compared with saline-injected individuals, and no statistical differences in the LH responses due to social status. This experiment demonstrated that the hypothalamus and pituitary, the two upstream components of the HPG axis, are fully functional in non-breeding helpers.

When T levels were examined further by comparing breeders and helpers, T levels of helpers showed significant changes both during the course of a breeding season and between years (Fig. 8.3). Despite helper males having statistically lower T levels than breeder males, the absolute levels and the seasonal patterns of changes in T levels of helpers suggested that their testes are functional. Thus, helpers appear to have functional HPG axes, a conclusion supported by the finding that testes volumes of helper males during the breeding season, although half the size of breeder male testes, are an order of magnitude greater than fully regressed testes (Schoech et al. 1996a). Field observations confirm that non-breeding helpers are reproductively capable, both in cases where one-year-old males pair with females and

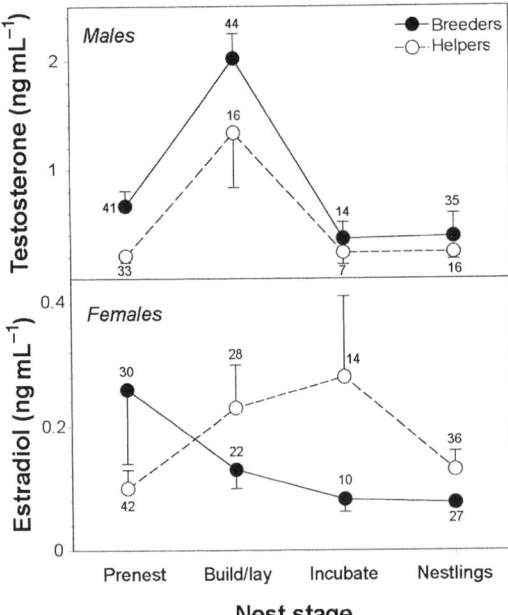

Figure 8.3. Plasma levels of T in male (upper panel) and E_2 in female (lower panel) breeder and helper Florida scrub-jays. Data are presented as means ± SE; sample sizes are shown above error bars or next to data points. Adapted from Schoech *et al.* (1996a); used with permission from Springer-Verlag.

produce offspring (Woolfenden and Fitzpatrick 1984; Webber and Cox 1987; Schoech *et al.* 1996a) and in cases where a male non-breeding helper occupies a breeding vacancy and within days his mate produces eggs.

Similar to males, female helpers in 1989 exhibited consistently low E_2 levels compared to breeders (Schoech *et al.* 1991). Following the rationale set forth above for males, LH levels, basal and in response to a GnRH challenge, were subsequently examined to compare the HPG axes of breeders and helpers. Basal LH levels of helper females were statistically equivalent to those of breeder females and, as was true for males, levels also differed between years (Schoech *et al.* 1996a). Additionally, helpers and breeders responded equivalently to a GnRH challenge. Thus, like their male counterparts, female helpers have functional hypothalami and pituitaries.

Interestingly, when E_2 titers were compared over the additional years of study, no statistical differences between E_2 levels of breeders and helpers were found, although there was a significant interaction between status and nest stage whose cause remains unknown (Fig. 8.3). Given that circulating plasma E_2 in females of either status is almost certainly of ovarian origin, the similarity in absolute E_2 levels between helpers and breeders makes it highly likely that female non-breeding helpers have fully functional HPG axes and are capable of reproducing when granted the opportunity.

Determination of ovarian follicle development by laparotomy found that follicle diameters in non-breeding helpers during the breeding season were significantly less than those of female breeders but three times larger than those of fully regressed females. Thus, despite being reproductively inactive, helpers undergo significant ovarian seasonal growth, and non-breeding female helper Florida scrub-jays are clearly reproductively capable.

Is CORT responsible for the reproductive quiescence of helpers? After controlling for handling time, CORT residuals did not differ between male breeders and helpers and there were no changes across the different stages of the breeding season (Schoech *et al.* 1991). CORT residuals of female breeders were marginally higher than those of female helpers, a finding in the opposite direction to that expected if CORT is involved in reproductive quiescence.

The follow-up study from 1992 to 1994 measured basal CORT levels by collecting all samples within one minute of capture. Findings were generally similar, except that no differences in CORT levels of breeders or helpers of either sex were detected (Schoech *et al.* 1997). Examination of the rate and pattern of change of plasma CORT levels with time were also made, and despite a profound corticosterone response to capture and handling, there were no differences attributable to sex or status. Thus, CORT does not appear to play a role in reproductive inactivity of Florida scrub-jays. These data, in conjunction with the sex steroid hormone findings, do not support the hypothesis that non-breeding helpers are reproductively suppressed. Instead, these birds are apparently capable of breeding, but generally lack the ecological opportunity to do so.

What role, if any, does prolactin play in mediating helping behavior? Irrespective of status or sex, PRL levels are low early in the season and increase to maximal levels during the incubation and nestling care stages of the reproductive cycle (Schoech *et al.* 1996b). Breeders also have higher PRL levels than non-breeding helpers and females have higher plasma PRL titers than

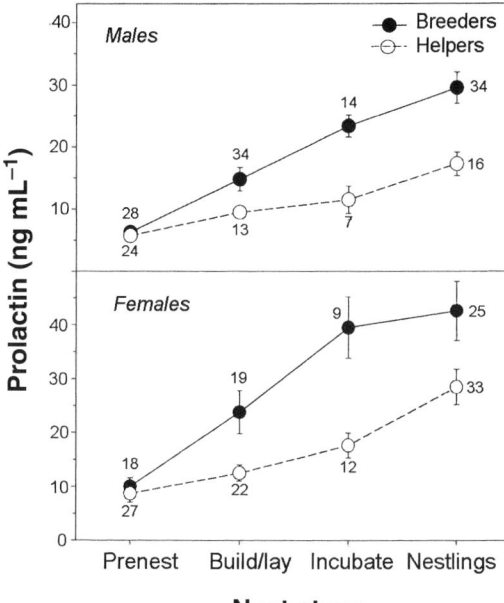

Figure 8.4. Plasma levels of PRL in male (upper panel) and female (lower panel) breeder and helper Florida scrub-jays. Data are presented as means and standard errors. Sample sizes are shown above error bars or next to data points. Adapted from Schoech et al. (1996b); used with permission from Elsevier Ltd.

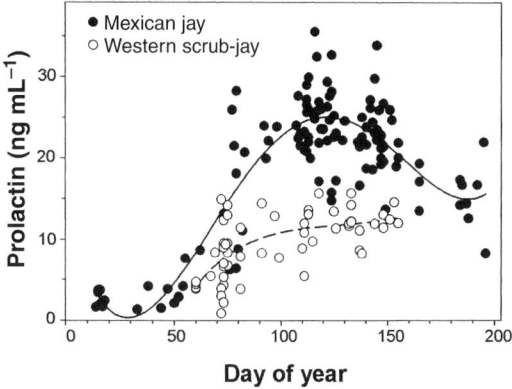

Figure 8.5. Changes in plasma PRL through the breeding season in Mexican jays and western scrub-jays. Data presented are from all birds regardless of sex, age, or breeding status. Trend lines are 4th-order regressions for each species. From Brown and Vleck 1998; used with permission from Oxford University Press.

same-status males (Fig. 8.4). Both of these differences may be attributable to either the role of PRL in brood-patch formation (Jones 1969, 1971) or the greater tactile and visual stimulation of incubating breeder females compared to males that do not incubate.

The finding that non-breeding helpers have elevated plasma PRL levels when nestlings are being provisioned suggests that PRL plays a role in mediation of alloparental care. This hypothesis is further supported by the existence of a significant correlation between feeding rates and plasma PRL levels, although this finding did not hold when the analysis was restricted to breeders (Schoech et al. 1996b).

Mexican jay

This congener of the Florida scrub-jay is found in oak-dominated montane environments of the southwestern United States and northern and central Mexico. In contrast to singular breeding Florida scrub-jays, the social system of Mexican jay is more complex, with plural breeding of up to five females within groups of up to 23 individuals (Brown 1994; Brown and Brown 1990). Furthermore, Mexican jays regularly engage in extra-pair copulations (Bowen et al. 1995; Li and Brown 2000), whereas Florida scrub-jays do not (Quinn et al. 1999).

Recent work on this species has focused on the role of PRL in mediating helping behavior. Brown and Vleck (1998) reasoned that if elevated PRL levels have evolved to enhance the expression of parental and alloparental behaviors, then there should be higher overall PRL levels in a species with helpers than in a closely related species without helpers. They also predicted that if helping is a selected trait in the Mexican jay, then plasma PRL levels of non-breeding helpers should increase independently of the stimulus of begging young.

On average, PRL levels in Mexican jays are more than double those of the western scrub-jay, a monogamous non-cooperatively breeding congener, thus supporting their first prediction (Fig. 8.5). The second prediction was also supported: PRL increases in non-breeder and breeder Mexican jays were similar in that levels reached their zenith during the incubation period, well before stimuli from nestlings could result in elevations in PRL secretion by helpers. Although the stimulus of the nest and eggs might explain elevated PRL titers in breeding females, this is not likely to be true for non-breeders who do not attend nests until young hatch. Similar results have been reported for Florida scrub-jays by Schoech et al. (1996b).

A detailed comparison within the genus *Aphelocoma* revealed that T levels in western scrub-jays were maximal for a brief period in March and, although absolute T levels in Mexican jays were comparable, their T levels were elevated over a longer period (Vleck and Brown 1999). These authors hypothesized that the differences in T profiles stem from fundamental differences in the social systems of the species, with western scrub-jays having a brief period of elevated T levels coinciding with territory establishment and courtship. In contrast, the period of elevated T in Mexican jays is more prolonged, similar to polygynous species in which males interact with multiple females (Wingfield *et al.* 1990; Beletsky *et al.* 1995).

When grouped by age or breeding status, within-species comparisons of T levels revealed links between T and social behaviors that may relate to the breeding system of Mexican jays. When the effect of nest stage was examined, T levels were highest in males, irrespective of status, prior to the time when nestlings or fledglings were present. The marked decline in T levels when young are present is consistent with the finding that elevated T levels are incompatible with parental or alloparental behaviors. Vleck and Brown (1999) conclude that regardless of whether an individual male in a group is the nest owner, it is to his benefit to provide nestling care given the possibility that he may have either fathered one or more of the nestlings or be closely related to the nestlings (Li and Brown 2000). Non-nesting jays in a group provide one-half of a brood's food, and by maximizing alloparental care males may increase their direct or indirect fitness gains. Under this scenario it may be in a male's best interests to have reduced T during the time when nestlings are present.

White-browed sparrow-weaver

This inhabitant of semi-arid areas of tropical Africa lives in groups of two to 11 birds that hold year-round territories. In each group of birds, a breeding female is paired with a breeding male and they are helped by related and unrelated ("invader") conspecifics. Related birds provision the young and defend the territory while unrelated birds simply defend the territory. Unrelated helpers may benefit by acquiring breeding territory or by enlisting the help of other group members during future breeding attempts.

Breeding males have significantly larger testes and greater body masses than related and unrelated helper males. However, fat scores and CORT titers do not differ between males of different status. Breeding males have the highest levels of T and LH, and their T levels are maximal midway through the second of two breeding periods (Wingfield *et al.* 1991).

Similarly, female breeders have larger ovarian follicles and heavier body masses than helpers. Related helpers and breeding females have similar levels of LH, and these are higher than in unrelated helpers. Luteinizing hormone is generally high during breeding and barely detectable during non-breeding in all birds, irrespective of status. Interestingly, T is similar in helper and breeder females while E_2 is undetectable in all females except breeders immediately before egg-laying.

To determine whether white-browed sparrow-weavers have fully functional HPG axes, Wingfield *et al.* (1991) challenged males and females with a cGnRH (chicken GnRH) injection. Unfortunately, because these birds were from outside the study population, social status was unknown. In males, this resulted in a dramatic increase in LH within 2 minutes followed by a slight increase in T after 10 minutes. Females responded more noticeably than males with an even larger increase in LH. In both males and females, LH and T returned to basal levels within 30 minutes.

Wingfield *et al.* (1991) compared CORT levels of breeders and helpers, and found that titers were similar in all birds irrespective of sex or social status. They concluded that the low levels of reproductive hormones in helpers do not result from their being "psychologically castrated" through the actions of CORT.

In contrast to most other species, levels of T appear to be unrelated to aggression. DHT, another androgen that might mediate aggression, was undetectable in all samples (Wingfield *et al.* 1991). Companion studies examined how aggression was mediated in this species. In one experiment, group social structure was destabilized following temporary removal of the breeding male (Wingfield *et al.* 1992). Intense aggression, at times lasting for days, resulted as a replacement male tried to establish himself as the breeder. Instead of the expected elevation of both T and CORT after removal of breeding males, only LH increased in replacement males and it declined to basal levels once dominance hierarchies were established and aggression subsided.

In a second experiment, Wingfield and Lewis (1993) simulated territorial intrusions by presenting four or five birds in a cage while simultaneously playing a recorded group chorus. Territory holders showed a high degree of aggression upon simulated intrusion as they head-bobbed and carried grass around in flight. The authors once again found that LH, but not T, appeared to regulate aggressive behavior in this species. Breeding pairs were the only birds to approach the pseudo-intruders, while non-breeding birds of the groups provided them with vocal support. Titers of LH were equivalent in birds of different social status within control groups but breeding females showed the highest levels in groups challenged by intruders.

In summary, T does not appear to play a role in the mediation of aggressive behavior during territorial disputes in white-browed sparrow-weavers. Instead, LH appears to mediate such behavior. Whether this divergence from patterns found in most avian species can be attributed to white-browed sparrow-weavers being a cooperative breeder or a tropical dweller remains unclear. Since group members are territorial year-round, territorial behavior may be independent of regulation by T (Wingfield and Soma 2002). Additionally, traditional views of "psychological castration," mediated through stress-induced suppression of helpers by breeding birds, do not hold for this species, as CORT levels were similar across all ages and status categories. Rather, data from this species lend support to the suggestion of Reyer *et al.* (1986) that helpers of some species may choose their social status, thereby gaining benefits of group living such as access to resources or increased protection from predation, rather than being directly suppressed by more dominant conspecifics.

Australian magpie

A large (300 g), black and white crow-like species, Australian magpies inhabit open savannah woodland throughout most of Australia. Males within groups can be polygynous and a dominance hierarchy exists for both sexes. Dominant females harass subordinates trying to nest, resulting in the failure of most attempts (Carrick 1963). In contrast to females, which are behaviorally prevented from breeding, non-breeder males are thought to be physiologically compromised. For example, subordinate males, yearling males, and males on poor quality territories or in nomadic flocks have markedly smaller testes than breeders on good territories (Carrick 1972).

In adult (>3.5 yrs) males, the seasonal patterns of LH and T are similar to that seen in most seasonally breeding species, peaking just prior to egg-laying, and absolute levels are equivalent in breeders and non-breeders. Schmidt *et al.* (1991) postulated that non-breeding adults are capable of breeding but are prevented from doing so by dominants. Unfortunately, a number of methodological problems render this interpretation questionable.

Nonetheless, it is apparent that adults and subadults can be either breeders or non-breeders, although first-year birds never breed; adult non-breeders have levels of LH and T that are similar to those of adult breeders; and subadult breeders have higher LH and T than similar-aged non-breeders. With the exception of first-year birds, age per se does not explain the observed differences in reproductive status, although it may be a contributing factor.

Differences in the endocrine titers between magpies of different ages and social status are most likely attributable to dominance hierarchies that exclude many from reproductive opportunities. However, because subadult males that experience stimulatory interactions with females are physiologically comparable to adult male breeders, they are clearly physiologically and behaviorally capable of breeding.

Harris's hawk

Harris's hawk is one of only two species of birds of prey in which cooperative breeding is common. In Arizona, where the endocrine research on this species has been performed, breeding group size ranges from two to seven individuals. The breeding pair of birds (the alpha birds) build the nest, incubate the eggs, and feed and brood the nestlings. The alpha female performs the majority of the incubation and rarely leaves the nest, where the alpha male and other group members (the helpers) feed her. Helpers can be either males or females, and are divided into beta helpers and gamma helpers. The former are always in adult plumage and are unrelated to the alpha pair while the latter are usually juvenal-plumaged offspring of the alpha pair. Alpha males often hunt with helpers for food and, if successful, helpers transfer food to one of the alpha pair for delivery to nestlings. All members of a group defend the nest against predators and against encroachment by conspecifics (Bednarz 1995).

To determine whether helpers were physiologically capable of breeding, Mays et al. (1991) collected blood samples to compare LH, T, E_2, and CORT profiles of alpha birds with same-sexed beta and gamma helpers. They found that both breeder and helper adult-plumaged males had elevated T and LH levels during the nest building period compared to levels during other stages, whereas T and LH levels in juvenal-plumaged males were consistently lower. This suggests that most adult-plumaged male helpers are physiologically able to breed, a result consistent with the observation that beta males regularly attempt to copulate with breeder females (Dawson and Mannan 1991). That females are usually uncooperative in such attempts presumably assures paternity for the alpha male. Mays et al. (1991) suggested that breeders manipulate beta males by allowing them access to the nest and that this may contribute to their achieving full breeding condition. Beta males occasionally incubate, feed young more, and spend more time harassing predators and in cooperative hunting (Dawson and Mannan 1991). Thus, the benefits to the alpha pair of showing more tolerance to beta than to gamma males may be considerable.

In contrast, both adult- and juvenal-plumaged female helpers had lower titers of T, LH, and E_2 compared to breeder females (Mays et al. 1991). The authors reasoned that adult female helpers pose more of a threat to group integrity than adult male helpers, since if a female helper successfully breeds then the extended group will have more chicks that require provisioning. Consequently, female helpers that could breed may be behaviorally suppressed by the dominant alpha female. Alternatively, since juvenal-plumaged helpers are generally closely related to the breeders, subordinate females may refrain from breeding due to incest avoidance (Chapter 9).

In a companion study, Vleck et al. (1991) examined correlations between progesterone, PRL levels, and behavior. They found no relationship between progesterone and either incubation or care of nestlings, although female breeders had elevated progesterone levels during the nest-building stage. Prolactin levels increased significantly in both males and females during incubation, but decreased soon after hatching.

Perhaps the most intriguing finding of their study was that PRL levels of adult-plumaged male helpers peaked during the nestling stage, when they provide far more food for the chicks than do breeders of either sex. Additionally, during this time PRL levels of adult-plumaged male helpers are higher than those of male breeders and comparable to those of female breeders who attend the nest continually when nestlings are present. Vleck et al. (1991) suggest that PRL facilitates nurturing behaviors of beta male helpers directed towards chicks to which they are not related, and thus may play a critical role in the maintenance of cooperative breeding in this species. Unfortunately, without parentage and provisioning data, these conclusions remain speculative.

Bell miner

This Australian species resides in loose colonies of 20 to 200 birds within which individual breeding females establish exclusive territories. Breeding and non-breeding males help at one or more nests and there may be multiple female breeders within a colony. Based upon behavioral observations, the breeding system is believed to be monogamy (Clarke 1988).

Poiani and Fletcher (1994) measured total androgen levels (T + DHT) and the degree of gonadal development and found that breeders had higher androgen levels and larger testes than non-breeding helpers. They also noted a marked increase in androgen levels with increased age in males prior to reaching sexual maturity at approximately nine months, at which point they are capable of breeding. When sexually mature non-breeders (SMNB) were subdivided into younger (9–18 months) versus older (>34 months) birds, there were no differences in androgen levels, although very limited data on testicular function suggest that older SMNBs may have fully functional testes while younger SMNBs do not. Poiani and Fletcher (1994) interpret their findings as evidence that helpers are "reproductively suppressed," although whether the delayed maturation is imposed or voluntary is debatable. Imposed delayed maturation (IDM) is expected to result from low-ranking helpers being dominated by breeders. In contrast, voluntary delayed maturation (VDM) would occur as a result of a decision by non-breeders to avoid potential costs associated with elevated T. Poiani and Fletcher (1994) postulated that the lower testicular development in younger SMNBs is due to the lower cost of remaining reproductively quiescent among younger birds, but younger SMNB males are generally more closely related to the breeding female than older SMNB males, and thus incest avoidance may be playing a role in causing the observed differences.

Poiani and Fletcher (1994) suggest that IDM could occur when dominant birds aggressively punish young birds in order to prevent plural breeding within the breeders' nesting area and to minimize paternity loss due to helpers engaging in extra-pair copulations. In contrast, VDM may be responsible for older SMNB male helpers that have low androgen levels coupled with spermatogenic gonads. This strategy might enable these older helpers to gain copulations without suffering the consequences likely to result from an open challenge of the male breeder. It may be that this dichotomy is an oversimplification and that examination of CORT titers would enable resolution as to how much delayed maturation is voluntary or imposed (Mays *et al.* 1991; Schoech *et al.* 1991; Wingfield *et al.* 1991).

Red-cockaded woodpecker

To determine whether non-breeding helpers were reproductively competent, in this genetically monogamous species, Khan *et al.* (2001) compared T levels of male breeders and helpers, the vast majority of which are males. They found no differences in plasma T levels between male breeders and helpers and the parallel profiles from these two groups peaked during the copulation period when levels were as much as three times higher than during the prebreeding and nestling care stages (Fig. 8.6). The equivalent T levels suggest that helpers are reproductively competent and comparable to breeders but are excluded behaviorally from access to females during their fertile periods. This is consistent with observations of helpers occupying a vacant breeding position and breeding immediately (Walters *et al.* 1988) and of one-year-old birds successfully reproducing (Walters 1990).

Further insight into the dynamics of behavioral suppression can be gleaned from the finding that male helpers residing in a group in which they are not related to the breeding female had higher T levels than helpers sharing a territory with their mother (Khan *et al.* 2001). The authors point out that T levels in the unrelated helpers may be elevated in part due to increased agonistic interaction with the breeding male who is presumably protecting his reproductive interests. Although

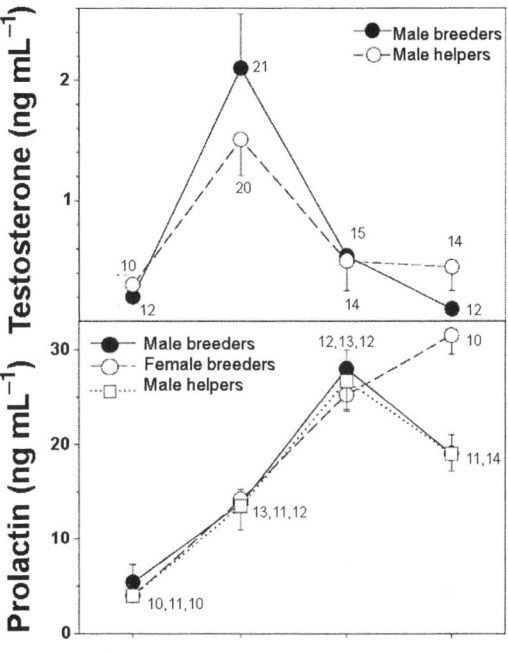

Figure 8.6. Changes in plasma T in male (upper panel) and PRL in male and female (lower panel) red-cockaded woodpeckers during the breeding season. Data are presented as means and standard errors. Sample sizes are shown next to data points. When data points are too close together to distinguish from one another, sample sizes are listed in the order: female breeder, male breeder, and helper. Adapted from Khan *et al.* (2001); used with permission from Elsevier Ltd.

this hypothesis is supported by behavioral observations (Walters *et al.* 1988; Daniels and Walters 2000b), these findings could also be explained either by a lack of stimulation between the helper and the female (Poiani and Fletcher 1994; Schoech *et al.* 1996a) or, once again, by incest avoidance.

Khan *et al.* (2001) also measured plasma levels of PRL to assess its relationships to parental and alloparental behaviors. Among both breeders and helpers of both sexes, PRL levels were low during the pre-breeding stage, increased through courtship and copulation, and reached levels approximately five times higher than prebreeding levels during laying and incubation. Interestingly, whereas PRL levels of male helpers and breeders decreased from this point and were lower while provisioning nestlings than during laying and incubation, those of female breeders were at their highest during the nestling provisioning period.

Since males in this species incubate, the authors speculate that the role of PRL in brood-patch formation and maintenance may explain the patterns observed through incubation. Why PRL levels subsequently continue to elevate in breeding females but not males remains unknown.

The authors also explored whether provisioning rates were correlated with plasma levels of PRL, as found in Florida scrub-jays. Although male and female breeders provisioned at similar rates that were greater than those of helpers, there were no relationships between provisioning rate and PRL titers. Paradoxically, the trend in breeders was toward less feeding with higher PRL.

Superb fairy-wren

Breeding groups of this small Australian passerine range in size from a single pair with no helpers to groups consisting of one female and five males (Rowley 1965). Approximately 60% of the groups consist of a dominant "breeder" male and younger male helpers (Mulder *et al.* 1994). Despite having a social mate, the social system is exceptionally promiscuous. Males without helpers have a two-fold higher rate of paternity in their own nests than dominant males in groups with helpers (Dunn and Cockburn 1996) and female promiscuity abounds, with up to 95% of broods containing nestlings sired by males outside of the social group (Mulder *et al.* 1994).

Male T profiles showed two distinct periods of increase, although the second was of much greater magnitude. First, T levels increased months before the breeding season and were significantly correlated with the molt into breeding plumage (Peters *et al.* 2000). Second, T levels increased during the breeding season and were maintained at elevated levels for an extended period of time (Peters *et al.* 2001). This latter finding is consistent with polygynous species that have a longer time period during which they may experience additional mating opportunities, increased agonistic male–male interactions, or both (Wingfield *et al.* 1990). Peters *et al.* (2001) conclude that the pattern of circulating T in male superb fairy-wrens is mostly driven by the former mechanism, that is, stimulation of the HPG axis due to high rates of extra-pair solicitation and copulation. This conclusion is supported by a significant positive correlation between T levels of males and the total number of fertile females in the study population, independent of whether their own mates are fertile.

As is the case in many of the species examined in this chapter, dominant males have higher T levels than helpers (Peters *et al.* 2001). Consistent with the "challenge hypothesis" (Wingfield *et al.* 1990), T levels are higher in dominant breeder males that share a territory with helpers than in males that breed without helpers, the latter of which have T levels that are equivalent to those of helpers. T levels are clearly affected by multiple aspects of the social environment, such as inter- and intra-sexual interactions in aggressive and sexual contexts.

One of the more intriguing findings of Peters *et al.* (2001) was that T levels did not decrease with the onset of paternal care. Peters *et al.* (2002) found that endogenous T levels differed between males of all statuses (dominant > pair > helper) and declined gradually as the sampling period progressed, but were not correlated with either age or condition. Unfortunately, provisioning rates were not determined, and thus their conclusion that endogenous T levels do not affect provisioning in males requires confirmation.

Focal watches of T-implanted males that found a three-fold reduction in provisioning rates further confound our understanding of the role of T in this species (Peters *et al.* 2002). How are the suppressive effects of exogenous T explained in light of the co-occurrence of elevated endogenous T levels during the parental care period? The authors conclude that maximal mean endogenous T titers of <0.5 ng ml^{-1} are at a level that does not interfere with nestling provisioning, yet allows extra-group courtship. In contrast, T levels that resulted from T implants, although within the range of the maximum levels recorded in the species, were on average over an order of magnitude higher than those of controls (Peters *et al.* 2002).

Peters (2003) examined the role of T in mediating the trade-off between parental and courtship behaviors by implanting males in unassisted pairs with T and using focal watches to construct time–activity budgets. T-treated males spent less time on paternal behaviors and more time on sexual behaviors than controls. Surprisingly, given that male courtship behavior during the nestling care period is typically directed towards extra-group females, T-treated males directed sexual behaviors towards their mates while failing to increase extra-group courting. These findings show that T plays a role in mediating courtship in superb fairy-wrens. However, less clear is why T-implanted males failed to increase extra-group courtship efforts. Perhaps, as argued by Peters, there are no differences in extra-pair courtship because endogenous T levels are always sufficient to motivate this most conspicuous male behavior.

CONCLUSIONS

Reaching definitive conclusions about universal hormonal mechanisms responsible for cooperative behaviors is problematic. Much of this difficulty can be traced to the great variance in social systems that fall under the umbrella of cooperative breeding. Can one expect shared endocrine mechanisms between species such as red-cockaded woodpeckers, which are both socially and genetically monogamous, and superb fairy-wrens, which are exceptionally promiscuous? Even when making comparisons within the genus *Aphelocoma*, the differences in social structure between Mexican jays and Florida scrub-jays are striking.

Another factor that makes drawing definitive conclusions difficult is that virtually all of the research carried out to date is correlational. There are three steps that must be taken before it is possible to conclude that a given hormone causes a behavioral or physiological response. First, one typically notes a correlation between the hormone and the behavior. Second, removal of the hormone results in cessation of the behavior. Third, replacement of the hormone reinstates the behavior. With few exceptions, most studies of cooperative breeders have yet to progress past the first step. Additionally, the above studies have investigated only nine of the hundreds of cooperatively breeding bird species, clearly an insufficient representation to allow discussion of "universal mechanisms." Nonetheless, we conclude with an attempt to draw out some of the common themes from the work discussed above.

Prolactin

The link between PRL and alloparental care is intriguing and merits further study. Vleck *et al.* (1991) found that the subgroup of helpers (beta males) in Harris's hawks that provisioned nestlings the most also had the highest levels of PRL. Similarly, Schoech *et al.* (1996b) found that Florida scrub-jay non-breeders that were active allofeeders had higher levels of PRL than those that did not provision young, and also determined that nest stage was a better predictor of PRL levels in helpers that either photoperiod or endogenous rhythms. Further, the increase in plasma PRL levels in Florida scrub-jay helpers began prior to when helpers are allowed at the nest by breeders, indicating that increased PRL levels are in anticipation of rather than in reaction to the stimulus of nestlings begging. These results, along with additional work on Mexican jays and red-cockaded woodpeckers, suggest an endocrine mechanism that has evolved in close conjunction with helping behavior.

Endocrinology and the unselected hypothesis

This leads directly to consideration of how endocrine studies contribute to the debate concerning the evolutionary origins of helping behavior. Jamieson and Craig (1987a) suggested that helping, like parental behavior, is a stereotypical response to auditory and visual stimuli from young birds. As such, they argued that rather than being under genetic control, helping behavior in cooperatively breeding birds does not differ from parental behavior and is an unselected by-product of philopatry and group living (the "unselected hypothesis"). This implies that endocrine responses are a consequence rather than a cause of helping behavior. For example, Jamieson and Craig (1987a) state "it is not hormones that control behavior, but rather it is behavior in association with certain stimuli that induces hormonal secretion." Unfortunately, the relationships between hormones and behavior are not so simple. We would suggest that an illuminating exercise for anyone who believes the first part of the above quote would be to study the copulatory behavior of a castrated house pet.

Is PRL a cause of or a response to helping behavior? The data from a number of studies show that PRL rises prior to the expression of allofeeding in the two cooperatively breeding jay species examined, that it is several times higher in Mexican jays than in western scrub-jays, and that it is higher in Florida scrub-jay helpers that provision young than it is in those that fail to help at the nest. These findings all support a role for this hormone in mediating helping behavior. If helping behavior is essentially the same as parental behavior, as proposed by the unselected hypothesis, then there should not be such a pronounced difference in PRL levels between cooperative and non-cooperative breeding congeners. While these findings support PRL as a potential mechanism mediating alloparental care, it is not possible to differentiate between PRL as a mediator of parental versus alloparental behaviors. Even if experimental removal of PRL resulted in a diminution of alloparental care and replacement of PRL restored the behavior, we would be greatly surprised if the same experiment did not similarly affect parental behavior.

Corticosterone

Elevated levels of CORT can negatively impact the reproductive axis at multiple levels (Pottinger 1999) and the increased stress of subordinate status can result in elevated CORT levels (Wingfield and Farner 1983). Consequently, assuming that non-breeding helpers are subordinate to breeders, it has been hypothesized that elevated CORT levels cause the reproductive quiescence of non-breeding helpers. This hypothesis has been examined in Harris's hawks (Mays *et al.* 1991), Florida scrub-jays (Schoech *et al.* 1991, 1997), and white-browed sparrow-weavers (Wingfield *et al.* 1991). However, contrary to predictions, plasma levels of CORT in helpers were equivalent to those found in breeders in two of the three species, the exception being the Florida scrub-jay. Given that, in contrast to mammalian cooperative breeders (Creel 2001), there is no substantive evidence from avian cooperative breeders that dominants are stressed as a result of group living, this counters the hypothesis that CORT levels play a key role in the reproductive suppression of helpers in birds.

This is perhaps not surprising given that most such groups have relatively stable, established dominance hierarchies that are often characterized by minimal aggression. It is interesting that despite considerable evidence that establishment of rank between individuals in groups that do not breed cooperatively most frequently results in elevated levels of CORT in the subordinate (Creel 2001), in white-browed sparrow-weavers, the only cooperatively breeding bird in which this has been investigated, CORT levels were unaffected (Wingfield *et al.* 1992). Thus, it appears that the lack of reproductive activity by non-breeding helpers cannot be attributed to the effects of CORT.

Reproductive hormones

Consideration of hormones of the reproductive axis yields few clear-cut interspecific commonalities although male helpers of most species have lower levels of T than breeders. The two primary factors that play a role in this difference are the age of the helper and whether the helper is the offspring of the opposite-sexed breeder in the group.

That age must be considered is emphasized by the findings that younger males generally have lower T (or total androgen) levels than older birds with few exceptions. Most notably, there were no age effects upon T in superb fairy-wrens in which all males are reproductively active, irrespective of status (Peters *et al.* 2001). Of the remaining seven species for which data exist, only in Harris's hawks and Mexican jays, in which one-year-old birds do not breed, was age found to be an overriding factor in low T levels or reproductive quiescence of helpers. In the other five species – pied kingfisher, Florida scrub-jay, bell miner, Australian magpie, and red-cockaded woodpecker – the researchers concluded that, while a contributing factor, age per se could not explain the differences between breeders and helpers. Observations that helpers readily became breeders when the opportunity arose support this conclusion.

Relatedness to the breeders, in particular the opposite-sexed breeder, is a factor in the quiescence of the reproductive axes of helpers in most, if not all, species (Chapter 9). However, relatedness does not always affect relative titers of some reproductive hormones. For example, in Florida scrub-jays and white-browed sparrow-weavers, two species that have both male and female helpers, only in females were there found to be effects of relatedness, with helper females living in groups in which they were not descendants of the breeder male having higher E_2 and LH levels than helper females that were living with their fathers.

While the mechanisms underlying the reduced HPG axis function of related helpers remain obscure, three alternative, but not necessarily mutually exclusive, hypotheses have been suggested. First, helpers living with their parents are subject to inbreeding avoidance. Second, helpers remain reproductively inactive because up-regulation of the HPG axis would result in elevation of T levels that could interfere with the expression of alloparental behaviors, thus lowering indirect fitness gains. This scenario would most likely apply only to males and relies on T inhibiting alloparental care as it does parental care. Third, in the absence of intra-pair interactions the reproductive axis of a helper is not stimulated. This is supported by studies that found endocrine differences in helpers living with their parent as compared with those living with an opposite-sexed breeder.

Final thoughts

While there appear to be some common trends in the endocrine physiology of cooperative breeding, many of the above caveats make drawing definitive conclusions impossible. We emphasize that endocrine studies of cooperative breeding are in their infancy and we reiterate our hope that others will join us in our efforts to gain a better understanding of this intriguing social system in all of its complexities.

ACKNOWLEDGMENTS

We thank Anne Peters and Suzanne Oppenheimer for access to unpublished data, Carol Vleck and Jerram Brown for providing published data, and Al Dufty for commenting on the manuscript. Kate Eldridge provided editorial comments. The authors were supported in part by the National Science Foundation.

9 • Incest and incest avoidance

WALTER D. KOENIG
University of California, Berkeley

JOSEPH HAYDOCK
Gonzaga University

Cooperative breeders live in groups that usually, although not always, consist of close relatives. Following Hamilton's (1964) rule, high relatedness serves to increase fitness by facilitating kin selection and the evolution of cooperative behaviors such as are frequently observed within societies of cooperative breeders. On the other hand, if high relatedness extends to opposite-sex breeders, there is the risk of inbreeding and its attendant genetic problems leading to significantly lowered fitness. How cooperative breeders resolve these conflicting selective pressures has long been recognized as a dilemma and has served as an important focus of research in this field.

Determining how societies resolve the potential problem of inbreeding has turned out to be unexpectedly difficult. First, there are semantic problems. Inbreeding is both a relative and a hierarchical phenomenon, and it is likely that inbreeding may be avoided at some levels but not others (Dobson *et al.* 1997). At one extreme, mates may share no known relatives going back several generations and can be considered "outbred" in that they share fewer genes identical by descent than two random individuals in the population. At the other extreme, mates may be close relatives such as siblings or a parent and offspring. Unless all genetic variability in the population has been previously lost, such individuals will be genetically more similar to each other than two randomly chosen individuals in the population, and thus their offspring will have an elevated proportion of homozygous loci identical by descent. Otherwise, the relatedness of individuals falls in between these extremes, in which case whether they are considered inbred or not can often depend on what level of inbreeding one focuses on.

For our purposes, this problem is operationally resolvable by restricting our interest to matings between close relatives. Those between first-order relatives, for which the coefficient of relatedness, r, between mates is on average ~0.5, are often referred to as "incestuous matings," while those between more distantly related individuals still sharing a common ancestor within the last two to three generations ($0.0625 < r < 0.5$) may arbitrarily be considered "inbred matings." Inbreeding between relatives more distantly related than this ($r < 0.0625$) are generally undetectable genetically due to the very small increase in homozygosity expected in the inbred offspring, and observationally because large numbers of known individuals need to have been followed for multiple generations.

Difficulties in determining the extent of inbreeding and inbreeding avoidance do not end once operational definitions of inbreeding and incest have been chosen (Pusey and Wolf 1996). Until the advent of DNA fingerprinting, authors had to infer the existence of incest from patterns of dispersal and demography, a procedure that has turned out to be embarrassingly misleading in several cases when extra-pair matings turned out to be common in groups that otherwise appeared to involve incest. Even today, with powerful molecular techniques at our disposal, limitations imposed by the inheritance process make the detection of incest difficult in cooperative breeders (Cockburn 1998; McRae and Amos 1999; Parker *et al.* 1999), potentially forcing workers to abandon the attempt to unambiguously determine parentage and instead rely on measures of genetic similarity to assess the extent and consequences of inbreeding (Richardson *et al.* 2001, 2002). As a result, despite considerable advances in understanding the mating system of many cooperative breeders, there remains much controversy as to the extent and importance of inbreeding in highly social species.

One additional problem worth mentioning is that in contrast to many areas of behavioral ecology where

---

Ecology and Evolution of Cooperative Breeding in Birds, ed. W. D. Koenig and J. L. Dickinson. Published by Cambridge University Press.
© Cambridge University Press 2004.

theory has provided valuable insights that have helped guide and interpret the findings of field workers, theory has been of little use in deciding, a priori, what to expect in terms of the extent of inbreeding and incest in cooperative breeders. On the one hand, the potentially deleterious effects of inbreeding, primarily via the phenomenon of inbreeding depression, are well documented empirically, at least under laboratory or captive conditions, and reasonably well understood theoretically despite debate over its precise genetic basis (Hedrick 1994; Lacy and Ballou 1998). However, detecting inbreeding depression in the wild has proved difficult, largely because inbreeding is rarely observed. Consequently, a variety of plausible counterarguments can be made supporting other hypotheses for a low rate of incest other than incest avoidance.

First, as mentioned above, Hamilton's (1964) theory of kin selection and inclusive fitness opened the door for the idea, apparently confirmed so dramatically by the correlation between haplodiploidy and eusociality in insects, that the strikingly social behaviors observed in cooperative breeders should be (perhaps even must be) associated with the kind of high genetic relatedness obtainable only through regular incest. Although later, more reasoned consideration of this idea rejected the hypothesis that inbreeding favors kin selection (Dawkins 1979), the logical inference that the two phenomena might be linked has been difficult to dismiss entirely.

Second, a case can be made that regular inbreeding reduces its own potentially adverse fitness consequences by purging the genome of severely deleterious recessive alleles (Barrett and Charlesworth 1991). Indeed, systems of regular inbreeding are not uncommon in many taxa, notably insects. Thus, cooperative breeders, by engaging in a regular system of inbreeding, may have succeeded in largely eliminating the detrimental consequences that inbreeding inflicts on most other vertebrates.

Third, parallel to the potential costs associated with inbreeding ("inbreeding depression"), there are potential costs of outbreeding ("outbreeding depression"). This has led to the concept of "optimal outbreeding," an extensive discussion of which is provided by Shields (1982), who argued that maintaining coadapted gene complexes yields benefits to inbreeding that outweigh its costs. Empirical support for this effect and for outbreeding depression is minimal (Pusey and Wolf 1996), but the possibility that it is important in some species remains.

And finally, as vividly pointed out by Shields (1987), reproductive competition can often provide an alternative explanation for the apparent rarity of incest observed in most natural populations. The bottom line of this and the prior counterarguments, reiterated more than once in the literature, is that there are both potential costs and benefits of inbreeding, and thus whether or not a particular population engages in inbreeding or incest avoidance cannot necessarily be predicted a priori (Bengtsson 1978; Waser et al. 1986).

With this combination of logistic and theoretical ambiguity, it is not surprising that the role that incest and inbreeding play in cooperative breeders has proved to be highly controversial and something of a minefield of failed inference. Although this problem is not restricted to cooperative breeders, it is particularly prominent in these species because most helpers come from within the group and thus the potential for incest and inbreeding is particularly high (Brown 1978). There are two primary controversies. The first is over the extent of incest and inbreeding. Occasional cases of inbreeding certainly occur, and most will agree that such cases may be more frequent in cooperative breeders than in less social species as a passive consequence of the high degree of philopatry and viscous population structure exhibited by social species. What is controversial is a relatively small subset of species for which incest and inbreeding are thought to be a regular part of the mating system, occurring with sufficient frequency to significantly influence behavior and population structure, as is the case in many social insects. And, if incest or inbreeding does turn out to be a regular event in even one species of vertebrate, why is it so rare in all the others?

This leads to the second controversy, which involves the mechanisms that lead to the rarity of incest in the vast majority of cooperative breeders where it is not a regular part of the mating system. Specifically, is incest rare because it is actively avoided, or is its rarity an epiphenomenon of some other behavior? And if the former, what exactly are the mechanisms of incest avoidance?

Related to this second controversy is the issue of inbreeding depression mentioned above. Is inbreeding depression detectable and significant in cooperative breeders, and thus an important selective force in the evolution of inbreeding avoidance? Or do cooperative breeders avoid the potential detrimental effects of inbreeding, and if so, how?

Table 9.1. *Evidence for and against inbreeding depression in cooperative breeders*

Species	Evidence	Comments	Reference
Mexican jay	Smaller brood size and lower survival of inbred offspring	Not backed by genetic data	Brown and Brown 1998
Pukeko	No clear inbreeding depression	Unconfirmed	Craig and Jamieson 1990
Common moorhen	Incestuous matings produced eggs with significantly lower hatchability	Backed by genetic analyses	McRae 1996b
Green woodhoopoe	Low hatchability correlated with high incidental frequency of inbreeding	Circumstantial	Ligon and Ligon 1990b
Green woodhoopoe	No evidence for decreased fledging success or survivorship of putatively inbred offspring	No genetic data; behavioral evidence only for parentage	Du Plessis 1992
Red-cockaded woodpecker	Lower hatching rate of inbred offspring and lower survival and recruitment of inbred fledglings	Large sample sizes; backed by indirect genetic evidence of strong monogamy	Daniels and Walters 2000a
Acorn woodpecker	High cost of failing to breed incestuously	Indirect estimate of minimum cost of inbreeding	Koenig *et al.* 1999
Dwarf mongoose	No decrease in reproductive success or survivorship	Backed by band-sharing but not by direct parentage analysis	Keane *et al.* 1996
Mashona mole-rat	Very low reproduction and weaning success among inbred pairs	Laboratory study	Greeff and Bennett 2000

CASE EXAMPLES

In the vast majority of cooperative breeders, incest occurs occasionally but is rare or at least rarely observed (Brown 1987; Cockburn 1998). Here we bypass these species and instead focus on cases where it has been proposed that the frequency of inbreeding and incest are sufficiently high to play an important role in their mating system. These include the Mexican jay, pukeko, Australian fairy-wrens (*Malurus* spp.), green woodhoopoe, Seychelles warbler, and common moorhen. We also discuss two parallel cases among cooperatively breeding mammals, those of the naked mole-rat and the dwarf mongoose. Evidence for and against inbreeding depression in many of these species is summarized in Table 9.1.

Mexican jays: the danger of demographic inference

Mexican jays are cooperative breeders in which groups contain up to several social pairs of breeders nesting separately along with a variable number of helpers that may feed at several nests within the group (Brown and Brown 1990). Based on the apparent rarity of dispersal and the frequent retention of offspring within social units, Brown (1974) inferred the existence of regular incest and inbreeding in Mexican jays, leading to an estimated average relatedness of birds within groups (r) of 0.8, which "virtually confirms for one population the prediction that can be made from Hamilton's theory that r should indeed be high in communal breeders that behave altruistically" (Brown 1974, p. 75). Shortly thereafter, based on additional data indicating more dispersal than previously suspected, Brown (1978, p. 141) rescinded this statement, concluding that "although some individuals do breed in their natal unit, incest has not been observed any more frequently in *Aphelocoma* than in noncommunal species." More recent work has substantiated this assertion (Brown and Brown 1998) and in addition found that Mexican jays exhibit high levels of extra-pair paternity (Li and Brown 2000).

Brown's (1974) originally flawed conclusions were based on one of the better long-term datasets available at the time. Furthermore, failing to appreciate the bias toward philopatry and short-distance dispersal inherent in most dispersal data is by no means limited to those studying cooperative breeders (Koenig et al. 1996). We emphasize it here not to be critical; indeed, Brown's willingness to reinterpret his results upon further study was exemplary. Rather, it provides a clear warning of the dangers of inferring incest from demographic data that has unfortunately not always been heeded by subsequent workers.

Pukeko: incest down under?

Another prominent early example of putative incest in a cooperative breeder is the case of the pukeko, a polygynandrous moorhen found in New Zealand. Based on a combination of social bonds, observed copulations, and actual egg-laying by females, Craig and Jamieson (1988) reported that the vast majority of birds were retained in their natal territory as adults, leading to a 77% frequency of incestuous matings, mostly between fathers and daughters (50% of 34 incestuous pairings) but also between mothers and sons (35% of incestuous pairings) and siblings (15% of incestuous pairings). Despite this remarkably high frequency of apparent incest, no obvious inbreeding depression was observed (Craig and Jamieson 1990). Craig and Jamieson (1990) suggested this was due to several possible mechanisms for reducing the disadvantages of incest, including the relatively high fecundity of pukeko, possible gamete selection by females, and the possibility that the adverse genetic consequences of incest had been eliminated by regular inbreeding in the past.

More recent work employing molecular techniques to determine parentage has unfortunately been performed on different populations from that originally studied by Craig and Jamieson and has only incidentally addressed their assertions regarding the extent of inbreeding. Nonetheless, to the extent that they can be used to test Craig and Jamieson's (1988) assertions, these studies offer little support for them. Lambert et al. (1994) found considerable mixed parentage in polygynandrous groups and suggested that a high level of band sharing within groups was "a likely consequence of strong philopatry and inbreeding" (p. 9641), but they had insufficient background demographic data to document any incestuous matings. Furthermore, two observations reported in the paper suggest at least the possibility of inbreeding avoidance. First, "in none of the groups did individuals hatched during the course of this study later reproduce" (p. 9644), a result consistent with young avoiding incest with their known or probable parents. Second, and more subtly, Lambert et al. (1994) claim to have been able to unambiguously determine parentage within groups, which, as discussed by Cockburn (1998), McRae and Amos (1999), and Parker et al. (1999), is unlikely if incestuous matings (either father–daughter or mother–son) were involved. Consider a pair with a son from a prior year: if the father sires all offspring (no incest), it will usually be possible to exclude the son as the parent. However, barring mutation, all genetic material in the son is also present in either his mother or his father, and thus, if the son incestuously sires the offspring, then all paternal genetic material passed on to the offspring will be present in both the son and the father and it will generally not be possible to exclude either as the potential sire. Thus, successful paternity assignment is itself indirect evidence that incest was uncommon in the population they studied.

A parallel molecular study on a third population (Jamieson et al. 1994) similarly reported no cases in which parentage was ambiguous, again suggesting that incest was rare. Unlike the population originally studied by Craig and Jamieson (1988), habitat saturation appeared to be low, allowing considerable dispersal of young and resulting in groups containing birds that were not closely related.

Given these more recent results, the most that can be said concerning Craig and Jamieson's (1988) hypothesis that pukeko regularly engage in close inbreeding is that it remains unconfirmed and is at least not a general phenomenon in this species. Again, inference based on demographic data was apparently misleading, but in this case even subsequent molecular studies have thus far failed to clarify the issue. Whether this is because of

Fairy-wrens: independence between the social and genetic mating system

Birds in the genus *Malurus* are a stunning group of small Australian birds, all 13 species of which are cooperative breeders (Rowley and Russell 1997). Extensive studies have been performed on two species, the splendid fairy-wren and, more recently, the superb fairy-wren. The saga of incest and inbreeding in fairy-wrens began with a note on the composition of breeding groups by Rowley *et al.* (1986) that detailed an unprecedented rate of incest in the splendid fairy-wren based on social pair-bonds. Specifically, of 136 pairs, 29 (21.3%) consisted of birds apparently related to each other by $r = 0.5$, including 13 mother–son, 7 father–daughter, and 9 full-sibling pairs. If this were not enough, its denouement was even more spectacular: demolishing the ubiquitous assumption that mating occurred within the group, allozyme studies revealed an unprecedented rate of extra-pair fertilizations (EPFs), reducing, and perhaps even eliminating, the apparently high rate of inbreeding indicated by the observed social relationships within the population (Brooker *et al.* 1990).

Subsequent studies on the superb fairy-wren using more sensitive molecular techniques have confirmed virtual independence between the social and genetically effective mating systems. Almost all broods contained extra-group young, and 76% of offspring were sired by extra-group males (Mulder *et al.* 1994), the highest incidence of EPFs described for any species. Although there is little evidence that this extraordinary rate of extra-group mating is directly related to incest avoidance (Mulder *et al.* 1994), there would appear to be no doubt that one consequence is that incest within social groups is significantly reduced below what it appears to be based on group composition. Dunn *et al.* (1995), for example, found that helpers occasionally sired young within their own groups, but never when they were living with their mother.

The result is that, as in Mexican jays and pukeko, inbreeding is apparently much less frequent than originally inferred and not a regular part of the mating system. One can hardly fault Rowley *et al.* (1986) for their incorrect inference: not only did they make the traditional assumption that mating would be within the social groups, but the high incidence of close relatives as social mates was extraordinary and completely unprecedented. We are fortunate that the techniques for determining parentage followed so quickly so as to clarify this bizarre situation, which otherwise could easily have misdirected our view of inbreeding in cooperative breeders for years.

Although the independence between the social and genetic mating systems is extraordinary in the fairy-wrens, several other species have recently been shown to exhibit very high frequencies of extra-group matings. Among birds, Li and Brown (2000) recorded extra-pair young in 63% of Mexican jay broods, fathered almost entirely by other males within the same social unit, and Richardson *et al.* (2001) documented 40% extra-group paternity in Seychelles warblers. Similar examples are now being revealed among cooperatively breeding mammals, including Ethiopian wolves, in which 70% of copulations are performed by extra-pack males (Sillero-Zubiri *et al.* 1996), and pilot whales, where incest within large pods of closely related animals is apparently obviated by males mating outside their natal pod (Amos *et al.* 1993). Although in none of these cases has it been determined that extra-pair matings are directly related to incest or incest avoidance, the incidence of incest is clearly reduced as a consequence of this behavior in many species, including the fairy-wren, Ethiopian wolf, and pilot whale.

Green woodhoopoes: incidental inbreeding?

Green woodhoopoes live in socially monogamous groups containing as many as 10 or more non-breeding helpers (Ligon and Ligon 1990). Two independent studies have reported contrasting results regarding the frequency of incest. Ligon and Ligon (1990b), studying a population near Lake Naivasha in Kenya, found incest to be rare: during more than six years of their study, they "never recorded a pair bond between close relatives in the same flock, such as siblings or parents and offspring" (Ligon and Ligon 1990b, p. 48). However, as in many cooperative breeders, dispersal is apparently limited and the Ligons speculated that the independent dispersal of close relatives to the same (non-natal) group could lead to sufficient incidental inbreeding to cause the unusually high degree of hatching failure observed within the population (Ligon and Ligon 1988). On the other hand, cases of incidental inbreeding were rarely observed. In their words, "given the conservative, short-distance movement by both males and females, the fact that close relatives occasionally form pair bonds is not surprising. What is surprising is that we have recorded it so infrequently" (Ligon and Ligon 1990b, p. 48).

Given the rarity with which incidental inbreeding was actually observed, it would seem premature to conclude that it is the cause of the dramatically low rate of hatchability seen in the Lake Naivasha population. Indeed, given the rarity with which inbreeding was documented, it is possible that inbreeding is avoided even outside natal groups. More likely, however, this population of woodhoopoes may be an example of what appears to be a relatively common situation in cooperative breeders and other highly social species: incest is strongly avoided at the within-group level, but not at the level of the subpopulation within which much dispersal apparently takes place (Dobson et al. 1997).

These results contrast with those of Du Plessis (1992), who reported frequencies of incestuous pairing of 15% and 29% in two populations of green woodhoopoes in South Africa. Additional work has moderated these estimates to some degree, but still some 15% of matings are estimated to be among first-order relatives in these populations (M. Du Plessis, personal communication). Based on strong mate-guarding, long and obvious copulation behavior, and aggressive territoriality, Du Plessis (1992) argues cogently that extra-pair matings are unlikely to counter these apparent cases of incest. Nonetheless, given the situation of the fairy-wrens, this is clearly an assumption that needs to be confirmed. If incest is indeed common, it apparently entails no cost on either fledging success or survivorship (Du Plessis 1989), suggesting that inbreeding depression is either absent or that the potential costs of inbreeding are counterbalanced by the risks of dispersal (Du Plessis 1992). These results, if not contradictory, at least leave considerable room for doubt as to the actual extent of and consequences of inbreeding in this species.

Naked mole-rats: incest underground?

One mammal that exhibits many parallel features with the avian examples already discussed is the naked mole-rat. This cooperatively breeding, subterranean rodent lives in colonies of up to 300 individuals, the vast majority of which are reproductively suppressed by the single dominant breeding female (Sherman et al. 1991; Faulkes and Bennett 2001). Extremely high intracolony relatedness based on genetic analyses (Faulkes et al. 1990b; Reeve et al. 1990), combined with behavioral evidence from laboratory populations (Sherman et al. 1991), suggests that inbreeding occurs regularly, despite the occurrence of active incest avoidance in other species of mole-rats (Burda 1995; Greeff and Bennett 2000; Faulkes and Bennett 2001; Herbst and Bennett 2001). Indeed, intracolony relatedness has been estimated in some groups to be as high as 0.8, enough to be an important factor in explaining their highly developed sociality (Reeve et al. 1990). Given their fossorial habits and poor thermoregulatory abilities, this has plausibly been thought to be an adaptive response to the high costs of dispersal.

As with several of the prior species, however, subsequent research indicates that the situation is more complicated than originally envisioned. Several lines of evidence now indicate that outbreeding is not only more common than originally thought but preferred over inbreeding. Most dramatic is the morphologically and physiologically distinct dispersal morph discovered by O'Riain et al. (1996, 2000b). Although the dispersal morph appears to be rare, it clearly indicates that dispersal and outbreeding are important. More recent work has supported this conclusion, both in the form of laboratory studies indicating that animals prefer non-kin as potential mates (Clarke and Faulkes 1999; Cisek 2000) and in the form of data from wild populations documenting dispersal up to 2.6 kms (Braude 2000; O'Riain and Braude 2001).

Based on these new data, O'Riain and Braude (2001) suggest that the high relatedness values obtained by Reeve et al. (1990) were due primarily to recent common ancestry rather than a propensity to inbreed per se, and Braude (2000, p. 7) concludes that "inbreeding is not the system of mating for this species and outbreeding is probably frequent." Although much work remains to be done on this fascinating species, it would appear likely that incest, although perhaps unusually common, is not the regular part of the mating system of naked mole-rats that earlier work suggested.

Dwarf mongooses: inbreeding without cost?

Dwarf mongooses, a small African carnivore, live in packs of 2–21 individuals within which only the oldest, most dominant pair generally breed. Young of both sexes frequently disperse to obtain breeding positions, but nonetheless, based on pedigrees, 14% of 241 offspring were apparently the result of incestuous matings between full siblings or a parent and offspring, suggesting a fairly high level of inbreeding (Keane et al. 1996). Detailed behavioral analyses further revealed no tendency for dispersal to be related to incest avoidance, and litter counts generated no evidence of either a survival or reproductive disadvantage to inbred pairs or their incestuous offspring compared to non-inbred pairs (Keane et al. 1996).

The apparent result is a mating syndrome of pervasive, mild inbreeding considerably less extreme than that originally suggested for species such as Mexican jays and naked mole-rats, but still much higher than virtually all other known vertebrates and, equally significantly, much higher than would be achieved by inbreeding avoidance of the sort practiced by most other cooperative breeders. This conclusion was backed up to some extent by genetic band-sharing data, but direct genetic estimates of relatedness were not made. Consequently, an alternative explanation is that many of the incestuous matings, inferred assuming that dominants parent all offspring, instead involve reproduction by subordinates, which are in fact known to reproduce frequently in this population (Keane et al. 1994). Confirmation of incest using more sensitive genetic techniques is clearly desirable.

Seychelles warblers: island incest following a bottleneck

The above examples might lead one to conclude that there remain no cases in which the mating system of a cooperative breeder involves more than a modest incidence of inbreeding and that, with the exception of species like green woodhoopoes and dwarf mongooses where crucial genetic work has yet to be done, the controversy over whether incest is avoided in cooperative breeders or not has been resolved. In fact, this is almost, but not quite, the case.

We consider there to be two possible exceptions with sufficient data to warrant discussion. The first is the extraordinary case of the Seychelles warbler. Endemic to the Seychelles islands in the western Indian Ocean, Seychelles warblers were reduced to 26 individuals on a single island in 1959. Since then, the population has recovered and has been intensively studied by Komdeur and his colleagues. Groups consist of a breeding ("primary") pair along with a variable number of helpers, many but not all of which are fledglings from prior years that have not successfully obtained breeding vacancies of their own. Although originally thought to be primarily non-breeders, nearly half of helper ("secondary") females have recently been found to lay eggs, producing 15% of offspring in the population (Richardson et al. 2001, 2002). Among both primary and secondary females that breed, extra-pair matings are common, resulting in 32% extra-pair paternity of eggs laid by primary and 58% extra-pair paternity of eggs laid by secondary females (Richardson et al. 2001).

This high rate of EPFs, combined with relatively low levels of genetic polymorphism most likely stemming from the earlier bottleneck, make determining the extent of incest and the existence of incest avoidance particularly challenging. However, evidence from patterns of genetic similarity indicates that the secondary female's relatedness to the primary male has no effect on whether she breeds or not (there was no significant difference between the relatedness indices between primary males and secondary females that bred and that did not breed), or on whether she obtains an EPF or not (secondary females whose eggs were fertilized by an extra-pair male were not significantly more closely related to the primary male than those whose offspring were sired by the primary male) (D. Richardson, personal communication). Secondary females were more likely than primary females to lay eggs sired by extra-group males, but given the complexity of the situation, it is unclear to what extent this lowers the overall incidence of inbreeding.

The final outcome of this complex system is not yet clear. Incest avoidance may yet play a central role (Chapter 5). However, it is possible that incest occurs relatively frequently through reproduction by secondary females. Thus far there appears to be no obvious mechanism at work to reduce the frequency of incest beyond, perhaps, the fact that fewer than half the secondary females, many of which are closely related to the primary male, lay eggs and breed. Furthermore, there appears to be no relationship between either relatedness of parents or offspring heterozygosity and offspring survival, suggesting an absence of inbreeding depression (D. Richardson, personal communication). Although such an absence would be extraordinary, it could be explained by the hypothesis that the deleterious effects of inbreeding were largely eliminated during the genetic bottleneck suffered by the population but a few generations ago. Given the central role that this species has played in recent advances in the field of cooperative breeding, resolving these issues should prove particularly illuminating.

Common moorhens: incest as making the best of a bad job?

The second exception is the common moorhen, a primarily monogamous species but one in which mating groups sometimes consist of more than one male and female (polygynandry) and brood parasitism is common (McRae and Burke 1996). Using minisatellite DNA markers, McRae (1996b) investigated the mating system of this species and reported that in groups where a daughter was present along with her parents, the father mated incestuously with his daughter and sired all her eggs. Hatchability of incestuous eggs laid by such "junior" females was significantly lower than both non-incestuous eggs laid by "senior" females in the same nest and eggs laid by non–incestuous secondary females, indicating inbreeding depression on the order of 42–45%. McRae (1996b) suggests that two factors result in this situation. First, the reproductive options for young moorhens are limited, making it preferable for some to delay dispersal and remain in their natal groups. Second, moorhens are at an early stage of evolving sociality and thus have yet to develop mechanisms to avoid incest.

How frequent are incestuous matings? Based on 228 group-years of moorhens over a 3-year period, 162 (71%) were monogamous and an additional 20 (9%) were cooperatively polygynous containing two or more adult males but only one female (McRae 1996b). Thus 46 (20%) contained two or three adult females, apparently representing 34 different sets of birds. Of these 34 sets, younger females were known to be daughters of the core pair in 18 (53%) cases, were unrelated or related in some other way in 9 (26%) cases, and were of unknown relatedness in 7 (21%) cases. Excluding these latter cases and assuming that incest occurs whenever daughters are living in groups with their parents, the overall annual incidence of groups in which incest occurs can be estimated as $0.2 \times (18/[34 - 7]) = 0.13$. Adding the six additional cases of incest McRae (1996b) found within monogamous pairs yields a minimum estimated annual frequency of incest in the population of 16%, a value close to the rate of incest originally suggested by Rowley *et al.* (1986) for superb fairy-wrens and far higher than the incidental rate of incest confirmed in any other vertebrate (Ralls *et al.* 1986).

Thus, the rate of incest would appear to be singularly high. However, in lieu of information on how long communal nesting has occurred in this population, it is not possible objectively to evaluate McRae's (1996b) hypothesis that mechanisms to avoid inbreeding have not had sufficient time to evolve. Certainly one would expect selection against incest to be strong and incest avoidance mechanisms to develop quickly, given the high inbreeding depression indicated by the low hatchability of incestuous eggs.

McRae (1996b) argues that the stage for incest is set by the limited options available to late-hatching daughters. She further provides evidence that mothers do not incur a cost from allowing their daughters to breed communally with them, but that daughters would do better, and in fact prefer, to breed on their own when they can. In those cases when they are unlikely to be competitive for mates, they "make the best of a bad job" by delaying dispersal, remaining in their natal group, and (apparently) breeding incestuously with their father. This scenario matches the predictions of theoretical studies (Bengtsson 1978; Waser *et al.* 1986) and is intuitively appealing: even in cases when inbreeding depression is high, breeding incestuously will surely be better than not breeding at all (Greeff and Bennett 2000). Unfortunately, this logic appears not to hold in at least one species, the acorn woodpecker, where offspring in similar circumstances do not attempt to breed and, more cogently, groups containing nothing but relatives often fail to breed at all rather than breed incestuously (Koenig *et al.* 1998, 1999).

Thus, a satisfying explanation for the moorhen situation remains elusive. Why is incest so common given the apparently high cost of inbreeding depression, and if it is a consequence of limited opportunities, why is incest not more common in other cooperative breeders, a large proportion of which live under conditions of comparably high ecological constraints? One possibility is that incest is unrelated to cooperative breeding, as suggested by Cockburn (Chapter 5). In any case, common moorhens may be the only known non-insular species of bird that appears to commonly engage in incest as a normal part of its mating system.

MECHANISMS OF INCEST AVOIDANCE

Considered in conjunction with the majority of cooperative breeders for which incest is rare, the above examples support the conclusion that inbreeding is, with only a few notable and several unconfirmed exceptions, low and unlikely to be playing an important role in the evolution of sociality among the vast majority of cooperative breeders. However, it does not follow that incest is actively avoided in species where it is rare. Rather, low frequency of incestuous matings can often be attributed to alternative hypotheses, including various demographic constraints (Rowley *et al.* 1993) and reproductive competition (Shields 1987). These alternative hypotheses, combined with logistic challenges of testing inbreeding avoidance (Pärt 1996) and the difficulty of obtaining unambiguous parentage data, have made confirmation of active incest avoidance in cooperative breeders at least as challenging as demonstrating the frequency of incest.

Unfortunately, there have been relatively few attempts to test alternative explanations for an observed low frequency of incest. Here we discuss three species where this issue has been specifically addressed, the long-tailed tit, the red-cockaded woodpecker, and the acorn woodpecker. We then follow up with a summary of the various mechanisms of incest avoidance that have been observed or proposed in cooperative breeders.

Long-tailed tits: avoiding incest by divorce

Long-tailed tits initially breed independently, but males are relatively philopatric and frequently act as helpers at their parents' nest if their own nesting attempt fails (Gaston 1973). Since the annual mortality rate of adults is high, this results in a potentially significant risk that, in the absence of some incest avoidance mechanism, a surviving breeder female that has successfully fledged young might end up breeding incestuously with a philopatric son following the death of her mate. Hatchwell *et al.* (2000) examined this situation and found that pairs breeding successfully were significantly less likely to remain together in the following year than unsuccessful pairs, a result contrary to the commonly observed pattern that divorce is either unrelated to prior breeding success or more frequent following a failed nesting attempt (Ens *et al.* 1993). Hatchwell *et al.* (2000) hypothesize that the uniquely high rate of divorce observed among successful pairs serves as an incest avoidance mechanism, a conclusion supported by the finding that female divorcees did not subsequently mate with a male member of their family flock and by the failure of the observed pattern of divorce to match the predictions of two alternative hypotheses.

Unfortunately, since few birds remain faithful following successful nesting, the hypothesis that incest would be more likely if birds remained with their partners following a successful breeding season is inferential. Furthermore, divorce in long-tailed tits is not particularly costly because, unlike most cooperative breeders, there are few constraints on independent breeding in this species (Hatchwell 1999). Thus, the high rate of divorce following successful breeding could be due to some other factor not considered by Hatchwell *et al.* (2000).

On the other hand, analogous situations in which animals abdicate a breeding position rather than stay and risk committing incest have been observed in other, more typical cooperative breeders where ecological constraints are high (Walters *et al.* 1988; Koenig *et al.* 1998). Although such cases are not ironclad, opting out of an (apparently) secure mating arrangement or breeding position when the probability of incest is high would appear to provide strong support for active incest avoidance when it occurs.

Red-cockaded woodpecker: dispersal and incest avoidance

Red-cockaded woodpeckers, now an endangered inhabitant of pine forests in the southeastern United States, frequently live in groups containing a small number of male offspring (helpers) from prior years. Based on a uniquely extensive data set acquired by following more than 200 groups in a 1100 km^2 area of North Carolina, Walters *et al.* (1988) and Daniels and Walters (2000b) documented circumstantial evidence for incest avoidance. This included a lack of parent–offspring incest within groups and cases of female breeders abdicating from groups in which they were established following the death of their mate when remaining offspring were their sons, but not when such helpers were not their sons.

A detailed analysis of inbreeding depression and the effects of incest avoidance on dispersal presented by Daniels and Walters (2000a) found the cost of inbreeding to be high due primarily to lower hatching success and lower survival and recruitment of fledglings. However, the effect of potential incest on female dispersal was otherwise modest: among females living on their natal territory, 18% remained rather than dispersed if all related group males were gone and there was no potential for incest, representing a small but significantly greater proportion than the 2% remaining when related males were present. Otherwise, females did not disperse further when there was a larger proportion of closely related males living nearby, nor did they avoid nearby vacancies when closely related males were present. In fact, the majority of females dispersed only 1–3 territories from their natal group, almost completely overlapping the range within which related males were often present. Incest avoidance was one, but not necessarily even the most important, factor influencing patterns of breeding female dispersal (Daniels and Walters 2000a).

Once again a likely explanation for these patterns is that incest avoidance is well-developed at the within-group level, but not at the larger subpopulation level within which most dispersal takes place (Dobson et al. 1997). Daniels et al. (2000) go on to explore the potential for these patterns to result in the accumulation of considerable inbreeding in red-cockaded woodpecker populations over time using a spatially-explicit model, an exercise with considerable potential ramifications for conservation of this species (see Chapter 12).

Acorn woodpeckers: incest avoidance at a cost

Acorn woodpeckers live in polygynandrous groups containing not only cobreeding males and joint-nesting females but offspring from prior years that delay dispersal and serve as non-breeding helpers. Cobreeders are almost always close relatives, either brothers, sisters, or some combination of parents and their (same-sex) offspring. Consequently, within-group relatedness is very high, with most individuals being siblings or offspring of everyone else within the group.

This results in a situation in which, in lieu of mechanisms preventing it, incest would almost certainly be common, especially following reproductive vacancies occurring after the death of breeders. However, in the vast majority of cases, breeder males are unrelated to the breeder females and incest does not occur. How is this accomplished?

Koenig and Pitelka (1979) and Koenig et al. (1984) originally suggested that reproductive roles in acorn woodpecker groups were determined by three rules that acted to reduce and largely avoid inbreeding. First, birds are reproductively suppressed while living in their natal group by the presence of a related breeder of the opposite sex. Thus, sons are non-breeding helpers that do not participate in reproduction if their mother (or other breeding female present when they were born), is still present in the group. Following her disappearance and replacement by a (presumably) unrelated female from elsewhere, sons can inherit breeding status within their natal group and cobreed with their father. Second, once birds disperse, they become (potential) breeders. And third, birds disperse alone or in unisexual groups only, never in bisexual groups. As a consequence of these behaviors, incest is uncommon in the population, with Koenig et al. (1984) describing 10 known or probable cases resulting from a combination of philopatry and breakdowns of the (putative) incest avoidance mechanisms.

The conclusion that these behaviors constitute active incest avoidance was subsequently challenged by Craig and Jamieson (1988) and Shields (1987) on the basis that the proposed mechanisms did not exclude alternative explanations for the observed patterns. Shields (1987) made a particularly strong case, pointing out that the acorn woodpecker data presented by Koenig et al. (1984) can be plausibly interpreted as supporting a stronger role for reproductive competition determining reproductive roles within acorn woodpecker groups than incest avoidance, a conclusion made more cogent by the striking within-group reproductive competition documented by Mumme et al. (1983a) and Hannon et al. (1985). Shields (1987) went on to suggest two contrasts to discriminate between these alternatives with respect to causing the observed patterns of dispersal and suppression of breeding.

The resolution of this controversy was forced to wait over a decade until molecular techniques allowed us to re-examine the occurrence of incest and to perform the strong inference tests needed to discriminate between the incest avoidance and reproductive competition hypotheses suggested by Shields (1987). Focusing on what happens following reproductive vacancies, Koenig et al. (1998) found support for both incest avoidance and reproductive competition playing important roles in determining reproductive behavior within groups. Supporting incest avoidance, offspring are significantly more likely to remain in their natal group and subsequently breed following vacancies of the opposite sex. However, many offspring left their natal group following vacancies of the opposite sex, and of those that remained, half did not breed in their natal groups, supporting a critical role for reproductive competition as well.

Of particular interest are groups with offspring of the same sex as a reproductive vacancy. Vacancies in such groups are filled more slowly than if helpers of the same sex are not present, suggesting that the helpers somehow deter immigrants from replacing the missing breeders. More cogently, such groups frequently fail to attempt breeding and suffer reduced reproductive success for each of the three years following the vacancy compared to groups in which either no non-breeding helpers or only helpers of the opposite sex are present. Koenig et al. (1999) analyzed this situation in detail and estimated that 9–12% of the overall reproductive potential of the population was lost due to such cases, which, since groups contain adult birds of both sexes, are attributable to incest avoidance. Thus, not only does incest avoidance appear to be strong in this population, but it is achieved at a relatively high cost, indirectly demonstrating that inbreeding depression must be considerable.

The result of these behaviors is that incest is rare. Of 400 offspring genetically fingerprinted, only 14 (3.5%) were apparently the product of incestuous matings (Dickinson et al. 1995; Haydock et al. 2001), a value only slightly higher than the 2.3% average inbreeding observed in 14 species (excluding the splendid fairy-wren and an earlier estimate for the acorn woodpecker) compiled by Rowley et al. (1993). All 14 of these young were produced by two groups lacking an unrelated adult breeder male and were the result of mother–son (nine young) and sibling (five young) incest (Haydock et al. 2001).

Thus incest, although rare, does occur. This raises a particularly vexing question: why do bisexual acorn woodpecker groups lacking a breeder of one sex, faced with the alternative of incest, not even attempt to breed? For example, of 14 groups in which no breeder males were present, 10 (71%) failed to breed altogether. Given that breeding incestuously, even faced with high inbreeding depression, is likely to be better than not breeding at all (Bengtsson 1978; Waser et al. 1986; Greeff and Bennett 2000), it is difficult to understand why under these circumstances groups do not breed incestuously. The identical situation arises in Damaraland mole-rats, where both wild and captive colonies in which the breeding female has died will fail to breed, sometimes for years, until a new, unrelated individual is introduced (Jarvis and Bennett 1993; Bennett and Faulkes 2000), and in meerkats, where extensive periods of reproductive quiescence were observed among surviving groups of relatives following the disappearance of dominants of either sex (O'Riain et al. 2000a).

Incest avoidance mechanisms in cooperative breeders

Table 9.2 summarizes the various incest avoidance mechanisms documented thus far in cooperative breeders. Even in species where inheritance of the natal territory is common, a majority of birds usually disperse and, as is generally true for birds, dispersal is frequently female-biased (Greenwood 1980). Such dispersal is a major means by which incest is reduced in cooperative breeders. However, alternative explanations for sex-biased dispersal, particularly reproductive competition (Moore and Ali 1984), have precluded the conclusion that patterns of dispersal have evolved as an incest avoidance mechanism per se. Most likely incest avoidance is but one of several functional consequences of dispersal in most cooperative breeders (Daniels and Walters 2000a).

There are, however, at least two types of dispersal in several cooperative breeders that are less readily explained by reproductive competition and thus are likely to be adaptations for incest avoidance. First is increased probability of dispersal by helpers following the death of breeders of the same sex, rather than the opposite sex, as has been documented in acorn woodpeckers (Koenig et al. 1998) and Florida scrub-jays (Balcombe 1989). In acorn woodpeckers, for example, dispersal following vacancies of the same sex is more common for both sexes but particularly important to helper males, only 31% of which remain in their natal group through the breeding season following the death of their father (a same-sex vacancy) compared to 61% following the death of their mother (an opposite-sex vacancy)(Koenig et al. 1998).

As pointed out by Balcombe (1989), aggression by and competition from unrelated replacements (stepparents) provides an alternative explanation for increased dispersal by helpers of the same sex once replacement from outside the group has occurred. Thus, what provides compelling evidence for incest avoidance

Table 9.2. *Mechanisms of incest avoidance in cooperative breeders*

Mechanism	Example	Reference
Dispersal		
From natal territory	Most species, facilitated by sex-biased dispersal	
More likely if opposite-sex breeder related	Acorn woodpecker	Koenig *et al.* 1998
	Red-cockaded woodpecker	Daniels and Walters 2000b
	Florida scrub-jay	Balcombe 1989
Abdication of breeding position	Red-cockaded woodpecker	Walters *et al.* 1988; Daniels and Walters 2000b
	Acorn woodpecker	Koenig *et al.* 1998
Dominants suppress offspring reproduction		
In both sexes	Naked mole-rat	Faulkes and Bennett 2001
In opposite-sex offspring	Acorn woodpecker	Koenig *et al.* 1998, 1999
	White-browed scrubwren	Magrath and Whittingham 1997; Whittingham *et al.* 1997
	Stripe-backed wren	Rabenold *et al.* 1990; Piper and Slater 1993
	Bicolored wren	Haydock *et al.* 1996
	Superb fairy-wren	Dunn *et al.* 1995
	Meerkat	O'Riain *et al.* 2000a
	Damaraland mole-rat	Bennett *et al.* 1996; Cooney and Bennett 2000
	Common mole-rat	Burda 1995
	Mashona mole-rat	Herbst and Bennett 2001
Other parental behaviors		
Divorce following successful breeding	Long-tailed tit	Hatchwell *et al.* 2000
Extra-group mating	Pilot whale	Amos *et al.* 1993
	Ethiopian wolf	Sillero-Zubiri *et al.* 1996
	Meerkat	O'Riain *et al.* 2000a
	Superb fairy-wren	See text
Kin recognition	(No unambiguous cases)	

is not so much the difference in helper dispersal following the filling of same-sex vacancies, but the fact that such vacancies are generally filled from outside the group to begin with. This is because in the absence of incest avoidance, helpers should be capable of inheriting and breeding following the death of their same-sex parent, and should therefore only be kept from doing so when outcompeted and expelled by an older or larger coalition of unrelated birds (Koenig *et al.* 1998).

The second kind of dispersal event that provides particularly persuasive, if circumstantial, evidence for incest avoidance is the abdication of breeding positions by breeders when faced with the situation when all group members of the opposite sex are close relatives. Given that ecological constraints are high in both acorn woodpeckers and red-cockaded woodpeckers, where this behavior has been observed, abdication strongly suggests that the costs of inbreeding are sufficiently high to warrant what in some cases appear to be extreme measures designed to avoid it.

The second major category of behavior that serves to reduce inbreeding in cooperative breeders is the suppression of offspring reproduction in cases where incest might otherwise occur. In naked mole-rats, reproductive suppression in both sexes appears to be imposed by the female breeder, but in most species the key interaction is with parents of the opposite sex. In acorn woodpeckers, for example, both sexes can inherit and breed in their natal group, but only after the death and replacement of related breeders of the opposite sex (Koenig et al. 1998). Similarly, male helpers have been found to potentially inherit and breed in the natal group in several other cooperative breeders, but only when their mother has died and been replaced by a new, unrelated female. A similar effect also occurs in several cooperatively breeding mammals and has been demonstrated experimentally in three species of mole-rats (Burda 1995; Cooney and Bennett 2000; Herbst and Bennett 2001). The mechanism by which reproduction is suppressed, and the extent to which it is imposed by dominant control rather than self-restraint, is often not known but probably varies, as suggested by depressed levels of reproductive hormones in some species but not in others (O'Riain et al. 2000a).

Beyond these two major categories of incest avoidance mechanisms, at least two additional behaviors have been documented, including divorce following successful breeding, discussed above in long-tailed tits, and extra-group mating in situations where potential mates within the group are relatives. The latter appears to be common in several group-living mammals including pilot whales, Ethiopian wolves, and meerkats. Although such extra-group mating is common in several cooperatively breeding birds, it has yet to be demonstrated to be specifically related to incest avoidance (Mulder et al. 1994).

Recognition of relatives is basic to these mechanisms of incest avoidance. However, recognition is not known to entail anything beyond association. In other words, offspring treat all individuals present in their natal group when they were born as close relatives and avoid mating with them, at least as long as they live in the offspring's natal group; more complex mechanisms of kin recognition such as phenotype matching or recognition alleles (Holmes and Sherman 1983) have not been demonstrated to occur. However, few attempts appear to have been made to detect such behaviors, with the notable exception of Daniels and Walters' (2000a) study showing that female red-cockaded woodpeckers do not appear to recognize or adjust their dispersal behavior depending on the distribution or number of male relatives in nearby territories. As a consequence, incidental incest occurring because a related male and female dispersed independently to the same territory appears to occur regularly, if rarely, in this and several other species. Apparently such incidental inbreeding has not been sufficiently common to provide strong selection for more sophisticated kin recognition mechanisms, but additional study in this area is clearly warranted.

CONSEQUENCES OF INCEST AVOIDANCE BEHAVIORS

Incest avoidance and reproductive skew

Although largely beyond the scope of this chapter, the importance of optimal-skew theory based on concessions among potential cobreeders is an important issue in cooperative breeders (Chapter 10). Optimal-skew theory has been used to explain many basic aspects of cooperative breeding, one of which is the general pattern of non-breeding by female offspring living in matrifilial societies, including the majority of species considered here (Reeve and Keller 1995).

Incest avoidance provides an alternative explanation for this phenomenon (Emlen 1996). Reeve and Keller (1996) dismissed this possibility on the basis that females should have easy access to extra-pair mating opportunities involving unrelated males, and thus should not be constrained by the lack of potential mates within their social unit. Although this does appear to be true in some social species (Table 9.2), it is not in others. In acorn woodpeckers, for example, there appears to be no mechanism by which a breeder female could restrict her daughters from laying eggs communally beyond the normal egg destruction observed in all joint-nesting associations. Nonetheless, joint nesting between mothers and daughters has only been observed to occur following the death and replacement of the breeder male, when incest is no longer an issue. Thus incest avoidance, and not concessions between joint-nesting females, determines whether or not helpers in this species breed. Similarly, other characteristics of acorn woodpecker behavior are consistent with the hypothesis of incest avoidance determining reproductive roles within groups (Koenig *et al.* 1998).

One solution to this dilemma is to add incest avoidance as a new factor potentially influencing reproductive skew (Haydock and Koenig 2003). There are two concerns with this approach. First, skew models are traditionally focused exclusively on within-sex interactions (Keller and Reeve 1994). Thus, adding incest avoidance, which is only one intersexual phenomenon that may influence reproductive partitioning and provide alternatives to skew theory, may not be technically difficult but is conceptually awkward. Second, in most cases, if incest avoidance were included, it would have to be sufficiently strong that breeding by helpers would not be predicted to occur. This would effectively reduce the analysis to one excluding helpers altogether, which is tantamount to how most authors are currently dealing with this issue.

Demographic consequences of incest avoidance

Inbreeding entails at least two potential costs. First, inbreeding depression is a potential problem for the viability of species like red-cockaded woodpeckers that live in small isolated populations and exhibit highly viscous patterns of dispersal (Daniels *et al.* 2000). Inbreeding depression is also a potential problem for any population that suffers a bottleneck. If the cause of the bottleneck is not resolved, inbreeding depression is likely to facilitate the demise and ultimate extinction of the population. If the cause is resolved and the population recovers, the long-term effects of inbreeding are unknown. In theory, however, it should alter the selective benefits of certain behaviors. For example, by purging the genome of deleterious recessive alleles, inbreeding depression and the benefits of inbreeding avoidance might be significantly reduced. Whether this is responsible for some of the more extraordinary behaviors currently observed in the Seychelles warbler, including its apparent lack of incest avoidance, is unknown but seems plausible.

Incest avoidance may, however, have significant adverse effects on productivity of populations well before inbreeding depression becomes an issue by restricting available mates. In humans, simulations by Hammel *et al.* (1979) suggest that the costs of even quite restrictive incest taboos are likely to be small if the population is at least several hundred individuals in size, but as the population decreases to under 50, even incest avoidance at the level typically observed in cooperative breeders (prohibition of mating within the nuclear family) leads to population decline unless individuals engage in behaviors leading to the production of offspring outside of normal marriage. Such a demographic situation would clearly result in pressure to change the prevailing level of incest avoidance.

In acorn woodpeckers, cases in which groups fail to breed because of a reproductive vacancy are fairly common. In cases where such groups contain related birds of both sexes, this decline in fertility is attributable to incest avoidance. Based on computer simulations, Koenig *et al.* (1999) estimated the decrease in potential population growth attributable to these events to be between 9 and 12%, tantamount to a decline in overall population growth (r values) of the order of 1.8 to 2.3% year^{-1}. This demonstrates that the potential demographic consequences of incest avoidance can be considerable in cooperative breeders where groups are often small and relatedness high. Although this cost may often be cryptic, it is always present, depressing population growth to an extent that could clearly facilitate population decline and extinction. For acorn woodpeckers, this is unlikely to be true in much of the Pacific coast but may be a contributing difficulty in the southwestern United States, where populations are more demographically fragile and in some cases dependent on immigration for their persistence (Stacey and Taper 1992; Ligon and Stacey 1996).

The bottom line is that inbreeding and inbreeding avoidance may play critical roles in population declines and conservation. Specifically, inbreeding avoidance places restrictions on mating that can depress population growth rates and facilitate population declines, leading to small populations where inbreeding depression may add further blocks to recovery.

CONCLUSION

Although inbreeding and inbreeding avoidance are important in virtually all vertebrates, they are of particular importance in cooperative breeders where individuals live in relatively small groups and relatedness is typically high. In the vast majority of cooperative breeders, these phenomena appear to be important in determining reproductive roles, influencing reproductive skew, and generally shaping reproduction within groups to an extent comparable to, and possibly even exceeding, that due to reproductive competition (Koenig *et al.* 1998).

Although short-distance dispersal makes incest beyond the natal group a significant possibility in many of these species, there is no evidence as yet to indicate that individuals recognize relatives with which they are unfamiliar or avoid incest following dispersal. However, more work in this area is clearly warranted.

Throughout the history of study on cooperative breeders, claims have been made that incest was common and inbreeding avoidance absent. With the accumulation of more data and the development of genetic techniques to determine parentage, the majority of these claims have proved wrong or at best premature. There remain, however, a handful of cooperatively breeding birds and mammals for which inbreeding avoidance appears to be absent and incest appears to be sufficiently common that it may constitute a normal part of their mating system. One of these, the insular Seychelles warbler, has suffered a population bottleneck recently that may have played an important role in eliminating the usual selective benefit of incest avoidance. However, no such convenient, albeit ad hoc, explanation exists for the apparently high levels of incest observed in the common moorhen. The primacy of incest avoidance in most cooperative breeders makes its apparent absence in these few all the more perplexing. We can hope that continuing long-term studies combined with improved and simplified parentage analyses will eventually provide an explanation. Until then, these species remain outliers with an important message yet to be decoded.

ACKNOWLEDGMENTS

We thank all the field assistants who helped monitor the Hastings acorn woodpecker population over the years, our colleagues Janis Dickinson, Ron Mumme, Mark Stanback, and Frank Pitelka for their assistance, and Jeff Walters for comments on the manuscript. This work was supported by the National Science Foundation, most recently by grant IBN-0090807.

10 · Reproductive skew

ROBERT D. MAGRATH
Australian National University

ROBERT G. HEINSOHN
Australian National University

RUFUS A. JOHNSTONE
University of Cambridge

Cooperatively breeding birds span a huge range of social organization, defined in part by the number of individuals of each sex and their reproductive roles within the group (Brown 1987; Stacey and Koenig 1990a; Cockburn 1998). In some species or groups, individuals of one sex share reproduction roughly equally, while in others a single individual monopolizes reproduction by that sex. Reproductive skew theory attempts to understand this variation in the partitioning of reproduction among individuals of the same sex. Such variation can be quantified as "reproductive skew," ranging from 0 (egalitarianism) to 1 (monopolization). A key assumption of many models of optimal reproductive skew is that dominant individuals control the reproduction of subordinates, but that in some circumstances allow subordinates to share reproduction as a way of enlisting their cooperation (Vehrencamp 1979, 1980, 1983a, 1983b; Emlen 1982b). Dominants therefore determine the partitioning of reproduction and the magnitude of skew through the size of the reproductive "concession" granted to subordinates. Reproductive skew models were originally designed to explain variation among species, but are also applicable to variation within species and even within single populations (Emlen 1995; Reeve *et al.* 1998).

Within the last decade there has been a resurgence of interest in reproductive skew, stimulated in part by the ability to use molecular methods to determine parentage and thereby quantify skew. The renewed interest is revealed in the development of many new and refined models of optimal skew (Keller and Reeve 1994; Keller and Chapuisat 1999; Johnstone 2000; Magrath and Heinsohn 2000; Reeve 2000) and critiques of the models (Emlen 1995, 1996, 1997a; Emlen *et al.* 1998; Clutton-Brock 1998; Reeve *et al.* 1998; Johnstone 2000; Magrath and Heinsohn 2000; Reeve and Keller 2001).

We follow recent reviews in recognizing two major classes of reproductive skew model: transactional and compromise (Johnstone 2000; Reeve 2000; Reeve and Keller 2001). In transactional models, individuals "trade" direct reproduction to achieve individual benefits and group cooperation. The original "concession" models of Vehrencamp and Emlen are transactional, because dominants allow some subordinate reproduction in return for group stability and cooperation. In contrast, compromise models are based on competition to maximize reproductive share, potentially without regard to group stability or cooperation. We also outline recent synthetic models (Johnstone 2000; Reeve 2000) showing that these contrasting views of how individuals interact are not mutually exclusive.

In this chapter we first briefly review models of skew and then consider whether these models are useful in understanding reproductive sharing among male birds. We then provide a verbal and graphical explanation of transactional, compromise, and synthetic models of skew, after which we consider the potential influences of incest and female choice. Next we explain the difficulty of testing models, illustrated by case studies, and finally we consider the prospects for further progress and provide our general conclusions.

We focus on reproductive sharing among males in order to minimize overlap with Chapter 11, which considers sharing among females. This restriction imposes little constraint on the discussion of theory because most models are applicable to either sex, and for completeness we include model variants specific to females. However, males and females differ in important ways, and measuring male skew requires molecular evidence, so our assessment of evidence and prospects are explicitly limited to males.

Ecology and Evolution of Cooperative Breeding in Birds, ed. W. D. Koenig and J. L. Dickinson. Published by Cambridge University Press.
© Cambridge University Press 2004.

TRANSACTIONAL MODELS

The common feature of transactional models of skew is that they assume individuals limit their own share of reproduction in return for stable cooperation. However, different models assume different types of interaction between dominants and subordinates and take into account varying ranges of behavioral options. Here we emphasize simpler models, which encapsulate the major features of all such models, but also list some variations of these basic models. We will not use the terms "reproductive skew" or "optimal reproductive skew" to refer to a particular type of model, since the degree of reproductive sharing is what all models try to explain, and all hypotheses could be cast in an optimality framework.

Concession model

The concession model is the original "optimal skew" model developed by Vehrencamp (1979, 1980, 1983a, 1983b). It assumes that dominants have perfect control over the amount of reproduction gained by subordinates, and that subordinates either stay in the group and cooperate or leave the group. The model predicts the minimum amount of reproduction offered by the dominant that only just compensates the subordinate for staying in the group. The model predicts the dominant's optimal "staying incentive" (Reeve and Ratnieks 1993) or magnitude of "concession" (Clutton-Brock 1998), and therefore the "optimal skew" from the perspective of the dominant. The model considers a group with one dominant and one subordinate and assesses the combined effects of the relatedness between the dominant and subordinate, ecological constraints on dispersal of the subordinate, and the effect of the subordinate's presence on the group's reproductive success.

The model predicts that a dominant will concede a smaller share of reproduction when the subordinate is more closely related, has less chance of reproducing independently, and has a larger effect on the group's productivity. Thus reproductive skew is predicted to be greatest, and the society most "despotic," when the subordinate is a close relative, with little prospect of breeding independently, who greatly improves the group's success. At the other extreme, reproductive sharing will be roughly equal, and the society most "egalitarian," when the subordinate is unrelated, has a good chance of reproducing elsewhere, and has a limited effect on group productivity. The effect of kinship arises because the closer the relatedness, the greater the indirect benefit to the subordinate arising from the increased productivity of the group, and therefore the lower requirement for the dominant to offer direct reproduction to entice the subordinate to remain in the group. Stronger ecological constraints mean that a smaller staying incentive is sufficient to retain the subordinate in the group since it has poor prospects elsewhere. The group-productivity effect is counterintuitive and arises because the lower the subordinate's effect, the lower the indirect fitness that can be obtained by helping relatives and therefore the greater the share of direct reproduction the dominant must offer to entice the subordinate to stay. Nonetheless if the dominant has nothing to gain from the subordinate's presence it will offer no concessions and the subordinate should leave the group. Intermediate levels of relatedness, constraint, or contribution to group success mean that skew can range anywhere between 0 and 1.

Restraint model

The restraint model uses the same general framework as the concession model, but assumes that the dominant has the ability only to evict a subordinate and not to control the subordinate's share of reproduction while it is a member of the group (Johnstone and Cant 1999). Thus the subordinate is selected to take as great a share of reproduction as is tolerated by the dominant. The dominant will tolerate a subordinate's presence only if its inclusive fitness is higher than it could achieve by breeding alone. The subordinate must therefore exercise restraint or it will be evicted from the group.

Predictions about the subordinate's share of reproduction from the restraint model are opposite to those of the concession model. The subordinate takes a larger share of reproduction, and skew will therefore usually decline, when it is more closely related to the dominant, it faces greater ecological constraints, and it has a larger effect on the group's productivity. The two models make opposite predictions because the same factors that make association profitable for the subordinate, and thus reduce the staying incentive it requires, also make association profitable for the dominant, and thus increase the level of subordinate reproduction that it will tolerate. Reproduction lost to a close relative has a smaller effect on the dominant's inclusive fitness than reproductive share lost to a non-relative, so it will tolerate

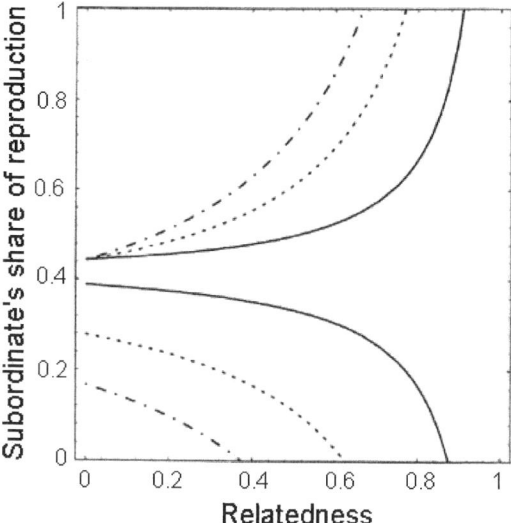

Figure 10.1. Combined effects of concession and restraint models, which delimit the range of sharing still allowing group stability. The figure shows the minimum (concession) and the maximum (restraint) share of reproduction that the subordinate can acquire in a stable group as a function of r, the relatedness between the dominant and subordinate. These lower and upper limits are shown for three different values of x, the expected reproductive success of a subordinate who disperses to breed independently: 0.7 (solid lines), 0.5 (dotted lines) and 0.3 (dot-and-dashed lines), all relative to an established lone breeder. In all three cases k, the productivity of the association relative to that of a lone breeder, is equal to 1.8. From Johnstone (2000); used with permission from Blackwell Verlag.

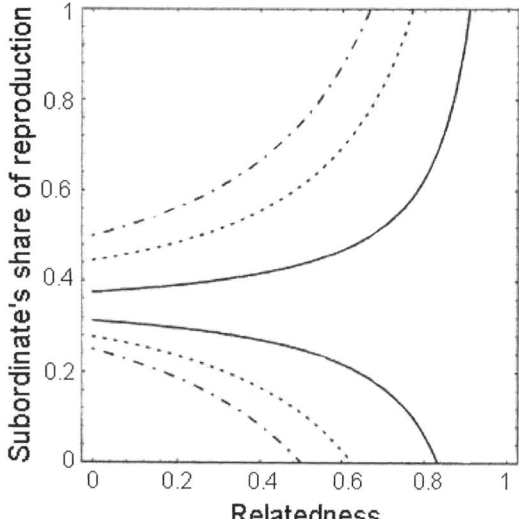

Figure 10.2. Combined effects of concession and restraint models, which delimit the range of sharing still allowing group stability. The figure shows the minimum (concession) and the maximum (restraint) share of reproduction that the subordinate can acquire in a stable group as a function of r, the relatedness between dominant and subordinate. These lower and upper limits are shown for three different values of k, the productivity of the association relative to that of a lone breeder: 1.6 (solid lines), 1.8 (dotted lines), and 2.0 (dot-and-dashed lines). In all three cases x, the expected reproductive success of a subordinate who disperses to breed independently relative to an established lone breeder, is equal to 0.5. From Johnstone (2000); used with permission from Blackwell Verlag.

greater reproduction by close relatives. The inclusive fitness cost of evicting a subordinate when it has little chance of finding a breeding vacancy is greater if the subordinate is a closer relative, so it will tolerate greater reproduction by the subordinate under severe ecological constraints. Finally, a dominant will tolerate greater reproduction by a subordinate if it has a greater effect on the group's productivity.

Concession and restraint: the zone of stability and conflict

The concession and constraint models together define the boundaries of group stability. Subordinates will leave groups if they do not get the minimum share of direct reproduction required under the concession model, and will be evicted by the dominant if they take a greater share than that allowed under the restraint model. Figures 10.1 and 10.2 show the predicted minimum and maximum share of reproduction by a subordinate, as defined by the concession and restraint models, as a function of relatedness and ecological constraints (Fig. 10.1), and relatedness and the subordinate's effect on group productivity (Fig. 10.2). The figures show that as relatedness (r) increases, the success of a dispersing subordinate (x) decreases, and the effect of a subordinate on the group's productivity (k) increases, so the minimum reproduction the dominant must offer to ensure group stability declines, while the maximum reproduction a subordinate can "get away with" increases. They also show that the range of subordinate reproductive sharing allowing group stability increases with relatedness, ecological constraints, and magnitude of beneficial effect on group productivity. This makes intuitive sense:

cooperatively breeding groups are predicted to be most stable when consisting of relatives facing severe ecological constraints that are much more productive together than when breeding alone. However, this range of group stability also defines a behavioral zone of conflict in interactions among individuals (Reeve 2000). We discuss this zone in more detail below.

Modifications

The concession and restraint models together define the major features of transactional models, but there are several important modifications which affect reproductive sharing and reproductive skew. Here we provide a brief overview of some other transactional models, indicating in what ways they modify the predictions of concession and restraint models.

Competitive ability, fighting, and peace incentives
Under the concession model, a dominant provides a subordinate with just sufficient reproduction to compensate it for remaining in the group, and its share of reproduction can be much lower than the dominant. Consequently the subordinate could potentially gain higher reproductive success by fighting to take over as the dominant. Reeve and Ratnieks (1993) modified the concession model to include the relative fighting ability of the subordinate, and predicted that the dominant can offer a "peace incentive" to prevent the subordinate from fighting. The peace incentive is in addition to the staying incentive of the concession model. Their model predicts that the dominant will offer greater peace incentives, thereby decreasing reproductive skew, when the subordinate is of higher competitive ability. This result is important because a subordinate's competitive ability is likely to covary with group type in cooperatively breeding birds. Although there has been no formal model, it is also plausible that dominants may be more wary about attempting to evict a more competitive subordinate, who may therefore be able to exercise less "restraint" and take a greater share of reproduction.

Bribery
Within the range of group stability (or "zone of conflict") defined by concession and restraint models, individuals may potentially compete to obtain their greatest share of reproduction. Such competition potentially reduces the group's reproductive productivity. Under these circumstances, individuals may "bribe" each other by offering a greater reproductive share in return for cooperation and the resulting greater group productivity (Reeve and Keller 1997). Such bribery is another form of social transaction and can potentially reduce reproductive skew.

Cost of reproduction for females
It can be optimal for a dominant female to allow a related subordinate to lay eggs in her nest if the cost of laying eggs is an accelerating function of clutch size (Cant and Johnstone 1999). Because it is more costly to lay each successive egg, there comes a point when the net benefit to a dominant of laying another egg is less than that of allowing a relative to lay eggs at a lower cost. The effect of a cost of laying on reproductive skew depends on other assumptions. If the subordinate is constrained to remain in the group and contributes nothing to parental care, dominants will always grant closer relatives a greater share of reproduction, as in the restraint model. However, if subordinates boost brood productivity and can disperse, reproductive skew initially increases with relatedness as in the concession model, but above a threshold relatedness, skew declines with relatedness. The model highlights a potential difference between males and females, because only females are likely to suffer an accelerating cost of producing successive young (Cant and Johnstone 1999). Further discussion can be found in Jamieson (1999).

Relatedness symmetry
Asymmetrical relatedness introduces an important complication to the original concession model (Reeve and Keller 1995, 1996). For example, in a trio consisting of a mother and daughter with an unrelated male, each female is related to the offspring of the other by the same amount ($r = 0.25$), and relatedness is symmetrical. However, in a nuclear family consisting of a mother and daughter with the daughter's father, the mother is less closely related to offspring produced by the daughter ($r = 0.25$) than the daughter is to offspring produced by her mother ($r = 0.5$), and relatedness is asymmetrical. Other things being equal, the daughter will require, and the mother will offer, fewer concessions under these conditions. Furthermore, if dominant control is exercised through the threat of eviction, the daughter must exercise greater restraint.

The above argument would seem to be applicable to males as well: in a nuclear family consisting of a father and son with the son's mother, the father should be less closely related to offspring produced by the son than the son is to offspring produced by his father. The original argument, however, assumes that the daughter mates with an outside, unrelated male while raising young within her home group. If she were to mate with her father, there would also be a difference in the relatedness of each female to her own offspring (mother–offspring = 0.5, daughter–offspring = 0.75), canceling out the asymmetry in relatedness to each other's offspring. The assumption that a daughter avoids incest and mates outside the group is reasonable in many cases (see Chapter 9), but the equivalent assumption makes no sense in the context of skew among males, because reproductive sharing by males within a group is not affected by paternity they gain outside the group. Relatedness asymmetry is thus likely to be relevant only to skew among females.

Bidding

If a subordinate can potentially join more than one group, dominants might enter a "bidding war" to gain the services of the subordinate (Reeve 1998). In this situation, the subordinate's share of reproduction will be higher than in the concession model. Overall, the model predicts that skew should be low independent of relatedness and should decline as the subordinate's contribution to group productivity increases.

Manipulation

In the concession model a dominant can entice a subordinate to remain in the group by offering a share of reproduction such that the subordinate's inclusive fitness is the same whether it stays or attempts to breed independently. The dominant increases its concession to match increasing prospects of independent breeding. Crespi and Ragsdale (2000) provided a Machiavellian twist to this equation by suggesting that dominants might increase a subordinate's likelihood of staying by deliberately reducing its chance of breeding independently! For example, a dominant might harass a subordinate, or perhaps feed it poorly during development, to reduce its ability to compete for independent breeding situations. If the subordinate has little prospect of breeding independently it will then "choose" to stay in the group, ultimately as a consequence of the dominant's manipulation. The result is that this model predicts stable monopolization of breeding under a much greater range of relatedness and ecological conditions than the concession model.

Social queuing and adult survival

Previous models of reproductive skew have considered only current reproduction within a group, but many cooperatively breeding birds are long-lived and subordinates may gain the delayed benefit of inheriting dominance within the group (Wiley and Rabenold 1984; Emlen 1999). Thus the benefits of group membership may include future reproduction, not just current direct and indirect fitness. Kokko and Johnstone (1999) put skew theory into a life-history framework in a dynamic model looking at the probability of inheriting dominance status upon the death of the current dominant. They extended the basic concession model by adding the probability of survival of the dominant and subordinate from one breeding season to the next.

Social queuing dramatically increased reproductive skew because a dominant needed to offer fewer concessions when a subordinate can inherit its position. Whenever a subordinate had similar or higher survival than the dominant, an increase in adult survival led to a greater chance of inheritance and hence a reduced need for concessions. This was not always true if subordinates had lower survival. Subordinate survival usually had a much greater effect on skew than relatedness, and groups in which the dominant monopolized reproduction were often stable, even when the dominant and subordinate were unrelated. If there was an effect of relatedness, it was towards greater skew with higher relatedness, as in the concession model. Dominants would even tolerate a small reduction in group productivity if the subordinate was a relative as long as the subordinate's probability of independent reproduction was sufficiently low, illustrating the idea of "parental facilitation" proposed by Brown and Brown (1984). Similarly, the dominant tolerated some reduction in group productivity if it experienced increased survival when a subordinate was present, even if they were unrelated. Ragsdale (1999) also modeled delayed benefits, using a different approach, and similarly found that future benefits permitted higher skew and group stability in a greater range of conditions.

Multi-member groups

Most models of reproductive skew assume that only two individuals compete to share reproduction. Reeve and Emlen (2000) extended the transactional approach to consider groups of arbitrary size, assuming that subordinates are identical in relatedness and prospects of successful dispersal, and that group productivity is a decelerating function of group size. They showed that in a "saturated" group, one that has reached the point where subsequent joining by subordinates is no longer beneficial, staying incentives may be largely insensitive to relatedness, although the saturated group size itself, which influences total skew, does depend on relatedness.

Johnstone *et al.* (1999) modified a concession model to examine reproductive sharing among a dominant and two subordinates, who are themselves equally subordinate but can differ in their relatedness to the dominant and to each other. The model also allowed group productivity to be a decelerating or accelerating function of group size: that is, the second subordinate may have a smaller or larger effect than the first.

The specific predictions of the model cannot be summarized simply, because the effects of relatedness on skew depended on other variables. For example, skew could increase or decrease with the relatedness between subordinates and the effect was stronger when the subordinates were more closely related to the dominant. Whether skew increased or decreased depended on whether reproductive productivity was, respectively, a decelerating or accelerating function of group size. An important conclusion was that the reproductive share granted to one subordinate can be affected by characteristics of the other subordinate and the specific effect of group size on reproductive productivity.

COMPROMISE AND SYNTHETIC MODELS

Compromise models of skew differ from transactional models in that individuals compete to gain direct reproduction without regard to group stability. Individuals do not engage in "social contracts" over reproduction in order to maintain group stability and the observed skew is a compromise between the optima of dominants and subordinates (Johnstone 2000). Nonetheless, reproductive skew need not be complete because dominants may have incomplete control of subordinate reproduction.

Tug-of-war model

Reeve *et al.* (1998) developed a general "tug-of-war" model in which competition for reproductive share comes at the price of reduced group productivity. They assumed that a given group productivity is divided between a dominant and subordinate, but that there is a cost, in terms of reduced group productivity, when resources are devoted to competition rather than reproduction. Depending on the specific model, dominant individuals were assumed either to be more efficient at using resources or to have access to greater resources; in both cases they had some advantage in competition with subordinates. Nonetheless, both the dominant and subordinate face a trade-off between maximizing group productivity and maximizing their share of direct reproduction. Group productivity was assumed to increase linearly with the amount of resources that are not devoted to competition.

Overall, these tug-of-war models predict that the relative competitive ability of the subordinate is the major determinant of skew. The less competitive the subordinate compared to the dominant, the smaller the share of reproduction it is able to claim, and thus the higher the skew. The effect of relatedness contrasts with the concession model, as skew is either unaffected by relatedness or declines with increasing relatedness. Assuming that dominance is defined by the efficiency of using resources, kinship has little influence on the partitioning of reproduction. As relatedness increases, both individuals reduce their efforts to claim reproduction for themselves, to an extent roughly inversely proportional to their competitive ability. The end result is that while aggression declines, the level of skew remains roughly constant or declines very slightly. Assuming instead that dominance is defined by access to resources rather than efficiency of using them, skew is predicted to decrease markedly with relatedness, at least when the dominant and subordinate are not close kin. The reason is that at low levels of relatedness, the entire share of reproduction that the subordinate controls is devoted to the struggle over reproduction. Within this part of the parameter range, variation in skew is thus solely the result of variation in the dominant's level of effort, which tends to decline with relatedness, as in the "efficiency" version of this model.

Several other less general models also examine the consequences of dominants not having control over subordinate reproduction. For example, Cant (1998) examined reproductive sharing among females contributing to a joint clutch. In the model, which assumed that additional breeding females do not increase group productivity, the dominant has no direct control over how many eggs the subordinate will contribute to their joint clutch, yet reproductive skew was predicted to increase with relatedness among females, as in the concession model. This is because the indirect fitness cost of reducing the dominant's success as the clutch size exceeds the optimum will prompt the subordinate to restrain her reproduction more when the dominant is a closer relative.

Compromise within a transaction framework: the synthetic model

Although transactional and compromise models are based on entirely different assumptions about the ways individuals interact, recent models by Johnstone (2000) and Reeve (2000) suggest that there can be compromise within a transactional framework. The threat of subordinate departure, considered in the concession model, and the threat of eviction, considered in the restraint model, together circumscribe a range within which groups can be stable despite conflicts among individuals. Reeve (2000) calls this a "zone of conflict" or "window of selfishness." In other words, in groups in which dominants do not have complete control, subordinates can compete for a share of reproduction up to but not exceeding the maximum defined by the restraint model, without the risk that they will be evicted.

The synthetic model developed by Johnstone (2000) predicts that both competitive ability and kinship will affect skew (Fig. 10.3). More competitive subordinates acquire a larger share of reproduction, as in the tug-of-war model, so that reproductive skew is lower when subordinates are more competitive. Skew is affected by kinship in complex ways. At low levels of relatedness ($r < 0.4$), both the threat of departure and the threat of eviction exert an influence on the division of reproduction, so that both individuals require some share of reproduction. Within this range, the subordinate's share of reproduction tends to decrease with relatedness, particularly when it is relatively uncompetitive. This reflects the decline in the minimum concession it requires as relatedness increases. At intermediate levels of relatedness ($0.4 < r < 0.7$), only the

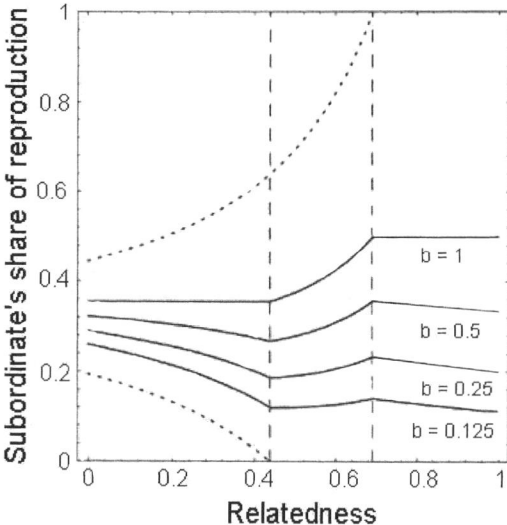

Figure 10.3. Predicted share of subordinate reproduction from Johnstone's (2000) synthetic model. The dotted lines show the minimum and the maximum share of reproduction that a subordinate may acquire in a stable group as a function of relatedness between dominant and subordinate. The minimum share follows from the concession model, while the maximum share follows from the restraint model. Together they delimit the "zone of conflict" in which reproductive competition can occur without affecting group stability. The solid lines show the actual share the subordinate is predicted to acquire in the synthetic model for several different values of b, the competitive ability of the subordinate relative to that of the dominant. For all the curves shown, k (the productivity of the association relative to that of a lone breeder) is equal to 1.8, and x (the expected reproductive success of a subordinate who disperses to breed independently) is equal to 0.35. The vertical dashed lines delimit regions in which, moving from left to right: (1) both the threat of eviction and the threat of departure constrain the division of reproduction, (2) only the threat of eviction constrains the division of reproduction, and (3) neither the threat of eviction nor of departure impose any restrictions on skew. From Johnstone (2000); used with permission from Blackwell Verlag.

threat of eviction continues to exert an effect because the subordinate's minimum share has dropped to zero: that is, the dominant need not offer a concession. Within this range, the subordinate's share of reproduction tends to increase with relatedness, particularly when it is a relatively strong competitor. This reflects the increase in the maximum share it may safely claim as relatedness increases (the "restraint" effect). At high levels of relatedness ($r > 0.7$), both the threat of departure and the threat of eviction become irrelevant, and the model yields the same solution as the basic subordinate-inefficiency tug-of-war. Skew therefore shows little change, or a slight decrease, with kinship. It is unlikely, however, that such high levels of relatedness occur regularly in cooperatively breeding birds: groups commonly contain a range of relatedness from non-relatives ($r = 0.0$) to first-order relatives ($r = 0.5$), but the infrequency of incest (Chapter 9) means that higher relatedness values are probably rare.

Skew is also influenced by group productivity and the possibility of independent reproduction. As with the impact of relatedness, the effects of these parameters vary in relation to kinship. At low levels of relatedness the synthetic model yields similar predictions to the concession model, so that the subordinate's share of reproduction decreases with increasing group productivity and increases with the opportunity for independent breeding. At intermediate levels of relatedness, when only the threat of eviction remains relevant, the predictions of the synthetic model match those of the restraint model, in which the subordinate's share of reproduction increases with group productivity and decreases with the opportunity for independent breeding. Finally, at high levels of relatedness, the model reduces to the basic tug-of-war, and neither group productivity nor the opportunity for independent breeding influence skew.

In summary, the predictions of the synthetic model approximate those of concession, restraint, or tug-of-war models depending upon the values of the relevant parameters, so that patterns of skew may primarily reflect the threat of departure by subordinates, the threat of eviction by dominants, or the struggle over reproduction within the group.

OTHER INFLUENCES ON REPRODUCTIVE SHARING

Female choice and sexual conflict

Reproductive skew among males can at least partly reflect female control. This possibility was identified in early discussions of reproductive skew by both Vehrencamp (1979, 1980) and Emlen (1982b), yet has only recently been subject to formal modeling (Cant and Reeve 2002). Here we consider why females might attempt to influence, or even totally control, reproductive sharing by males, and outline Cant and Reeves' models. The possibility of female control contrasts with previous models that assume that reproductive skew results entirely from social transactions or compromise about reproduction within a sex.

Cant and Reeve's (2002) "work incentive" model addresses a situation in which females benefit from manipulating reproductive sharing among males. This is likely to be important in many species and is well illustrated by dunnocks, in which groups are composed of unrelated adults. In a polyandrous trio of this species, whether a beta male helps feed nestlings is determined by whether he copulates with the female, so it is in the female's interest to copulate with both males (Davies 1992). On the other hand, it is in the alpha male's interest to try to stop the beta from copulating with the female. Thus there is conflict between the sexes and low reproductive skew is partly or primarily the result of female control.

The work-incentive model assumes that females will allocate paternity so as to maximize the total care provided by two males. The model assumes that the amount of offspring care provided by each male is a function of his share of paternity, and that groups are stable regardless of reproductive share. It predicts that when the beta male's genetic quality is lower than that of the alpha, the beta's share of paternity decreases with relatedness, as also predicted by the concession model. When the males are identical in quality, or the female cannot distinguish genetic quality, she allocates paternity equally. As the relative cost of care to the beta male increases, the female increases his share to compensate him for this cost, and therefore skew declines. Importantly, the model predicts conflicts of interest among all parties, including between the alpha male and the female.

Cant and Reeve (2002) also develop a "staying incentive" model based on transactional models of skew. They assume that the female has total control over the allocation of paternity but that the alpha male will evict the beta if the latter exceeds a threshold share of paternity, as in the restraint model. Predictions of the model depend on whether the alpha male benefits from sharing reproduction. When both the female and the alpha male benefit from the presence of the beta, predictions are similar to the concession model: for example, skew increases with relatedness. However, when only the female benefits, the predictions are similar to the restraint model, and skew declines with increasing relatedness. If the alpha male has partial control over beta reproduction, groups are likely to be unstable, and so real groups should be characterized by a lack of conflict between the alpha male and female and generally follow the predictions of the concession model. The predictions of the staying-incentive and work-incentive models are then mostly similar except for this lack of conflict between the alpha male and female, and the lack of an effect of ecological constraints under the work-incentive model, which assumes that group stability is unaffected by skew.

A female's perspective on reproductive skew among males might also involve other genetic costs and benefits. For example, females might benefit from genetic diversity among their offspring and so mate with more than one male when the males are unrelated to each other. Similarly, although a subordinate male may gain some net benefit from offspring of an incestuous mating with his mother even if those offspring are of lower fitness than offspring of the dominant, it is less likely to be in the interests of the female to raise her son's inbred offspring in comparison with the dominant's outbred offspring. Females could therefore attempt to increase skew among males by avoiding mating with close relatives. Cant and Reeve's (2002) staying-incentive model explicitly excludes the issues of incest and relatedness asymmetry, and thus has restricted application to societies forming through natal philopatry.

Incest avoidance

A subordinate or breeding female may avoid reproduction if this would entail incest (Emlen 1995, 1996). There is evidence of incest avoidance and possibly inbreeding depression in some cooperative breeders (Chapter 9). If incest is avoided and the only potential mates are close relatives, there is no need for the dominant male to constrain reproduction by the subordinate, and there should be neither risk of eviction nor competition with the dominant. Incest avoidance might therefore result in complete skew when other models would predict reproductive sharing, and reproductive monopolization may not be due to manipulation, transactions, or competition among males. Given that cooperatively breeding groups often form through natal philopatry of young, incest avoidance is likely to have a major affect on reproductive sharing.

Incest has not yet been incorporated in any formal models, but it would be useful to do so. We expect that the concession model will, paradoxically, predict lower skew if the subordinate is related to the resident female and their inbred offspring are of low quality. This is because, whatever the share of reproduction required by the subordinate in order to stay, a reduction in the value of his offspring due to inbreeding depression means that he will then require a bigger share to compensate him for not dispersing. Active incest avoidance by subordinates might then provide evidence against concession models.

CONFOUNDING VARIABLES AND THE PROBLEM OF TESTING THE MODELS

In this section we highlight some of the difficulties in interpreting the evidence in support of skew models, focusing on patterns of skew and relatedness in males; several other authors have reviewed the empirical evidence advanced in support of skew models (Keller and Reeve 1994; Emlen 1995, 1997a; Reeve *et al.* 1998; Reeve 2000; Reeve and Keller 2001). In our view, there is as yet no species for which there is unequivocal support for or rejection of any specific transactional, compromise, or synthetic model of skew among male birds. There is no species for which all relevant parameters have been measured precisely, so it has not proved possible to test quantitative predictions about skew, and qualitative predictions of whether skew should increase or decrease have not allowed discrimination among models. More fundamentally, we still face the challenge of testing the assumptions on which the models are built. In this section we focus on the problems of trying to distinguish among models of skew, after which we consider case studies that illustrate these problems. Our aim is to stimulate research by clarifying the challenges we face, rather than to stifle research by raising obstacles.

We focus in particular on predictions regarding relatedness and skew. This is because it has proved easier to quantify relatedness than the effect of subordinates on group productivity, ecological constraints, or competitive ability. Relatedness is also easy to compare among individual groups, populations, and species. Of the additional parameters, the effect of subordinates on group productivity would seem straightforward to quantify as the number of extra young produced as a result of a subordinate's presence. However, it has proved notoriously difficult to test whether the presence of subordinates affects group productivity, let alone specifying the precise shape of the relationship (Magrath and Yezerinac 1997; Cockburn 1998). In addition, subordinates could affect a dominant's fitness in other ways, such as through increased survival, providing further challenges to the field biologist. Variation in ecological constraints has also allowed useful tests of skew models, but is difficult to measure precisely in part because it is a dynamic property of life-history rather than an externally imposed constraint (Kokko and Lundberg 2001). Furthermore, different individuals will have different abilities or opportunities to compete for independent breeding vacancies. For example, Florida scrub-jay subordinates remain longer as helpers if they are less competitive in acquiring breeding vacancies (Marzluff et al. 1996). If a greater ability to compete for a share of reproduction within a group also implies a greater ability to compete for breeding vacancies, it may be difficult to distinguish between the effects of competition and dominant control, since more competitive subordinates within a group may be granted greater concessions because they have greater opportunity to disperse, not because they are more effective at competing for reproductive share.

The greatest challenge in distinguishing among models in cooperatively breeding birds is that a correlation between relatedness and skew can arise because some other variable covaries with relatedness (Emlen 1996, 1999; Magrath and Heinsohn 2000). Furthermore, such confounding variables can lead to a convergence of qualitative predictions from quite different models.

The problem of confounding variables afflicts all field observational studies, regardless of the question, but is particularly serious in the interpretation of patterns of skew in cooperatively breeding avian societies based on natal philopatry of young. Natal philopatry will result in a nuclear family consisting of the dominant pair and their son (or daughter), but subsequent deaths and social rearrangements can lead to groups with variable relatedness. If the female dies or disperses and is replaced by an unrelated immigrant, then a "stepmother" group is formed. Similarly, replacement of the dominant male can lead to a "stepfather" group, assuming an unrelated immigrant replaces the dominant, or relatedness can be maintained if an older brother (for example) inherits dominance. Another change among breeders can result in an "unrelated" group in which subordinates are unrelated to either dominant. Thus replacement of individuals can result in groups in which subordinates can be siblings, half-siblings, nieces, nephews, or even unrelated to offspring of the dominant pair. Groups formed through natal philopatry pose difficulty for testing skew models because any effect of relatedness on reproductive skew can be confounded by relative competitive ability of subordinates, incest avoidance (Emlen 1996), and most likely female mate choice. Relatedness asymmetry can also lead to a convergence of predictions (Emlen 1996), at least for females. We consider each in turn.

The relative competitive ability of subordinate compared to dominant males is likely to covary with their relatedness to the dominant pair for three reasons (Emlen 1996; Clutton-Brock 1998). First, assuming no death or dispersal of group members, a male that remains on his natal territory will be with his mother and father. As time passes, it is likely that one or both of the dominants will be replaced. Thus, because of the way groups with lower relatedness form, subordinate males will be older, on average, when with an unrelated female. Second, individuals that force themselves into groups of unrelated individuals are likely to be competitively superior to those that stay on the natal territory. Third, individuals that have been unsuccessful in gaining paternity may eventually leave groups for that reason, so that, as time goes by, remaining subordinates are on average more competitive and therefore gain a greater share of paternity. In the meantime, group composition may have changed, so that they are (incidentally) less related to the dominants. It is even conceivable that the opposite effect could occur if better competitors are more likely to leave the group and acquire breeding vacancies, leaving less competitive subordinates at home.

The problem with covariation of relatedness and competitive ability is that almost all transactional and compromise models predict that more competitive subordinates will either take or be given a greater share of reproduction. The restraint model is an exception if competitive subordinates suffer reduced ecological constraints on independent breeding, but even there it is possible that evicting a more competitive subordinate will be more difficult. Overall, competitive ability is likely to render qualitative predictions inadequate for distinguishing among models and might explain many apparent effects of relatedness on skew.

Incest avoidance predicts the highest skew in nuclear families because subordinates choose not to mate with their mother (Emlen 1995, 1996). Thus, high skew in nuclear families could arise from incest avoidance, the dominant offering few concessions to subordinates (a transactional model), or the subordinate being a poor competitor compared to the dominant (a compromise model or a peace incentive in a transactional model). Although there is good evidence of incest avoidance in many species, it is not easy to be confident that incest is avoided precisely because transactional and compromise models provide alternative explanations for subordinates not mating with mothers. Furthermore, incest apparently occurs regularly in a few species (Chapter 9), so the importance of incest avoidance should be assessed on a case-by-case basis.

Female control could further confound correlations between reproductive skew and relatedness. First, females may avoid incest, even if males do not, again producing high skew in nuclear families. Second, if females choose males on the basis of genetic quality they may avoid young males, who are likely to be related subordinates, thereby increasing reproductive skew when males differ in competitive ability and subordinates are more closely related, as also predicted by concession models. In some circumstances, skew may even be decreased when the beta has lower genetic quality (Cant and Reeve 2002). Third, females might benefit from mating polyandrously to increase genetic diversity, and they are likely to benefit more when males are unrelated, again as predicted by the concession model. Finally, females may mate with subordinates to enlist their help with care of offspring (Cant and Reeve 2002). Again, such an enticement is more likely when subordinates are not closely related to the brood and gain less indirect fitness by helping. In sum, female control could produce many patterns associated with relatedness and male competitive ability that are also predicted by skew models.

Relatedness asymmetry is correlated with mean relatedness of a subordinate to the offspring of the dominant pair. In nuclear families mean relatedness is 0.5, in stepmother groups it is 0.25, and in unrelated groups it is 0. However, as discussed above, the relatedness of the dominant and subordinate to each other's offspring is potentially asymmetrical in nuclear families but symmetrical in other types of groups. The effect of asymmetry is to make the predictions of different models more similar, at least for females. For example, relatedness asymmetry leads to a prediction of high skew in nuclear families in tug-of-war models despite the general prediction that skew will not be affected or will decline with relatedness. To this extent the predictions of the concession and tug-of-war models converge.

CASE STUDIES

We now illustrate the difficulties and opportunities in testing models of reproductive sharing by taking four case studies. Together they illustrate many of the possibilities and problems of testing models of skew using natural variation, as they are taxonomically diverse and have contrasting social organizations. We emphasize reproductive skew among males even though three of the species can have multiple breeders of each sex.

White-browed scrubwrens: simple groups formed by natal philopatry

White-browed scrubwrens are typical cooperative breeders in that groups usually form through natal philopatry of males, so that subordinates are usually closely related to one or both members of the dominant pair. However, because of death and dispersal, only 44% of beta males were related to both dominants, with 38% related to the alpha alone, 16% related to neither dominant and 2% related only to the female (Magrath and Whittingham 1997). Social groups can contain more than one subordinate male, but additional subordinates never gain paternity (Whittingham *et al.* 1997), so we focus on alpha and beta males only.

The pattern of reproductive sharing and relatedness in scrubwrens is consistent with the concession model but is also consistent with other models, including an influence of incest avoidance and female choice (Fig. 10.4). First, beta males never gained paternity when with their mother, but sometimes did so when with an unrelated female. This is consistent with the concession model, incest avoidance, and almost any model incorporating the effect of competitive ability. Under the concession model, the alpha male need provide the fewest concessions to entice the beta to stay when he is most closely related to the alpha male's offspring, as is the case in a nuclear family. In other groups, the alpha male's offspring will be less closely related and so the alpha male should offer greater concessions and skew will decline. The pattern is also consistent with incest avoidance by the beta male, assuming he avoids mating with his mother.

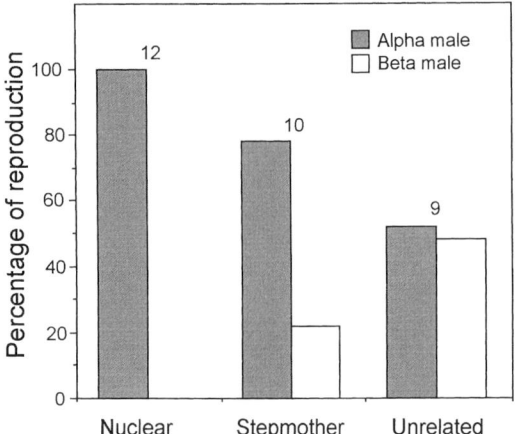

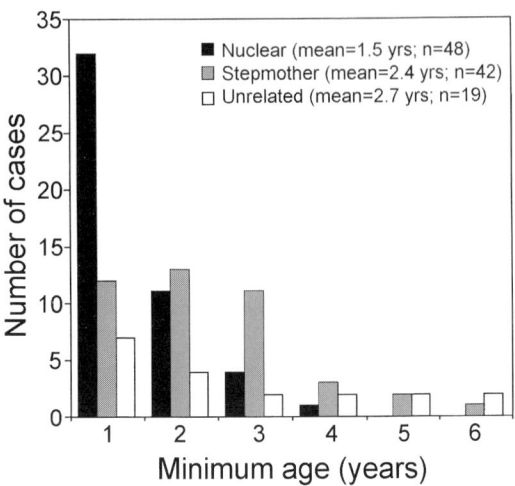

Figure 10.4. Beta male white-browed scrubwrens gain a greater share of paternity when less closely related to the dominants. In "nuclear" groups the subordinate is with his mother and father, in "stepmother" groups he is with his father and an unrelated immigrant female, and in "unrelated" he is unrelated to either dominant. Sample sizes are numbers of broods, but percentages were calculated on total nestlings fingerprinted (34 nestlings from nuclear groups, 26 from stepmother groups, and 24 from unrelated groups). Five nestlings resulting from extra-group paternity were excluded before calculating percent sharing. From Magrath (1999), based on data in Whittingham et al. (1997).

Figure 10.5. The average age of beta male white-browed scrubwrens varies according to their relatedness to the dominants, as a consequence of groups forming through natal philopatry. Beta males are youngest, on average, when still with their mother and father (nuclear groups), and older when with an unrelated female (stepmother groups) or with two unrelated dominants (unrelated groups). Sample sizes are group years. From Magrath (1999).

Differences in competitive ability could also explain the pattern, because on average beta males are younger in nuclear families, both in absolute terms (Fig. 10.5) and relative to the age of the alpha male (Magrath 1999). Second, in groups with an unrelated breeder female, sharing appears to be more equitable when the beta male is with an unrelated alpha male than when he is with his father. Incest is not an issue in this comparison, so a larger sample would provide a promising test of the concession model. However, beta males could be more competitive in unrelated groups, in which they tend to be older, compared with stepmother groups (Magrath 1999). Furthermore, such males can occasionally force their way into unrelated groups (Magrath and Whittingham 1997), and thus may be competitively superior individuals regardless of age.

Another complexity with assessing models of skew in scrubwrens is that groups are only more reproductively successful than pairs when the breeding female is a yearling (Magrath 2001). Given the high survival of adults, only 25% of groups contain a yearling female in any one year, so for most groups in most years, group size has no detectable effect on productivity. Furthermore, subordinates do not significantly affect the survival of dominants. The lack of any obvious benefit for dominants in most groups is a problem because most models of skew assume that subordinates lead to increased reproductive success, because a dominant should only offer a concession in return for a benefit gained from the subordinate's presence. Thus the simplest test of different models would require restricting analyses to sharing among males in groups with yearling females, which has thus far not been possible.

Finally, the pattern of reproductive sharing is also consistent with female control over male reproduction. First, females may avoid incest even if their sons do not. Second, if females seek multiple paternity to increase genetic variability among their young, they will have the most to gain from multiple paternity when the sires are genetically different. Female choice could thus explain the most equitable sharing of paternity in groups in which the males are unrelated. Finally, subordinate male scrubwrens provide more care to nestlings if they have paternity (Whittingham and Dunn 1998), and so again females may seek copulations with beta males particularly when such males have the least indirect fitness to gain via the alpha male's paternity. The end result is that female control may lead to high skew in nuclear families because of incest avoidance by the female, and least skew in unrelated groups, so that the ranking of skew becomes the same as for the concession model and any model incorporating the effect of male competitive ability.

In summary, overall correlations between relatedness of males and reproductive skew could not be used to test among models. Comparisons among groups in which the female is unrelated to the males avoid the confounding effect of incest but only partly solve the effects of differences in male competitive ability and female choice. Furthermore, comparisons are best restricted to groups with yearling females because the assumption that subordinates increase reproductive success does not appear to hold for groups with older females. Overall, it has not yet been possible either to accept or to refute alternative explanations for the observed pattern of reproductive sharing. We expect that most of these problems with testing models of skew in scrubwrens will apply to many cooperatively breeding birds.

Acorn woodpeckers: multiple breeders, incest avoidance, and constrained random mating

Breeding groups of acorn woodpeckers in California can have multiple breeders of each sex and a polygynandrous mating system. Groups typically contain coalitions of from one to seven (usually one to three) cobreeder males competing for mating with one or two (rarely three) joint-nesting females, in addition to non-breeding helpers of both sexes that are offspring from previous broods (Haydock and Koenig 2002). Incest is rare and there is good non-experimental evidence that it is avoided (Koenig et al. 1998; Haydock et al. 2001; Chapter 9). Males have a lower chance of dispersing to a breeding vacancy and thus appear to face more severe ecological constraints than females. Males also appear to derive greater fitness from being a member of a cobreeding coalition than do females. Given these differences between males and females, dominant males should need to offer fewer concessions to subordinate males than dominant females would need to offer subordinate females.

As predicted from the concession model, reproductive skew was greater for cobreeder males than for cobreeder females, and for males the observed magnitude of skew was even roughly that predicted quantitatively by the concession model, but Haydock and Koenig (2002) provide an alternative explanation. Because of the low number of eggs in a clutch, Haydock and Koenig (2002) compared observed skew against null models. Males had higher skew than expected from a null model of random reproductive success, whereas females had more equitable reproduction than expected by chance. However, Haydock and Koenig argue that the skew observed among males arose simply from the pattern of non-independence of paternity within broods. Supporting this hypothesis is the finding that skew among males did not differ from a random model incorporating the observed incidence of 72% broods sired by a single male. Furthermore, there was no evidence of stable dominance in either sex, with success at gaining paternity in one brood a poor predictor of success in further broods. Nor was there a detectable correlation of reproductive success with age, weight, or condition.

Such an absence of dominance, if true, violates an assumption of both transactional and compromise models of skew. It remains possible, however, that dominance may be stable within each breeding attempt, but that males burn out after short periods of dominance, as if they are engaged in a tug-of-war (S. Vehrencamp, personal communication). It is also possible that variation in success among broods is the result of a long-term transaction among males, rather than the lack of consistent dominance, although sharing in the long-term appears to be more equitable then predicted by the concession model.

Examining the effect of ecological constraints provides another way to test models of skew in acorn woodpeckers. Emlen (1984) showed that yearlings were more likely to remain on the natal territory in years when ecological constraints were greater, resulting in larger group sizes. Within larger groups, behavioral evidence suggests that some males do not attempt to breed, so it is possible that reproductive skew is greater. Thus, one could argue that greater ecological constraints have lead to greater skew, as predicted by concession models (S. Vehrencamp, personal communication). However, it is also possible that larger groups are not comparable with smaller groups, and so this is not a good test (W. Koenig, personal communication). Certainly, we consider that it is more convincing to test predictions of ecological constraints using groups of the same size.

Overall, Haydock and Koenig (2002) argue that while the sex difference in skew is consistent with the concession model, other evidence is contrary to the predictions and assumptions of both the concession and restraint models. Their work implies an absence of dominance within sexes, which challenges a fundamental assumption of most skew models, and suggests that the mechanics of sperm competition may have a major influence on skew within broods. Furthermore, individuals go to great lengths to avoid incest, as if this were a rule of thumb, not a strategy based on the reproductive payoffs of incestuous versus non-incestuous mating, suggesting that incest avoidance could not easily be incorporated into skew models. While Haydock and Koenig (2002) reject concession and restraint models, they did not assess other models of skew, nor provide direct evidence that dominance is truly lacking. Finally, variation in ecological constraints provides another possible way to test models.

Pukekos: geographic variation in ecological constraints

The pukeko is a rail with geographically variable group composition and mating behavior. A population studied by Craig and Jamieson (1990) on the North Island of New Zealand had low adult mortality and limited options for juvenile dispersal, and groups were consequently composed of close relatives. Groups were permanently territorial and were composed, on average, of 3.3 breeding males, 1.8 breeding females, and 1.9 non-breeding helpers, who were not observed to copulate and had underdeveloped gonads. By contrast, a population studied in the South Island consisted of small groups of unrelated adults, including 1.5 males, 1.3 females, and few helpers. This population re-established territories each year, had relatively high adult mortality, and offspring dispersed independently, breeding as yearlings (Jamieson et al. 1994; Jamieson 1997). All birds appeared to be capable of reproduction. In both populations birds had a clear dominance hierarchy. The different ecological constraints and group composition both suggest that, under the concession model, North Island birds should have higher reproductive skew than South Island birds.

Consistent with the concession model, South Island males did appear to have a more egalitarian mating system than North Island birds (mean skew 0.25 ± 0.08, $n = 12$, and 0.58 ± 0.12, $n = 6$, respectively), although the difference was not statistically significant, perhaps reflecting the small sample size. Dominance interactions were not obvious during copulations, whereas they were during interactions over food or territories, suggesting that dominant individuals might be "conceding" reproduction to subordinates, as required by the concession model. Furthermore, subtle differences in behavior suggested greater reproductive dominance in the North Island population, again as predicted by the concession model. Finally, in support of an assumption of the concession model, male coalitions had higher reproductive success than solitary males (J. Quinn and I. Jamieson, unpublished manuscript).

As suggested by Jamieson et al. (1994) and Jamieson (1997), however, the concession model is not the only possible explanation of their results. Females may influence reproductive skew among males because they potentially benefit from male incubation and care of young. Furthermore, dominant males may be unable to control subordinate reproduction, particularly in the South Island where subordinate males are "coalition partners" and may be more equally competitive and harder to control. In addition, it appears that males cannot accurately predict paternity from copulation rates or dominance (Jamieson et al. 1994), in which case the concession model may not yield a stable solution (Kokko 2003).

Overall, the egalitarian mating system in South Island birds is consistent with expectations from concession models, and the tolerance of subordinate copulation suggests some form of social transaction. However, there is no direct evidence that dominant males can control subordinate reproduction, particularly if females favor equitable reproduction among males. The fact that dominant North Island males are also tolerant of subordinate reproduction, despite stronger ecological constraints, argues that dominants may not be able to control subordinate reproduction. Finally, this pioneering study was carried out before the development of additional transactional and compromise models of skew, and the results do not exclude other models.

White-winged choughs: obligate cooperation and annual variation in constraints

White-winged choughs are large, highly social passerines that live and breed in groups of up to 20 individuals.

Group members contribute to all aspects of parental care, from building nests to care of fledged young. Cooperative breeding is obligate, as pairs cannot breed alone and even trios are rarely successful. Furthermore, the number of young fledged increases linearly in groups of four or more (Rowley 1978; Heinsohn 1992) because additional individuals provide more food and reduce starvation among nestlings (Boland *et al.* 1997a). Chough society has a hierarchical structure in which coalitions of relatives can disperse together and join with other coalitions to form groups, and reproductive sharing is affected by both dominance within coalitions and the success of other coalitions (Heinsohn *et al.* 2000). The social rearrangements following increased mortality caused by a severe drought have revealed the role of coalitions in reproductive sharing and group stability (Heinsohn *et al.* 2000).

Reproductive sharing in chough groups changed dramatically after a drought, during which the death of breeding females led to dispersal of coalitions and formation of new social groups (Heinsohn *et al.* 2000). DNA fingerprinting showed that before the drought a single female and male monopolized reproduction and were assisted by their grown offspring. Groups were composed of a small number of coalitions, usually only one coalition of each sex, headed by each breeder. During the drought, death of the female breeder often led to dispersal by coalitions that subsequently joined to form amalgamated groups consisting of from two to seven coalitions. Within these new groups, reproduction was usually shared among coalitions, but with only one breeder per coalition, such that reproductive sharing was dependent more on the number of coalitions than on group size (Fig. 10.6).

The change from high skew before the drought to low skew after it is consistent with the concession model. Before the drought, groups were composed of close relatives, each additional group member increased the production of offspring, and individual choughs had few dispersal options (Heinsohn 1992). Under these conditions, the concession model predicts high skew, which is consistent with the observed reproductive monopolization. The drought allowed new dispersal opportunities, leading to groups composed of unrelated coalitions and increased reproductive sharing. Again, this is consistent with the concession model, which predicts greater sharing when dispersal is easier and relatedness is reduced. Finally, the

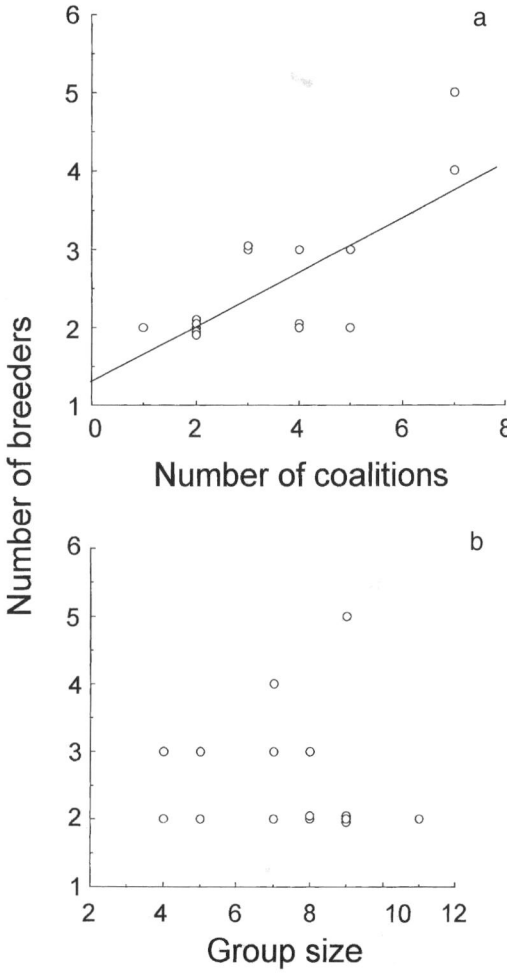

Figure 10.6. Reproductive sharing in white-winged choughs depends on the number of coalitions of relatives in the groups. The figures show number of breeders after the drought according to (a) the number of coalitions of relatives in the group, and (b) the size of the group. From Heinsohn *et al.* (2000); used with permission from the Royal Society of London.

high skew before the drought appears not to be due to incest avoidance, as incest was recorded in two groups, repeated over multiple years (Heinsohn *et al.* 1999, 2000), showing that it is not an absolute constraint on reproduction.

Reproductive sharing within groups appears to be consistent with the synthetic models of skew, although no such model has tackled the kind of hierarchical social system observed in white-winged choughs. Reproduction within coalitions was always monopolized by a

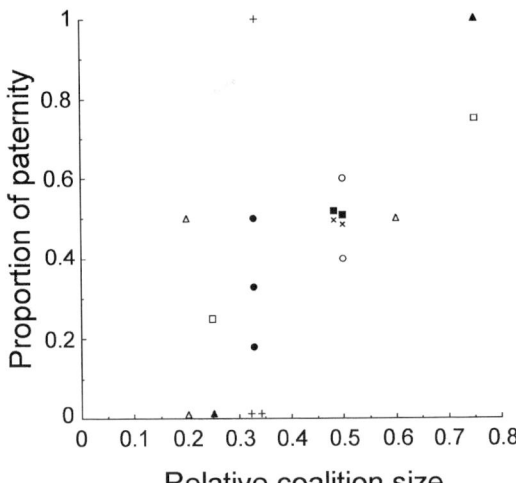

Figure 10.7. Larger coalitions gain a greater proportion of paternity in white-winged chough groups. The figure shows the proportion of paternity attained by 19 male coalitions versus the relative size of their coalition in the group. The data are restricted to the eight stable groups with two or more competing male coalitions. Relative coalition size equals the size of the coalition divided by the total males in the group. Symbols are unique for each of the eight groups.

single individual, presumably the most dominant, which is consistent with the close relatedness among coalition members. By contrast sharing among coalitions within the same group was more equitable, and the reproductive share of the breeder in each coalition was related to the number of members it contained relative to other coalitions (Fig. 10.7). If one assumes that larger coalitions are more competitive, then sharing among coalitions is related to competitiveness of the coalition and is therefore potentially a "tug-of-war."

Patterns of group stability also appear consistent with transactional and synthetic models of skew. Before the drought groups consisted of fewer coalitions, most of which contained a successful breeder, whereas after the drought groups had multiple coalitions with multiple parentage, together with individuals that were not successful. Thus it is possible that there is a period after group formation in which coalitions and individuals jostle for a share of reproduction, and those that fail leave to try breeding elsewhere. This suggests that group stability is closely related to the extent of reproductive sharing, as assumed by transactional and synthetic models.

Overall, reproductive sharing in white-winged choughs is related to the hierarchical structure of groups, with reproductive sharing among but not within the coalitions of relatives that make up breeding groups. Furthermore, the heads of larger coalitions gain a greater share of reproduction. Individuals in these coalitions thus face the dual optimization problem of having sufficient members to compete successfully against other coalitions for breeding vacancies while assuring they gain a worthwhile share of either direct or indirect fitness for themselves within their coalition. Such complexity has never been encapsulated in formal models of skew, but the results suggest that such an approach would be valuable.

PROSPECTS

Many qualitative patterns of reproductive sharing among males are consistent with one or more models of reproductive skew, but qualitative patterns are unlikely to discriminate among models or to rule out alternatives such as female choice and incest avoidance. Female choice is likely to prove important in explaining reproductive skew among males, as there is strong theoretical and empirical evidence that female birds in general can choose among males and benefit from multiple mating, including some evidence that they can affect reproductive sharing among males within social groups (Chapter 5). There is also growing evidence that one or both sexes avoid incest in some species (Chapter 9), sometimes to a greater extent than seems plausible from the cost of inbreeding depression (Koenig et al. 1998). Nonetheless, the equivocal support for skew models may reflect the difficulty of testing them, not problems with the models themselves, and evidence for competing explanations is also often equivocal.

We believe that testing assumptions rather than just predictions is the most important way to make progress in understanding patterns of reproductive sharing (Johnstone 2000; Magrath and Heinsohn 2000). It can be difficult or impossible to distinguish among models on the basis of their qualitative predictions, and yet the assumptions of models differ substantially, as do the proximate causes of reproductive skew. Precise measures of all parameters in all models would allow quantitative predictions and therefore stronger discrimination among models, but measuring all parameters precisely may prove impossible. Another way to make progress could be to focus on predictions about behaviour of competing individuals and group stability, rather than primarily focusing on reproductive sharing (Johnstone 2000; Cant and Reeve 2002), but here we focus primarily on the importance of testing assumptions, and their likely validity.

Control by dominants and knowledge of sharing

Transactional models of skew assume that dominants control subordinate reproduction, either directly or through the threat of eviction from the group, and that such control is exercised within the constraint of maintaining group stability. So far, however, there appears to be no direct evidence that subordinate dispersal can be manipulated by reproductive concessions made by dominants, either among male birds or in vertebrates more generally (Clutton-Brock 1998). There is also no direct evidence that dominants have precise control over the magnitude of skew. What behavioral mechanisms would allow such precise knowledge and control of subordinate reproduction? One problem of particular importance to vertebrates is that small brood sizes mean that a precise partitioning of reproduction may not be possible or may require keeping track of reproduction over several breeding attempts (Tsuji and Tsuji 1998; Haydock and Koenig 2002). Similarly, there is mixed evidence on whether dominants could evict subordinates if they wished, and none that eviction can follow a subordinate exceeding a threshold share of reproduction.

Recent theoretical work suggests that it is essential to determine how well individuals can assess reproductive share. Kokko (2003) modified a concession model and showed that there was no stable reproductive share if subordinates could not accurately assess their share of reproduction within the breeding group. Evolutionary "knowledge" of the average share in the population was insufficient for stability.

There is mixed evidence that dominants can potentially control either reproductive sharing or group membership. First, the pattern of group membership and reproductive sharing in dunnocks suggests limited control by dominants (Davies 1992). In this species adults within a breeding group are unrelated, and dominant males do not benefit from the presence of subordinates. Thus, skew models predict that groups should not be stable. Nonetheless, dominants are often unable to prevent other males from becoming members of the breeding group, suggesting that they are unable to evict subordinates, contrary to the assumption of the restraint model. Dominant males are also generally unable to prevent reproduction by subordinates, contrary to the assumption of concession models.

Second, in contrast to dunnocks, pied kingfisher males do appear to have control over group membership, and only allow unrelated males to join breeding groups, and assist with the provisioning of young, when feeding conditions are poor and assistance is valuable to the dominant (Reyer 1990). However, such control of group composition is not exercised through control of reproductive sharing. Third, white-fronted bee-eater males, which nest in colonies, can disrupt the breeding attempts of sons or other close relatives to recruit them as helpers (Emlen and Wrege 1992). This suggests that manipulation of independent breeding may increase group stability and reproductive skew as suggested by Crespi and Ragsdale (2000), at least in colonial breeders. Finally, low-quality Florida scrub-jay subordinates remain on their natal territory longer than high-quality birds (Marzluff et al. 1996), supporting the possibility that dominant territory breeders could increase group stability by manipulating the quality of subordinates.

The mechanisms of control, and possibilities of knowledge, are likely to differ between the sexes (Chapter 11). For example, female acorn woodpeckers and groove-billed anis remove eggs from a communal nest until they themselves start laying, thereby potentially exercising some degree of control over reproductive skew. The eggs of different female ostriches are different in appearance, allowing knowledge and female control even in a mixed clutch. Knowledge of reproductive share, and possibilities for control, may be more indirect among males. Males probably cannot identify kin within a brood, and sperm competition is more indirect than egg-tossing. For example, although copulation frequency is a good guide to paternity in some species (Davies 1992), it is not in others (Jamieson et al. 1994).

Group size and productivity

Most models assume that group productivity increases with the addition of subordinates, and models involving more than one subordinate also make assumptions about the shape of the relationship between group size and productivity. Despite these assumptions, it is often difficult to demonstrate whether group size affects productivity, let alone to determine the shape of the relationship. If groups do not increase success, they are stable in only very restricted circumstances and relatedness will have little effect on skew (for example, $k = 1.6$ in Fig. 10.2). In fact, about one-third of studies of cooperative breeders have detected no group-size effect (Cockburn 1998; Hatchwell 1999), and most studies reporting a positive correlation are not backed up with experiments or other robust evidence (Cockburn 1998). Clearly it is important to measure the effect of subordinates on productivity in order to assess the relevance of models and to make quantitative predictions. As discussed earlier, it is also important to measure other potential benefits to dominants of retaining subordinates within groups.

A further complication is that group size may have different effects on group productivity in different circumstances. For example, the benefit of having subordinates may be greater under poorer environmental conditions or when breeders are younger (Reyer 1990; Magrath 2001). A sample of 11 species showed that in good conditions groups had a mean of only 1.3 times the success of pairs, yet 2.3 times the success in poor conditions (Magrath 2001). This complication creates difficulties in measuring group-size effects, but also potentially allows within-population tests of models. Paradoxically, for example, the concession model predicts that dominants will offer a higher share of reproduction when related subordinates have a lower effect on group productivity.

Finally, a tacit assumption of most models is that the benefits of group living follow automatically and predictably from "convincing" a subordinate to remain in the group. However, subordinates may differ in the amount of care provided or even in whether they provide care at all (Heinsohn and Cockburn 1994; Boland et al. 1997b; Magrath and Whittingham 1997). Furthermore, both the decision to help and the magnitude of help can be influenced by kinship (Komdeur 1994a; Russell and Hatchwell 2001), suggesting yet another variable that confounds any association between relatedness and skew.

Kin discrimination

All models of skew assume that individuals have perfect knowledge about the kinship of other group members. This raises the issue of what mechanisms are involved and what precision of recognition is possible (Komdeur and Hatchwell 1999). Although there have been a few studies suggesting that individuals may be able to recognize kin without prior association (Bateson 1982; Petrie et al. 1999), most kin recognition in birds appears to rely primarily on a "rule of thumb" based on prior association (Komdeur and Hatchwell 1999; Hatchwell et al. 2001b; Chapter 9). For example, an adequate, albeit imperfect, rule for avoiding incest by males in species with a single female in a breeding group might entail no more than recognizing whether the resident female has been replaced or not. By contrast, quantitative predictions of skew models rely on very precise estimation of kinship, such as between fathers and uncles, fathers and older brothers, and brothers and half-brothers. Unfortunately, there may be no mechanism by which this could be routinely achieved. Realistic predictions of skew models rely on realistic assumptions about the mechanism and precision of estimation of kinship; such assumptions should be included in models of skew. As shown by Kokko (2003), there may be no stable level of skew if individuals lack knowledge of reproductive success in current nests.

Benefits of grouping and cooperation

Most models of skew assume that the benefits of group membership for dominants and subordinates are based on the current share of reproduction and group productivity. However, there are numerous other potential benefits to subordinates that may affect reproductive sharing and could potentially be included in models of skew. Perhaps most importantly, subordinates can ultimately accede to dominance status within the group (Wiley and Rabenold 1984; Emlen 1999), and transactional skew models by Kokko and Johnstone (1999) and Ragsdale (1999) suggest that subordinates will accept a smaller share of reproduction, or even no reproduction, when they are compensated by the long-term benefit of inheriting breeding status. Other benefits include increasing skill (Komdeur 1996) or the chance of local dispersal (Ragsdale 1999). From the dominant's point of view, increasing a relative's skill at breeding, or even just allowing them a safe haven while they mature, may also be an indirect fitness benefit, and so a dominant may tolerate subordinates even if they reduce group productivity. This benefit to the dominant is similar to the idea of parental facilitation (Brown and Brown 1984).

Incest

Models of skew have thus far assumed that there is no constraint on incest, and there are mixed views on whether avoidance of incest should be considered separately or included in skew models. We suggest that the best way to incorporate the effects of incest on reproductive sharing may depend on the basis for decisions about inbreeding and who makes those decisions. The impact of inbreeding depression on the sharing of reproduction could potentially be incorporated in the models of skew outlined above (Vehrencamp 1983a; Emlen 1999; Johnstone 2000). For example, the reduced fitness prospects of inbred young would lead to a devaluation of group productivity, in proportion to the share of reproduction that the subordinate obtains, which would presumably affect optimal skew. However, incorporating the effect of incest in this way assumes that individuals take the quantitative effects of incest into account when assessing reproductive payoffs. However, at least in some species, individuals appear to avoid incest almost completely, even if this leads to long periods without any reproduction (Koenig et al. 1998; Chapter 9). Thus, individuals may follow a crude rule of thumb rather than assess the magnitude of inbreeding depression, reducing the value of including incest avoidance in a general model. Consequently, it may make more sense simply to exclude some individuals from the set that potentially reproduces (Koenig et al. 1998; Chapters 9 and 11).

A second problem of incorporating the effects of incest in a synthetic model is that either or both sexes may avoid incest. As noted above, a female may avoid incest even if it were adaptive for a subordinate, and conflicts between the sexes have only just begun to be incorporated into skew models.

Female control of skew among males

Most models of skew have ignored the problem of mate choice and assume that reproductive sharing is the outcome of transactions or compromise entirely among members of one sex. This assumption is almost certainly false. There is good evidence that females can benefit from, and at least partly control, reproduction among male birds generally, and the same appears to be true within groups of spotted hyenas (Engh et al. 2002). Cant and Reeve's (2002) model shows that female interests can affect reproductive skew among males, so the female's role in controlling male skew within cooperatively breeding groups should be assessed empirically. Their models also present new empirical challenges regarding the genetic quality of males, the female's ability to assess genetic quality, and the need to measure the costs and benefits of parental care for each individual within a group.

CONCLUSIONS

What have models of reproductive skew contributed to the study of cooperatively breeding birds? Models of skew have provided explicit and sometimes surprising predictions about reproductive sharing and group stability, and, like all valuable models, have raised questions of interest and highlighted what we do not know. It was a surprising prediction, at least at the time of publication of the original concession model, that dominants might exercise more severe control over closer relatives and that dominants should allow greater concessions when subordinates have a lower effect on group productivity. Transactional and compromise models have also highlighted two general and contrasting views of social interactions, involving exchange of benefits rather than merely competition, which has stimulated other models of social behavior such as negotiation over extra-pair paternity (Shellman-Reeve and Reeve 2000).

These models have also highlighted areas for further research on the control dominants have over subordinate reproduction and group membership, the abilities and precision of kin recognition, and the role of the opposite sex in determining reproductive sharing. Models of skew provide another reason for increasing our knowledge of the effect of group size and helping on group productivity and how this varies among and within species. Although specific models of skew may be based on incorrect assumptions, they are all founded on the fundamental assumption that natural selection and inclusive fitness are important in social evolution, and are the only models so far to make quantitative predictions about reproductive sharing. Regardless of the current empirical support or lack thereof, it makes no sense to reject models of skew as a whole; however, it makes a lot of sense to question and test the specific assumptions on which they are based.

ACKNOWLEDGMENTS

We are grateful to Walt Koenig and Janis Dickinson for inviting us to contribute this chapter, and to Walt Koenig, Hanna Kokko, Ian Jamieson and Sandy Vehrencamp for insightful comments on a draft. We emphasize, however, that not all of our views or interpretations are shared by these people: a sign, we think, of a healthy and dynamic area of research. RDM and RGH would like to thank the Australian Research Council for supporting our work on the social behavior of birds, which sparked our interest in reproductive skew, and Steve Emlen and Morné Du Plessis, who fanned the flames by inviting us to participate in a symposium on the subject.

11 · Joint laying systems

SANDRA L. VEHRENCAMP
Cornell University

JAMES S. QUINN
McMaster University

Joint nesting is a relatively rare form of cooperative breeding in which two or more breeding group members of the same sex contribute genes to a clutch of eggs and cooperate in the care of young (Brown 1987; Vehrencamp 2000). Traditionally, joint nesting referred to multiple-female clutches. However, with the development of DNA techniques for assigning paternity, a growing number of cooperative species with shared-paternity clutches have been discovered. Joint-female (or communally laying) systems and joint-male (or cooperatively polyandrous) systems exhibit many important differences. Nevertheless, several avian joint-female species are also characterized by the presence of two or more adult males who share paternity to some degree. Here we focus on the diversity of joint-female systems, referring the reader to Chapter 10 and other reviews for discussions of breeding systems with male cobreeding (Faaborg and Patterson 1981; Hartley and Davies 1994; Ligon 1999).

Most joint-female species are non-passerines. By contrast, helper-at-the-nest species, as well as cooperatively polyandrous species, are found among both the passerines and non-passerines. There may be a good explanation for this pattern. Communally laying species all share one important feature: males make a large contribution to incubation and care of the young. In some joint-female species males perform all of the incubation and subsequent care, whereas in others the males perform more than half of the incubation, including nocturnal incubation.

In a survey of the phylogenetic origins of communal-laying species, all were found to arise in taxa with a history of strong male incubation (Vehrencamp 2000). Although some male passerines contribute up to 50% of diurnal incubation, they do not possess brood patches and have never been known to incubate at night (Bailey 1952; Skutch 1957, 1976; Ball 1983). Communal laying is therefore probably rare among passerines because females are the primary incubators in this taxon. There are, of course, many species with a strong paternal care role that are not communal nesters. Male parental care is thus a necessary, but not a sufficient, preadaptation for joint-female nesting. One of the issues we address is the possible reason for this evolutionary constraint.

Beyond this one common characteristic, communally laying species differ in virtually all other aspects of their biology. The most conspicuous difference lies in their mating systems. Some communal nesters are socially monogamous, with breeding groups comprised of joint-nesting pairs. Others contain multiple males who compete for access to females in cooperatively polyandrous or polygynandrous breeding units. Finally, some joint-nesting breeding units contain a single male, but even these species differ in the duration of association between the male and the joint-laying females, and range from stable polygynous harem units to sequential polygynandrous units with sometimes no contribution by females to incubation or care of the young. Therefore, a second issue we examine here is the role of conflicting male and female interests in the evolution of joint laying by females.

A long-standing factor that has been implicated in the evolution of cooperative breeding in general is ecological constraints on independent breeding (Emlen 1982a; Emlen and Vehrencamp 1983; Koenig *et al.* 1992; Chapter 3). Classical concessions models of reproductive skew (Chapter 10) predict that skew should be high under conditions of strong ecological constraints, not low as in true joint-laying species (Emlen 1982a; Vehrencamp 1983a; Keller and Reeve 1994). Some, but by no means all, joint-laying species do seem to be

---

Ecology and Evolution of Cooperative Breeding in Birds, ed. W. D. Koenig and J. L. Dickinson. Published by Cambridge University Press.
© Cambridge University Press 2004.

characterized by ecological constraints. We therefore attempt to evaluate whether these species more appropriately meet the assumptions of alternative skew models such as the eviction model (Johnstone and Cant 1999), the high-cost-of-reproduction model (Cant and Johnstone 1999), or the tug-of-war model (Reeve et al. 1998). For further details of these models, see Chapter 10.

A final consideration is whether joint nesting by females arises as a consequence of conspecific brood parasitism (CBP). Conspecific brood parasitism occurs when a female lays eggs in a conspecific female's nest and leaves without providing parental care (Yom-Tov 1980; Andersson 1984; Petrie and Møller 1991). The asymmetry in parental effort is assumed to benefit the parasite while imposing some level of cost on the host. Mutual cooperation, on the other hand, involves joint laying and shared parental effort, and is assumed to result in a net reproductive benefit to both females relative to each nesting solitarily. Recent models by Zink (2000) and Andersson (2001) specify the conditions under which females might choose to nest solitarily, parasitize another female's nest, or breed cooperatively with another female; these models are discussed in greater detail below.

In this chapter we review the major taxa of joint-female nesters, summarizing their mating systems, parental care allocation, nesting ecology, and fitness consequences of different group compositions. In the discussion we examine whether the conditions under which a brood parasite should stay and help are met in different joint-nesting species. We then attempt to summarize the key factors favoring the evolution of communal nesting and speculate on the most appropriate skew model for each taxon given existing information.

REVIEW OF JOINT-FEMALE NESTING SPECIES

Ratites

Mating system

Ratites are the most ancestral extant group of birds. Sibley and Ahlquist (1990) found that the flightless ratites (order Struthioniformes) are a monophyletic group that is closely related to tinamou (order Tinamiformes). The mating system of flightless ratites ranges from monogamous in kiwis (Reid and Williams 1975; Taborsky and Taborsky 1991, 1992), monogamous or sequentially polyandrous in cassowaries (Crome 1976), to polygynous and sequentially polyandrous in greater rheas (Bruning 1974), and polygynandrous in ostriches (Bertram 1992). Emus are generally socially monogamous in natural settings, but rarely engage in sequential polyandry and promiscuity (Coddington and Cockburn 1995). Whether female emus sometimes lay in more than one nest is unresolved. To date, DNA fingerprinting has been restricted to analyses of domestic ratites and a recent study of four communal ostrich clutches (Kimwele and Graves 2003).

Territoriality in ratites is variable. Ostrich males defend a large territory against other males but allow access to their nest by all females (Bertram 1992). Territoriality by males was also reported in kiwis (Taborsky and Taborsky 1991), but not in emus (Coddington and Cockburn 1995). Rheas are usually not territorial although the nest-owning male becomes aggressive towards both sexes in the vicinity of his nest as egg-laying progresses (Handford and Mares 1985). Cassowaries of both sexes defend territories in the non-breeding season.

With the exception of ostriches, ratite males perform all of the parental care. Lone females or groups of females typically mate and then lay eggs in a sequential series of nests, leaving the male to incubate the joint clutch. Ostriches have a complex breeding system, comprising a single major hen that bonds with the male, lays most or all of her eggs in his nest, and participates in incubation and fledgling care, and several minor hens that lay eggs in multiple nests without contributing to parental care (Hurxthal 1979; Bertram 1992). Recent microsatellite DNA analysis suggests that major hens sometimes have more eggs incubated in other nests than in their own (Kimwele and Graves 2003). Ostrich major hens help with diurnal incubation, while the male performs the nocturnal incubation.

Tinamou inhabit a range of habitat types and show a wide variety of breeding systems including polygyny plus sequential polyandry as well as monogamy (Handford and Mares 1985). DNA analyses of tinamou have not been reported yet. Although poorly studied in general, most species are thought to be solitary except during the late breeding season when family groups with a single adult male can be seen (Sick 1964). Both sexes of the ornate tinamou, a mountain grassland species, are territorial (Pearson and Pearson 1955) while the brushland tinamou inhabits lowland woodland and scrub, males living solitarily on a loosely defined home range

(Lancaster 1964a). Male slaty-breasted tinamou set up territories into which they attract a temporary harem that lays and then moves on to another male's nest (Lancaster 1964b).

Nesting behavior
The nesting habits of ratites are also quite variable. Nests range from a bulky irregular platform of sticks and leaves in cassowaries, through burrow-bound nests in kiwis, to simple scrapes on the ground in ostriches and rheas. Eggs vary from white in ostriches to avocado-green in emus. There is a huge range in the relationship between egg and body mass with the largest, the ostriches, laying eggs that represent only 1.5% of female mass while the much smaller kiwis lay eggs that are up to 25% of the maternal mass (Bertram 1992). The small ratio of egg to adult body mass in large ratites allows the incubation of large clutches. Joint laying by multiple females is therefore more likely to occur in larger-bodied species.

Competitive interactions among laying females have been noted in several joint-nesting ratite species. Greater rhea hens visiting nests while the male incubator was absent were observed to roll eggs into or out of the nest and may have been skewing the incubated clutch (Bruning 1974). Handford and Mares (1985) described a report in which three species of tinamou were observed moving eggs with their bills, in some cases moving eggs out of the nest. Whether the movements of rhea or tinamou eggs benefited the female moving the eggs was not determined. In ostriches, the major hen can apparently distinguish her own eggs from those of other females and rolls some of the eggs that are not her own to an outer ring that is not incubated (Fig. 11.1). This activity results in an incubated clutch that generally includes all the eggs laid by major hens (except one in nest C of Fig. 11.1) and is typically limited to 20 or fewer eggs in total. The skew in egg ownership also benefits the nesting male because he apparently has much greater paternity confidence in eggs laid by the major hen than in those laid by minor hens.

As far as is known, most male ratites assume virtually all care of the eggs and fledglings. In greater rheas, breeding males feed very little during 40 days of incubation (Bruning 1974) and spend more time vigilant and less time feeding when caring for chicks (Fernández and Reboreda 1998). The major exceptions are kiwi females that help dig the nest burrow and female ostrich major hens that take on a significant share of incubation and guard the chicks with the male.

Fitness effects
Because of the transient nature of groups in many ratite taxa, effects of group size on survival are probably minimal. The number of females laying eggs in a nest probably has the greatest influence on fitness. The main factor determining hatching success in the large-bodied ratites seems to be the number of eggs that can be successfully incubated. This number is quite large and exceeds the number of eggs that can be laid by a single female (at two-day intervals) within a time frame that does not unreasonably extend the laying period and associated exposure of the eggs to excess heat and predators. In ostriches, daily nest predation rate during laying is 6% and egg viability drops sharply after 15 days of exposure, during which time a female can only lay eight eggs (Bertram 1992).

The maximum number of eggs an ostrich can effectively incubate is 20. About half usually belong to the major hen and the territorial male (probability of paternity with major hen estimated behaviorally at 92%). In another study four clutches genotyped using microsatellite DNA revealed that the territorial male fathered an average of approximately 70% of incubated eggs laid by the major hen (Kimwele and Graves 2003). The other half of the clutch belongs to two or more minor hens; about one-third of these eggs are sired by the territorial male (Bertram 1992; Kimwele and Graves 2003). Males attempt to mate with as many minor hens as possible, and undoubtedly sire some of the eggs laid in other males' nests. Minor hens apparently prefer to mate with territorial nesting males, especially those in whose nests they are laying, and refuse matings by other males (Hurxthal 1979). In the greater rhea hatching success increases with clutch size up to about 30 eggs (Fig. 11.2), followed by diminished hatching success with further increases (Fernández and Reboreda 1998).

Conclusions
With the exception of ostriches, communal nesting ratites lay eggs in the nest and leave the care to the male. The small egg size relative to body size allows the incubation of a large clutch. A single female cannot lay the number of eggs that can be successfully incubated in a reasonable period of time, leaving room for multiple females to contribute eggs to a joint clutch.

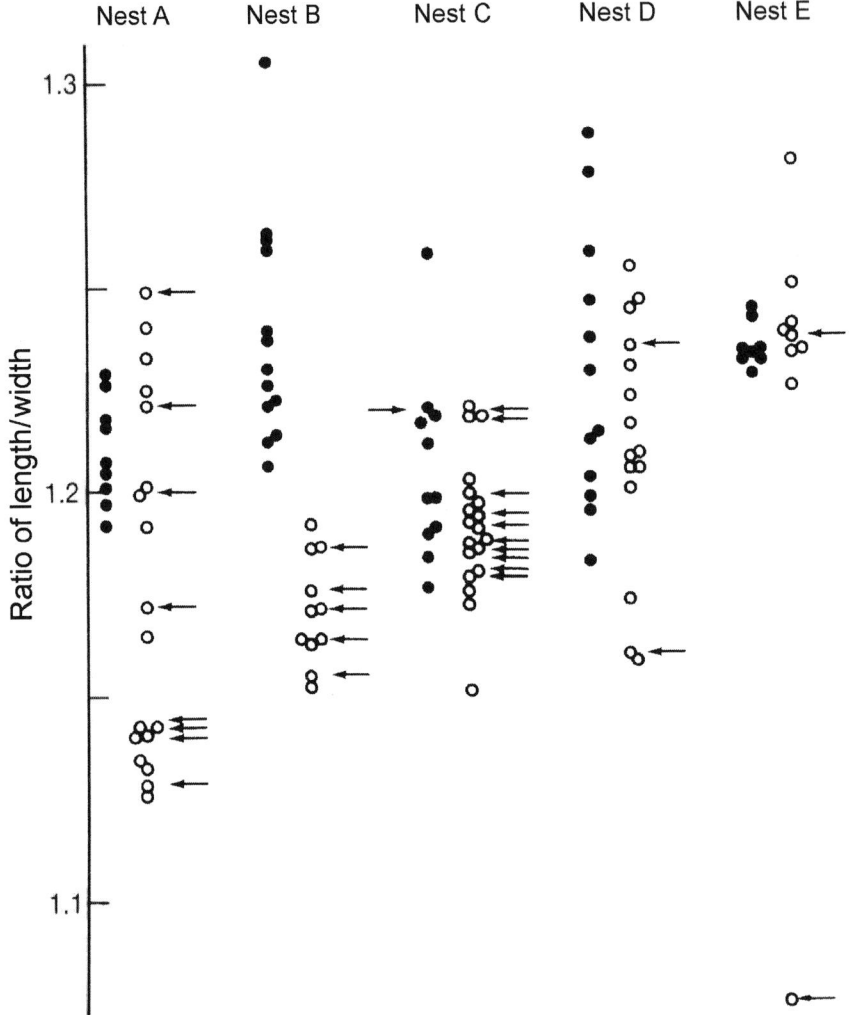

Figure 11.1. The shape of major and minor hens' eggs in five ostrich nests. Closed circles are major hen eggs, open circles are minor hen eggs. Arrows indicate eggs that were expelled. Expelled eggs are almost exclusively those of minor hens, despite the similarity of their shape to major hen eggs. Major hens are believed to use a combination of egg surface appearance, size, and shape to distinguish their eggs from those of other females. From Bertram (1992); used with permission from Princeton University Press.

In rheas, sole male parental care appears to be very costly for the male, and less than 20% of males appear to attempt to breed during a breeding season (Fernández and Reboreda 1998). Those that do breed benefit from large clutches of up to 30 eggs. The availability of breeding males appears to be very limited and females may have few breeding opportunities. The laying by multiple females in a single nest does not appear to be costly to females until clutch size exceeds 30 eggs.

Ostriches are the only ratites in which brood parasitism is feasible. We consider minor hens parasitic because they leave the care of their eggs to another female. However, the costs to the host appear to be small and outweighed by the benefits. Because females can apparently recognize their own eggs (Bertram 1992), they can adjust the egg composition of the incubated clutch to maximize their own fitness, and lose essentially no eggs to this selection process. Territory-holding males sire most of the

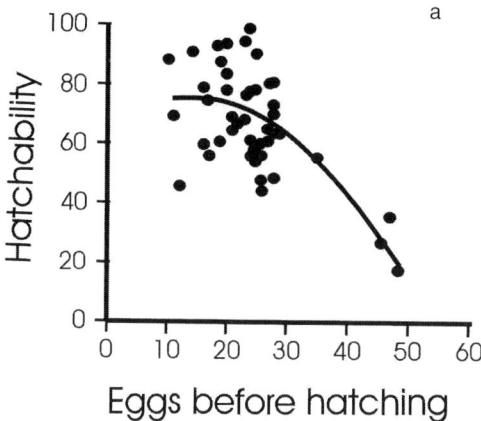 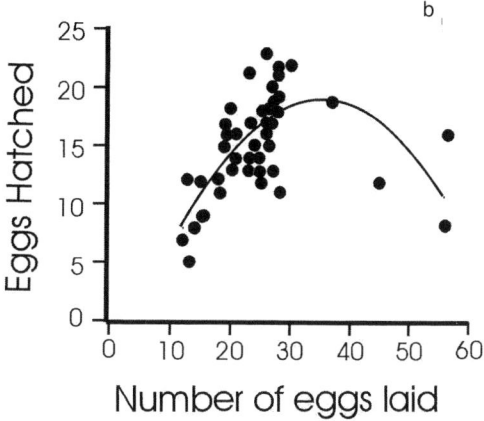

Figure 11.2. Greater rhea egg hatchability. (a) Percent hatchability of eggs shown as a function of the number of eggs before hatching. Hatchability drops sharply when over 30 eggs are laid in one nest. (b) Number of eggs hatched as a function of the number of eggs laid. Numbers of hatched eggs increased until up to 30 or 35 eggs laid and then decreased. From Fernández and Reboreda (1998); used with permission from *The Auk*.

eggs of the major hen, and also sire some of the eggs laid by the parasitic minor hens in their own nest and other males' nests. The parental birds also benefit from the presence of additional young via the predator dilution effect (Treisman 1975). The benefit of large broods of mobile young is so great that the synchronous broods on adjacent territories become merged into crèches, driven apparently by kidnapping of the young of subordinate pairs by the dominant pair in a neighborhood (Hurxthal 1979).

Magpie goose

Mating system

This species is the sole member of the family Anseranatidae, and the sister taxon to the rest of the avian order Anseriformes (geese and ducks) (Harshman 1994). It is therefore a fairly old lineage, close to the common ancestors of the ratites, galliforms, and the first birds. Although goose-like in appearance, this species possesses clawed toes and males are significantly larger than females (2.8 kg versus 2.0 kg) (Marchant and Higgins 1990). The birds walk and run well on land.

The magpie goose mating system is unique among waterbirds, with the predominant reproductive unit consisting of stable polygynous trios of one male with two females. One female is dominant over the other, but the subordinate female maintains a long-term association with the dominant breeders and ascends to the dominant position if the dominant female dies (Marchant and Higgins 1990). Both females contribute eggs to the clutch and participate in nest construction, incubation, and defense. Studies of these birds in captivity also indicate that auxiliaries of both sexes may be associated with these breeding units and assist with some aspects of nest defense and incubation (Marchant and Higgins 1990; Whitehead 1999). DNA studies indicate that cobreeding females are more closely related to each other than to females in different groups (Horn *et al.* 1996). Males are closely related to their male neighbors, hinting at the possibility that these birds exhibit the usual avian pattern of male philopatry, rather than the female philopatry typical of other waterfowl. The close female relatedness suggests that female relatives disperse in groups. Finally, there is anecdotal evidence of conspecific brood parasitism, although CBP is difficult to distinguish from cooperative cobreeding in this species (Whitehead and Tschirner 1991).

Nesting behavior

Magpie geese nest in seasonal flood plains, making them very difficult to study in the wild. Thousands of breeding groups may nest synchronously in colonies, with nests in high-density areas separated by as little as 5 m. Nests are floating platforms in dense reeds built up with rushes and other vegetation in shallow water. Eggs are slightly smaller relative to body size than expected from allometric equations for waterfowl (Whitehead and

Tschirner 1990a). Clutch size for a single female is about 8–10, and maximum clutch size for communal groups is about 16. Description of egg disappearance by Frith and Davies (1961) suggests that egg removal by co-breeding females occurs. True predation from snakes, crows, dingos, eagles, *Varanus* lizards, and water rats is high. Males perform the greater share of incubation and defend nests vigorously against potential predators. In captivity males incubate largely at night, but in the wild they appear also to incubate diurnally, allowing females to forage (Whitehead 1999). Male weight declines significantly during incubation (Whitehead 1999).

The incubation period is relatively short for such a large bird, 25–31 days. Hatchlings and adults remain in the vicinity of the nest for about four days until all eggs have hatched, and the family then leaves and roams throughout the marsh. The precocial goslings both feed themselves and receive provisioned food from the adults. They grow extremely rapidly, as there is a premium on becoming volant before the flood plain dries out (Whitehead and Tschirner 1990b; Whitehead and Saalfeld 2000). The young remain with the adults for at least a year, and in captivity yearlings sometimes join other breeding groups as auxiliaries (Marchant and Higgins 1990).

Fitness effects
Unfortunately there are no data available on reproductive success as a function of group size. The average incubated clutch size of groups with more than one laying female does not appear to be twice that of a single female's clutch, so it is likely that per capita number of eggs incubated decreases for multi-female groups. In a two-year study by Whitehead (1999), there was no evidence that birds showing a stronger attachment to or defense of the nest site during researcher nest visits were more likely to hatch a brood. Nevertheless, survivorship of eggs and goslings might be enhanced by more attending parents, and especially by more experienced and aggressive parental males.

Conclusions
Like swans and geese, magpie geese mate for life and males play a very important role in incubation and offspring care. Their close phylogenetic affinity to the ancestral stock of birds, including taxa with sole male incubation such as ratites, megapodes, and button quail, may explain the origin of the high level of paternal care. Large body size relative to egg size may facilitate the incubation of enlarged multi-female clutches as in ratites. However, ecological constraint in the form of a very brief time window for reproduction seems to have favored provisioning of young and permanent polygamous harems rather than the temporary intersexual associations seen in most ratites. Moreover, males reach maturity much later than females and are likely to suffer higher mortality as a result of the risky aspects of their parental role, so the adult sex ratio is likely to be skewed in favor of females. Polygynous females may have a lower mean annual reproductive success than monogamous females, but their probability of acquiring a high-quality male mate may be enhanced by dispersing in sib groups. It remains to be seen whether dominant females own a larger proportion of the communal clutch.

Pukeko, Tasmanian native hen, and common moorhen

Mating system
Pukeko (known as purple swamphen outside of New Zealand), along with the common moorhen and Tasmanian native hen, are members of the family Rallidae, clustering by DNA–DNA hybridization quite distinctly from the cranes and their allies, other members of the order Gruiformes (Sibley and Ahlquist 1990). Pukeko inhabit open areas associated with permanent or seasonal wetlands and are omnivorous, feeding mostly on aquatic and semi-aquatic plants but often provisioning chicks with insects. They have mate-sharing by both sexes along with a range of mating systems within the same population: monogamy, polygyny, polyandry, and polygynandry (Jamieson 1997; J. Quinn and I. Jamieson, unpublished data). DNA analyses revealed polygynandrous groups in both the South Island (Jamieson *et al.* 1994) and the North Island (Lambert *et al.* 1994) New Zealand study areas. Dominance status and copulation rates did not reveal consistent relationships with paternity in either study (Jamieson *et al.* 1994; Lambert *et al.* 1994). Pukeko maintain territories that are year-round in the North Island study areas (Craig and Jamieson 1990), but only maintain territories during the breeding season in the Otokia study site on the South Island (Jamieson *et al.* 1994). Territory defense is predominantly the task of breeding male group members (Craig 1979; Jamieson *et al.* 1994). Breeding groups from the North Island are made up of related cobreeders (Lambert *et al.* 1994), in contrast with South Island pukeko groups in which groups are smaller and most adults are unrelated to other

group members (Jamieson *et al.* 1994). Intergroup brood parasitism has not been observed.

Tasmanian native hens feed on seeds and leaves of many species in winter and spring, and on seeds and insects in the summer. The mating system of Tasmanian native hens exhibits mate-sharing by both males and females, with monogamy, polyandry, polygyny, and polygynandry all occurring in a single population (Goldizen *et al.* 1998). Slightly over half of groups are monogamous pairs and multi-female groups are relatively uncommon (Ridpath 1972; Goldizen *et al.* 2000). DNA analysis suggests that multi-male groups are genetically monogamous (Gibbs *et al.* 1994). However, a reconsideration of this study suggests that the groups examined probably contained only one breeding male with others being young birds still in their natal group, and thus polyandry probably does, in fact, occur in multi-male groups (Goldizen *et al.* 2000). Tasmanian native hens maintain year-round territories with all members defending the territory (Ridpath 1972), and brood parasitism has not been reported (Ridpath 1972; Gibbs *et al.* 1994; Goldizen *et al.* 1998, 2000). Breeding groups are made up of related cobreeders (Goldizen *et al.* 2000).

Common moorhens use a wide range of freshwater wetlands associated with emergent vegetation (del Hoya *et al.* 1996). Moorhens are most commonly monogamous, but polyandry and polygyny were noted occasionally (McRae 1995). DNA analysis allowed assignment of all chicks to the socially monogamous pairs, and thus extra-pair fertilization is apparently uncommon. However, McRae and Burke (1996) identified brood parasitism as a relatively common strategy. Male breeders, and to a lesser extent female breeders, defend territories (Petrie 1984) and females in multi-female groups are typically closely related (Gibbons 1986), often producing a mother–father–daughter trio (McRae 1996b).

Nesting behavior
Pukeko build nests hidden within vegetation often just above the water or floating (del Hoya *et al.* 1996) and female cobreeders typically lay their eggs in the same nest. Egg color pattern allows separation of eggs by maternity (J. Haselmayer, J. Quinn, and I. Jamieson, unpublished data). Reproductive skew is greater in North Island than South Island groups whether non-breeding helpers are included or not (Jamieson 1997). Furthermore, the maximum egg production by one of the cobreeding females, presumably the subordinate, is apparently suppressed in North Island groups, possibly through dominance interactions including mounting behavior by dominant females (Jamieson 1997). In South Island cobreeding groups, both females lay large clutches leading to larger total clutch size compared with North Island nests. North Island pukeko that copulate subsequently incubate eggs (Craig and Jamieson 1990), while almost all South Island pukeko males incubate eggs (Jamieson *et al.* 1994). Non-breeders are less likely to encounter the clutch and showed little incubation behavior.

Within three days of hatching, young are led from the hidden nest and all group members participate in brooding and feeding (Craig and Jamieson 1990). Pukeko chicks are precocial and can leave the nest within hours of hatching (Craig and Jamieson 1990) and begin feeding themselves at 10–14 days, feeding independently by 25–40 days (del Hoya *et al.* 1996).

Tasmanian native hens build nests hidden within vegetation, often over water, and female cobreeders typically lay their eggs in the same nest. Egg color patterns only allowed separation of eggs by maternity when one female laid unusual white eggs (Goldizen *et al.* 1998, 2000). Although no skew data have been reported, there is indirect evidence suggesting that females have skewed egg laying. One female cobreeder engaged in more than 80% of observed copulations in seven of nine group years, presumably because of dominant female suppression of subordinates (Goldizen *et al.* 2000). Furthermore, mean clutch sizes of polygynous (about 8 eggs) and polygynandrous (about 6) groups are not twice the size of monogamous (about 5) and polyandrous (about 5) groups (Fig. 11.3a; Goldizen *et al.* 1998). Males and females share in all aspects of parental care (Ridpath 1972; Gibbs *et al.* 1994). Chicks leave the nest one to two days after hatching and are capable of feeding themselves after one to two weeks, with parental supplementation until eight weeks of age (del Hoya *et al.* 1996).

Common moorhens nest above the water in shallow bowl nests that may be elaborately constructed of twigs, reeds, rushes, and sedges (del Hoya *et al.* 1996). Eggs are distinctively patterned, allowing the detection of communal joint nesting as well as CBP (Gibbons 1986; McRae 1997). Eggs in the nest may be destroyed by group males or one of the breeder females. Work by McRae (1996b) has shown that successful joint nesting requires females to lay relatively synchronously; when not synchronized, eggs are destroyed. Egg-laying in joint nests is skewed in favor of the "senior" female, which does as well in communal nests as do monogamous females. "Junior" females lay fewer eggs and

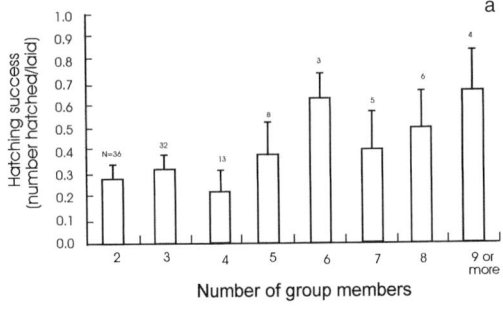

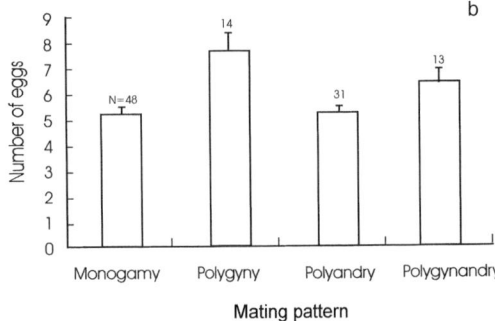

Figure 11.3. Reproduction versus group size and mating pattern in Tasmanian native hens. (a) Hatching success as a function of group size ($\bar{x} \pm$ SE; $N =$ number of group-years). (b) First clutch size as a function of mating pattern ($\bar{x} \pm$ SE; $N =$ number of group-years with known mating pattern and known size of first clutch). From Goldizen et al. (1998); used with permission from Blackwell Publishing Ltd.

have reduced hatching success, probably because they often mate with their own fathers and suffer inbreeding depression. However, senior females do not overtly suppress their daughters' breeding attempts.

In general, adult females that assist their parents with brood care also attempt to breed, but they still help if unsuccessful at breeding. Nest building and incubation are performed by both sexes, with the male taking the night-time stint (Petrie 1983). Hatchlings remain in the nest for one to two days, swim well by the third day, and dive after about a week. Young are fed and cared for by adult group members and brood division may occur. They are self-feeding at 21–25 days of age and are fed by adults for up to 45 days (del Hoya et al. 1996).

Fitness effects
Studies of North Island pukeko suggest that dominance status of breeding males does not affect copulation or fertilization success in some populations (Lambert et al. 1994). North Island pairs were found to have higher reproductive success than groups, although limited samples and habitat variables as well as group membership stability may explain these differences (Craig 1980). North Island groups were typically large relative to groups on the South Island and were made up of related adults (Craig and Jamieson 1988; Lambert et al. 1994) that included sexually mature non-breeders (Jamieson 1997).

J. Quinn and I. Jamieson (unpublished data) found that unrelated male pukeko formed coalitions to acquire high-quality breeding territories, which generally yielded greater cumulative reproductive success than breeding singly on low-quality territories during a five-year study at Otokia on the South Island. Although females do not contribute much to territory defense, subordinate females may gain reproductively by joining another female and nesting communally on a high-quality territory. However, the established female gains no apparent benefit by sharing her nest with an unrelated subordinate, and suffers costs in terms of reduced per capita hatching success in the enlarged clutch. This conflict of interest between males, who benefit from additional females by having two rather than one mate laying eggs, and females, who suffer reduced hatching success of their eggs, seems to have been resolved in the favor of males and subordinate females.

To summarize, North Island pukeko have higher longevity and reduced dispersal options compared with those at Otokia on the South Island, leading to larger groups that tend to be made up of related adults (Jamieson 1997). These features are associated with elevated levels of skew in the North Island populations, even when non-breeding females are excluded from the analysis, presumably due to the suppression of subordinate female laying (Jamieson 1997).

Multi-female groups of Tasmanian native hens produced larger clutches, and hatching success per egg generally increased with the number of group members (Fig. 11.3b; Goldizen et al. 1998), presumably because of increased vigilance and nest defense by larger groups. Communal clutches were only about 40% larger than single-female clutches. This suggests that suppression of subordinate hen egg-laying may have occurred, although this has not been verified by DNA analyses. Many reproductively mature one- or two-year-old Tasmanian native hens were not observed to copulate and probably acted as helpers (Goldizen et al. 2000).

Communal nesting in the common moorhens studied by McRae (1996b) was usually the result of delayed dispersal of females that remained on natal territories and bred with their fathers. Multi-female moorhen groups in this high-density breeding population produced more eggs than monogamous pairs, and senior female breeders were as successful as monogamous females at laying eggs, hatching, and production of independent young. Junior females produced and hatched fewer eggs, but did not raise significantly fewer young than senior or monogamous females. Senior breeding females did not suffer reduced fitness when their daughters laid communally in their nests, perhaps because of increased parental care and age-related differences in clutch size. However, junior breeders did more poorly than they may have as monogamous breeders and left the communal group to become monogamous in two cases when a mate became available (McRae 1996b).

Conclusions

In general, it does not appear that the pukeko or Tasmanian native hen social system involves brood parasitism, since cobreeders always contribute towards parental care. Rather it appears that restricted dispersal options for young in the North Island pukeko populations and the Tasmanian native hens have resulted in a skewed mating system that typically includes helpers and suppression of egg-laying. Subordinates in these systems may gain both direct and indirect fitness by remaining and helping to raise young. However, dominant female pukeko from the Otokia population appear to be compromised in terms of hatching success. Helpers are not common in this system, and there appears to be no suppression of egg-laying. Clutch sizes of communal nests are large and prone to hatching failure. Perhaps the reduced longevity and increased dispersal options have led to a conflict between males, which gain from communal clutches, and dominant females, which lose in terms of hatching success. The end result is a system in which dominant females cannot always prevent subordinate females from joining the group, perhaps because of undetected protection of these females by group males, and cannot recognize and expel subordinates' eggs (J. Haselmayer, J. Quinn, and I. Jamieson, unpublished data).

Common moorhens engage in both communal breeding, usually mother–father–daughter groupings, and brood parasitism, typically of unrelated breeders (McRae and Burke 1996). Brood parasitism typically involves the evening laying of a few parasitic eggs before the parasite begins laying in her own nest, or following the loss of her nest (McRae 1996a, 1997). McRae (1996a) suggests that communal breeding strategies may have facilitated the evolution of brood parasitism by increasing the likelihood of accepting the eggs of another female where it imposes no cost, and subsequently of accepting brood-parasitic eggs also, and by increasing the likelihood that offspring delaying dispersal would settle near their natal territory and thus increase the chance that a brood parasite is related to its host. Alternatively, restricted availability of breeding opportunities and delayed dispersal may have set the stage for a brood parasite with no option but to remain with the host, leading to communal nesting.

Acorn woodpecker

Mating system

Woodpeckers, along with barbets and the brood-parasitic honeyguides, are members of the order Piciformes. Most barbets and woodpeckers breed as territorial mongamous pairs, but a few species contain helpers-at-the-nest. In all woodpeckers and the single barbet species (Restrepo and Mondragón 1998) observed to date, the male undertakes nocturnal incubation. Woodpeckers are a sister taxon to the intraspecifically brood-parasitic honeyguides, and this cluster is a sister taxon to the barbets, so male nocturnal incubation may be the ancestral trait in this order (Sibley and Ahlquist 1990).

Among woodpeckers, cooperative breeding is relatively rare and largely limited to the melanerpine group. Only in the acorn woodpecker and perhaps some of the tropical members of this genus are females known to nest jointly.

Acorn woodpecker groups in high-density populations may contain 1–6 cobreeding males, 1–3 cobreeding females, and 0–10 non-breeding helpers. Breeding males thus usually outnumber breeding females, although groups with two females and one male occur. The mating system of this species has been called "opportunistically polygynandrous" (Koenig and Stacey 1990). About 40% of groups consist of a monogamous pair, 38% contain one female and two or more males (polyandry), 16% contain two females and multiple males (polygynandry), and 5% contain one male and two females (polygyny). Thus, only 21% of groups include joint-laying females. Cooperating females are almost always related, usually full- or half-sibs or occasionally a mother and her daughter. Cooperating males are also usually close relatives. Breeding males are not related to the breeding female(s), however, because of incest avoidance (Chapter 9). There is no evidence of extra-group fertilization or conspecific brood parasitism.

Acorn woodpeckers are highly territorial and usually resident throughout the year. The key adaptation that enables them to persist on their territories through the winter is their notable habit of storing acorns in several large, conspicuous granaries. Granaries are vigorously defended by all group members. High-density populations in California are usually saturated, so a large fraction of the surviving offspring of both sexes from the prior breeding season remain on the parental territory as non-breeding helpers. Sons are more likely to ascend to breeding status within the group, becoming breeders when their mother dies and is replaced by an unrelated female, while daughters are more likely to disperse to fill breeding vacancies outside the group. Competition for vacated breeding slots is intense and characterized by power struggles lasting days (Koenig 1981). Duration and size of such struggles are both positively correlated with granary size (Hannon et al. 1985). Moreover, dispersers of both sexes are more successful if they compete for these slots as a sibling coalition. Female fights are especially intense, and single female breeders can sometimes be evicted from their territories by invading sibling groups. About 63% of joint-nesting females achieve their breeding position by dispersing in a sibling coalition (Koenig and Mumme 1987).

Nesting behavior

As in all woodpeckers, acorn woodpeckers excavate and nest in tree cavities. Predation on the nests is generally low, but groups will usually attempt a second brood if the first is lost or if the acorn crop is very good. Joint-breeding females generally lay synchronously and exhibit strategies that guarantee cooperation and equitability. If females start laying in different nests, each may remove and destroy the eggs of the other, so the only way females can cobreed is to simultaneously use the same cavity. If one female begins laying before the other, the second female removes and destroys the prior eggs of the first layer. One effect of egg destruction is to reduce skew in egg ownership (Mumme et al. 1983a; Koenig et al. 1983, 1995). Reproductive skew between two female breeders is consequently very low, with the more successful female owning an average of 58% of the eggs, not significantly different from half (Haydock et al. 2001). The order of laying can vary from one nesting attempt to another, so over time females can expect to achieve similar reproductive success. Joint laying by more than two females is rare.

All potentially reproductive males (those not related to the breeding female) attempt to follow and mate-guard the female during her fertile period, but fighting and dominance interactions among the males are evident (Stacey 1979b; Joste et al. 1982; Mumme et al. 1983b; Hannon et al. 1985). Paternity studies indicate significant skew in fertilization success, with the more successful male on average siring 77% of the young in a given nest (Haydock et al. 2001). In groups with more than two males, it is rare for a third male to achieve any paternity. The male that achieves the larger share of paternity is not necessarily the older or larger male in the group; if anything, he appears to be lower in weight and condition (binomial test, $P < 0.02$: data from Haydock and Koenig 2002, Table 3). Alpha males thus may pay a cost for their reproductive access to the female. In some instances of second nesting attempts, the beta male may achieve the higher paternity, suggesting that one male cannot consistently monopolize reproduction (Haydock and Koenig 2002).

Female breeders perform most of the diurnal incubation, split equitably between them. Males also incubate during the day, with breeders participating more than non-breeding helpers (Mumme et al. 1990). Only males incubate at night. Sometimes this duty is shared by more than one male, while in other cases one male takes sole responsibility (Joste et al. 1982). All parents and helpers feed the nestlings, with breeding females contributing the most, breeding males a bit less, and helpers the least (Koenig and Mumme 1987; Mumme et al. 1990).

Fitness effects

In general, groups with an additional male, female, or helpers produce more offspring in a season than solitary pairs. However, the increase is at best 20 to 50%, so the per capita reproductive success in groups is lower than the success of pairs (Fig. 11.4). Adult survival is primarily affected by the amount of acorn stores on the territory. Females suffer higher mortality during the breeding season and have lower annual survivorship than males. Female survival also decreases as the number

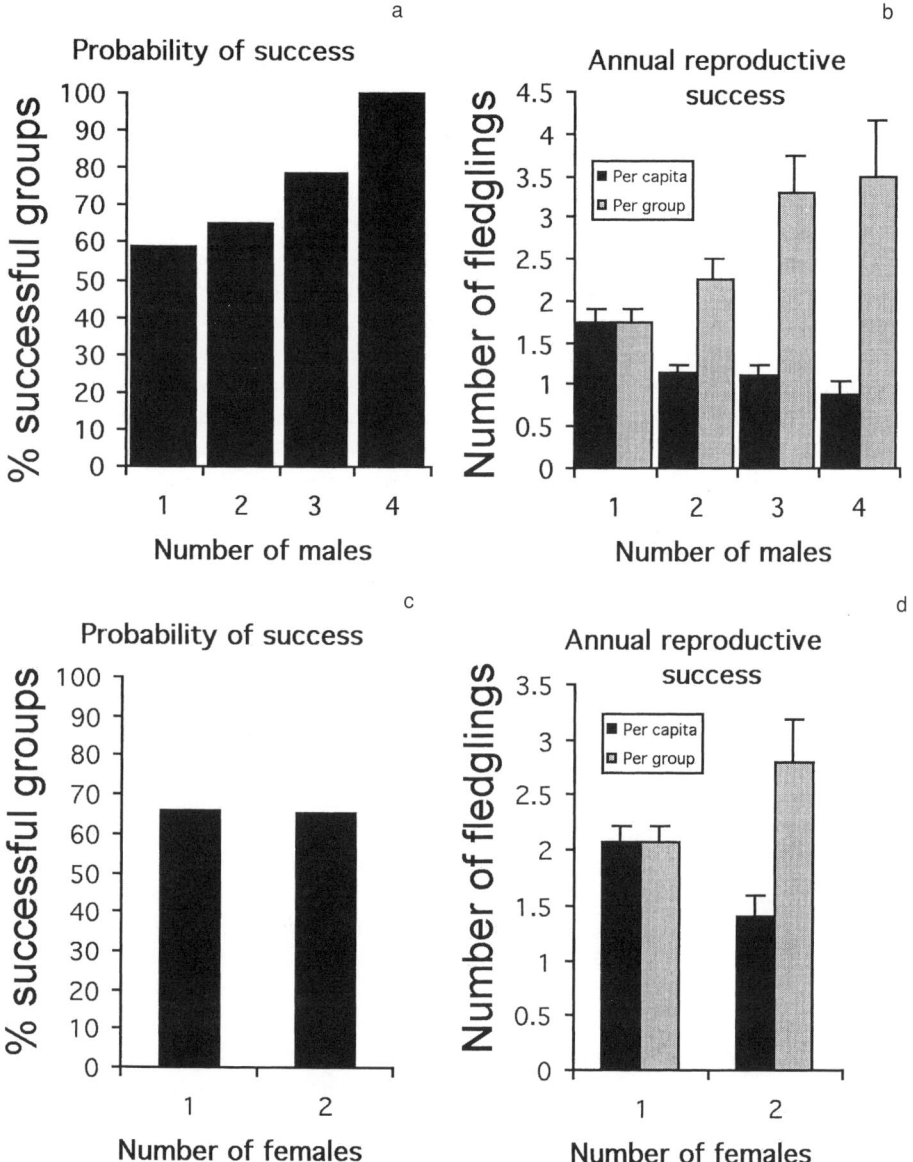

Figure 11.4. Effects of group size on reproductive success in the acorn woodpecker. (a) Probability of successful nesting as a function of number of males. (b) Per-group and per-male annual reproductive success for males. (c) Probability of successful nesting as a function of the number of females. (d) Per-group and per-female annual reproductive success for females. Redrawn from Koenig and Mumme (1987); used with permission from Princeton University Press.

of female cobreeders increases, possibly because of undetected dispersal events (Haydock and Koenig 2003). Established joint-nesting females therefore have lower lifetime reproductive success than established single females. The advantage to joint breeding may therefore arise only from increasing the probability of successfully winning a power struggle for a breeding vacancy, and perhaps in maintaining ownership of that breeding slot against invading female groups (Koenig and Mumme 1987).

Conclusions

Joint nesting in female acorn woodpeckers cannot be characterized by conspecific brood parasitism. Cobreeding most often occurs as a result of two females dispersing from their natal group together and winning a power struggle for a vacant slot on another territory. Occasionally daughters remain and cobreed with their mothers following replacement of their fathers, and in other cases females disperse and join a female relative who has succeeded in acquiring a breeding slot. Habitat saturation and constraints on independent breeding are clearly important causal factors in the formation of cooperative groups of relatives in both sexes. However, cobreeding females may sustain an uneasy truce in which each would prefer the other to leave.

Another level of complexity in this system is the potential for conflict between the sexes. A group of males with a female vacancy may encourage the acceptance of a female duo over a single female, since the males' per capita reproductive success is higher. Once a group has bred successfully for several years and accumulated young males, those males would potentially benefit by fostering the eviction of their mother, especially if she is alone, so that they can become breeders with a new female. Finally, when groups with a breeder vacancy contain helpers of the same sex as the missing bird, a conflict arises between the helpers and the breeders of the opposite sex (Koenig et al. 1998). Thus, intersexual conflict plays an important role in this system.

Crotophagines

Mating system

All four species of the Crotophaginae (order Cuculiformes) are communal nesters (Davis 1942). Cuckoos are renowned for their unusual breeding systems, which include territorial monogamous pairs, polyandry, and obligate and facultative interspecific brood parasitism in addition to communal nesting. In the non-parasitic monogamous species, which are believed to represent the ancestral cuculiform breeding system, males are the primary incubating sex, performing all nocturnal and at least half of the diurnal incubation. This sex bias in parental effort has been proposed to act as a constraint or pre-adaptation for the evolution of the other unusual breeding systems found in this taxon (Ligon 1993, 1999).

Three species have been reasonably well studied: the groove-billed ani, smooth-billed ani, and guira cuckoo. All occur in the New World tropics and subtropics and inhabit moderately open, secondary growth habitat. Breeding groups are composed of approximately equal numbers of male and female adults plus a variable number of non-breeding helpers, which usually are retained offspring from prior broods. Groups defend all-purpose territories against each other. Breeding adults pair off, with males vigorously defending their mates against other males. Pairs often separate themselves from the rest of the group, especially during laying, and can be observed allogrooming and contact roosting throughout the year.

Breeding units thus appear to be composed of cooperating monogamous pairs. However, the birds do not pair for life, and alliances can change with the addition of new group members. When extra females are present, they typically lay fertile eggs so polygamy can occur. Cases of CBP from extra-group individuals have been observed in both ani species (Loflin 1983; Vehrencamp et al. 1986) and verified with genetic data in smooth-billed anis (Blanchard 2000). Conspicuous pairs are not evident in the guira cuckoo and some instances of polygamy were detected with DNA analysis, so the mating system in this species remains unclear (Quinn et al. 1994; R. Macedo, personal communication).

Same-sex birds in multi-pair units are generally unrelated to each other. Birds disperse at about nine months of age in groove-billed anis, just prior to the onset of the breeding season (Bowen et al. 1989). Roaming solitary birds can often be seen and heard at this time as they approach prospective groups, sit in the top of a tree, and give a special joining call (Bowen 2002). Smooth-billed ani offspring tend to remain longer in their natal groups, leading to larger overall group sizes and more non-breeding helpers (Loflin 1983). Genetic analyses on nine partially sampled smooth-billed ani groups were also consistent with low levels of relatedness among group members and occasional retained offspring (Blanchard 2000).

Crotophagines exhibit the typical avian dispersal pattern of greater female than male movements. Essentially all females eventually disperse from their natal units and join new groups. A few males are recruited into their natal group as breeders, but most males also disperse and become established in groups, often close to their natal territory (Bowen et al. 1989).

The three well-studied species vary in group size and the level of organized cooperative breeding among the pairs. The groove-billed ani exhibits the smallest group sizes (mean 4.2, range 2–8) and the most organized level of cooperation (Koford et al. 1990). Modal group size is four individuals, consisting of two breeding pairs with no helpers. Single pairs are relatively common, and three-pair groups occur primarily in high quality habitat. Groups larger than six or seven tend to be unstable. Extra birds may be offspring from late broods of the prior year who eventually disperse or they may be peripheral or parasitic pairs who leave after a failed nest attempt. Groups with an odd number of adults are much rarer than even-numbered groups. Occasionally nests of groove-billed anis are abandoned because all of the eggs are tossed out. This occurs either because the laying females are strongly out of synchrony or because an extra-group pair has attempted to join by parasitizing the clutch.

Smooth-billed ani groups tend to be larger (mean 6.7, range 2–17 in Florida; mean 3.8 to 6.2 in different years in Puerto Rico), but frequently contain non-breeders, and the modal number of laying females per group is still usually two (Loflin 1983; G. Schmaltz, personal communication). The incidence of non-monogamous mating, CBP, and breeding disorganization may be somewhat higher in this species than in groove-billed anis (Davis 1940; Loflin 1983; Blanchard 2000).

Guira cuckoos also tend to occur in large groups (mean 6.7, range 2–13) and have the least-well-organized form of cooperative nesting (Macedo 1992). DNA fingerprinting analyses on several partially sampled groups suggest that polygamous mating is widespread (Quinn et al. 1994). Up to seven females may lay in a single nest, and the average number of eggs laid per female (typically 1–3) is much smaller than in the anis (typically 4–6) (Cariello et al. 2002). Nest abandonment is common and may be caused by a sabotage strategy, to force renesting, on the part of individuals who did not contribute eggs to the clutch (Macedo 1992). In addition, adults have been observed to remove and kill nestlings, and patterns of nestling loss indicate that this practice occurs frequently (Macedo and Melo 1999; Macedo et al. 2001).

Nesting behavior
Nests are bulky, open-cup structures placed in dense, often thorny, trees or bushes. Each pair takes turns constructing the nest, with the male bringing twigs to the female waiting at the nest, who puts them in place (Skutch 1959). Green leaves are continuously brought to the nest during incubation and tucked under the eggs. The breeding females in a group rarely begin laying synchronously, and the order in which they start to lay affects the number of eggs they lose. Early eggs are tossed out of the nest, or in the case of smooth-billed anis, sometimes covered with leaves and sticks (Loflin 1983; Quinn and Startek-Foote 2002). The final number of eggs a given groove-billed ani female contributes to the incubated clutch tends to be greater for the last-laying female, but the skew in egg ownership is usually low and less than what it would have been without egg-tossing or burial. First-laying ani females strategically increase their egg size, which may increase nestling survival and further eliminate any reproductive skew (Vehrencamp et al. 1986). In unusually large groups of smooth-billed anis, additional eggs may become accidentally buried during incubation; which eggs are buried at this stage seems to be random with respect to egg ownership and laying order (Blanchard 2000). The eggs of the crotophagines are not only exceptionally large for the bird's body size (Lack 1968; Macedo 1992) but they also have extremely thick shells to withstand layering and frequent turning in the nest. For clutch sizes up to about 14–15 eggs (above the leaf lining), hatch rates are extremely high (Vehrencamp 2000).

All breeding group members incubate, but relative effort differs greatly. The behaviorally dominant male performs most of the nocturnal incubation and up to half of the diurnal incubation; he also vigorously defends the nest against potential predators. Among ani females the last bird to start laying contributes less than the other females (Köster 1971; Vehrencamp 1977). Contributions to nestling provisioning tend to be more evenly distributed among the adults, and juvenile helpers also provision nestlings. In 68% of groove-billed ani groups, the dominant male is mated to the last-laying female, but in the remaining groups he is mated to an early-laying female who usually manages to achieve equal or greater egg ownership by laying a very large clutch (Vehrencamp et al. 1986). Female age and body size were positively associated with incubated clutch size. In a few cases where one female in a two-pair group disappeared and a new female joined, the surviving female became mated to a different male, suggesting that the dominant male is able to predict and acquire the female most likely to gain the reproductive skew advantage.

Fitness effects

Figure 11.5 illustrates the effects of group size on nest predation rate, annual reproductive success, adult survival, and lifetime reproductive success in the groove-billed ani. Two-pair groups can raise about twice as many offspring as single pairs, but larger groups are not as efficient because some of the late-hatching young are outcompeted and trampled beneath their older nestmates (Vehrencamp 1978; Koford et al. 1990). Nest predation rates do not differ significantly for different group sizes. Breeding season survival increases with increasing group size, especially for females. Combining annual reproductive success with survival probabilities leads to a clear reproductive superiority of two-pair groups over single pairs (Vehrencamp et al. 1988). For males, survival is largely a function of dominance status, with single-pair males and dominant group males suffering higher breeding season mortality (19%) than beta group males (10%).

Since dominant males may have sired the greater share of eggs in the nest, fitness of dominant and subordinate males is probably approximately equal, and very similar for males in two-pair and single-pair groups. Loflin (1983) obtained virtually identical results in the smooth-billed ani, with two-female groups having the highest per-female annual reproductive success (Quinn and Startek-Foote 2000). He also believed that survivorship increased for adult birds in larger groups as a result of vigilant sentry behavior and shared parental care.

Conclusions

Communal nesting in the crotophagines is both egalitarian and mutually beneficial to group members relative to single-pair breeding, at least for intermediate-sized breeding units. Cobreeding in these birds does not depend on habitat saturation to retain birds on their natal territories. Although good-quality habitat may be saturated, there is adequate availability of vacant territories for both anis and guiras (Koford et al. 1986; Macedo and Bianchi 1977). Brood parasitism may sometimes be used as a mechanism to gain entry into a group, but birds that successfully insert eggs into nests are subsequently allowed to join the group and contribute their share of the parental care. In guiras, the larger number of females contributing eggs to a clutch relative to the number of males, coupled with late egg-tossing and infanticide, may be indicative of a fairly high level of brood parasitism in this species.

PASSERINES

Helper species with occasional cobreeding

In helper-at-the-nest species with female helpers, extra eggs attributable to the helpers are sometimes observed. The incidence of such helper egg-dumping is often extremely rare (Florida scrub-jay, Quinn et al. 1999), while in others it is low but persistent, on the order of 10% of nests (Vehrencamp 2000). In the magpie-jay, a system with only female helpers, up to two additional eggs were laid in groups containing older helpers (Langen 1996a). In the Arabian babbler about 10% of nests contained eggs of two females (Zahavi 1974). Recent work indicates that cobreeding occurs only when a new group has been established. In subsequent nests, one female becomes the primary breeder (Lundy et al. 1998). In the white-winged chough, Rowley (1978) reported an incidence of double clutches of about 16%. Such clutches had low hatching success because the mud nests of this species are not large enough to handle enlarged clutches. Subsequent studies on this species discovered that cobreeding was more likely to occur when new groups form (Heinsohn 1992, 1995; Heinsohn et al. 2000). During a particularly severe drought, about 20% of the population died and large groups split into fragments of same-sex siblings. These sibling groups merged with opposite-sex fragments from elsewhere, and cobreeding occurred among both the males and females. Single pairs are incapable of successfully reproducing in this species because provisioning is very costly, so cooperation may be mutually beneficial during these population crashes.

Among helper-at-the-nest species, the Seychelles warbler exhibits the highest level of cobreeding by helpers. The primary population of this rare, endemic species on Cousin Island has recovered from a severe population bottleneck and now completely saturates the island. It was initially described as a helper-at-the-nest species in which the helpers are usually female offspring from the prior year (Komdeur 1994b). Pooled over all habitats, 27% of all breeding units contain helpers, but the fraction of units with helpers increases in higher-quality habitats. Helpers significantly increase the reproductive success of the breeding pair, but productivity is not doubled by the presence of a helper. The

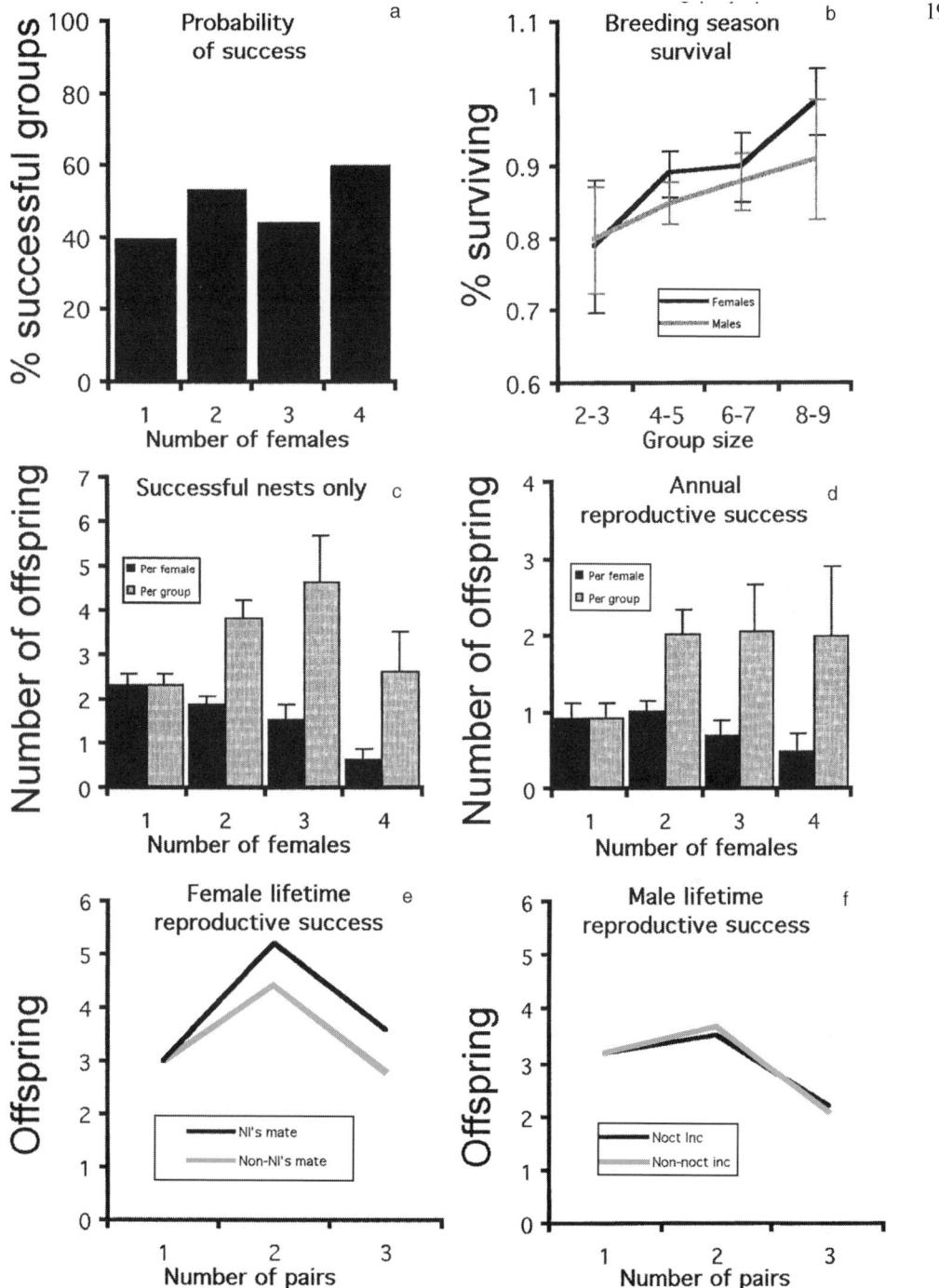

Figure 11.5. Effects of group size on components of reproductive success and survival in the groove-billed ani. (a) Probability of successful nesting as a function of the number of breeding females. (b) Breeding-season survivorship of adult males and females as a function of group size. (c) Number of offspring surviving to one month of age in successful nesting attempts (1 or more juveniles produced) per group and per female as a function of number of breeding females. (d) Annual reproductive success per group and per female. (e) Estimated female lifetime reproductive success for mate of nocturnal incubator and mate of non-incubators. (f) Estimated male lifetime reproductive success for nocturnal incubator and non-nocturnal incubator (assuming a bird remains in the same role its entire life). Redrawn from Vehrencamp et al. (1988) and Koford et al. (1990).

normal clutch size for a female is one, but two-egg and (rarely) three-egg clutches are also observed. In a recent study of maternity and paternity using DNA profiling, 44% of female helpers were found to have contributed an egg to the primary female's clutch, and 15% of all offspring produced were attributed to helper females (Richardson *et al.* 2001). About 40% of the young in this study were sired by breeding males on other territories, suggesting that both the helper and primary females sought extra-group copulations. The helper females assist the primary female with incubation as well as nestling care. On average, primary females benefit from the presence of female helpers and preferentially produce female offspring when living on high-quality territories (Komdeur *et al.* 1997). This species thus represents an unusual example of intermediate reproductive skew involving a mixture of helpers-at-the-nest and cooperative joint breeding without a significant contribution of the male to incubation.

Taiwan yuhina

Yamashina (1938) first described communal nesting in this timaliine babbler. He observed five nests, which contained eggs of two or three distinct types and were attended by 4–8 birds of both sexes. More recent work is unpublished, but we are able to provide a very brief summary based on meeting abstracts, notes, and personal communications with S.-F. Shen and H.-W. Yuan. Sixty percent of groups contain an even number of birds and equal ratios of males and females, which form clearly associated pairs. Modal group size is four. Per capita reproductive success is higher for even-sized groups, and is highest for groups of four and six relative to single pairs. One pair appears to dominate the others and contributes more parental effort. Although the dominant female performs much of the incubation, some but not all of the other group members also incubate. In some groups, the alpha male performs the greatest proportion of diurnal incubation, and males may occasionally incubate at night. Female reproductive skew is presumably low, since the number of egg types corresponds with the size of the group. The birds typically disperse and breed in non-natal groups at one year of age, and thus are neither habitat-saturated nor related to their same-sex cobreeders. This species therefore seems to have a breeding system much like that of the anis, with joint-nesting pairs that perform better than single pairs. Although females are usually the primary incubating sex, the alpha male has a large role in incubation for a passerine. It remains to be seen whether reproductive skew favors the dominant pair, and whether the true joint nesting in this species represents a major exception to the male incubation rule or is driven by the relatively strong paternal effort.

DISCUSSION

Conspecific brood parasitism and the evolution of joint nesting

Except for ratites and tinamou, with their ancestral state of all females laying and leaving parental care to the male, all other known joint-nesting species evolved from an ancestral state of single nesting females who care for their own young with male assistance. Several of the species described in this chapter show evidence of CBP, and in crotophagines and pukeko brood parasitism may be a mechanism for joining established breeding groups. It is therefore reasonable to ask whether cooperative joint nesting evolved via an intermediate step of conspecific brood parasitism, which generates joint-clutch experiments that are then subjected to natural selection. Specifically, parasites must be selected to stay and care, and hosts must be selected to accept the newcomer.

CBP is most common and conspicuous in species with precocial self-feeding young, especially the Anseriformes and Galliformes, but is also found in some semiprecocial species (Laridae, Rallidae) and some altricial species (Columbiformes, Cuculiformes, Passeriformes) (Yom-Tov 1980; Petrie and Møller 1991). These semiprecocial and altricial species are not only characterized by the need for offspring provisioning, but many are also colonial breeders. Although female parental care is the dominant pattern among the species with CBP, several are biparental, and a few exhibit greater male than female incubation effort, including the greater roadrunner (Miller 1964; Vehrencamp 1982), yellow-billed cuckoo (Nolan and Thompson 1975), American coot (Lyon 1993a, 1993b), and common moorhen (Gibbons 1986; McRae 1996a). Detailed studies on a few species with frequent CBP show that the parasitic females are often young floater females without a territory or mate, females that have lost their nests, mated territorial females who lay in a neighbor's nest before producing their own clutch, or relatives of the host female (Brown 1984; Gibbons 1986; Møller 1987; Evans 1988; Lyon 1993a;

McRae 1996a, 1997; McRae and Burke 1996; Andersson and Åhlund 2000).

According to the models of Zink (2000) and Andersson (2001), a female that has laid eggs in a conspecific nest is usually better off deserting rather than staying to help raise the offspring, because she then has the ability either to start her own nest or parasitize additional nests. However, if the advantages of leaving are reduced or the benefits of staying are increased, cooperation could be the better strategy.

Several factors could reduce the advantages of leaving:

(1) The availability of additional breeding sites is low. This factor is a strong deterrent to leaving in species with all-purpose territories, especially in habitat-saturated populations, since vacant, good-quality territories are in very short supply and highly contested. Extreme shortage of territories is certainly characteristic of acorn woodpeckers and North Island pukeko, while South Island pukeko and anis may experience a shortage of high-quality territories.
(2) Potential mates for a second nesting attempt are in short supply. If females require the assistance of a male for territory defense, nest construction, incubation, and/or offspring care, they will be strongly restricted in their ability to start a second nest by the availability of unmated males. Coupled with the spatial or temporal constraints of leaving outlined above, the need for biparental care probability explains the much lower incidence of CBP in altricial species.
(3) The time available for initiating or parasitizing a second nest is limited. Highly seasonal and synchronous breeding in a population would severely restrict the ability to find additional nests to parasitize and reduce the success of one's own delayed nest (Yom-Tov 1980). This factor appears to be an important incentive for staying in the magpie goose.

Several factors could also operate to increase the benefit of staying:

(1) Offspring survival is significantly improved if the parasite stays to help raise them. This factor is obviously most important for species in which the young require extensive provisioning, and is likely to play a role in all of the communal species except ratite and tinamou species. Provisioning of young clearly separates the communal magpie goose from the rest of the Anseriformes in which parasitic females always desert.
(2) A high level of relatedness among the females increases the benefit of staying and cooperating. Andersson's CBP model and empirical data on white-fronted bee-eaters (Emlen and Wrege 1986) suggest that females are more likely to stay and cooperate if they are related to the host female because of inclusive fitness benefits. Zink's model suggests that females should be less likely to parasitize relatives because he assumes parasitism is costly for the host. Cobreeding females are related in the magpie goose, acorn woodpecker, Tasmanian native hen, common moorhen, North Island pukeko, white-winged chough, jays, and Seychelles warbler, but not in the ratites, South Island pukeko, anis, or Taiwan yuhina. Relatedness therefore shows no consistent association with joint nesting, even when the reproductive fitness costs are considered. In the magpie goose, acorn woodpecker, and white-winged chough, sibling females may benefit from dispersing together. In the moorhen, Seychelles warbler, and other helper species, cobreeding results from retained daughters who are already cooperating as helpers.
(3) Females gain access to high-quality males by first parasitizing and then remaining with them (Vehrencamp 2000). This factor is likely to be most important for species with a significant paternal care role, and is potentially the key one that explains why communal laying is much more likely to occur in species with a strong male parental care role. This idea has been neither modeled nor demonstrated with empirical data in birds. However, it is well established in fish with male care of offspring that females are strongly attracted to males demonstrating high parental care abilities in their courtship displays and to males that have previously acquired eggs from other females (Knapp and Kovach 1991; Forsgren *et al.* 1996; Ostlund and Ahnesjo 1998).

Finally, host females must be selected to accept the parasite. The most important factor favoring acceptance must be the ability to incubate an enlarged clutch successfully, which reduces the cost of parasitism to the host. In ratites, incubation of multiple clutches is possible because of the relatively small egg size, and in

Seychelles warbler it is possible because of the small clutch size. Anis have evolved special adaptations for incubating large clutches. In pukeko and acorn woodpeckers, on the other hand, hatching success declines with the addition of a second female so the reproductive costs of joint nesting are higher.

A second factor leading to host acceptance is male control over female access to the nest, or male promotion of parasites who copulate with them. This factor could also explain the widespread occurrence of joint nesting in taxa with strong male parental care.

Multiple contexts for the evolution of joint nesting

Our review has revealed some important associations between mating systems, parental care patterns, ecological contexts, fitness benefits of cooperative nesting, conflicting interests between the sexes, and within-sex degree of skew. Joint nesting has clearly evolved in several contexts. The primary factors that we believe are important for the evolution of joint nesting in each taxon are summarized in Table 11.1 and discussed below, along with some speculation as to the most appropriate skew models for each system.

For the ratites, the key factors favoring communal laying are sole male parental care and the ability of the incubator to cover a clutch that is larger than what a single female can lay. Large body size relative to egg size facilitates the incubation of large clutches, and precocial development of young permits single-parent care of fledglings. Males benefit by attracting multiple females to their nest, and actively prevent females from attempting to roll each others' eggs out of the nest by completely controlling access to the nest. Females benefit, up to a point, from the presence of other females' eggs. The result is a polygynandrous mating system characterized by simultaneous polygyny and sequential polyandry, and very low skew among cobreeding females.

This mating system is a win–win situation for both sexes, allowing both males and females to maximize mating opportunities and reproductive success (Vehrencamp and Bradbury 1984). Females do not participate in cooperative brood care so transactional skew models are not appropriate. The best characterization of this system is "beneficial sharing" in which it is neither profitable nor possible for a dominant female to skew the clutch in her own favor. In ostriches, on the other hand, one female does take control of the nest but subordinate hen laying is solicited. By recognizing her own eggs, the major hen retains only enough of the subordinate females' eggs to maximize her own reproductive success. Subordinate females gain no benefit from staying or cooperating and leave to lay in other nests. A bidding game model could be appropriate in this situation.

We lack much information on the magpie goose, but the dominance of one female and the hint of egg-tossing suggest that some reproductive suppression is occurring. There appears to be a strong temporal constraint on the breeding period and an advantage to having multiple provisioners for the offspring. High-quality parental males may also be a limiting resource for females. A concessions model might apply to this system, with ecological constraints favoring retention of female offspring and moderately high levels of skew. However, these birds are long-lived and the subordinate does remain to help, so the "hopeful reproductive" model could also apply (Kokko and Johnstone 1999).

In the simultaneously polygynandrous species, including the acorn woodpecker, pukeko, and Tasmanian native hen, cooperative polyandry, with multiple males and one female, is a common breeding unit. With the exception of South Island pukeko, these species are all characterized by ecological constraints on territory availability, so offspring are often retained, cobreeders of the same sex are related, and per capita reproductive success declines with increasing group size (see supersaturation hypothesis in Chapter 3). Cooperative male defense of resources is advantageous. Males have thus evolved the strategy of sharing reproductive access to the breeding female, but their relative success (skew) may be dependent on dominance behavior. The fact that reproductive skew increases with increasing ecological constraints indicates that some elements of the transactional skew models are operating. Behavioral observations, however, suggest that a tug-of-war model might be most appropriate. The addition of another female benefits the males but not the females, so there is a conflict of interest between the sexes. Males, who perform a significant amount of the parental care, should encourage the joining of multiple females and prevent females from ejecting each other. Cobreeding females may then minimize their loss from the low per capita breeding success by evolving mechanisms to reduce the skew between them. This situation, with

Table 11.1. *Summary of breeding unit characteristics of species described in text*

Species	N breeding males[a]	N breeding females[a]	Mating system[b]	Females related	Helpers present	Incubation by[c]	Evidence of conspecific brood parasitism	Chick development	Ecological constraint	Origin of female cobreeding[d]
Greater rhea and tinamou	1	1–6	Sequential polygynandry	No	No	M only	n/a	Precocial	None	Male PC + large clutch
Ostrich	1	3–6	Sequential polygynandry	No	No	M and major hen	Yes	Precocial	None	Male PC + large clutch
Magpie goose	1	2–3	Harem polygyny	Yes	Yes	A	Possibly	Semi-precocial	Temporal constraint	Male PC + EC
Pukeko (North Island)	1–3	1–2	Cooperative polygynandry	Yes	Yes	M (night) A (day)	No	Semi-precocial	Territories	Male PC + EC
Pukeko (South Island)	1–3	1–2	Cooperative polygynandry	No	Rare	M (night) A (day)	No	Semi-precocial	Low	Male PC + EC
Tasmanian native hen	1–3	1–2	Cooperative polygynandry	Yes	Yes	M (night) A (day)	No	Semi-precocial	Territories	Male PC + EC
Moorhen	1	1–2	Monogamy	Yes	Yes	M (night) A (day)	Yes	Semi-precocial	Territories	Male PC + EC
Acorn woodpecker	1–4	1–2	Cooperative polygynandry	Yes	Yes	M (night) A (day)	No	Altricial	Territories	Male PC + EC
Groove-billed and smooth-billed anis	1–3	1–3	Monogamy	No	Some	M (night) A (day)	Yes	Altricial	Low	Male PC + group benefit
Guira cuckoo	1–4	1–4	Unknown	No	Yes	Unknown	Yes	Altricial	Low	Male PC + group benefit
Taiwan yuhina	2–3	2–3	Monogamy	No	Some	Mostly alpha female+male	Unlikely	Altricial	Low	Group benefit
Seychelles warbler	1	1	Monogamy	Yes	Yes	F only	Unlikely	Altricial	Territories	EC + female helpers
White-winged chough	1–3	1–2	Monogamy	Variable	Yes	F (night) A (day)	Unlikely	Altricial	Territories; provisioning	EC + female helpers
White-throated magpie-jay	1	1	Monogamy	Yes	Yes	Breeding F only	Unlikely	Altricial	Territories	EC + female helpers

[a]Refers to typical numbers of males and females but not the extremes.
[b]Mating system refers to social mating system.
[c]M = male; F = female; A = all.
[d]Male PC = male parental care; EC = ecological constraint.

opposite-sex control of group membership, remains to be modeled.

The socially monogamous species (anis and Taiwan yuhina) are the only ones in which per-female (and per-pair) annual reproductive success is equal or slightly larger for cooperative groups compared to single nesting pairs. In both of these systems, the modal group size of two-pair groups coincides with the peak in per-pair reproductive success. Slightly larger groups occur, and their success is not as high as the modal group size but still about as good as single pairs. Much larger groups suffer significantly lower success and are rare. Selective pressure clearly favors group formation from the perspective of both sexes, and conflict of interest between the sexes is minimal. Conflict of interest within the sexes is low but not absent, resulting in a slight degree of skew.

In both systems, a dominant pair performs the majority of incubation. Although the subordinate pair appears to be usurping the parental effort of the dominant pair, relations between the pairs are apparently friendly and no obvious attempt is made to eject them. This observation implies that the benefits of cobreeding exceed the dominants' cost of ejecting the subordinates. The subordinate birds may experience subtle costs and benefits from their association with the dominant pair. In anis, the subordinate male avoids the risks and energetic costs of nocturnal incubation, but his mate may lose some of her eggs to egg-tossing. The beta Taiwan yuhina female may gain a similar benefit of reduced parental effort. For both systems, it is crucial to determine whether mating is truly monogamous, or whether another subtle benefit and perhaps a driving force for joint nesting is increased mating opportunities for the alpha male and access to a high-quality mate for the beta female. Finally, acceptable habitat is not limited, so both sexes disperse and cobreeders are unrelated. These species appear to fit the classical concessions skew model with unrelated individuals quite well. However, the laying order, egg-tossing, and clutch adjustment strategies among female anis could also fit a tug-of-war model. The two models could be distinguished on the basis of whether or not the degree of skew increased with increasing ecological constraints (true only for the concessions model), or with the augmentation of one female's competitive ability (true only for the tug-of-war model).

Occasional joint nesting in the passerine helper species (jays, white-winged chough, and Seychelles warbler) occurs in the context of strong ecological constraints and the prevalence of female helpers. Why these species preferentially retain female helpers, rather than the more typical pattern of male retention, is not clear. A small fraction of the helpers manage to insert an egg into the clutch despite strong suppression by the dominant breeding female. Incest avoidance cannot explain the typically high degree of skew in these species, because subordinate females have access to male mates who are not their fathers. The incubation-clutch-size constraint of added subordinate female eggs may be much less in the Seychelles warbler, with their normal clutch of one egg, which could explain the higher incidence of helper laying in this species. In the white-winged chough, cobreeding increases when ecological constraints are relaxed and same-sex group members are less related. These findings are consistent with the concessions model.

In conclusion, conspecific brood parasitism may well be the proximate mechanism by which joint brood experiments are generated in some species. True joint nesting, with cooperative brood care by cobreeding females, can only evolve and become stable if females benefit from joining another female, and neither female is then selected to leave or sabotage the clutch. A large role of the male in parental care, plus at least one other factor that increases the cost of leaving or increases the benefit of staying, seems to be required. These specific conditions appear to be uncommon in nature, but at least four different contexts have been outlined here. Thus female joint nesting is rare, but highly variable in terms of inter- and intra-sexual strategies and relative fitness effects.

ACKNOWLEDGMENTS

We thank Ian Jamieson, Walt Koenig, Joey Haydock, Sheng-Feng Shen, and Hsiao-Wei Yuan for allowing us to summarize unpublished data. M. Andersson provided many useful insights into the dynamics of his CBP model. Comments on earlier drafts of the manuscript by Walt Koenig and Joey Haydock greatly improved its quality.

12 · Conservation biology

JEFFREY R. WALTERS
Virginia Polytechnic Institute and State University

CAREN B. COOPER
Cornell Laboratory of Ornithology

SUSAN J. DANIELS
Virginia Polytechnic Institute and State University

GILBERTO PASINELLI
University of Zurich

KARIN SCHIEGG
University of Zurich

The primary objective of conservation is to preserve biodiversity. Biodiversity encompasses not only distinct life forms such as species and subspecies, but also unique adaptations such as cooperative breeding. Cooperatively breeding birds exhibit a variety of distinctive traits that render some species unusually vulnerable to, or resistant to, habitat loss, degradation, and fragmentation, and to the problems inherent to small populations. Especially relevant are extreme philopatry, sensitivity to habitat quality, and the presence of large numbers of non-breeding adults (helpers). To our knowledge, no one has previously assessed how cooperative breeders as a group are faring against the threats to their continued existence they currently face. In this chapter we conduct such an assessment and examine the interaction between the distinctive features of cooperative breeders and the various threats to biodiversity.

In the absence of a body of previous work, we take a simplistic approach and look for broad, general patterns rather than attempting complex analyses. The existing data simply will not support the latter, compelling us to propose hypotheses rather than test them. We fully recognize that, given the diverse ecologies of cooperatively breeding birds (Chapter 5), there will be numerous exceptions to any generalization made about the group. We also recognize that many generalizations will apply to some of the diverse array of cooperative breeding systems but not others, for example to systems with singular breeders assisted by natal helpers but not to systems with plural breeders. Nevertheless, we believe that the features on which we focus are common enough that pointing out the consequences of these features will promote conservation of the fascinating adaptation of cooperative breeding. To illustrate our ideas, we rely heavily on the two cooperatively breeding species we have studied, the red-cockaded woodpecker and brown treecreeper.

In order to carry out our assessment, we first assembled a list of cooperatively breeding species to examine. We used Brown's (1987) list as a starting point, and consulted recent literature to remove and add species to the list. We only included species in which cooperative breeding was well documented as a regular rather than occasional event. This procedure, for example, resulted in our excluding most of the species listed as "unclassified and miscellaneous" on Brown's (1987, Table 2.2) list. Also, we only included species for which sufficient information was available to assess the features we wished to discuss. The 127 species we considered, listed in Appendix 12.1, can be regarded as a sample of relatively well-studied cooperative breeders. Our list obviously is more restricted than other recent ones. For example, Arnold and Owens (1998) considered 308 of 9672 extant bird species to be cooperative breeders, while Ligon and Burt (Chapter 1) list 357 cooperatively breeding species. Scientific names of the species we discuss are also listed in Appendix 12.1.

THE CONSERVATION STATUS OF COOPERATIVELY BREEDING BIRDS

As of August 2001, the IUCN Red List (IUCN 2000) listed 1192 avian species as critically endangered, endangered, or vulnerable (12% of extant species). None of the 127 cooperative breeders in our sample was listed as critically endangered or endangered, and only five were considered vulnerable (4% of our sample). The pattern was similar in the only bird family (Corvidae)

Ecology and Evolution of Cooperative Breeding in Birds, ed. W. D. Koenig and J. L. Dickinson. Published by Cambridge University Press.
© Cambridge University Press 2004.

represented by a sufficient number of species in our sample to merit examination. Thirteen jays and crows (11% of extant corvid species) appear on the Red Book lists, as does one of 16 corvid species (6%) included in our sample. This crude assessment suggests that cooperative breeders are not over-represented among the world's most endangered bird species, and may even be under-represented. However, the Red Book lists are biased toward poorly known species, whereas our sample is biased toward well-known species. Hence ours may not be a fair sample for this assessment.

The Red Book lists are dominated by island endemics. Three of the five cooperative breeders listed as vulnerable are island endemics (Seychelles warbler, Galápagos hawk, and Maui alauahio). If cooperative breeders are especially vulnerable to human impacts, they may be over-represented among the many island forms already extirpated (Burney et al. 2001), and therefore under-represented among those that remain.

We therefore examined the prevalence of cooperative breeders among endangered and threatened birds in two continental areas, one (United States and Canada) inhabited by relatively few cooperative breeders, the other (Australia) by many. The Red Book lists include 21 species breeding in the United States and Canada, representing 3% of the extant breeding bird species in these countries. Of 15 cooperatively breeding species in our sample occurring in the United States and Canada, two (13%) are included in the Red Book lists (red-cockaded woodpecker, Florida scrub-jay). The Red Book lists include 22 species breeding in Australia, which represents 4% of this country's extant breeding bird species. None of the 35 Australian cooperative breeders in our sample is included in the Red Book lists. Combining the data from the two continents, 3–4% of their extant breeding bird species appear on the Red Book lists, compared to 5% of our sample of cooperative breeders from these continents. Again, there is no indication that cooperative breeders tend to be over-represented among endangered and threatened bird species.

CAUSES OF EXTINCTION

In assessing the relative vulnerability of cooperative breeders to various threats to biodiversity, we use the conceptual framework of declining-population and small-population paradigms of Caughley (1994). The small-population paradigm is concerned with problems inherent to small populations, and the declining-population paradigm with processes by which populations are driven toward extinction by external agents. We first address agents of population decline.

The majority of recent and imminent extinctions of avian species involve persecution by humans and/or impacts of associated exotic species, primarily on islands. There is no obvious reason that cooperative breeders would be especially subject to these processes, unless they were disproportionately represented on islands, and indeed we found no indication that cooperative breeders are especially persecuted by humans or especially sensitive to exotics. We focus instead on the question of whether cooperative breeders are particularly vulnerable to three other processes that increasingly threaten species globally, that is, three likely agents of future extinctions: habitat loss, habitat degradation, and habitat fragmentation.

HABITAT LOSS

There is no question that the primary threat to biodiversity currently is habitat loss, and all species are vulnerable to this threat. Rabinowitz et al. (1986) provide a convenient way to assess relative vulnerability to habitat loss by defining rarity categories. This framework has three dimensions, geographic range (narrow versus broad), habitat use (generalists versus specialists), and population size (everywhere small versus somewhere large). Seven of the eight categories created represent rarity, and vulnerability to habitat loss, in some sense. Only habitat generalists with large population sizes and broad geographic ranges are not rare (Rabinowitz et al. 1986), and only species in this category are broadly resistant to habitat loss.

Although we have not conducted a systematic analysis, a cursory assessment of the species in our sample suggests some hypotheses about the relative vulnerability of cooperative breeders to habitat loss. First, although there are numerous glaring exceptions such as the American crow and laughing kookaburra, it follows from the theoretical idea that the evolution of cooperative breeding is linked to sensitivity to habitat quality (Stacey and Ligon 1991; Emlen 1991; Koenig et al. 1992; Walters et al. 1992b) that cooperative breeders will tend to be habitat specialists. Based on this supposition, we hypothesize that cooperative breeders will be disproportionately affected by complete elimination of particular habitat types.

We found no indication that cooperative breeders tend to more often have either narrow or broad ranges compared to non-cooperative species. It does appear, however, that cooperative breeders tend to have populations that are somewhere large, as opposed to everywhere small. This pattern is consistent with observations made in some of the first studies of cooperative breeders that resulted in the concept of habitat saturation (Selander 1964; Brown 1974; Stacey 1979a; Koenig and Pitelka 1981; Emlen 1982a; Woolfenden and Fitzpatrick 1984). Based on this supposition, we hypothesize that cooperative breeders will tend to be relatively resistant to initial reductions in habitat area, and therefore to be among the species that persist when habitat is reduced but not completely eliminated.

Finally, neither cooperatively breeding species nor habitat loss is evenly distributed over the planet. Hence the vulnerability of cooperative breeders also depends on the degree of correspondence between the two distributions, as well as the distribution of conservation efforts. Cooperative breeders constitute a larger portion of the avifauna in the *Eucalyptus* woodlands of Australia than anywhere else (Brown 1987), and perhaps next largest in the savannas of Africa. Thus, conservation of savanna ecosystems is of particular importance to cooperative breeders. In Australia, conservation efforts are focused on rainforest, while *Eucalyptus* woodland communities, despite their great extent and variety, are poorly conserved (Specht 1981; Yates and Hobbs 1996). Rapid human population growth in Africa makes conservation particularly challenging, but the savannas where cooperative breeders are especially concentrated are not among the areas where habitat loss is most extreme currently (Brooks and Thompson 2001).

Cooperatively breeding birds, like all organisms, are threatened by habitat loss. As a group they may differ slightly from non-cooperative species in their response to habitat loss, with an unusually high proportion of species being relatively resistant to initial habitat reduction, and an unusually high proportion being relatively sensitive to elimination of particular habitats. Overall, though, habitat loss does not appear to represent an appreciably greater or lesser threat to cooperative breeders than to other birds.

HABITAT DEGRADATION

Habitat degradation, although as pervasive as habitat loss and habitat fragmentation, generally does not receive as much attention as the other two processes. Habitat degradation encompasses a variety of phenomena, such as introduction of exotics, loss of large, old trees in forests and woodlands, and altered species composition in grasslands due to grazing of domestic animals and altered fire regimes. It follows from the theoretical idea that the evolution of cooperative breeding is linked to sensitivity to habitat quality that cooperative breeders would be unusually vulnerable to habitat degradation.

The validity of this generalization depends on how widely applicable this hypothesis about the evolution of cooperative breeding is. In North America, where much of the work that supports the evolutionary hypothesis has been done, the generalization appears valid. The Florida scrub-jay (Woolfenden and Fitzpatrick 1984) and red-cockaded woodpecker (Conner *et al.* 2001) are two conspicuous examples of North American cooperatively breeding habitat specialists that are sensitive to habitat degradation. But even in North America there are notable exceptions to the generalization, such as the American crow.

The evolutionary hypothesis appears to be less applicable in other regions, notably the southern temperate regions of Africa and Australia where cooperative breeders are most prevalent (Ford *et al.* 1988). Habitat degradation is thought to be a primary cause of declines of woodland birds in Australia, but there is no indication that cooperative breeders are disproportionately represented among the affected species (Recher and Lim 1990; Robinson and Traill 1996; Ford *et al.* 2001). We conclude that cooperative breeders probably are disproportionately represented among avian species that are particularly sensitive to habitat degradation, but this relationship is not consistent enough that habitat degradation represents a universal threat to cooperative breeders.

HABITAT FRAGMENTATION

Species differ in their sensitivity to fragmentation (Walters 1998), and in which of the several different mechanisms responsible for adverse effects of fragmentation impact them. Changes in microclimate that penetrate from abrupt edges into habitat fragments can alter the ecology of those fragments, producing adverse effects on some animal and plant species (Franklin and Forman 1987; Chen and Franklin 1990). This type of edge effect appears to be only a minor factor in the responses of bird species to fragmentation. In contrast, another type of edge effect, elevated rates of nest predation and nest parasitism along edges, is the primary mechanism responsible for declines of many bird species in fragmented forest habitats in eastern North America (Robinson *et al.* 1995; Faaborg *et al.* 1998). In these systems, populations of nest predators and nest parasites may be sufficiently elevated in fragmented landscapes to reduce nesting success even in large fragments. This mechanism apparently is characteristic of only some ecosystems: for example, it does not seem to occur in the Australian *Eucalyptus* woodlands in which many cooperative breeders dwell (Ford *et al.* 2001; Zanette and Jenkins 2000). Even where it does occur, there is no reason to suppose cooperative breeders are any more or less likely to be affected than other species.

This is not true of the final mechanism responsible for adverse effects of habitat fragmentation: disruption of dispersal due to isolation of remnant habitat patches. Thus we devote much of the remainder of this chapter to a discussion of this mechanism. Our hypothesis is that dispersal patterns differ between cooperative and non-cooperative species in ways that make the former more likely to be adversely impacted by isolation effects, and that cooperative breeders are more likely to be sensitive to habitat fragmentation as a result. We will first review dispersal patterns of cooperative breeders, and then discuss their relationship to isolation effects.

Dispersal patterns among cooperative breeders

Dispersal patterns of some cooperative breeders are similar to those of non-cooperative species. These include species in which helpers are non-natal, or are offspring from early broods that assist in raising later broods within a breeding season but disperse before the subsequent one. In these systems young leave their parents within their first year to eventually acquire a breeding position elsewhere, in some cases spending one or more breeding seasons as a non-breeder. In Galápagos hawks, for example, juveniles do not remain with their natal group, but are instead aggressively expelled from the natal territory at age 3–5 months and join non-territorial, non-breeding populations (DeLay *et al.* 1996). A similar description applies to a great many non-cooperative species.

Systems in which natal helpers remain for one or more breeding seasons are a different matter. Zack (1990) pointed out that in these systems, helpers not only delay dispersal, but also alter their dispersal behavior. Helpers wait to inherit the breeding position on the natal territory or locate one on a nearby territory during forays from the natal territory. Brown (1987) termed this dispersal strategy stay-and-foray (SAF), and contrasted it with the more typical strategy among birds of permanently emigrating to search for vacancies, which he termed depart-and-search (DAS). The SAF strategy results in a distribution of dispersal distances that is unusually highly skewed toward very short distances, that is, extreme philopatry (Zack 1990). This is due not only to helpers becoming breeders on their natal territory, but also to the short dispersal range of individuals employing SAF compared to those employing DAS.

The concept of contrasting SAF and DAS dispersal strategies is an accurate description of the behavior of red-cockaded woodpeckers (Walters *et al.* 1992b). In this species many males and a few females exhibit SAF, whereas some males and most females exhibit DAS, departing the natal territory during their first year. The dispersal range of individuals employing SAF is quite limited, as helpers normally move no more than three territories from their natal site, and often inherit the natal territory. Some individuals employing DAS move only short distances as well, but others move long distances (Walters 1990; Daniels 1997).

This conceptual framework for dispersal is generally applicable to cooperative breeders in which helpers are retained on their natal territory; some exceptions are discussed elsewhere (Chapter 5). It is well documented in a large number of species that helpers become breeders on their natal territory, either by inheriting the breeding position or, in some species, by acquiring a portion of the natal territory through territorial budding, and that dispersal by helpers is limited to the immediate vicinity of the natal territory (Table 12.1). Even if those individuals that become breeders on their natal territory are excluded, median dispersal distances are generally only one or two territories from the natal territory, compared to typical values of four to eight territories for non-cooperative species exhibiting DAS (Arcese 1989; Nilsson 1989; Payne 1991). Thus, the thesis that SAF is associated with short dispersal distances is well

Table 12.1. *Documented dispersal characteristics of some cooperatively breeding birds*

Species	Bud[a]	Bias[b]	Short[c]	Long[d]	References
Hoatzin	X	F			Strahl and Schmitz 1990
Acorn woodpecker		F	X		Hooge 1995; Koenig *et al.* 2000
Red-cockaded woodpecker	X	F	X	X	Walters 1990; Walters *et al.* 1992b; Daniels and Walters 2000a
Green woodhoopoe		N	X		Ligon and Ligon 1990b
Arabian babbler		F	X		Zahavi 1990
Florida scrub-jay	X	F	X	X	Woolfenden and Fitzpatrick 1984; Breininger *et al.* 1995; Thaxton and Hingtgen 1996; Stith *et al.* 1996
Mexican jay		N	X		Brown 1994
White-throated magpie-jay		M			Langen 1996b
Brown treecreeper	X	F			Noske 1991; Cooper 2000
Superb fairy-wren		F			Mulder 1995
Splendid fairy-wren		F	X	X	Rowley and Russell 1990; Russell and Rowley 1993a, 1993b
Red-winged fairy-wren		F	X		Russell and Rowley 2000
Galápagos mockingbird		F	X	X	Curry and Grant 1989, 1990
Stripe-backed wren		F	X	X	Rabenold 1990; Piper *et al.* 1995
Bicolored wren		F	X		Rabenold 1990; Haydock *et al.* 1996

[a] Some helpers acquire territories by budding.
[b] Bias in dispersal distance: F = females move longer distances; M = males move longer distances; N = no bias.
[c] Helpers disperse short distances (generally ≤ 3 territories).
[d] Some individuals disperse long distances (> 5 territories).

supported. The behavior associated with SAF has been documented also, and matches Brown's (1987) conceptualization well. Woolfenden and Fitzpatrick (1984) described the forays of Florida scrub-jay helpers attempting to locate breeding vacancies, and Hooge (1995) used radio-tracking to document and describe forays by acorn woodpecker helpers.

There is also evidence of occasional long-distance movements in many cooperatively breeding species (Table 12.1), suggesting that some individuals employ DAS. In some species, such as the splendid fairy-wren (Russell and Rowley 1993a) and stripe-backed wren (Piper *et al.* 1995), some young depart their natal territory in their first year. In the stripe-backed wren, as in the red-cockaded woodpecker (see above), long-distance dispersal is associated with these juvenile birds rather than with helpers (Piper *et al.* 1995), suggesting the juveniles employ DAS. In these species young birds appear to make an irreversible choice between the two dispersal strategies in their first year (Walters *et al.* 1992b), based on social and ecological factors in the natal territory and immediate neighborhood (Pasinelli and Walters 2002).

In other species, an individual may remain as a helper for a period, and then adopt DAS. For example, female Florida scrub-jays typically remain on their natal territory at age one, but may depart and move long distances in their second year (Woolfenden and Fitzpatrick 1984; Stith *et al.* 1996). Perhaps they are not actively attempting to disperse in the first year, and defer their selection of dispersal strategy until their second year. Extended stays as helpers and frequent short-distance dispersal suggest that some female scrub-jays employ SAF, whereas occasional long-distance movements suggest others employ DAS. It may be that in some species individuals employ SAF for a period, and then switch to DAS. This conjecture would be supported if long-term helpers engaged in long-distance movements or became floaters. Switching from DAS to SAF would be evidenced by the return after a prolonged absence of a former fledgling or helper. Both types of event appear to be rare. We therefore suspect that generally individuals commit to a particular strategy rather than switching between them.

The degree of skew toward short distances in the distribution of dispersal distances in a particular species depends on the proportion of individuals that employ SAF rather than DAS. This often differs between the sexes, and typically it is males in which SAF is more common, and thus male dispersal distances that are shorter (Table 12.1). There are several exceptions, however, including species in which the sexes disperse similar distances and those in which females disperse shorter distances.

The best available data suggest fascinating variation among species in how often SAF is employed, and by which individuals. In the splendid fairy-wren, as in the red-cockaded woodpecker, some individuals of both sexes employ DAS, but unlike in the woodpecker, the frequency of DAS is low in females as well as males (Rowley and Russell 1990; Russell and Rowley 1993a). In the superb fairy-wren (Mulder 1995) and the hoatzin (Strahl and Schmitz 1990), apparently all males employ SAF and all females DAS, after a delay of a year spent on the natal territory. In the white-throated magpie-jay, all females adopt SAF and all males DAS (Langen 1996b). In the Galápagos mockingbird (Curry and Grant 1990), brown treecreeper (Walters *et al.* 1999; Cooper and Walters 2002a), and Florida scrub-jay (Woolfenden and Fitzpatrick 1984; Breininger *et al.* 1995; Stith *et al.* 1996), all males employ SAF whereas females use both strategies. Finally, in the green woodhoopoe (Ligon and Ligon 1990b) and red-winged fairy-wren (Russell and Rowley 2000), apparently all individuals practice SAF.

Such variation is not surprising. The best-developed theories about the evolution of cooperative breeding are those that address the retention of young on their natal territory. The evolutionary question, expressed as "why delay reproduction and dispersal?" by Brown (1987, p. 63) and "why stay?" by Emlen (1984), can also be expressed as "why employ SAF?" The theoretical answers to this question (Emlen 1991; Stacey and Ligon 1991; Koenig *et al.* 1992) are based on costs and benefits that are expected to vary among species, between the sexes within a species, and even among individuals within a sex (Walters *et al.* 1992b). For example, in red-cockaded woodpeckers, benefits of SAF to males exceed those to females because male dominance and inbreeding avoidance mechanisms preclude females from inheriting the breeding position on their natal territory (Walters *et al.* 1992b). Thus, variation among species of cooperative breeders in dispersal patterns may be a function of variation in the relative fitness benefits of SAF and DAS. This variation provides an opportunity to test the theory.

Whatever the origin of this variation, it will be expressed in differences in dispersal-distance distributions. There are of course many additional sources of variation that contribute to differences between species, expressed especially in differences in distances moved by individuals employing DAS. For example, in the red-cockaded woodpecker median natal dispersal distance of females employing DAS is only two territories from the natal site (Daniels and Walters 2000a), whereas the median distance for female superb fairy-wrens employing DAS is 12 territories (Mulder 1995). There is no indication that distances moved by individuals employing DAS vary more among cooperative breeders than among non-cooperative species, however. The consistent difference between non-cooperative species and cooperative breeders in which young are retained on the natal territory as helpers is the preponderance of short dispersal distances associated with SAF in the latter. We now examine how this feature might affect response to habitat fragmentation.

Dispersal behavior and response to habitat fragmentation

Disrupted dispersal is one of the major causes of decline of fragmented bird populations (Stouffer and Bierregaard 1995; Matthysen 1999; Walters *et al.* 1999). Theoretical modeling suggests dispersal success in fragmented populations depends on factors such as reproductive rate, habitat distribution, dispersal behavior, and how a species interacts with the intervening matrix between habitat patches (With and King 1999; Fahrig 2001). There is nothing distinctive about habitat distribution among cooperative breeders that would make them any more or less sensitive to habitat fragmentation than other species. Their generally low reproductive rates may make them somewhat more vulnerable to fragmentation effects. Also, cooperative breeders may be vulnerable to habitat fragmentation by virtue of being, in most cases, permanent residents rather than migrants. Many migrants engage in prospecting behaviors prior to dispersal that may make them less sensitive to matrix barriers than non-migrants (Reed et al. 1999). However, we suggest that it is primarily the unusual dispersal behavior of cooperative breeders that makes them more likely to be affected by habitat fragmentation, and to respond to it in a slightly different way, than non-cooperative species.

The point is perhaps clearest if one imagines the habitat matrix as a significant barrier to movement. If this is true, then only a small fraction of individuals attempting to disperse between habitat patches will be successful. In patches that are large enough to hold many groups, filling of breeding vacancies will be facilitated by the presence of helpers, and the groups within such patches therefore will be unusually persistent. The situation is very different in small patches containing only one or a few groups that are beyond the dispersal range of helpers in other patches. Assuming incest avoidance mechanisms exist (Chapter 9), filling of breeding vacancies in these patches will depend heavily on individuals employing DAS moving between patches. Since only a fraction of the population adopts DAS, a much smaller number of individuals will be available to fill breeding vacancies than in non-cooperative species. In the extreme case, there may be no dispersers of one or both sexes available. Thus we hypothesize that cooperative breeders will tend to be more sensitive to patch isolation, and to be more strongly affected by patch size, compared to non-cooperative species. We predict they will fare better than non-cooperative species in relatively large, isolated patches, and worse in relatively small, isolated patches.

As the permeability of the matrix to dispersers increases, the proportion that successfully move between patches increases, but the number still may not be sufficient to maintain the population, depending on the fraction of individuals that employ DAS. Indeed, modeling studies with red-cockaded woodpeckers indicate that the spatial distribution of territories strongly affects population behavior, even if one assumes that the matrix does not impede movement at all (Walters et al. 2002; Schiegg et al. 2002). In these modeling exercises all females are assumed to employ DAS. However, in this species, as in many other cooperative breeders, females do not hold territories alone. Thus isolated territories are often abandoned because of lack of replacement of the more philopatric males, only a fraction of which employ DAS (Schiegg et al. 2002). In fact, in the model runs breeder male replacement in small populations was almost entirely by helpers, and almost never by individuals employing DAS.

Both species of cooperative breeders that we have studied are sensitive to habitat fragmentation due to disruption of dispersal. Fragmentation of the pine savannas in which red-cockaded woodpeckers live causes small populations to decline (Conner and Rudolph 1991), and isolated territories at the edges of large populations are abandoned because breeding vacancies are not filled (Walters et al. 2002). Brown treecreepers are sensitive to fragmentation of the eucalypt woodlands in which they live (Barrett et al. 1994; Cooper and Walters 2002b), and there is compelling evidence that this is due to disruption of dispersal. Specifically, dispersal of females between isolated habitat patches is insufficient (Walters et al. 1999; Cooper and Walters 2002a; Cooper et al. 2002a), due partly to the fact that some females employ SAF, but also to a matrix that apparently is inhospitable to this species (Cooper et al. 2002b). In the case of brown treecreepers the isolation problem is exacerbated by male philopatry, which forces females to disperse out of their natal patch in order to avoid pairing with a related male. This will be true of many cooperative breeders (Chapter 9). Consequently, declining brown treecreeper populations are characterized by the presence of many unpaired territorial males in isolated habitat patches (Walters et al. 1999; Cooper et al. 2002a).

Disruption of dispersal by habitat fragmentation applies primarily to those species with natal helpers that exhibit SAF, and thus cooperative species exhibiting other social systems are not expected to be especially sensitive. Is sensitivity to habitat fragmentation the rule among species with natal helpers that employ SAF? There is some evidence that Florida scrub-jays are affected by habitat fragmentation due to disrupted dispersal (Stith et al. 1996; Thaxton and Hingtgen 1996; Breininger 1999). Generally, however, there is little empirical information. Certainly we do not expect all such species to be sensitive to habitat fragmentation. The impact of habitat fragmentation will depend on distances between patches relative to helper dispersal distance, the fraction of the population employing DAS, the impact of the matrix on movement, and the size of remaining habitat patches. We expect some species to be unusually resistant to effects of habitat fragmentation when remaining patches are large and distances between patches small. When dispersal can occur through SAF, populations will be resistant, and when DAS is required, they will be sensitive.

SMALL POPULATIONS OF COOPERATIVE BREEDERS

The small-population paradigm refers to problems inherent to small populations, regardless of what agents of decline caused them to become small (Caughley 1994). These problems relate to the ability of a species to persist once it is reduced to small populations, and to recover once the factors that led to its decline have been addressed. These problems include effects of demographic stochasticity, environmental stochasticity, and catastrophes on population persistence, and adverse effects of inbreeding depression and loss of genetic diversity (Shaffer 1981, 1987). Small populations of cooperative breeders differ from those of non-cooperative species with respect to both population dynamics and genetic structure in ways that are important to their conservation. We first address population dynamics, and then genetic structure.

Population dynamics

Demographic and environmental stochasticity produce fluctuations in population size that can bring small populations to low levels from which they cannot recover. Catastrophes represent an extreme version of the same process. Cooperative breeders are no more or no less prone to demographic stochasticity, environmental stochasticity, and catastrophes than other species, but they are not affected by these phenomena in quite the same way as other species due to the existence of helpers. Helpers, by providing a pool of replacement breeders, buffer the effects of mortality and productivity on breeding population size. If mortality is high and productivity low, it is the number of helpers, not the number of breeders, which is reduced. This has the effect of dampening fluctuations in the breeding population due to demographic and environmental stochasticity, and perhaps even those due to catastrophes. Environmental stochasticity is often expressed in annual variation in survival and productivity. In cooperative breeders this variation will translate into annual variation in the size of the helper class, while the number of breeders remains relatively constant.

This effect has been demonstrated in modeling studies of red-cockaded woodpeckers (Letcher et al. 1998; Walters et al. 2002; Schiegg et al. 2002). These studies suggest that because of the buffering effect of helpers, small populations of cooperative breeders are better able to persist in the face of demographic and environmental stochasticity than small populations of other species. In species with natal helpers that employ SAF, however, this will be the case only if territories are aggregated so that breeding vacancies that arise are within the dispersal range of helpers in other groups. Territory inheritance by helpers provides some buffering even in isolated territories, but the full effect requires dispersal between territories by helpers. Thus, we hypothesize that the viability of small populations of many cooperative breeders will be unusually sensitive to the spatial distribution of territories, but where territories are highly aggregated even very small populations will be remarkably persistent. The extent to which this is true of a particular species will depend on the proportion of individuals of each sex that are helpers, the dispersal range of helpers, and the proportions of helpers that gain breeding positions through territory inheritance and dispersal.

Floaters may provide the same sort of buffer against the effects of demographic and environmental stochasticity in some non-cooperative species that helpers do in cooperative breeders (Smith 1978). However, it is not clear how widespread or how effective buffering due to floaters may be in non-cooperative species. Therefore we hypothesize that when populations are reduced in a region, cooperative breeders are likely to be among the species that are best able to persist.

It is difficult to obtain empirical evidence of the relative stability of small populations of cooperative breeders, as it is usually unclear whether the small-population paradigm alone, or the declining-population paradigm as well, applies to a particular case (Caughley 1994). Certainly researchers have long commented on the stability of the populations of cooperative breeders they study, many of which are rather small. But some populations fluctuate greatly, for example the acorn woodpecker population studied by Stacey and Taper (1992). Unfortunately, more empirical data likely will become available as more cooperative breeders become endangered and reduced to small populations.

Genetic structure

The extreme philopatry exhibited by many cooperative breeders has consequences for their population genetic structure. In species in which many individuals of one or both sexes employ SAF, neighborhoods will be characterized by high genetic relatedness. The clan structure exhibited by some species represents an extreme form of this. Clans, first described in white-fronted bee-eaters (Hegner and Emlen 1987; Emlen 1990), denote a higher order of social structure linking sets of breeding groups that often represent extended families. White-fronted bee-eater clans defend feeding territories, but are not segregated at breeding colonies, whereas in several other species clans occupy neighboring breeding territories. In brown treecreepers, for example, the related males within a clan occupying neighboring territories feed young on each other's territories, but feeding across clan boundaries does not occur (Noske 1991; Cooper 2000; Doerr and Doerr 2001). In many cooperative breeders, then, members of the less philopatric sex must locate mates outside of their neighborhood, or outside of their clan, to avoid inbreeding. In species in which both sexes are highly philopatric, the potential for inbreeding is especially great.

Inbreeding is considered elsewhere (Chapter 9). Here we reiterate a few major points relevant to the vulnerability of small populations of cooperative breeders to inbreeding depression. Behavioral mechanisms to avoid close inbreeding are common among cooperative breeders, reducing the frequency of inbreeding considerably from what it otherwise might be. Recognition appears to be based on association, that is, individuals refrain from mating with familiar family members. Still, mating with familiar relatives occurs occasionally in many species, and regularly in a few.

There is also the possibility that individuals that obtain mates within their neighborhood will pair with an unfamiliar relative that dispersed from their natal territory prior to their birth, or with a descendent of such an individual. Daniels and Walters (2000a) showed that female red-cockaded woodpeckers can essentially eliminate this possibility by dispersing more than three territories from their natal site, yet do not usually move this far. We conclude that instances of close inbreeding are more common among cooperative breeders than among other species, and that relatedness within neighborhoods is generally higher, conclusions supported by the modeling results of Stevens and Wiley (1995).

One might imagine that cooperative breeders, because of a long history of at least somewhat elevated levels of inbreeding, would be less susceptible to inbreeding depression than non-cooperative species. This may be true of some species, such as the Seychelles warbler (Chapter 9), but clearly is not true of others. Studies of both Mexican jays (Brown and Brown 1998) and red-cockaded woodpeckers (Daniels and Walters 2000a) revealed inbreeding depression manifested in higher rates of hatch failure and reduced post-fledging survival of juveniles, the typical effects observed in birds (Keller 1998; Daniels and Walters 1999).

There are no empirical studies that address the impact of inbreeding on small populations of cooperative breeders. Some of the indirect evidence reviewed above suggests that impacts may be large (high relatedness within neighborhoods, high frequency of incestuous matings), and some of it suggests that impacts may be small (inbreeding avoidance behavior, prior history of inbreeding). Perhaps susceptibility of small populations of cooperative breeders to adverse effects of inbreeding depression will prove to be more variable, rather than greater or less, compared to susceptibility of non-cooperative species. Thus, although isolated small populations of cooperative breeders may be unusually stable demographically, it is not clear that they are any more or less vulnerable genetically.

THE SYMBOLIC VALUE OF COOPERATIVE BREEDERS

Not all species whose conservation is an issue receive equal attention. Public interest and conservation efforts mostly focus on charismatic vertebrates. Even among these, certain species receive particular attention because they are thought to have strong interactions with many other species in their community ("keystone species") or to be particularly sensitive to factors that threaten many species in their community ("indicator species"). Cooperatively breeding birds often are relatively charismatic, and as a result their symbolic value in conservation is high. There is no reason to presume that they are particularly likely to be keystone species, but because of the habitat specificity of many, they may be especially likely to be indicator species. Certainly they are likely to be "umbrella species," that is, species whose conservation results in the protection of other, less charismatic species within their communities.

In North America, the Florida scrub-jay is the highest-profile indicator species for the scrub habitat community in which it lives, and the same can be said for the red-cockaded woodpecker and its longleaf pine savanna community, as well as the acorn woodpecker and its oak savanna community. The Florida scrub-jay and red-cockaded woodpecker are habitat specialists, and are highly sensitive to habitat quality (see above). Similarly, in Australia, two cooperative breeders, the brown treecreeper and hooded robin, can be viewed as indicator species for *Eucalyptus* woodland communities in northeastern New South Wales (Barrett et al. 1994). However, another cooperative breeder, the noisy miner, is a clear indicator of degraded habitat in this same system (Ford et al. 2001). Some cooperative breeders are symbolic of the plight of the habitat on which they depend, whereas others are favored by the influence of humans on landscapes. Whichever is the case, group living makes them conspicuous and captures the attention of the public. As a result, one can expect cooperative breeders often to be in the forefront where conservation is concerned. It therefore pays to be aware of their special attributes with respect to conservation.

CONCLUSIONS

Cooperative breeders face the same threats to their continued existence as other birds, and like those others, exhibit considerable variability in how they are impacted by these threats. With respect to some threats, for example their susceptibility to introduced predators and competitors, there is nothing distinctive about cooperative breeders. But with respect to many other threats, cooperative breeders are distinctive. Because many are habitat specialists, they often will be particularly sensitive to habitat loss and habitat degradation. Because of their unusual dispersal behavior, many cooperative breeders will be sensitive to habitat fragmentation, but also relatively persistent when reduced to small populations. Because of their charismatic nature, those in trouble will often be in the public eye, which will promote their conservation and funding for their study. They thus provide fertile ground not only for the sociobiological research for which they are known, but also for research on basic problems in conservation.

ACKNOWLEDGMENTS

We thank Rob Heinsohn for helpful comments, and the editors for their tireless efforts to make this a better paper. We also are grateful to the many students, technicians, and collaborators who have contributed to the red-cockaded woodpecker and brown treecreeper projects, especially Hugh Ford, Jay Carter, and Phil Doerr.

APPENDIX 12.1. COOPERATIVELY BREEDING SPECIES CONSIDERED IN THIS STUDY

Family	Common name	Scientific name
Struthionidae	Ostrich	*Struthio camelus*
Anseranatidae	Magpie goose	*Anseranas semipalmata*
Picidae	Red-cockaded woodpecker	*Picoides borealis*
	Acorn woodpecker	*Melanerpes formicivorus*
Lybiidae	D'Arnaud's barbet	*Trachyphonus darnaudii*
	Red-and-yellow barbet	*T. erythrocephalus*
Ramphastidae	Toucan barbet	*Semnornis ramphastinus*
Bucconidae	White-fronted nunbird	*Monasa morphoeus*
Bucorvidae	Southern ground-hornbill	*Bucorvus leadbeateri*
Phoeniculidae	Green woodhoopoe	*Phoeniculus purpureus*
Todidae	Puerto Rican tody	*Todus mexicanus*
Halcyonidae	Blue-winged kookaburra	*Dacelo leachii*
	Laughing kookaburra	*D. novaeguineae*

Appendix 12.1. (cont.)

Family	Common name	Scientific name
Cerylidae	Pied kingfisher	*Ceryle rudis*
	Micronesian kingfisher	*Halcyon cinnamomina*
Meropidae	European bee-eater	*Merops apiaster*
	Red-throated bee-eater	*M. bulocki*
	White-fronted bee-eater	*M. bullockoides*
	Carmine bee-eater	*M. nubicus*
	Rainbow bee-eater	*M. ornatus*
	Little green bee-eater	*M. orientalis*
Coliidae	Speckled mousebird	*Colius striatus*
Opisthocomidae	Hoatzin	*Opisthocomus hoazin*
Crotophagidae	Smooth-billed ani	*Crotophaga ani*
	Groove-billed ani	*C. sulcirostris*
	Guira cuckoo	*Guira guira*
Psittacidae	Eclectus parrot	*Eclectus roratus*
Psophiidae	Pale-winged trumpeter	*Psophia leucoptera*
Rallidae	Common moorhen	*Gallinula chloropus*
	Dusky moorhen	*G. tenebrosa*
	Tasmanian native hen	*G. mortierii*
	Purple gallinule	*Porphyrula martinica*
	Pukeko	*Porphyrio porphyrio*
Laridae	Brown skua	*Catharacta lonnbergi*
Accipitridae	Galápagos hawk	*Buteo galapagoensis*
	Harris's hawk	*Parabuteo unicinctus*
Acanthisittidae	Rifleman	*Acanthisitta chloris*
Tyrannidae	White-bearded flycatcher	*Phelpsia inornata*
Climacteridae	Red-browed treecreeper	*Climacteris erythrops*
	Black-tailed treecreeper	*C. melanura*
	Brown treecreeper	*C. picumnus*
	Rufous treecreeper	*C. rufa*
Maluridae	Purple-crowned fairy-wren	*Malurus coronatus*
	Superb fairy-wren	*M. cyaneus*
	Red-winged fairy-wren	*M. elegans*
	Blue-breasted fairy-wren	*M. pulcherrimus*
	Splendid fairy-wren	*M. splendens*
Meliphagidae	Noisy miner	*Manorina melanocephala*
	Bell miner	*M. melanophrys*
Pardalotidae	White-browed scrubwren	*Sericornis frontalis*
	Yellow-rumped thornbill	*Acanthiza chrysorrhoa*
	Striated thornbill	*A. lineata*
	Buff-rumped thornbill	*A. reguloides*
Petroicidae	Hooded robin	*Melanodryas cucullata*
Pomatostomidae	Hall's babbler	*Pomatostomus halli*
	New Guinea babbler	*P. isidorei*
	Chestnut-crowned babbler	*P. ruficeps*

(cont.)

Appendix 12.1. (cont.)

Family	Common name	Scientific name
	White-browed babbler	*P. superciliosus*
	Grey-crowned babbler	*P. temporalis*
Laniidae	Yellow-billed shrike	*Corvinella corvina*
	White-crowned shrike	*Eurocephalus anguitimens*
	Grey-backed fiscal shrike	*Lanius excubitoroides*
Corvidae	White-winged chough	*Corcorax melanorhamphos*
	Apostlebird	*Struthidea cinerea*
	Varied sitella	*Daphoenositta chrysoptera*
	Florida scrub-jay	*Aphelocoma coerulescens*
	Mexican jay	*A. ultramarina*
	Unicolored jay	*A. unicolor*
	White-throated magpie-jay	*Calocitta formosa*
	Azure-winged magpie	*Cyanopica cyana*
	Formosan magpie	*Urocissa caerulea*
	American crow	*Corvus brachyrhynchos*
	Northwestern crow	*C. caurinus*
	Tufted jay	*Cyanocorax dickeyi*
	Beechey jay	*C. beecheii*
	Bushy-crested jay	*C. melanocyaneus*
	San Blas jay	*C. sanblasianus*
	Green jay	*C. yncas*
	Plush-capped jay	*C. chrysops*
	Yucatan jay	*C. yucatanicus*
	Pinyon jay	*Gymnorhinus cyanocephalus*
	Brown jay	*Psilorhinus morio*
	Black-faced woodswallow	*Artamus cinereus*
	White-breasted woodswallow	*A. leucorhynchus*
	Australian magpie	*Gymnorhina tibicen*
	Ground cuckoo-shrike	*Coracina maxima*
	Chestnut-fronted helmetshrike	*Prionops scopifrons*
	Retz's helmetshrike	*P. retzii*
	White helmetshrike	*P. plumatus*
Picathartidae	Rufous rockjumper	*Chaetops frenatus*
Muscicapidae	Western bluebird	*Sialia mexicana*
	Anteater chat	*Myrmecocichla aethiops*
Sturnidae	Golden-breasted starling	*Cosmopsarus regius*
	African pied starling	*Spreo bicolor*
	Chestnut-bellied starling	*Lamprotornis pulcher*
	Galápagos mockingbird	*Nesomimus parvulus*
Sittidae	Brown-headed nuthatch	*Sitta pusilla*
	Pygmy nuthatch	*S. pygmaea*
Certhiidae	Fasciated wren	*Campylorhynchus fasciatus*
	Bicolored wren	*C. griseus*
	Gray-barred wren	*C. megalopterus*

Appendix 12.1. (cont.)

Family	Common name	Scientific name
	Striped-backed wren	*C. nuchalis*
	Band-backed wren	*C. zonatus*
	Spotted wren	*C. gularis*
	Black-capped donacobius	*Donacobius atricapillus*
Paridae	Black tit	*Parus niger*
Aegithalidae	Long-tailed tit	*Aegithalos caudatus*
	Bushtit	*Psaltriparus minimus*
Sylviidae	Seychelles warbler	*Acrocephalus sechellensis*
	Common babbler	*Turdoides caudatus*
	Arrowmarked babbler	*T. jardineii*
	Large grey babbler	*T. malcolmi*
	Black-lored babbler	*T. melanops*
	Brown babbler	*T. plebejus*
	Arabian babbler	*T. squamiceps*
	Jungle babbler	*T. striatus*
	Bare-cheeked babbler	*T. gymnogens*
	Blackcap babbler	*T. reinwardtii*
Passeridae	Alpine accentor	*Prunella collaris*
	Dunnock	*P. modularis*
	White-browed sparrow-weaver	*Plocepasser mahali*
	Grey-headed social-weaver	*Pseudonigrita arnaudi*
Fringillidae	Maui alauahio	*Paroreomyza montana*
	Medium ground-finch	*Geospiza fortis*
	Common cactus-finch	*G. scandens*
	Brown-and-yellow marshbird	*Pseudoleistes virescens*
	Bay-winged cowbird	*Molothrus badius*

13 • Mammals: comparisons and contrasts

ANDREW F. RUSSELL

University of Cambridge (Present address: University of Sheffield)

Working on birds has two major logistical advantages over mammals: their observation need not rely on habituation, and they can be individually marked more easily. It is therefore not surprising that despite the studies summarized in Solomon and French (1997), work on cooperative breeding in vertebrates has been markedly bird-biased (Emlen 1991), with the result that theory on the evolution of cooperative breeding has been primarily developed for birds (Chapters 1 and 3). Such theory has addressed three questions: (1) Why do individuals delay dispersal? (2) Why do they delay breeding? (3) Why do they help? The answers to these questions do not form a linear progression of results that will lead us to understand the evolution of cooperative breeding, for both birds and mammals may be cooperative without delaying either dispersal or breeding, and may be non-cooperative whilst delaying both. Nevertheless, these questions remain a useful framework for understanding cooperative breeding in the majority of cases (Emlen 1995), and will form the structure of this chapter.

DEFINITIONS AND TERMINOLOGY

Cooperative breeding arises when three or more individuals cooperate to provide care to offspring, and hence some individuals inevitably care for offspring that are not their own. Following Cockburn (1998), I encompass a broad number of circumstances in which individuals may knowingly provide care to others, including when individuals have not bred (helpers-at-the-nest) as well as when they have bred (plural, communal, cooperative polygamous systems).

In birds, cooperative species are usually characterized by individuals incubating and/or feeding non-descendant offspring. Cooperative species are more difficult to define in mammals because, in many, individuals only conduct activities that are ambiguously cooperative (Jennions and Macdonald 1994; Solomon and French 1997). Examples include cooperative defense of offspring (muskox, African elephant), predator surveillance systems (ground squirrels, primates), cooperative food acquisition (killer whale, spotted hyena) and allogrooming (primates). One reason for this may be that lactation effectively removes the potential for non-mothers to provision offspring in many species. I define cooperative species to be those in which individuals carry out activities that, at a potential cost to themselves, have the potential to improve the condition and survival of recipient breeders and/or non-descendant offspring. Such care includes feeding breeding females, allolactation, babysitting, carrying young while foraging, huddling during hibernation, pup-feeding and sentineling/alarm calling. Finally, I term the individuals that carry out these behaviors helpers, whether or not they have bred in a current attempt, and whether or not help has been shown to result in measurable benefits to the recipients.

As in birds (Arnold and Owens 1998), roughly 3% of mammals are cooperative, although this number will doubtless increase considerably with the widening of our definition of cooperative breeding (Cockburn 1998) and with improvements to our knowledge of mammalian systems. In addition, cooperative breeding in mammals has a number of similarities with that of birds (Table 13.1). Its distribution is unequal with respect to phylogeny (Chapter 1), being most prevalent in rodents, mongooses, dogs, and callitrichid primates. Most societies of cooperative mammals form through delayed dispersal of offspring (Emlen 1995), but delayed dispersal need not lead to cooperative breeding (Chapter 2). Finally, there is substantial diversity in mating system

Ecology and Evolution of Cooperative Breeding in Birds, ed. W. D. Koenig and J. L. Dickinson. Published by Cambridge University Press.
© Cambridge University Press 2004.

Table 13.1. *Social, reproductive, and cooperative characteristics of the main mammal species discussed in this chapter*

Species	Group size	Skew (female)	Fecundity	Cooperative activities	Helping sex
Naked mole-rat	290	High	3×10		Both
Damaraland mole-rat	40	High	3×4		Both, female biased
Common mole-rat	15	High	1.5×4		Both
Pine vole	10	Medium/high	3×5	Babysitting	Both
Prairie vole	20	Medium/high	3×5	Babysitting	Male
Mongolian gerbil	15	Medium	2×5	Babysitting	Both
Alpine marmot	15	High	1×4	Huddling	Male
Black-tailed prairie dog		Low	1×5	Sentineling	Female
Belding's ground squirrel		Low	1×6	Sentineling	Female
Meerkat	40	High	2×4	Allolactation, babysitting, pup-feeding, sentineling	Both, female biased
Banded mongoose	40	Low	2×3	Allolactation, babysitting, pup-feeding, sentineling	Both, male biased
Dwarf mongoose	20	High	2×3	Allolactation, babysitting, pup-feeding, sentineling	Both
African lion	15	Low	1×3	Allolactation	Female
Spotted hyena	80	Low/medium	1×1	n/a	n/a
Brown hyena	15	Medium	1×2	Allolactation, babysitting, pup-feeding	Both, female biased
African wild dog	50	High	1×10	Regurgitation, allolactation, babysitting, pup-feeding	Both
Gray wolf	10	High	1×6	Regurgitation, allolactation, babysitting, pup-feeding	Both, male biased
Jackals/coyote	5–15	High	1×4	Regurgitation, allolactation, babysitting, pup-feeding	Both, female biased
Cotton-top tamarin	10	High	1.5×2	Carrying, pup-feeding	Both
Common marmoset	15	High	2×2	Carrying, pup-feeding	Both, male biased
Golden lion tamarin	10	Medium/high	1×1.5	Carrying, pup-feeding	Both

Group size shows the maximum number excluding dependent offspring and is shown because the minimum for virtually all species is 2. Skew refers to females only (low 0–30%, medium 35–65%, high >65%), but is normally similar for males. Fecundity refers to average number of litters delivered per year and average number of pups per litter. In cooperative activities: huddling during hibernation provides thermo-regulatory benefits to offspring; babysitting provides general protection and serves to keep offspring in the area of birth; allolactation refers to individuals providing non-offspring with milk; regurgitation refers to cases where helpers provide food to breeders; offspring carrying is specific to callitrichids and allows mothers to forage efficiently; sentineling is where an individual stands on guard from a prominent point to look out for predators and alarm-calls at the sight of danger while the rest of the group forages more efficiently. Sentinel behavior and alarm-calling has been suggested to be selfish (Clutton-Brock *et al.* 1999b), but there is evidence that it may be personally costly (Sherman 1977), more common in the presence of pups, and beneficial to recipients (Manser 1999). Whether subordinates directly care for offspring in mole-rats is currently debatable. Biases in helper contributions by the sexes are not always known, and some biases may be slight.

and reproductive skew (Chapter 5), even within the same family (e.g. mongooses).

However, at least three differences are also apparent (Table 13.1). Although groups of mammalian cooperative breeders are sometimes characterized by small numbers of helpers, those of many species contain dozens and groups of one species (the naked mole-rat) may even consist of hundreds. Species of cooperative mammals are often also relatively fecund, with litters of a single offspring rare, and a few species producing litters of 10 or more. Mammals not only conduct a wide range of costly cooperative activities, but in many species each helper may perform several of them. For example, in the cooperative mongooses, helpers may carry out three or more costly activities including allosuckling, babysitting, pupfeeding, social digging, and sentinel behavior.

WHY DELAY DISPERSAL?

Most birds are territorial and breed in socially monogamous pairs (Lack 1968), so dispersal is often essential to securing a mate and a breeding position. In the majority of cooperative birds and mammals, cooperation arises when offspring delay dispersal and help their parent(s) to raise non-descendant kin (Emlen 1995). Thus, understanding the factors that promote delayed dispersal may hold the key to understanding the evolution of cooperative breeding. Because monogamy prevails in birds, delayed dispersal will usually be an inferior tactic to dispersal and independent breeding. However, because cooperative breeding may be the culmination of a series of individual decisions (Chapter 3), delayed dispersal should be regarded as an alternative to dispersing and floating rather than an alternative to dispersing and independent breeding. The question thus is, why do breeding constraints select for delayed dispersal in some species and dispersing and floating in others?

The answer to this question has centered on the possibility that species differ in their costs of dispersal and their benefits of philopatry. Delayed dispersal is frequently observed in saturated habitats, leading to the suggestion that dispersal costs mediated through habitat saturation have selected for delayed dispersal in birds (Chapter 3). However, experimental evidence shows that habitat saturation does not always lead to delayed dispersal (Smith 1978; Beletsky and Orians 1987; Carmen 2003), leading to the question, why should habitat saturation lead to delayed dispersal in some species and not others? Similarly, although delayed dispersal may be associated with nepotistic benefits (Chapter 2), it is unclear why such benefits should not be reaped by all species.

I make three points in this section. First, I suggest that habitat saturation appears to play a minor role in understanding delayed dispersal among mammals. Second, I outline the costs and benefits that have been proposed to account for delayed dispersal and group living among vegetarian and carnivorous mammals. Finally, I discuss how mammalian studies can help us to understand delayed dispersal in vertebrates. In particular, I highlight that although dispersal costs and philopatric benefits are inevitably important to understanding delayed dispersal, ultimately, delayed dispersal cannot occur unless group living is unconstrained; since, by definition, delayed dispersal cannot arise without group living.

Habitat saturation in mammals

While habitat saturation has been implicated in promoting delayed dispersal in some cooperative mammals (Messier and Barrette 1982; Doncaster and Woodroffe 1993; Creel and Macdonald 1995; Solomon and French 1997), evidence is generally weak. The key prediction of habitat saturation is that the creation of suitable breeding vacancies will lead to individual dispersal and independent breeding. This prediction has never been directly tested in mammals, but a number of observations suggest that dispersal decisions are often independent of habitat availability (Cheeseman *et al.* 1993; Clutton-Brock *et al.* 1999a; Creel and Creel 2001).

Habitat saturation in birds may be associated with territory inheritance by at least one sex (Zack 1990). Ironically, although habitat saturation has been suggested as a possible route to sociality in callitrichid primates (Goldizen and Terborgh 1989; Baker *et al.* 1993; Rylands 1996) and muroid rodents (Getz *et al.* 1993, 1994), inheritance of the natal territory is rare in both families, and in callitrichids there is no evidence that helpers are any more likely to inherit the natal territory than non-helpers (McGuire *et al.* 1993; Tardif 1997). Thus, there is circumstantial evidence to suggest that a saturated habitat has a minimal role to play in governing delayed dispersal within cooperative mammals, and its importance may be overstated because of its suspected influence in birds.

Delayed dispersal in vegetarians

Group living is extremely widespread in vegetarian mammals (e.g., sciurid and bathyergid rodents, ungulates, primates), and groups frequently comprise philopatric offspring, even in those families that are ostensibly non-cooperative. The main reason for the prevalence of group living in vegetarians is two-fold: first, their food tends to be found in large clumps, and second, they tend to be highly susceptible to predation. Thus, for many vegetarians, the relative costs of dispersal and independent living are likely to be high (Krebs and Davies 1993). Despite potential costs, individuals may often be constrained from group living (Clutton-Brock and Harvey 1977), and hence constrained from delayed dispersal. For example, group living is constrained in both ungulates and primates when individuals forage selectively on dispersed food items and when crypsis is the primary method of predator evasion (Jarman 1974; Clutton-Brock and Harvey 1977, 1978; Clutton-Brock 1989a). It is noteworthy that in such species, parents(s) may forcibly evict offspring from their territory soon after reaching independence (Jarman 1974; Clark 1978).

Food distribution and predation risk are also important in governing group living in sciurids (Armitage 1981) and mole-rats (Bennett and Faulkes 2000). In sciurids (ground squirrels, prairie dogs, marmots), the benefits of reducing dispersal and breeding in kin clusters (Dunford 1977) (kin-neighborhoods, Chapter 3) are well documented and include sophisticated nepotistic-mediated anti-predatory (Sherman 1977) and group defense systems, often achieved through complex mechanisms of kin recognition (Mateo 2002). In addition, however, the extent to which such species show delayed dispersal appears to depend upon ecological and life-history factors. For example, delayed dispersal is more common when the growing season is insufficiently long for offspring to reach adult body size in their first year and when thermodynamic benefits of communal hibernation are high (Armitage 1981, 1999; Blumstein and Armitage 1999).

Among mole-rats, delayed dispersal and sociality are common characteristics of those species that inhabit xeric regions of Africa, while solitary species are more common in mesic regions (Jarvis *et al.* 1994). The significance of this is apparently that xeric regions of Africa are characterized by dispersed clumps of plant tubers and high costs of digging in hard (naked mole-rat) or loose (Damaraland mole-rat) sand, except during short, unpredictable rains (Faulkes *et al.* 1997; Jarvis *et al.* 1998). Delayed dispersal therefore occurs in those species in which the benefits of group foraging for food items are high and the probability of dispersing, digging a new burrow, finding food, and finding mates is low during the brief rains (Jarvis *et al.* 1994, 1998; Spinks *et al.* 2000b). Such ecological costs and benefits may be a general phenomenon in subterranean rodents, for in the Americas, sociality in rodents may also be associated with desert environments (Lacey and Sherman 1997). Thus, in group-living mole-rats, delayed dispersal appears to be a consequence of both dispersal costs and group-living benefits, while in solitary species delayed dispersal may be constrained because food is not clumped, and hence offspring are forcibly evicted at independence (Bennett and Faulkes 2000).

Delayed dispersal in carnivores

In contrast to the situation with vegetarians, carnivorous mammals tend to be solitary, and correspondingly, offspring usually show immediate dispersal. The generality of this statement depends not so much upon the size of the predator as on the nature of the prey species and the method by which predators obtain their food. For example, small carnivores that feed diurnally on invertebrates, such as meerkats, dwarf and banded mongooses, typically live in groups that comprise philopatric offspring (Rood 1986). One reason for this is that insects can be a highly renewable resource (Waser 1981; Waser and Waser 1985), and so foraging competition does not select against group living (Creel and Macdonald 1995). In addition, predation risk is high among small diurnal species compared with nocturnal species, and hence delayed dispersal and group living is more beneficial (Waser *et al.* 1994). Interestingly, social mongooses tend to forage in the presence of sentinels (Rasa 1986; Clutton-Brock *et al.* 1999b), and increases in group size are commonly associated with increases in foraging efficiency and survival (Rood 1990; Clutton-Brock *et al.* 1999a).

In contrast, carnivores that forage on small vertebrates, such as mustelids, other mongooses, cats, and dogs chiefly show dispersal soon after attaining independence (Creel and Macdonald 1995). This is because small vertebrates do not renew themselves frequently enough to be found in large enough densities to support a group of carnivores (Waser and Waser 1985), although where rodents, for example, are found at unusually high densities for much of the year, carnivores feeding on small vertebrates like jackals and Ethiopian wolves can show delayed dispersal. Nevertheless, in such species, foraging tends to be conducted alone (Creel and Macdonald 1995), probably because obtaining vertebrate prey tends to rely more on an element of stealth that is difficult to achieve while hunting in groups.

Carnivores that specialize on large prey are more variable in their tendency to live in groups and to exhibit delayed dispersal. Benefits of group living include increased foraging efficiency (Gittleman 1989, Creel and Creel 1995), carcass defense (Kruuk 1975; Gorman *et al.* 1998), and offspring defense (Packer *et al.* 1990), but such benefits are not open to all species. Cats that hunt using stealth tend not to live in groups, and mothers abandon their offspring close to independence (Packer and Ruttan 1988). The exception to this is African lions, possibly because they may tackle prey larger than themselves and groups of lions are able to defend their kill against hyenas. The tendency for lions and hyenas to live in groups depends on their prey. In high (large) prey density areas, both can be found in large groups, and delayed dispersal is the norm, while in low (small) prey density areas, delayed dispersal is less extreme and group sizes tend to be substantially smaller (Gittleman 1989). Finally, African wild dogs, which chiefly forage on large prey, live in groups and show delayed dispersal, while other dogs such as coyotes and wolves only show delayed dispersal when they forage on large prey (Creel and Macdonald 1995). That dispersers tend to have lower survival than non-dispersers in carnivores (Lucas *et al.* 1997), with mothers typically abandoning offspring or driving offspring from their territory in those species in which individuals live alone, strongly suggests that delayed dispersal is constrained in non-social carnivores.

Synthesis: when to delay dispersal

Investigations into the causes of delayed dispersal in birds have centered upon two different approaches, the fine-scale within-species approach and the large-scale phylogenetic approach. Although within-species investigations have led to significant insights, they have not elucidated why delayed dispersal has evolved in particular species. Similarly, although large-scale phylogenetic analyses have been useful in showing some of the factors that are associated with immediate dispersal, they have been relatively unsuccessful in elucidating consistent correlates of delayed dispersal (Chapter 1). In contrast to these two approaches, studies on mammals have adopted a more "middle-of-the-road" approach, attempting to understand delayed dispersal within and between families of animals rather than within species or the whole class. Although this approach has not always been subjected to statistical tests, it has led to significant improvements to our understanding of delayed dispersal in mammals (Armitage 1981; Jarvis *et al.* 1994; Creel and Macdonald 1995; Lacey and Sherman 1997; Moehlman and Hofer 1997), and could be used to great effect in birds. Future work that is aimed at explaining delayed dispersal within phylogenetically, ecologically, and/or morphometrically similar species is likely to be extremely fruitful (Chapter 3).

Avian studies have generally viewed delayed dispersal as a function of net dispersal costs, mediated through ecological constraints on independent breeding. This consensus appears to have arisen because helping is generally the "best-of-a-bad-job" tactic in monogamous birds. However, delayed dispersal should be viewed as an alternative to dispersing and floating, rather than as an alternative to independent breeding. This is an important distinction, for when constraints exist on independent breeding, delayed dispersal need not preclude breeding any more than dispersing and floating. Understanding why some species delay dispersal while others show dispersal and floating is likely to be critical to advancing our understanding of the evolution of cooperative breeding, since in most cases immediate dispersal precludes cooperative breeding (Chapter 2). To this end, it may be more useful to investigate constraints on dispersal, such as whether or not the distribution of food within a territory is conducive to delayed dispersal in social species and whether food may facilitate dispersal in non-social species.

Observations in mammals suggest that delayed dispersal occurs when there are benefits (or at least reduced costs) to group living. Such an interpretation is supported by the fact that offspring of non-social species are often forcibly evicted from their natal territory by their parent(s) (Jarman 1974; Clark 1978; Faulkes and Bennett 2000). That birds may benefit from familiarity and nepotism (Chapter 2) and benefits of group living more generally (Chapter 5) suggests that delayed dispersal may be beneficial in many birds. The question then is, in the face of breeding constraints, why do so many birds disperse and float and so few delay dispersal?

Evidence from mammals suggests that the answer may lie in whether or not the type, size, and distribution of resources in combination with the risk of predation are conducive to group living (see above). Supporting evidence can be found in birds. For example, Zack and Ligon (1985) suggested delayed dispersal occurs in shrikes that live in high-quality habitats. E. M. Russell (2000) suggested that a lack of sufficient food year-round constrains delayed dispersal in temperate passerines, although in this case high fecundity and breeding success are also likely to constrain delayed dispersal because of intense competition following breeding (A. F. Russell 1999). Finally, delayed dispersal in the Seychelles warbler is only possible on territories of sufficient quality (Komdeur 1992). That group living may be associated with substantial anti-predatory and food-acquisition benefits (Krebs and Davies 1993), suggests that there is no a priori reason for the suggestion that delayed dispersal is the best-of-a-bad-job tactic and may even be superior to immediate dispersal when constraints exist on independent breeding. The bottom line therefore is that despite the inevitable importance of dispersal costs, if there are constraints on group living, delayed dispersal will not occur, and cooperative breeding becomes substantially less likely.

WHY NOT BREED?

Like theories on delayed dispersal, early models explaining who breeds in social groups of vertebrates were first proposed for cooperative birds (Chapter 10). Such models were based on the assumption that dominants had full control of the breeding rights over subordinates, and the latter thus bred only when both personally able to do so and permitted to do so by the dominant (concession models of reproductive skew: see Chapter 10). However, this paradigm appears to be particularly problematic in its application to social mammals, in which allowing a subordinate to breed is clearly costly and controlling subordinate reproduction is difficult (Clutton-Brock 1998).

Continuing with the foraging and competition theme from the previous section, I first review the role of food monopolization and fecundity on reproductive skew among females. Second, I discuss the proximate mechanisms of skew in social mammals. Third, I suggest why concessions ideas are unlikely to be a widespread model system for female mammals, but why they may be more applicable to males.

Reproductive skew among social mammals

The question of skew boils down to the net benefits that dominant individuals gain from suppressing subordinate reproduction and the net benefits subordinates gain from attempting to combat any suppression. Although exceptions occur (Cant and Reeve 2002; see Chapter 10), broadly, skew will be high (1) when resources are limiting and monopolizable, such that it is beneficial to control subordinate access to resources, and (2) when the relationship between the monopolization of such resources and reproductive success is significantly positive, such that controlling subordinate access leads to a net increase in dominant fitness.

In birds, understanding reproductive skew is largely confined to cooperative species, since few non-cooperative species live in social groups during the breeding season. This is not the case in mammals, and reproductive skew can equally be investigated in group-living non-cooperative and cooperative species.

Evidence from non-cooperative social mammals shows that low-skew, "egalitarian" societies tend to be found where resources are non-monopolizable and the relationship between monopolization and reproductive success is low. For example, low skew is typical among female ungulates and non-callitrichid primates in which unrelated mates are not limiting, food is found in large patches, and female fecundity is low (i.e., virtually always one offspring produced once per year) (Clutton-Brock 1989a). Similarly, skew is relatively low in female social carnivores such as African lions and spotted hyenas, in which cooperative foraging can yield prey large enough to support reproduction by many individuals and fecundity is also relatively low, although skew may increase in both species when group size and hence competition for food increases (Lewis and Pusey 1997; Packer et al. 2001). In contrast, the degree of skew among males in these species tends to be higher, most likely because the constraining factor for males is not food, but females, a more defendable resource in which an increase in access substantially increases reproductive success. The degree to which skew varies within males therefore depends on their ability to monopolize females, the behavior of females, and whether or not alliances of males are more able to obtain reproduction than singletons (Davies 1992; Cant and Reeve 2002; Hager 2003).

In cooperative breeders, the number and behavior of breeding females is likely to govern the degree of skew among males (Davies 1992; Cant and Reeve 2002), but for females, helpers are likely to constitute a further limiting resource. Therefore, female skew is likely to depend on the ability of dominants to control subordinate reproduction and the relationship between monopolization of helpers and reproductive success. Creel and Creel (1991) suggested that skew is highest among those carnivores in which the costs of breeding are highest and concluded that this is because costly reproduction for subordinates allows dominants to suppress their reproduction more easily. French (1997) used a similar argument to explain

Table 13.2. *Physiological similarities between subordinate and dominant males*

Species	Comparison	References
Damaraland mole-rat	LH, T (⇓ sperm)	Maswanganye et al. 1999; Bennett et al. 1993, 2000
Common mole-rat	LH	Spinks et al. 2000a
Zambian mole-rat	LH	Bennett et al. 2000
Mashona mole-rat	LH	Bennett et al. 1997
Meerkat	LH, T (testes)	O'Riain et al. 2000b ; Carlson et al. in press
Dwarf mongoose	LH, T (⇓ testes)	Creel et al. 1993
Gray wolf	LH, T (sperm)	Asa 1997
African wild dog	T	Creel and Creel 2001
Common marmoset	LH, T	Baker et al. 1993
Cotton-top tamarin	LH, T	Ginther et al. 2001
Golden lion tamarin	T	French et al. 1989

When considering those individuals of similar age, no differences were found between subordinate and dominant males in their levels of luteinising hormone (LH) or testosterone (T) in any species, suggesting no physiological suppression, although in some cases sperm levels were lower or testes smaller (indicated by ⇓).

why skew is high among more fecund tamarins and marmosets and lower in less fecund golden lion tamarins. Finally, Gilchrist (2001) calculated the costs of reproduction to be lower in low-skew banded mongooses than in high-skew meerkats and dwarf mongooses. Whether or not such relationships are causal, there is little doubt that among a large range of cooperative mammals the degree of reproductive skew is at least associated with the costs of reproduction.

However, these conclusions are likely to be incomplete because they only consider the relationship between skew and the ability of dominants to control the reproduction of subordinates, and do not include why dominants should want to suppress reproduction by subordinates when the costs of reproduction are high. The likely reason is that when the costs of reproduction are high, fecundity is also high, and the correlation between monopolization of helpers by a dominant and the latter's reproductive success is also likely to be high. Hence, both the ability of dominants to control subordinate reproduction and the relationship between control and reproductive success will select for high skew.

Mechanisms of reproductive skew

Reproductive skew is generally considered to be a consequence of suppression, but this terminology is unfortunate. In fact, skew may be a consequence of three non-mutually exclusive mechanisms: subordinate constraint, subordinate restraint, and dominant suppression. For example, in the absence of a stimulus from an unrelated mate, individuals may be constrained from breeding, or they may refrain from breeding if the probability of producing viable offspring is low. Although there are few data on inbreeding depression in cooperative birds (see Chapter 9) and mammals (Dietz and Baker 1993), it is clear in both classes that the frequency of subordinate breeding is considerably higher when in the presence of unrelated members of the opposite sex. In addition, studies in mammals show subordinate females may fail to reproduce in the absence of a dominant female if unrelated males are also absent, indicating that suppression is not always sufficient to explain reproductive skew (Burda 1995; Bennett et al. 1996, 1997; Clutton-Brock et al. 2001b).

Recently, studies have begun to investigate whether suppression is exercised behaviorally or physiologically. In birds, differences in levels of testosterone between dominant and subordinate males is variable, although in some cases this variation may be explained by differences in age and the presence or absence of unrelated females (Chapter 8). In contrast, among mammals, differences in hormone levels, testes size, and/or sperm quality between subordinate and dominant males are slight or non-existent, irrespective of whether or not unrelated females are accessible (Table 13.2). The notable exception is the naked mole-rat, in which

Table 13.3. *Evidence of physiological suppression among females*

Physiological suppression	Reproductive skew			
	Medium/high		Low	
No	Zambian mole-rat[a]	Related	Golden lion tamarin	Unrelated
	Mashona mole-rat[a]	Related		
	Common mole-rat[a]	Related	African lion[b]	Unrelated
	Meerkat[a]	Related	Banded mongoose[b]	Inbreeding
	Dwarf mongoose	Related	Spotted hyena[b]	Unrelated
	Gray wolf	Related		
	Common marmoset	Related		
Yes	Damaraland mole-rat	Related		
	Naked mole-rat	Inbreeding		
	Prairie vole	Unrelated		
	African wild dog	Unrelated		
	Common marmoset	Unrelated		
	Cotton-top tamarin	Unrelated		

Physiological suppression is defined as occuring when subordinates have lower levels of luteinizing hormone (LH) and/or estrogen (E) than dominants even in the presence of potential mates. In the absence of potential mates, results are shown only if no differences were found. Species are included only if individuals examined are of similar age. Related/unrelated refers to whether or not females commonly have access to unrelated males within their group (except where inbreeding appears common).
[a] lack of suppression shown by GnRH challenges (see Chapter 8).
[b] hormonal data currently lacking.

subordinate males have reduced levels of luteinising hormone (LH) and testosterone (T) as well as impaired spermatogenesis (Faulkes and Abbott 1991; Faulkes et al. 1991, 1994). That male hormone levels between subordinates and dominants do not differ at all in most mammals suggests that male reproductive skew in general is not dictated physiologically. This may be because suppression of physiology in males is difficult or costly, because behavioral suppression by simple mateguarding is generally effective, or because by physiologically suppressing males, dominants compromise a subordinate's tendency to help (Creel et al. 1993; Asa 1997).

While the differences between subordinate and dominant female birds in their levels of LH and estrogen may also depend on the presence of an unrelated male in the group, the situation with female mammals is far more variable (Table 13.3). Among carnivores and non-eusocial mole-rats, differences between subordinates and dominants in levels of sex hormones tend to be significant when males unrelated to subordinates are absent, but small or negligible when unrelated males are present, and small differences may be due to age and/or sampling during estrus (Creel et al. 1992; Carlson et al. in press). In these species, reproduction by subordinate females is apparently not suppressed physiologically, and experiments confirm this conjecture. For example, in meerkats and non-eusocial mole-rats, administering subordinate females in the absence of unrelated males with gonadotropin-releasing hormone (GnRH) gives rise to levels of LH comparable to that of dominants (Bennett et al. 1997, 2000; Spinks et al. 2000a; O'Riain et al. 2000a; Herbst and Bennett 2001).

However, in females of other mammals, physiological suppression appears to occur (Table 13.3). For example, in naked and Damaraland mole-rats, subordinate females have significantly lower levels of LH and estrogen than dominant females and have impaired ovulation and prepubescent ovaries (Faulkes et al. 1990a; Faulkes and Abbott 1997; Bennett et al. 1993, 1994). Furthermore, these hormonal differences are not reduced with a challenge of GnRH. Similarly, among cotton-top tamarins (French et al. 1984; Heistermann et al. 1989) and prairie voles (Carter and Roberts 1997), the presence of the dominant female causes abnormal ovulation and ovarian function even in the presence of unrelated males, a situation that is reversed by removal of the dominant female (Widowski et al. 1990). How dominants achieve physiological suppression is not known, but it is clearly not achieved by dominant-induced stress, at least as measured by cortisol, since in both birds and mammals subordinates generally have similar or lower levels of cortisol than dominants (Creel 2001; Carlson et al. in press; see Chapter 8).

Under what circumstances should dominants physiologically suppress subordinates? Answering this question depends on quantifying the relative costs to dominants of physiologically versus behaviorally suppressing subordinates (Chapter 8). This cost may include energetic costs to the dominant as well as effects that resulting changes in hormonal levels of subordinates may have for their subsequent contributions to cooperative efforts in the group (Asa 1997).

Three general points are appropriate. First, physiological suppression of subordinate males may be unusual. Second, among females, physiological suppression may be more likely when subordinates live in groups with unrelated males, and subordinate breeding is costly to the dominant. For example, subordinates appear to be physiologically suppressed when living in polyandrous groups but not in family groups in relatively fecund callitrichids like common marmosets (Abbott 1984; Hubrecht 1989), whereas in the less fecund golden lion tamarin physiological suppression does not occur irrespective of group type (French et al. 1992; French 1997). In naked mole-rats, physiological suppression may be important because inbreeding is common. That dominant females are aware of a subordinate's opportunity to breed is shown in experiments in pine voles, in which adding an unrelated male to groups causes dominant females to become aggressive towards subordinates (Brant et al. 1998). Finally, physiological suppression has not been shown to occur in carnivores. This may be due to a combination of unrelated males seldom being found in their groups, infanticide being a viable alternative, and an inherent costliness of subordinate breeding (Creel and Creel 1991, 2001; Moehlman and Hoffer 1997; Clutton-Brock et al. 2001b). Alternatively, a lack of physiological suppression may also indicate a potential for dominants to benefit from subordinate breeding under certain circumstances (Cant and Johnstone 1999).

Synthesis: theories explaining reproductive skew

Understanding reproductive skew in birds and mammals has generally concentrated on transactional models, which assume that dominants allow subordinates to breed to maintain their presence and help within the group. This is extremely unlikely to be true of most female mammals (Clutton-Brock 1998), but may be more true of males (Hager 2003). First, concession models are unlikely to explain female reproductive skew in non-cooperatively breeding social mammals, where non-breeders take no part in dominant breeding. Second, in mammals, it will seldom benefit a dominant female to allow a subordinate to breed in order to retain her in the group, for a subordinate will almost always produce substantially more offspring than she (the subordinate) can raise by herself. This is because, in contrast to many cooperative birds, females in cooperative mammals tend to be highly fecund (Table 13.1). Thus, allowing a subordinate to breed will have substantial negative effects on the helper : offspring ratio in cooperative mammals, leading to a reduction in the fitness of the dominant.

Third, concession models also predict that concessions are traded for philopatry, help, and/or subordination (Chapter 10). There is little evidence for this generally, and in meerkats subordinate breeding is not associated with any of these three life-history traits (Clutton-Brock et al. 2001b).

Finally, concession models predict that close kin should be less likely to breed in a group than more distant kin. This appears untrue among non-callitrichid primates (Hager 2003), coatis (Gompper et al. 1997), spotted hyenas (Engh et al. 2000), prairie voles (Firestone et al. 1991; Hodges et al. 2002), and meerkats (T. H. Clutton-Brock, unpublished data), which all show positive associations between a subordinate female's probability of breeding and its degree of relatedness to the dominant female. Indeed, in prairie voles, related pairs of females are likely to cobreed, while in unrelated pairs of females reproductive success is substantially reduced (Hodges et al. 2002) and one of the two will normally die (Firestone et al. 1991).

Positive relationships between relatedness and breeding probability are consistent with two other branches of skew models, incomplete control (Clutton-Brock 1998; Reeve *et al.* 1998; see Chapter 10) and nepotistic models (Wrangham 1980). Incomplete-control models are likely to predict the degree of reproductive skew among cooperative female mammals, in which dominant females commonly pay fitness costs when subordinates breed. In contrast, nepotism models may apply more frequently to non-cooperative species, such as non-callitrichid primates and spotted hyenas, in which some subordinate reproduction has little fitness costs to the dominant, and the dominant can increase inclusive fitness by selecting closer relatives to breed (Cant and Johnstone 1999). In contrast to the situation for females, concession models may more commonly explain reproductive skew among males in mammals, either because coalitions can be more effective at obtaining matings than singletons (van Hoof and van Schaik 1994; Hager 2003) or because females mate with multiple males in order to retain their assistance in obtaining paternal care (Davies 1992; Cant and Reeve 2002). In the former case, as predicted by concession models, skew is more egalitarian among groups of unrelated males (Packer *et al.* 1990; Keane *et al.* 1994), although such effects may be confounded by coalition size (Clutton-Brock 1998).

WHY HELP?

That so many mammal species commonly show delayed dispersal and group living raises the question as to why relatively few qualify as cooperative breeders. There may be three reasons for this. First, lactation may reduce the ability of individuals to provide care to offspring. Second, because so many mammals are polygynous, individuals may rarely be constrained from independent breeding, even in their natal group (Clutton-Brock 1989b), while dominants may not benefit from suppression (see previous section). Third, many mammals are vegetarian, which probably reduces the benefits that helpers could provide.

The effects that helpers have on their recipients' fitness are well documented in birds (Chapter 3), although the factors governing variation in helper contributions are not (Chapter 7). First, I document the effect of helpers on recipients and the fitness benefits gained in societies of cooperative mammals. Second, I review the evidence that help is costly and the factors that govern individual contributions to cooperation. My aim is to highlight the similarities and differences between birds and mammals, and to illustrate how mammal studies help us to understand helping behavior in general.

Effects of help on recipients

Avian studies show that helpers are typically associated with reductions in breeder investment and/or increases in offspring care (Hatchwell 1999). However, relatively few studies have investigated whether such effects translate into fitness advantages while controlling for confounding influences of individual quality, territory quality, and group size (Chapter 3). This is also the case among mammals (Jennions and Macdonald 1994; Solomon and French 1997). Nevertheless, those mammalian studies that have used experimental or multivariate analytical approaches to investigate helper effects have the potential to add significantly to our knowledge of the generality of such effects in cooperative vertebrates (Table 13.4).

Benefits to parents
In most mammals, increases in helper numbers are associated with reductions in parental investment (Solomon and French 1997). Among males, such reductions may be associated with subsequent increases in other activities that help to increase the reproductive success of a current breeding attempt, such as reducing the risk of predation or infanticide. Alternatively, males may seek to increase their reproductive success by attempting to increase extra-group copulations or personal survival. Whether males succeed in gaining benefits from reduced investment is currently unknown.

In mammals, females appear to be the primary beneficiaries of reduced investment, with helpers allowing females to forage longer and/or more efficiently. For example, among cooperative cetaceans, helpers guard offspring at the surface, allowing mothers to forage for longer (Whitehead 1996), while among callitrichid primates helpers carry offspring, allowing mothers to forage more efficiently (Tardif 1997). In birds, load-lightening is sometimes associated with significant increases in maternal survival and significant reductions in inter-clutch interval among multi-brooded species (Chapter 3). Whether or not helpers influence maternal survival in mammals is unknown, but helper removal experiments in prairie and pine voles suggest that helpers may allow females to have reduced inter-birth intervals (Solomon 1991; Powell and Fried 1992).

Table 13.4. *Effects of increasing numbers of helpers on their recipients and effects of kinship on contributions to cooperation*

Species	Effect on fecundity of dominant female	Effect on offspring	Effect of varying degrees of relatedness	Proposed personal benefits	References
Pine vole	⇓ inter-birth interval ± litter size	⇑ development ⇑ weight ± survival			Solomon 1991[a]
Prairie vole	⇓ inter-birth interval ± litter size				Powell and Fried 1992[a]
Mongolian gerbil	± inter-birth interval ± litter size	± offspring		⇑ opportunity to breed ⇑ experience	Salo and French 1989[a]
Alpine marmot		⇑ survival	⇑ weight loss ⇑ alarm calling		Arnold 1990a
Black-tailed prairie dog, Belding's ground squirrel					Hoogland 1995; Sherman 1977
Meerkat	⇓ inter-birth interval ⇑ litter sizes	⇑ growth ⇑ condition ⇑ survival	⇑ food to pups of dominant by males	⇑ survival ⇑ dispersal probability ⇑ breeding success	Clutton-Brock *et al.* 2001c[a]; Russell *et al.* 2003a; Young 2003
Banded mongoose		⇑ growth ⇑ condition ⇑ survival		⇑ survival ⇑ breeding success	Gilchrist 2001; Hodge 2003
Dwarf mongoose		⇑ productivity		⇑ survival	Creel *et al.* 1991; Creel and Waser 1994; Waser *et al.* 1995
African lion		± offspring	⇑ allosuckling		Packer *et al.* 1990
Brown hyena			⇑ food to pups of dominant by males		Owens and Owens 1984
African wild dog		⇑ survival		⇑ survival ⇑ dispersal probability	Creel and Creel 2001
Cotton-top tamarin		± survival		⇑ experience (females) ± inheritance ± advertisement	Tardif 1997; Washabaugh *et al.* 2002
Golden lion tamarin			⇑ carrying pups of dominant by males	⇑ experience (females) ± inheritance	Baker and Woods 1992

Helper effects on dominant females and offspring are shown only if an experimental or multivariate statistical technique was employed. Proposed personal benefits refer to those direct benefits that individuals may gain from helping, but since no study has investigated this effect properly the effects must be treated with caution.
⇑ = increase, ⇓ = decrease, ± = no effect.
[a] = experimental studies.

Helpers may also influence the fecundity of female breeders within reproductive attempts, an effect that has not been considered in birds, except where females mate polyandrously (Davies and Hatchwell 1992). In meerkats, for example, the number of individuals that help to raise a previous litter not only has a significant negative effect on inter-birth intervals, but also has a significant positive effect on the size of the following litter (Russell et al. 2003a). Such effects are likely to arise as a consequence of the effects of helper mediated load-lightening on maternal condition. Two pieces of evidence support this conjecture. First, helpers cause substantial load-lightening in meerkats. In the presence of helpers, dominant females do not contribute to babysitting and substantially reduce investment in pup-feeding (Clutton-Brock et al. 2000, 2001a), both of which are associated with significant growth costs to helpers (Russell et al. 2003b). In addition, the number of allolactators in meerkat groups also increases with the number of female helpers, and allolactating helpers are associated with considerable reductions in energy expenditure of dominant females during lactation (Scantlebury et al. 2002). Second, females helped by many helpers are significantly heavier at the conception of their following litter than those with fewer helpers, and female conception and litter size at birth are primarily a consequence of maternal condition (Russell et al. 2003a).

At present, it is unclear whether or not helpers are associated with significant increases in maternal fecundity in mammals generally. However, maternal condition is known to correlate with fecundity in a wide range of non-cooperative mammals (Clutton-Brock 1991), and positive relationships between helper number and maternal condition have been reported in other cooperative mammals, including cotton-top tamarins (Sanchez et al. 1999) and golden lion tamarins (Bales et al. 2001). Furthermore, in naked mole-rats, attaining the breeding position is associated with a dramatic increase in female body size, caused by elongation of the lumbar vertebrae (O'Riain et al. 2000b), while in meerkats, after controlling for age, females are significantly larger and heavier when they are dominant than when they are subordinate (A. F. Russell, unpublished data). Such effects raise the interesting possibility that load-lightening not only increases maternal condition, but also body size, allowing females to accommodate larger litters (Sherman et al. 1999; A. F. Russell, unpublished data).

Benefits to offspring

In birds, helpers commonly have a significant effect on the provisioning frequency, growth, and survival of dependent offspring (Chapter 3), and experimental and multivariate analytical techniques reveal similar effects in mammals (Table 13.4). Helper-removal experiments conducted in prairie voles, for example, show that helpers have significant effects on the developmental rate and ultimate weight of pups at weaning (Solomon 1991). In groups of banded mongooses, some pups from the same litter are raised by a personal helper (an "escort"), while others are not, allowing within-group comparisons of helper effects. Multivariate analyses controlling for differences in initial body weight show that escorted pups grow at a faster rate and survive better than non-escorted pups (Gilchrist 2001; Hodge 2003).

Finally, in meerkats, multivariate analyses show that helpers have significant effects on the biomass intake rate, daily weight gain and growth of pups (Clutton-Brock et al. 2001c; Russell et al. 2002). The results of these analyses were confirmed using novel experimental manipulations in which dependent offspring were temporarily removed from their own groups, causing increases in the helper : pup ratio, and added to other groups, causing decreases in the helper : pup ratios (Clutton-Brock et al. 2001c). The rate at which pups gained weight before and after the experiment was greater in the litters from which pups had been removed and lower in those litters to which pups had been added. That changing the helper : offspring ratio, but not helper number, influences the rate at which pups gain weight provides one of the clearest examples in any vertebrate that the relationship between helper number and offspring growth is causal.

Mammalian studies have also taken helper effects a step further than most bird studies by investigating whether helpers may have long-term effects on the offspring that they help to raise. For example, through increasing pup weight at weaning, helpers may cause long-term benefits in prairie voles, for pups that are heavy at weaning are preferred as social mates and have higher fecundity as breeders (Solomon 1993, 1994). Similarly, in banded mongooses, escorted pups are heavier at independence than non-escorted pups, and heavy females are likely to breed earlier as adults (Hodge 2003). Evidence in meerkats, in which helper number is known to account for most of the known variation in pup weights at independence (Russell et al. 2002), shows that individual weight at independence is associated with greater investment in helping (Clutton-Brock et al. 2002; Russell et al. 2003b), higher probability of surviving to reproducing age, greater probability of dispersing early (males) or breeding at a younger age (females), greater probability of attaining dominance (both sexes), and greater fecundity among breeding females (A. F. Russell, unpublished data).

Fitness benefits

There is little question that in many mammals helpers have dramatic effects on the reproductive success of their recipients, not only increasing their fecundity and breeding success, but also the breeding potential of their offspring. In addition, like birds, mammal helpers generally provide care to relatives (Emlen 1995), and may thus increase their overall fitness through indirect as well as direct pathways.

Indirect fitness benefits

Most cooperative vertebrates represent a poor testing ground for the relative importance of indirect fitness to the evolution of cooperative breeding, because philopatric individuals have little opportunity but to help kin. In other words, helping kin may be a consequence of philopatry and cooperation rather than its cause (Clutton-Brock 2002). A better test is to investigate whether, when having the choice, helpers prefer to direct their care toward more-related rather than less-related individuals, although a non-significant effect does not necessarily preclude the importance of kin selection (Keller 1997; Komdeur and Hatchwell 1999).

As with birds, there are a number of mammalian studies that provide evidence for kin preferences in helping behavior. Male brown hyenas (Owens and Owens 1984) and meerkats (Clutton-Brock et al. 2004) provide more food to closely related young than to more distantly related young, while male golden lion tamarins prefer to carry more-related offspring (Tardif 1997). Female African lions prefer to suckle the offspring of closely related mothers than those of more-distantly related mothers (Pusey and Packer 1994), while in Belding's ground squirrels (Sherman 1977) and black-tailed prairie dogs (Hoogland 1995), females with closely related kin present are more likely to utter costly alarm calls than those without kin present. Finally, in alpine marmots, Arnold (1990a) showed that individuals overwintering with closely related individuals lose more weight than those overwintering with less-closely related individuals. These last studies are of particular note because alarm-calling may have significant survival benefits for offspring in ground squirrels (Sherman 1977), and because Arnold (1990b) showed that in marmots offspring have higher overwinter survival if they had been helped by more-closely related individuals.

Direct fitness benefits

Helping to raise non-descendant young may also evolve if it improves individual fitness directly, and avian studies have identified five different ways in which this may be possible. First, by helping, individuals may gain immediate breeding rights in a current breeding event. However, in this case, individuals probably only help (i.e., raise offspring that are not their own) because they have gained reproductive success and cannot discriminate between offspring in broods of mixed parentage (Davies 1992). Second, individuals may help because of a requirement to pay rent while remaining philopatric. This remains a significant possibility, but no adequate tests of its predictions have been conducted in either birds or mammals. For example, helper removal experiments are not a good way of testing for payment of rent (Mulder and Langmore 1993) because they are not able to control for inevitable changes in dominance hierarchies or asymmetries of resident versus removed birds.

Third, helping may arise as an advertisement of quality (Zahavi and Zahavi 1997). That secondary helpers are more likely to gain a mate than non-helpers in pied kingfishers (Reyer 1990) and that potential breeders provision offspring more in the presence of competitors than in their absence in Arabian babblers (Ridley 2003) provides support for the advertisement hypothesis in birds. As yet, there is no comparable evidence in mammals (Baker et al. 1993; Tardif and Bales 1997). Fourth, studies in birds have found that helpers may gain experience allowing them to become better breeders than non-helpers (Komdeur 1996). In mammals, Salo and French (1989) showed that experimentally removed Mongolian gerbils are poorer breeders than those that remain in the presence of offspring. However, it is unclear why helpers should gain more experience than philopatric individuals, and such effects may arise due to differences in quality (Chapter 3) or changes in inherent aversions towards pups, a phenomenon so far shown to be common to rodents (Carter and Roberts 1997).

Finally, and perhaps most importantly, individuals may increase their direct fitness by helping to raise additional group members. This "group augmentation" hypothesis predicts that helpers increase group productivity and, through doing so, increase their probability of survival, dispersal, and ultimate breeding success. Recent theoretical advances show that group augmentation need not be subject to cheating, and that this is particularly true in the presence of kin (Kokko et al. 2001). That individuals may gain benefits from living in a group (Chapter 5) suggests that group augmentation may be a primary route through which helpers increase their fitness directly. However, tests of this idea are in their infancy, and studies have yet to examine the residual increase in direct fitness gained by helping to raise the productivity of the group. Nevertheless, a number of studies report findings consistent with the predictions of group augmentation.

Most importantly, studies of cooperative mammals commonly show positive associations between helper numbers and group productivity (Table 13.4), as well as group size and individual survival probability (Rood 1990; Clutton-Brock et al. 1999a; Creel and Creel 2001). Thus, by increasing group productivity, helpers may increase their probability of surviving to reproduce. Evidence from wild dogs and mongooses suggests that an individual's survival probability is a consequence of group size, and that individuals from larger groups disperse in larger coalitions, and thus, by helping to raise co-dispersers, individuals increase their probability of dispersing successfully (Rood 1990; Creel and Creel 2001; Young 2003). Indeed, this may be a general phenomenon in cooperative mammals, for in a large number of species individuals disperse in coalitions and there is evidence to suggest that larger units are more successful at becoming established than smaller units (Lucas et al. 1997; Lewis and Pusey 1997; Bennett and Faulkes 2000; Young 2003). The advantages of dispersing in coalitions are that individuals are likely to forage more efficiently and compete for space or access to groups more effectively (Packer and Pusey 1982; Clutton-Brock et al. 1999a; Creel and Creel 2001). Finally, by helping to raise group productivity, individuals will raise more helpers from which they may benefit in the future (Jennions and Macdonald 1994; Solomon and French 1997; Clutton-Brock et al. 2002).

Group augmentation also explains some behaviors that are difficult to explain by kin selection, such as helping by non-relatives, adoption, and kidnapping (Clutton-Brock 2002), and why small groups of Wied's black tufted-ear marmosets commonly accept immigrants, while large groups strongly repel them (Schaffner and French 1997). In addition, group augmentation can explain why in cooperative birds and mammals the philopatric sex generally contributes more to cooperation than the more dispersive sex, irrespective of which sex is philopatric. For example, in Seychelles warbler (Richardson et al. 2003) and white-throated magpie-jay (Langen and Vehrencamp 1999), females are philopatric and non-breeding females help more than males, while in most other cooperative birds males are philopatric and also the predominant helping sex (Cockburn 1998). In contrast, among mammals, females tend to be both the philopatric and the helping sex, but in those cases where males are more philopatric, males help more than females (Table 13.1).

Sex differences in contributions to cooperation with respect to philopatry make sense because philopatric individuals will benefit more from raising the size of their group in which they will remain and breed than dispersing individuals that remain only for a time before dispersing. Interestingly, in meerkats, females (the philopatric sex), provide more care to female offspring (the more active helpers), whereas male help is independent of offspring sex (Brotherton et al. 2001; Clutton-Brock et al. 2002).

Costs of helping

The more costly help is, the more cooperative breeding must involve the importance of indirect fitness benefits rather than direct benefits such as group augmentation. However, as is the case in birds (Chapter 4), few mammalian studies have investigated the costs of cooperation. In mammals, there is no question that helping is associated with substantial costs to condition, but there is as yet little evidence that helping is associated with energy-mediated survival costs (Arnold 1990a, 1990b; Tardif 1997; Russell et al. 2003b). It cannot be ruled out that survival costs would be apparent under certain conditions, but evidence so far suggests that survival effects are not found because helpers behave so as to minimize mortality risk. One exception to this is Belding's ground squirrels, where individuals that utter alarm calls may be more likely to be depredated (Sherman 1977).

For example, in callitrichid primates, helpers primarily care by carrying infants. Such behavior is associated with reduced caloric intake and reduced foraging in saddle-back tamarins (Goldizen 1987; Tardif 1997) and cotton-top tamarins (Price 1992), and reduced jumping distance in common marmosets (Schradin and Azenburger 2001). Although no study has investigated whether there are any fitness costs of such behavior, Tardif and Harrison (1990) and Price (1992) suggest that individuals might reduce potential fitness costs associated with carrying offspring by traveling less and remaining more hidden.

In meerkats, helping is associated with substantial growth costs (Russell et al. 2003b). However, there is no evidence to suggest that an individual's contribution to cooperation in a current event influences its probability of dying or dispersing by the following breeding attempt, or breeding in the following attempt. Furthermore, the costs of cooperation do not accumulate over many breeding events to affect mortality or dispersal, but do accumulate to have a negative effect on the probability that a subordinate female will breed.

There are at least three reasons why helping may have little effect on fitness costs in meerkats (Cant and Field 2001; Clutton-Brock et al. 2002; Russell et al. 2003b). First, helping may be condition-dependent. Second, helpers that invest heavily in a previous breeding attempt increase their foraging rate during the subsequent non-breeding season and reduce their investment in the subsequent breeding attempt, with the magnitude of this reduction depending on the inter-birth interval. Finally, helpers may reduce their investment later in life and before critical life-history phases. Thus, despite substantial short-term growth costs, evidence from meerkats suggests long-term fitness costs may be slight.

Contributions to cooperation

In Chapter 4, Heinsohn points out that we know little about those factors that influence individual contributions to cooperation in birds. This is also largely the case among mammals. For example, why cooperative breeding is so uncommon in non-callitrichid primates and spotted hyenas is unclear, especially given their tendency to live in stable family groups. Similarly, the apparent rarity of redirected care following breeding failure in mammals that commonly live in family groups or kin-neighborhoods is also difficult to explain. Our knowledge of the factors that are associated with individual contribution to cooperation in cooperative mammals is better understood. Primary among explanations of individual contributions to cooperation in mammals is unsurprisingly helper number, age, sex, and condition (Creel and Waser 1994; Lacey and Sherman 1997; Tardif 1997; Creel and Creel 2001; Clutton-Brock et al. 2002), and in some cases relatedness (see above). In addition, a significant factor that influences contributions to cooperation in rodents is prior exposure to young offspring (Carter and Roberts 1997), while in meerkats additional factors include previous contribution to cooperation and (for males) timing to dispersal (Clutton-Brock et al. 2002; Russell et al. 2003b). Finally, although individual contributions to cooperation vary in meerkats after controlling for differences in age, sex, and condition, those individuals that conduct much of one activity tend to conduct much of all others (Clutton-Brock et al. 2003), a result also suggested in naked mole-rats (Lacey and Sherman 1997).

The role that hormones may play in governing contributions to cooperation has become a relatively new field of research. Four hormones in particular have attracted attention: testosterone, estrogen, prolactin, and cortisol. Unfortunately, the relationship between different hormones is poorly understood, and experiments show that effects may vary depending when in the season they are administered (Roberts et al. 1996). In addition, whether hormones are the cause of behaviors or the effect is unknown.

Despite these problems, some patterns are emerging. First, high levels of testosterone are generally associated with low propensities to cooperate (Roberts et al. 1996), possibly because testosterone is associated with aggressive and sexual activity (Wingfield et al. 1990; Ketterson and Nolan 1994). However, in some cases high levels of testosterone are associated with high levels of paternal care, apparently because testosterone can be converted to estrogen, and estrogen levels may be positively associated with help (Trainor and Marler 2001, 2002). That high estrogen and low testosterone positively influences contributions to cooperation is shown in the Mongolian gerbil, in which fetal males positioned between two females tend to be feminized, show high levels of estrogen, low testosterone, and high levels of cooperation (Clark and Galef 2000). However, the hormone most universally associated with high levels of care is prolactin, with positive effects of prolactin on contributions to cooperation being evident in both birds (Chapter 8) and mammals (Dixon 1992; Mota and Sousa 2000).

Unfortunately, studies that have investigated the role of prolactin on contributions to cooperation have not investigated its role independently of either age or condition. Thus, such effects may not be additive to what we already know. Some interesting results are emerging from studies of male helpers in meerkats (A. A. Carlson, unpublished data). First, a helper's responsiveness to playback experiments of pup begging calls is significantly and positively associated with levels of prolactin and cortisol. Second, using multivariate analyses, it is clear that levels of prolactin and/or cortisol may have effects additional to helper number, age, and condition on contributions to cooperation. However, levels of prolactin and cortisol are correlative, and results suggest a dominant effect of cortisol, leading to the intriguing possibility that relationships between prolactin and cooperation may, in some species, be a consequence of cortisol. Although cortisol is generally viewed as being a stress hormone, in limited doses it may act as an anxiety hormone, leading to greater responsiveness of helpers to the needs of offspring, as has commonly been shown to be the case in humans.

Finally, hormone levels may be maternally derived (Place *et al.* 2002), leading to the possibility that mothers may be able to exert some degree of control over the subsequent helping tendencies of their offspring. Investigations of the effects that hormones have on helping tendencies either using experimental designs or multivariate analyses may elucidate some hitherto unknown features of societies of cooperative vertebrates.

Synthesis: comparisons of birds and mammals

Mammals represent an ideal testing ground for some of the theories that have been developed for birds. First, consider the observation that in birds males tend to be both the more philopatric and the predominant helping sex, which has led to the hypothesis that the philopatric sex may gain greater direct benefits by helping to raise further group members (Cockburn 1998). This hypothesis is supported in mammals, in which the philopatric sex tends to be female, and females are the predominant helping sex (Clutton-Brock *et al.* 2002, Table 13.1). Second, Cockburn (1998) suggested that positive relationships between productivity and helper number are stronger in those species in which both sexes help or just females help, although recent studies call this association into question (Magrath 2001; Hatchwell *et al.* 2003). In mammals, there is little evidence for this pattern, and correlations between helpers and productivity occur irrespective of helper sex (Tables 13.1, 13.4).

Third, avian studies suggest that fathers tend to be the sex that first reduces their investment in the presence of helpers (particularly when male mortality is high), but when mothers do reduce their investment, they may do so more fully than fathers (Hatchwell 1999). One explanation for this finding is that males are more likely to concentrate on enhancing their survival because they cannot be sure that they are the fathers of the current brood, although it cannot be ruled out that they use their extra time to carry out other activities such as territory defense. In mammals, both mothers and fathers reduce their investment, but mothers appear to do so more than fathers. One reason for this is that in mammals the costs of lactation are extreme (Clutton-Brock *et al.* 1989), and all individuals may benefit from mothers that are able to provide milk for their offspring and be in better condition to breed again earlier and produce more offspring. An analysis of mammals similar to that undertaken in birds by Hatchwell (1999) would elucidate some interesting generalizations concerning helper effects in cooperative vertebrates.

Finally, cooperative breeding in birds has been suggested to be most common in those species that have small clutch sizes (Arnold and Owens 1998). This suggests that, in birds, helpers allow mothers to reduce investment and increase longevity. This may contrast with the situation in mammals, in which cooperative species tend to be highly fecund. For example, cooperative mole-rats, dogs, and primates are more fecund than non-cooperative species in the same family. This point requires further investigation, but it appears that cooperative mammals may use helpers to increase productivity rather than personal survival.

CONCLUSIONS AND FUTURE RESEARCH

In virtually all cooperative birds and mammals, cooperative breeding arises when offspring delay dispersal and help to raise offspring that are not their own (Emlen 1995). Understanding delayed dispersal is therefore central to understanding cooperative breeding, and understanding delayed dispersal requires an understanding of group living. In contrast to suggestions in birds, habitat availability is probably a poor indicator of delayed dispersal in mammals. Instead, delayed dispersal appears largely to be a consequence of food type and distribution, as well as foraging style and predator evasion method (Jarman 1974; Clutton-Brock and Harvey 1977, 1978). Such findings have become apparent by comparing ecological and life-history correlates of delayed dispersal within families of mammals rather than studying their effect within species or within mammals as a whole. Similar approaches have seldom been adopted in avian studies. Future research in both classes may benefit substantially from two approaches. First, identifying when delayed dispersal and dispersal/floating tactics occur within species, with particular focus on territory quality and the survival of dispersers and delayers. Second, identifying causes of delayed dispersal in morphologically, ecologically, and/or phylogenetically similar species, again with a focus on food and offspring survival.

Reproductive skew varies significantly within and between species. Understanding the ultimate reasons for variation in skew among cooperative species may largely depend upon the ability of dominants to monopolize food, helpers, and/or mates, and the relationship between such monopolization and reproductive success. Where helpers have a significant effect on reproductive success, and helpers are limiting, dominants will be selected to suppress reproduction of subordinates. Conversely, helpers may be selected to "accept" suppression, especially when breeding is costly. I agree that testing the assumptions of skew models will be more illuminating than testing the predictions at this stage (Chapter 10). However, understanding the costs and benefits of subordinates breeding for dominants will also yield significant insights.

A significant number of mammalian studies have considered whether skew is achieved behaviorally or physiologically. Two patterns appear to be emerging. First, physiological differences between dominant and subordinate males appear to be generally lacking, indicating that reproductive skew in males is governed by behavior. Second, physiological differences among females appear to vary substantially, with some species showing no physiological differences and others showing dramatic differences. Understanding why there should be differences between the sexes as well as differences between species is likely to be an exciting and fruitful line of future research. One possibility is that physiological suppression arises when breeding by subordinates is seldom beneficial, whereas behavioral suppression may be more common when subordinate breeding is beneficial under certain circumstances. Although significant steps have been made in understanding the proximate mechanisms of skew, many studies have failed to investigate differences between dominants and subordinates with due consideration for differences in age or the presence and absence of unrelated mates. This lack of consistency between studies makes it difficult to make cross-study comparisons, and hence clouds our ability to make coherent generalizations.

In most birds and mammals, helpers direct their care towards relatives, and such observations support the hypothesis that kin selection may constitute a significant driving force in the evolution of cooperative breeding (Emlen 1997a). Recently, however, the importance of kin selection has been questioned on the grounds that helpers may not enhance productivity, frequently show no preference between different degrees of kin, may often preferentially help non-kin (Cockburn 1998), and may often help kin because they are philopatric rather than because of kin selection per se (Clutton-Brock 2002). There are a number of reasons for the persistence of this debate.

First, positive associations between helper number and reproductive success may be difficult to detect if they depend on helper effort rather than number (Innes and Johnston 1996), or if relationships between number and productivity are only apparent under certain circumstances (Chapter 3). Second, helping non-kin could be a secondarily derived effect arising after the initial evolution of cooperative breeding, which is known to be more likely in the presence of kin (Kokko et al. 2001). Indeed, there are extremely few species where all helpers direct care towards non-kin, and many of those may do so because they have reason to believe that they have gained direct fitness in a current brood (e.g., dunnocks). Third, a lack of preferential help for different degrees of kinship does not negate Hamilton's (1964) rule, and hence does not provide evidence against the importance of kin selection, assuming that no mechanism for discriminating among different level of kin exists (Keller 1997; Komdeur and Hatchwell 1999).

Fourth, direct fitness effects of helping are commonly confused with those of breeding. Helpers are, by definition, individuals that provide care to offspring that are not their own. Hence, direct fitness effects of helping must be calculated as the residual contribution to overall fitness resulting from increments to reproductive success that were themselves a result of helping. Studies that do not discriminate between fitness gained by helping and that gained by breeding will overestimate the relative importance of direct fitness to the evolution of cooperative breeding. Currently, the importance of direct fitness benefits is seen as a viable alternative when kin-selected ideas are insufficient. It is now time to test the predictions of direct fitness hypotheses, which have hitherto remained poorly tested.

There are three other areas of helper effects that remain severely under-represented in both birds and mammals. First, understanding the costs of cooperation is in dire need of attention, and an understanding of the adaptive nature of cooperative breeding relies on at least a basic understanding of costs. In addition, it may be that the costs (as well as benefits) of cooperation help to explain sex biases in helping, but evidence is so far generally lacking. Second, understanding individual decisions to help and contributions to cooperation is a field in which we know very little. Yet understanding cooperative breeding relies on our ability to identify the causes of such variation. Hormonal studies offer a new insight into explaining the variation, but studies have so far failed to control for a number of factors that correlate with both hormone levels and contributions to cooperation, including age, sex, and condition. Lastly, Hatchwell (1999) examined whether, in cooperative birds, helpers cause reductions in parental investment or additions to offspring provisioning, and found that the former arose when nestling starvation is rare. Repeating this type of analysis in mammals is likely to provide a useful test of the generality of these findings across cooperative vertebrates.

ACKNOWLEDGMENTS

I thank Walt Koenig and Janis Dickinson for allowing me onboard and for significantly increasing the clarity of my message. I also thank Tim Clutton-Brock, Ben Hatchwell, Andy Young, and Sarah Hodge for their countless discussions on cooperative breeding and their valuable comments on this chapter.

14 • Summary

STEPHEN PRUETT-JONES
University of Chicago

The existence of avian species in which social groups consist of more than a reproductive pair and individuals other than the parents feed offspring has been known for more than a century (Boland and Cockburn 2002), and active research on cooperatively breeding species has been ongoing since Skutch (1961) reviewed the important questions raised by such systems. Both the theoretical approach to evolutionary questions raised by cooperative breeding and the empirical investigation of such systems have changed greatly in these 40 years. In the first compilation of articles focused on cooperative breeding in birds Stacey and Koenig (1990a) presented summary chapters on long-term studies of cooperatively breeding birds, including several of the best-known species at the time. Research on those species yielded many of the data on which the theoretical foundation for our understanding of the evolution of cooperative breeding has been built.

The focus of Stacey and Koenig (1990a) on long-term behavioral and ecological studies was relevant because of the importance of such studies to identifying the costs and benefits of individual behaviors. Although the importance of long-term studies is no less significant today, it is also the case that with the advent of new techniques (particularly paternity analysis, phylogenetic methods, comparative analysis, and theoretical approaches such as reproductive skew theory), a questions-based approach to cooperative breeding has flourished and is reflected in the focus of this volume, in which conceptual issues relevant to the biology of all cooperatively breeding species, both birds and mammals, are examined.

In preparing this summary, I have chosen a novel, and I hope useful, approach to examining the diversity of cooperative breeding systems. Specifically, I try to identify and justify the most general and documented facts relating to the diversity of avian cooperative breeding systems. In essence, I have attempted to elucidate summary statements – 13 in all – that all researchers working on cooperative breeding in birds, or at least the majority, could agree on. I acknowledge that my choice of topics is based, in part, on my own interests and that other authors might choose different topics to emphasize. The chapters in this volume are the primary basis of the summary statements that I make here, but I draw on other literature as well.

(1) DIFFERENT AUTHORS DEFINE COOPERATIVE BREEDING DIFFERENTLY

Despite the generally accepted defining characteristics of cooperative breeding systems – the delayed dispersal of offspring and alloparental care by these philopatric individuals – different authors emphasize these characteristics differently in their definitions. Three examples illustrate this: Brown (1987) defines communal (cooperative) breeding as "a system of breeding that is characterized by the normal presence of helpers at some or all nests." Ligon and Burt (Chapter 1) define cooperative breeding as occurring when "non-breeding helpers occur within a social unit beyond the primary pair, irrespective of the presence or absence of potential breeders." Lastly, Cockburn (Chapter 5) defines it as "where more than two individuals combine to rear a brood of young."

Although arguably minor, the differences between these definitions illustrate two general points. First, although much of the theory of cooperative breeding focuses on factors promoting delayed dispersal, the primary identifying characteristic of cooperative systems is that individuals other than the breeding pair

---

Ecology and Evolution of Cooperative Breeding in Birds, ed. W. D. Koenig and J. L. Dickinson. Published by Cambridge University Press.
© Cambridge University Press 2004.

assist in the care of young. Second, the collection of species accepted as exhibiting cooperative behavior has moving boundaries, with the majority of typical cooperatively breeding species (in which parents are assisted by philopatric offspring) included in all of the definitions, but with other species excluded. For example, Ligon and Burt's definition excludes species such as the dunnock in which all of the group members reproduce (Davies 1990, 1992). In contrast, Cockburn's definition excludes species in which philopatric offspring do not assist the breeding pair such as the Siberian jay (Kokko and Ekman 2002). In the most inclusive list to date, Arnold and Owens (1998), adopting the definition of Emlen and Vehrencamp (1985) that cooperative breeding occurs whenever "more than two individuals rear the chicks at one nest," list a total of 308 bird species as being cooperative breeders (3.2% of the 9672 avian species listed in Sibley and Monroe 1990).

(2) THERE ARE DIFFERENT WAYS OF THINKING ABOUT THE "QUESTIONS" OF COOPERATIVE BREEDING

The focus of research, both theoretical and empirical, is at the outset dependent on how a research question is phrased and viewed in an evolutionary sense. This is certainly true of the field of cooperative breeding. Although the two defining characteristics, delayed dispersal and helping, are generally straightforward, there are different ways of expressing these characteristics as evolutionary questions or hypotheses. Emlen (1982a) articulated the important questions in cooperative breeding as "why do offspring remain on their natal territory?" and secondly "why do those individuals help?" Brown (1987) viewed the issue of delayed dispersal as comprising two components (delayed reproduction and delayed dispersal) and split the question of "why stay?" into "why delay reproduction?" and secondly "why stay?" The discovery of many systems in which the philopatric offspring do not delay their reproduction, or systems in which offspring that remain do not help (or help once they have dispersed), complicates the application of any single model of cooperative breeding, but also highlights the importance of addressing each of these questions independently.

The issue of dispersal has been subdivided even further. Thinking of dispersal in combination with mate searching, Brown (1987) expressed the problem as: Why in some species do individuals exhibit a strategy of stay-and-foray and in others disperse-and-search? Ekman et al. (Chapter 2) review data on species in which individuals exhibit one or both of these strategies. Zack (1990) further noted that in cooperative breeders, the entire pattern of dispersal is altered, not just delayed, with dispersal distances strongly skewed to short distances. This will vary with the species, and with sex within a species, but Zack (1990) correctly identified that the issue of dispersal can be viewed as two related questions of "why do individuals initially delay dispersal" and secondly "why do individuals disperse such a short distance once dispersal occurs?"

The importance of these different views, and of separating the general phenomenon of cooperative breeding into specific components, is two-fold. First, it highlights the important evolutionary questions that must be addressed in order to understand the origin and maintenance of cooperative breeding in individual taxa. Second, it identifies the hierarchical patterns of individual decisions that contribute to the diversity of cooperative breeding systems (Dickinson and Hatchwell, Chapter 3) and to the dynamics of familial structure in social species (Emlen 1991, 1995, 1997a).

(3) COOPERATIVE BREEDING IS NOT A MATING SYSTEM

Avian social mating systems are typically classified on the basis of two criteria – the number of mates that individuals of each sex have, and the existence or nature of a pair bond between mates (Ligon 1999). Most authors recognize four broad categories of social mating systems: monogamy (one, pair-bonded, social mate for both males and females), polygyny (males but not females have multiple mates, and males are pair-bonded with at least one female), polyandry (females but not males have multiple mates, and females are pair-bonded with at least one male), and promiscuity (males and/or females have multiple mates, with no pair bond); and two categories of reproductive mating systems: monogamy (one reproductive partner for both males and females) and promiscuity (males and/or females have multiple reproductive partners). There are many known variants of each of the above social systems, and there is also now a fifth distinctive social mating system recognized, that of polygynandry (multiple, pair-bonded mates for both males and females).

Because cooperative breeding is often defined by both life-history characteristics (timing of reproduction and dispersal) and an aspect of behavior (alloparental care), it cannot be said that cooperative breeding is a category of mating system or social organization. In fact, with the exception of lekking (as a variant of social promiscuity), all known social and reproductive mating systems have been identified in species categorized as cooperative breeders (Brown 1987; Ligon 1999; Cockburn, Chapter 5). Furthermore, across populations or across groups within a population, several social and reproductive mating systems can co-occur in one species. Examples include the dunnock (Davies 1985; Hartley and Davies 1994), pukeko (Jamieson 1997), and Tasmanian native hen (Goldizen et al. 1998).

In Brown's (1987) synthesis, cooperative breeding was referred to as a breeding system rather than a mating system, with which I would agree, and the 222 species known at that time to exhibit cooperative breeding were divided into 13 categories defined by the nature of territoriality (all-purpose, colonial, or neither), the number of breeders in a social group (singular versus plural), social mating system, and the identify of helpers (male versus female). The largest of Brown's categories was species that exhibited "unclassified and miscellaneous" breeding systems (76 species), an indication of both the diversity of behavior in cooperative breeding species and the amount of research remaining to be done on such species.

Cockburn (Chapter 5) takes a different approach, categorizing the "mating systems" of cooperatively breeding birds on the basis of both social and genetic relationships. In the 34 species for which Cockburn had detailed data on paternity and social behavior, he identified no less than 22 distinct systems, in nine broad categories! The main difference between the categorization of Brown (1987) and that of Cockburn, besides the inclusion of paternity data by the latter, was that Brown concentrated on the similarities between species whereas Cockburn focused on the differences. Cockburn's classification highlights the diversity of social and reproductive relationships among cooperatively breeding birds. Additionally, higher-level organizations of family structure can occur, such as clans of white-fronted bee-eaters (Hegner and Emlen 1987; Emlen 1990), brown treecreepers (Noske 1991; Cooper 2000; Doerr and Doerr 2001), and white-winged fairy-wrens (Rowley and Russell 1995). In my view, cooperative breeding is best viewed as a behavioral and ecological syndrome, an "unnatural collection of species" (Ligon 1999), that encompasses a diversity of both social and reproductive mating systems.

(4) THE IMPORTANCE OF "COOPERATION" VARIES ACROSS SPECIES, AND IS OBLIGATORY FOR VERY FEW

Various authors distinguish between obligate and opportunistic cooperatively breeding species (Dow 1980; Edwards and Naeem 1993; Chapter 1). By obligate cooperative breeders, these authors refer to species for which cooperative breeding is a normal and regular part of the species' social organization and reproductive behavior. In the vast majority of such species, however, pairs without helpers occur at varying frequencies in a population, and pairs are capable of successfully producing young without the aid of helpers.

There is an important distinction to be made between the species for which cooperation is truly obligatory and those for which it is not. This former group contains very few species, and only for one, the white-winged chough, has the obligate nature of assistance by auxiliary individuals been experimentally determined. In this species, a pair of birds (one male and one female) is incapable of successfully raising offspring to fledging and, in fact, the assistance of at least two helpers is required for successful reproduction (Rowley 1978; Heinsohn et al. 1988; Heinsohn 1991a, 1992; Boland et al. 1997a). This situation appears to arise because of a combination of diet (soil invertebrates) and habitat (areas with hard pan soil) that results in individuals – particularly young birds – having limited foraging success. Individual offspring do not attain foraging skills comparable to those of adults until they are four years of age (Heinsohn et al. 1988), and across groups, the rate of food delivery to nestlings does not level off until a group reaches a size of no less than 14 individuals (Heinsohn 1991a, 1992). Thus, the presence of helpers in white-winged choughs appears to be absolutely necessary for successful reproduction.

Other species for which assistance may also be obligatory are the yellow-billed shrike (Grimes 1980), brown jay (Williams et al. 1994), and apostlebird (A. Cockburn, personal communication), although experimental confirmation of these are needed. However unusual the white-winged chough, and possibly these other species, the evolution of cooperative breeding must have played a uniquely important role in the history of such species.

(5) DELAYED DISPERSAL AND HELPING BEHAVIOR ARE INDEPENDENT IN MANY SPECIES

The view of the connection between delayed dispersal and helping behavior has changed considerably over the last three decades. Because the species of cooperatively breeding birds that received extensive study early on were "typical" in the sense that the individuals providing alloparental care were generally philopatric offspring, the connection between delayed dispersal and helping appeared absolute. As more species received study, however, it became clear that there were species in which offspring delayed their dispersal but did not necessarily assist their parents (Gayou 1986; Veltman 1989; Birkhead 1991; Ekman et al. 1994), species in which offspring provided alloparental care after dispersing to other areas or territories (Curry and Grant 1990; Ekman et al. 1994; Dickinson et al. 1996), and species in which unrelated individuals provided alloparental care (Ligon and Ligon 1978a, 1983; Du Plessis 1993). Thus, delayed dispersal and helping are clearly independent in some species, and may be so in all cooperative breeders.

Delayed dispersal involves the evolution of behaviors in adults (increased tolerance, lack of aggression) and offspring (reduced aggression) as well as shifts in life-history traits in offspring (philopatry versus dispersal and postponement of reproduction). The evolution of helping behavior is independent of factors influencing delayed dispersal, and although the benefits of helping may be significant, they are neither necessary nor sufficient to account for delayed dispersal (Chapter 2). With respect to the relationship between helping behavior and delayed dispersal, the key question is whether in any species individuals delay their dispersal specifically for the benefits of providing alloparental care. Currently, the answer would appear to be no. Once individuals have delayed their dispersal, the loss of direct fitness benefits through individual reproduction may be mitigated by kin selection through individuals helping to raise close kin. Nevertheless, such benefits are a consequence of individual decisions made once their dispersal has been altered, not a cause of the decision to alter their dispersal.

(6) THERE ARE COSTS AND BENEFITS OF BOTH DELAYED DISPERSAL AND HELPING

In terms of current utility, delayed dispersal functions as an alternative to individuals confronted with constraints on dispersal (Emlen 1982a) or as the mechanism by which individuals obtain direct benefits (Stacey and Ligon 1991). The difference between these views of cooperative breeding has been extensively reviewed (Koenig and Pitelka 1981; Emlen 1982a; Koenig et al. 1992; Hatchwell and Komdeur 2000), but it is clear that delayed dispersal functions in a variety of ways in species in which it occurs.

Factors that have been shown to influence delayed dispersal include a shortage of nesting structures (Walters et al. 1992a), a shortage of mates (Pruett-Jones and Lewis 1990), limited dispersal options resulting from habitat saturation (Koenig et al. 1992), inadequate foraging skills by individuals (Brown 1985; Heinsohn et al. 1988), direct benefits, specifically inheriting part of the territory, that individuals may accrue by remaining on their natal territory (Woolfenden and Fitzpatrick 1984), increased chance of local dispersal (Ragsdale 1999), assistance in competing for reproductive vacancies (Hannon et al. 1985; Brown 1987), increased survivorship (Black and Owen 1987; Ekman et al. 2000; Green and Cockburn 2001; Kraijeveld and Dickinson 2001), protection from predators (Griesser 2003a, 2003b), an opportunity to improve skills (Komdeur 1996), inheriting dominance within the group (Wiley and Rabenold 1984), potential access to breeding opportunities (Wiley and Rabenold 1984; Stacey and Ligon 1987, 1991; Zack 1990; Komdeur 1996; Ragsdale 1999), increased access to food (Scott 1980; Barkan et al. 1986; Ekman et al. 1994; Pravosudova 1999), opportunities to attract mates (Kraaijevold and Dickinson 2001), and thermoregulatory advantages (Du Plessis, Chapter 7).

The above benefits of delayed dispersal can be either direct, such as a place to live if the habitat is saturated, or indirect, such as possible inheritance of part of the territory, and the benefits of remaining on the natal territory may be different than those resulting from the association with relatives (Chapter 2). If the benefits of extended association with relatives are significant, there may be competition among young for such opportunities (Black and Owen 1987).

The costs of delayed dispersal include competition for breeding vacancies (Koenig et al. 1995; Cockburn 1998), a reduced availability of unrelated mates (Walters et al. 1992b; Brown and Brown 1998; Ekman et al. 1999), and a risk of mortality while waiting for reproductive opportunities (Rabenold 1990; Russell and Rowley 1993b).

The costs and benefits of helping behavior have been reviewed extensively (Brown 1987; Koenig et al. 1992; Emlen 1982b, 1997a; Cockburn 1998) and will only be summarized here. These costs and benefits may vary with sex of the individual providing care (Hatchwell 1999; Cockburn 1998) and with both the physical environment (Reyer 1990; Magrath 2001) and the social environment (Reyer 1980, 1984). Helping may benefit the individual providing the care directly, or the parents directly, or both the helper and the parents simultaneously. Direct benefits to helpers include experience in selecting nest sites (Hatchwell et al. 1999) and tolerance of the philopatric individual by the parents (Mulder and Langmore 1993). Direct benefits to the parents can include energetic benefits or increased survivorship through load-lightening (Woolfenden 1975; Lewis 1982; Reyer and Westerterp 1985; Russell and Rowley 1988; Rabenold 1990; Crick 1992; Khan and Walters 2002; Heinsohn, Chapter 4). Benefits to both the helpers and the parents include reduced predation on the nest (Rabenold 1990), reduced risk of starvation of nestlings (Curry and Grant 1990), increased growth rates of nestlings (Ligon 1970; Woolfenden 1978; Reyer 1980; Dyer 1983), and increased production of young (Brown et al. 1982; Mumme 1992b; Komdeur 1996).

These benefits, however significant in some species, are not observed in all cases and there are many exceptions. For example, although the assistance of helpers increases the production of offspring in some species, there is no relationship between helpers and offspring production in others (Magrath and Yezerinac 1997; Magrath 2001; Hatchwell 1999).

Helping behavior also carries inherent costs. Although helpers are generally assisting in rearing close kin, and thus may receive inclusive fitness benefits, the individual that is providing alloparental care is (in most cases, and in some cases by definition) not reproducing on its own. Furthermore, although alloparental individuals generally direct their care toward close relatives (Curry 1988a; Emlen and Wrege 1988; Rabenold 1990; Komdeur 1994a; Dickinson et al. 1996), this is not always the case (Dunn et al. 1995). Helpers may occasionally receive reproductive benefits through helping comparable to those that they would obtain if they bred independently (Rabenold 1984; Bednarz 1988; Heinsohn 1991a), although in most cases they do not (Woolfenden and Fitzpatrick 1984; Koenig and Mumme 1987; Emlen and Wrege 1988, 1989; Reyer 1990; Stacey and Koenig 1990a; Dickinson et al. 1996; Dickinson and Akre 1998). More direct costs to helpers include lower survival or greater weight loss as a function of the amount of assistance delivered (Rabenold 1990; Heinsohn and Cockburn 1994). Heinsohn (Chapter 4) provides other examples of documented costs but also notes that the costs of helping behavior are not well studied.

(7) COOPERATION DOES NOT IMPLY LACK OF COMPETITION

Despite the cooperation exhibited by individuals in cooperative breeding species towards a common goal – the successful rearing of offspring – competition can and often does occur in virtually every possible circumstance involving same-sexed individuals. This competition can be subdivided into two general categories: that occurring between individuals within the same group, and that occurring between individuals in different groups.

Within groups, there can be competition among both females and males whenever more than one individual of the same sex occurs together. Furthermore, in joint-nesting species, in which more than one female lays eggs in the same nest, there may be competition among the breeding females. In such species, egg-tossing or egg destruction, in which one female destroys the eggs of another, is a common result, occurring in the guira cuckoo (Macedo and Bianchi 1997), groove-billed ani (Vehrencamp et al. 1986), acorn woodpecker (Mumme et al. 1983a; Koenig et al. 1995), and common moorhen (McRae 1996b). The competition among female moorhens can be sufficiently strong that unless the females lay synchronously, joint nesting is not successful (McRae 1996b). Similarly, in acorn woodpeckers females must lay synchronously in the same nest, otherwise egg destruction occurs (Mumme et al. 1983a; Koenig et al. 1983, 1995).

Among males there is competition within groups for dispersal opportunities as well as competition for reproductive opportunities. Whether all the males in a group attempt copulation with the resident female(s) often depends on relatedness (see below), but for those males that attempt copulation, competition will occur and can take several forms, including fighting and dominance interactions (Jamieson 1997; Nakamura 1998a, 1988b), mate-guarding of females (Joste et al. 1982; Mumme et al. 1983b; Hannon et al. 1985), attempts to stimulate the female to eject sperm from previous copulations (Davies 1983), forced copulations (Ewen and Armstrong 2000), and infanticide and nest destruction if males are unconvinced of paternity (Koenig 1990; Macedo et al. 2001). Differential parental care by males as a function of paternity (Burke et al. 1989; Li and Brown 2002) can also be viewed as a form of competition.

The competition among males within groups can potentially influence many aspects of group dynamics, including group composition and reproductive skew. There is, however, mixed evidence on each of these influences. In some species, dominant males can control group composition (Reyer 1990; Emlen and Wrege 1992) while in others they cannot (Davies 1992). There is currently no evidence that dominant males can control reproductive skew (Magrath et al., Chapter 10) and only equivocal evidence on whether interactions among males or female choice is the primary determinant of reproductive skew in cooperatively breeding species (Cant and Reeve 2002; Chapter 10).

Lastly, there can also be conflict between breeders and helpers over the level of alloparental care delivered. In superb fairy-wrens, breeders are apparently aggressive towards helpers if the latter do not remain consistently on the territory (Mulder and Langmore 1993). In the white-winged chough, helpers can deceive the breeders and directly consume the food that they appear to be delivering to nestlings (Boland et al. 1997b).

Between groups, there is a natural competition between individuals for territorial space and food resources. A circumstance characteristic of most cooperative species, the philopatry of offspring, leads to intense competition in some species between individuals of different groups for dispersal opportunities. For example, in acorn woodpeckers there are dramatic fights (power struggles) among females for dispersal and breeding opportunities, with coalitions of sisters fighting together being more successful at dispersing and filling vacancies than singletons (Koenig 1981; Hannon et al. 1985). In the New Zealand pukeko, unrelated males from different groups end up forming coalitions to defend and share quality territories in southern populations (J. Quinn and J. Jamieson, unpublished data) while related males within natal groups exhibit clear dominance hierarchies over food resources and territories in northern populations (Craig and Jamieson 1990; Jamieson et al. 1994; Jamieson 1997).

(8) COOPERATIVE BREEDERS SHARE MANY CHARACTERISTICS

Despite the diversity of cooperative breeding systems in birds, as a collection of species they often share similar ecological and life-history characteristics, over and beyond the defining traits of delayed dispersal and helping behavior. Similarities that have been recorded among cooperative breeders in at least certain geographical areas include year-round residency, high survivorship (Rowley 1968, 1976; Fry 1977; Brown 1987; Arnold and Owens 1998), possibly reduced clutch sizes (Brown 1987; Arnold and Owens 1998, 1999), prolonged dependence of offspring (Langen 2000), sensitivity to habitat quality and specialized patterns of habitat use (Walters et al., Chapter 12), and more common occurrence in less spatially occluded habitats (Cockburn 1996).

The similarities between many species of cooperatively breeding birds led to the expectation that there might be consistent ecological features or life-history characteristics that predicted the evolution of cooperative breeding (Rowley 1968; Ford et al. 1988; Arnold and Owens 1999). In general, this approach has highlighted the ecological correlates of cooperative breeding in specific geographical regions, but it has not provided a synthetic answer to the question of why some species exhibit cooperative breeding and others do not. For example, in Australia cooperative breeding is generally more common in aseasonal habitats (Rowley 1968; Ford et al. 1988) but in South Africa it is associated with seasonal habitats (Du Plessis et al. 1995). Brown's (1987) review suggested that cooperative breeders were more likely to be omnivorous, Emlen's (1982a) suggested that diet specialists were more likely to be cooperative breeders, and other studies have found no link between diet and cooperative breeding (Ford et al. 1988; Du Plessis et al. 1995; Arnold and Owens 1999; Langen 2000).

Neither the similarities in ecology (or in ecological constraints) nor life-history traits among cooperative breeders can predict which species do or do not exhibit cooperative breeding. This is particularly true when these relationships are viewed in the absence of phylogenetic information. Nevertheless, integrating the study of ecological interactions and life-history characteristics offers the best hope of understanding the selective pressures maintaining cooperative breeding in particular species (Hatchwell and Komdeur 2000).

(9) COOPERATIVE BREEDING IS ANCESTRAL IN MANY GROUPS

Cooperative breeding is certainly adaptive in many species in which it occurs. Nevertheless, this fact by itself is not evidence that cooperative breeding evolved as an adaptation in the specific lineages in which it plays such an important role. As noted by several authors in this volume, the identification of an important function of behaviors, such as cooperative breeding, is separate from an argument about the evolutionary origin of those behaviors (Reeve and Sherman 1993).

Cooperative breeding is non-randomly distributed across bird families (Arnold and Owens 1998). In many families, relatively few species and sometimes just one species exhibit cooperative breeding, whereas in some families most or all species breed cooperatively. Some of this variation is dependent on evolutionary history. Within the passerine parvorder Corvida, in numerous families all species exhibit cooperative breeding, whereas in the suborder Passeri there is no family in which all the species are cooperative breeders (Chapter 1). As suggested by this variation, recent comparative studies have shown that cooperative breeding is ancestral in many groups in which it occurs (Edwards and Naeem 1993; Ligon 1999), and Ligon and Burt (Chapter 1) suggest that cooperative breeding may be the basal condition in all species of Passeriformes. Additionally, cooperative breeding occurs in some non-passerine groups, and these are among the most ancient of all living neoavian birds (Chapter 1).

Throughout this volume, genera and families of birds are discussed that provide evidence of the variability in the appearance or loss of cooperative breeding. Examples include: (1) species complexes in which cooperative breeding is ancestral for the group but has been lost in particular species, such as scrub-jays of the genus *Aphelocoma* (Peterson and Burt 1992; Burt and Peterson 1993; Chapter 1); (2) species in which cooperative breeding is clearly derived and evolved in that specific lineage, such as the red-cockaded woodpecker (Walters 1990; Conner *et al.* 2001); (3) genera in which cooperative breeding is ancestral for the genus and has been maintained in all extant species, such as *Malurus* (Ligon and Burt, Chapter 1); (4) genera in which cooperative breeding is ancestral for the genus but has been lost in particular species, such as *Campylorhynchus* wrens (Farley 1995); (5) subfamilies in which cooperative breeding is ancestral and has subsequently been lost or maintained in genera and species in complex patterns, such as the Corvinae; and (6) Families in which cooperative breeding is ancestral for the family and has been either maintained or lost in particular genera, such as woodhoopoes (family Phoeniculidae), and bee-eaters (family Meropidae) (Ligon and Burt, Chapter 1).

(10) SEXUAL SELECTION CAN BE STRONG IN COOPERATIVE BREEDING SYSTEMS

Sexual selection occurs when there is variance in mating success among individuals and such variance is due to heritable variation (Andersson 1994). Sexual selection is most common among males, but can also occur between females. In cooperatively breeding species, there is the potential for sexual selection regardless of whether the reproductive mating system is strictly monogamous, as in the bell miner (Conrad *et al.* 1998; Painter *et al.* 2000), or highly promiscuous, as in fairy-wrens (Brooker *et al.* 1990; Mulder *et al.* 1994). The egalitarian nature of social behaviors, as well as the generally low levels of sexual dimorphism, in many cooperatively breeding species suggests that sexual selection may be limited in some taxa, but it certainly can occur in others.

With the advent of techniques allowing for identification of sires, our understanding of cooperative breeding systems, and all patterns of social organization in birds, has greatly expanded. As a result, it is now possible to quantify the reproductive skew among individuals and, indirectly, the importance of sexual selection. Not surprisingly, the results of such studies have been that extra-pair, but within-group, matings as well as extra-group matings can be common, both of which can contribute to variance in male mating success (Webster *et al.* 1995).

Fairy-wrens (genus *Malurus*) represent the most dramatic known example of extreme reproductive promiscuity coupled with social monogamy, both for cooperative breeding species in particular and birds generally. All *Malurus* are known to breed cooperatively (Rowley and Russell 1997), and for those species that have been studied, reproductive promiscuity through extra-group matings is the rule. In both superb fairy-wrens and splendid fairy-wrens as many as 75% of offspring are sired by extra-group males (Brooker *et al.* 1990; Mulder *et al.* 1994; Webster *et al.* 2003). In superb fairy-wrens, most females seek extra-pair matings (Dunn and Cockburn 1999), and there is strong unanimity in female choice, resulting in high variance in mating success among males. The interactions among males, in conjunction with female choice, affect the resultant distribution of matings. In superb fairy-wrens, males exert more effort in seeking extra-group matings if there are helpers in their group (Green *et al.* 1995). Also, females actively choose among males based on plumage status (Dunn and Cockburn 1999; Green *et al.* 2000). Subordinate male helpers contribute little to within-group paternity if those males are sons of the breeding female (Mulder *et al.* 1994). If, however, the helper males are unrelated to the breeding female, their share of paternity increases to more than 20% (Cockburn *et al.* 2003).

Reproductive promiscuity in fairy-wrens is associated with modification of their reproductive anatomy (Mulder and Cockburn 1993; Tuttle *et al.* 1996), high rates of sperm production (Tuttle and Pruett-Jones 2003) and extreme sperm counts (Tuttle *et al.* 1996). Sexual selection through reproductive promiscuity in fairy-wrens may also be responsible for patterns of plumage evolution in this group (A. C. Driskell and S. Pruett-Jones, unpublished data).

Extra-pair or extra-group matings in cooperatively breeding species have now been recorded in many species, including Mexican jays (Li and Brown 2000, 2002), tree swallows (Kempenaers *et al.* 1999), Seychelles warblers (Richardson *et al.* 2001), stitchbirds (Ewen and Armstrong 2000), and more. Although the reasons why females should mate outside the social group, or mate with multiple males, remain generally unresolved issues (Jennions and Petrie 2000; Tregenza and Wedell 2000), it is clear that female promiscuity is more correctly thought of as a rule rather than an exception in cooperatively breeding species, and apparently in birds in general.

(11) THERE CAN BE SIGNIFICANT INTERPOPULATION DIFFERENCES IN COOPERATIVELY BREEDING SPECIES

Patterns of social organization and mating systems in birds, including those that exhibit cooperative breeding, can vary significantly across populations (Ligon 1999). This variation is particularly important as it relates to testing hypotheses concerning the evolution of cooperative breeding. For the sake of brevity, I will use just two examples to illustrate this point, but there are other well-documented cases. The case of the acorn woodpecker illustrates how varying habitat in different areas influences the dynamics of cooperative breeding in each population. Superb fairy-wrens illustrate how different research teams studying similar populations of the same species in different localities can reach strikingly different conclusions.

The acorn woodpecker is distributed in oak woodlands in western North and Central America. Koenig and Stacey (1990) detail a comparison of the biology of this species at three sites: Hastings Reservation in California, Water Canyon in New Mexico, and the Research Ranch in Arizona. These sites vary in terms of the number of oak species present and in the variability of acorn production. At Hastings, there are five common species of oaks, with relatively low overall annual variability in acorn production, while at the Research Ranch there are only two species of oaks, with considerable annual variability in acorn production. Water Canyon also has just two species of oaks, but acorn production is less annually variable than at the Research Ranch.

Koenig and Stacey (1990) document that this variation in the availability and predictability of acorns influences virtually every aspect of the population biology and social behavior of acorns woodpeckers at the three sites. For example, at the Research Ranch, cooperative breeding was uncommon, with over 85% of groups breeding as simple pairs, whereas at Hastings, only 23% of groups bred as pairs. The population at Hastings exhibited the highest values for annual survivorship of birds, group size, group stability, and reproductive output, and the lowest values for annual turnover and the percentage of individuals dispersing each year.

In all cases, values for the Research Ranch contrasted the most with Hastings and those for Water Canyon were intermediate.

Koenig and Stacey (1990) use these comparisons to argue for the critical role that food resources and habitat variability play in influencing cooperative breeding in this species. They also argued that these differences illustrate how both ecological constraints and the intrinsic benefits of philopatry can influence delayed dispersal: at Hastings, ecological constraints appeared to be the most important factor leading to delayed dispersal, whereas at Water Canyon, and presumably the Research Ranch, the benefits of philopatry appeared to be more important.

The second example, the superb fairy-wren, regularly exhibits cooperative breeding throughout its range in eastern and southeastern Australia (Rowley and Russell 1997). During the austral breeding season of 1989, Ligon et al. (1991) studied superb fairy-wrens near Armidale, New South Wales, as did Pruett-Jones and Lewis (1990) in Canberra, 580 km away. Both research teams conducted field experiments to evaluate hypotheses for the evolution of delayed dispersal. At both localities the populations were relatively similar in terms of group size and composition. Nevertheless, the conclusions of the two studies were quite different.

In Pruett-Jones and Lewis's (1990) study, all but one auxiliary male (31 of 32) dispersed when provided with a dispersal and breeding opportunity, and the authors concluded that habitat saturation and a shortage of potential female mates explained delayed dispersal in the population they studied. In contrast, Ligon et al. (1991) found that the habitat was not saturated, there were available but unoccupied territories, and potential female mates were not limiting to males. There are several possible interpretations of the differences between these two studies. The two populations of superb fairy-wrens may, indeed, differ in terms of life-history and behavioral responses to habitat availability. Alternatively, some differences in the results may be due to the methodology employed by the two research teams. For example, Pruett-Jones and Lewis (1990) determined that females were limiting in October and November, after which time all dispersing females will normally have either found a breeding opportunity or died. Ligon et al. (1991) determined that there was an excess of females earlier in the season, in September, during which time females are normally dispersing. Ligon (1999) suggested that the differences in the conclusions of these two studies indicates that ecological constraints cannot be a widespread factor in the evolution or maintenance of cooperative breeding in the superb fairy-wren and that other factors such as phylogeny may be more important.

(12) INBREEDING IN COOPERATIVELY BREEDING SPECIES IS NO MORE INTENSE THAN IN OTHER SYSTEMS

The possibility of inbreeding seems particularly high in cooperative breeding systems, where groups often contain offspring from previous generations and dispersal distances tend to be short. Nevertheless, with a few exceptions, genetic analysis of paternity has revealed that incestuous matings are rare (Koenig and Haydock, Chapter 9). This includes cases in which previous claims of high levels of inbreeding are now known to be inaccurate, such as the superb fairy-wren (Rowley et al. 1986) and Mexican jay (Brown 1974, 1978; Li and Brown 2000). The exceptions in birds, in which inbreeding has been suggested or documented, include: (1) genetically unconfirmed observations of inbreeding in the green woodhoopoe (Du Plessis 1992); (2) relatively frequent inbreeding in the Seychelles warbler that may relate to a recent population bottleneck (Richardson et al. 2001); and (3) inbreeding associated with reduced hatchability of eggs and post-fledging survival of juveniles (inbreeding depression) in common moorhens (McRae 1996b), Mexican jays (Brown and Brown 1998) and red-cockaded woodpeckers (Daniels and Walters 2000a).

Various mechanisms act to reduce inbreeding in cooperative breeding systems (Chapter 9). Some mechanisms, such as extra-pair paternity and patterns of dispersal, may act to reduce incestuous matings, but it is not clear in any case whether these patterns evolved to have this effect. Other mechanisms, such as individuals not breeding in their group if a parent (or offspring) of the opposite sex is still present, and high rates of divorce between years (Koenig and Pitelka 1979; Koenig et al. 1998; Woolfenden and Fitzpatrick 1984; Stevens and Wiley 1995; Daniels and Walters 2000a; Hatchwell et al. 2000), do appear to have evolved specifically to reduce incestuous matings. The specific mechanisms by which individuals recognize their kin of the opposite sex are unknown.

(13) MOST COOPERATIVE BREEDERS STILL RESIDE IN AUSTRALIA

Despite the continuing discovery of cooperative breeding in birds around the globe, it remains a fact that on an absolute and relative basis, Australia supports the most known cooperatively breeding species. This prevalence of cooperative breeding in the Australian avifauna has been recognized for a long time (Rowley 1976; Dow 1980; Brown 1987), even if the reasons for it have not. An analysis of environmental factors does not explain the distribution of cooperatively breeding species in Australia (Dow 1980). There are several life-history correlates with cooperative breeding in Australia, such as non-migratory lifestyles, high survival, and low clutch sizes (Ford et al. 1988; Yom-Tov 1987), but by themselves these factors are insufficient to explain the occurrence of such systems there.

A recent comparative analysis (Cockburn 1996) suggests that the reason for this phenomenon may not be that complicated. The majority of Australian passerines belong to the parvorder Corvida, and this group evolved in Australia and radiated from there to other regions of the world (Sibley and Ahlquist 1985). Species within Corvida are particularly likely to exhibit cooperative breeding (Russell 1989; Edwards and Naeem 1993; Clarke 1995), with from 21% to 40% of Corvida in this category worldwide, excluding families in which males are emancipated from parental care such as birds of paradise (Cockburn 1996). Among clades of Corvida, the proportions of species that breed cooperatively are comparable both within and outside of Australia (Cockburn 1996).

The reason why species in Corvida are so likely to exhibit cooperative breeding remains an open question, and additional comparative analyses are clearly warranted. Nevertheless, the answer to the riddle of why there are so many cooperatively breeding species in Australia may be mostly historical, reflecting the phylogenetic fact that the Corvida lineages first evolved there and that cooperative breeding is a common feature of these species.

OTHER STATEMENTS

My choice of topics to focus on in this review has been motivated both by the existing literature and by personal interests. Additional generalizations are possible. Several examples are: (1) the apparently divergent hypotheses for the evolution of cooperative breeding can be viewed as complementary (Koenig et al. 1992; Hatchwell and Komdeur 2000); (2) the role of helpers in reproduction varies significantly, both within and among species (Smith 1990); (3) the factors responsible for the maintenance of cooperative breeding may be different than those responsible for the evolution of cooperative breeding (Chapter 1); (4) in most species, helpers eventually attain breeding status (Cockburn 1998); (5) the behavioral dynamics of cooperative breeders are dependent on the physiological mechanisms underlying specific behaviors (Chapters 7 and 8); and (6) the dynamics of social behaviors in cooperative breeding systems reflect a compromise between the reproductive interests of adults and helpers (Chapter 3). The reviews in this volume summarize many of the data that justify these statements.

SUMMARY

The diversity of avian cooperative breeding systems is the result of the interaction between evolutionary history, life-history traits, ecological relationships, and the costs and benefits of particular strategies in individual taxa. This interaction is dynamic and only predictable in hindsight once we know details of each of the component factors. The chapters in this volume are both excellent summaries of and introductions to the wealth of research on cooperative breeding systems, about which we know more than perhaps any other avian breeding system (Emlen 1997a). This latter view might suggest that the important questions have been answered, but as the chapters here illustrate, this is far from the case. The nature of the questions that are being addressed in cooperative systems may have changed, but fundamental issues remain unanswered. In particular, we are just beginning to understand the relative importance of phylogeny versus ecological factors in explaining the occurrence of cooperative breeding, the physiological basis of social behaviors of cooperative breeders, the causes of reproductive skew, the precise factors influencing dispersal and helping behavior in females, and the nature of female choice in cooperative breeders.

Cooperative breeding is a diverse syndrome, with significant intra- and interspecific variation in virtually every aspect of social behavior and demography. The recognition of this diversity has led to the realization that no single evolutionary model can explain either the evolution or the maintenance of cooperative breeding in all species. Phylogenetic relationships appear to be more important than ecological interactions as an explanation of cooperative breeding in some taxa, whereas in others the reverse is the case. While the earlier expectation that cooperative breeding would have a single, unitary evolutionary explanation has not been realized, the results of current research are exciting because they illustrate how dynamic the evolutionary process is and how complicated cooperative breeding really is.

ACKNOWLEDGMENTS

I wish to thank Walt Koenig and Janis Dickinson for inviting me to write this summary statement, and for their tolerance with my delays. Walt Koenig, Keith Tarvin, and Mike Webster provided important comments on an earlier draft of the manuscript.

Names of bird and mammal species mentioned in the text

BIRDS

Common name	Scientific name	Common name	Scientific name
Acorn woodpecker	*Melanerpes formicivorus*	Dunnock	*Prunella modularis*
Alpine accentor	*Prunella collaris*	Eclectus parrot	*Eclectus roratus*
American coot	*Fulica americana*	Emu	*Dromaius novaehollandiae*
American crow	*Corvus brachyrhynchos*	European bee-eater	*Merops apiaster*
Apostlebird	*Struthidea cinerea*	Florida scrub-jay	*Aphelocoma coerulescens*
Arabian babbler	*Turdoides squamiceps*	Forest woodhoopoe	*Phoeniculus castaneiceps*
Australian magpie	*Gymnorhina tibicen*	Galápagos hawk	*Buteo galapagoensis*
Beechey jay	*Cyanocorax beecheii*	Galápagos mockingbird	*Nesomimus parvulus*
Bell miner	*Manorina melanophrys*	Goldeneye	*Bucephala clangula*
Bicolored wren	*Campylorphynchus griseus*	Gray jay	*Perisoreus canadensis*
Black-billed woodhoopoe	*Phoeniculus somaliensis*	Great tit	*Parus major*
Black-eared miner	*Manorina melanotis*	Greater rhea	*Rhea americana*
Blue-footed booby	*Sula nebouxii*	Greater roadrunner	*Geococcyx californianus*
Blue tit	*Parus caeruleus*	Green jay	*Cyanocorax yncas*
Brown-headed nuthatch	*Sitta pusilla*	Green woodhoopoe	*Phoeniculus purpureus*
Brown jay	*Psilorhinus morio*	Groove-billed ani	*Crotophaga sulcirostris*
Brown skua	*Catharacta lonnbergi*	Guira cuckoo	*Guira guira*
Brown thornbill	*Acanthiza pusilla*	Hairy woodpecker	*Picoides villosus*
Brown treecreeper	*Climacteris picumnus*	Harris's hawk	*Parabuteo unicinctus*
Brushland tinamou	*Nothoprocta cinerascens*	Hoatzin	*Opisthocomus hoazin*
Bushtit	*Psaltriparus minimus*	Hooded robin	*Petroica cucullata*
Button quail	*Coturnix chinensis*	Hooded warbler	*Wilsonia citrina*
Cactus wren	*Campylorhynchus brunneicapillus*	Hoopoe	*Upupa epops*
		Island scrub-jay	*Aphelocoma insularis*
Carrion crow	*Corvus corone*	Jungle babbler	*Turdoides striatus*
Cassowary	*Casuarius casuarius*	Kiwi	*Apteryx mantelli*
Common babbler	*Turdoides caudatus*	Laughing kookaburra	*Dacelo novaeguineae*
Common moorhen	*Gallinula chloropus*	Lewis' woodpecker	*Melanerpes lewis*
Cooper's hawk	*Accipiter cooperii*	Long-tailed tit	*Aegithalos caudatus*
Crescent honeyeater	*Phylidonyris pyrrhoptera*	Magpie goose	*Anseranas semipalmata*
Downy woodpecker	*Picoides pubescens*	Maui alauahio	*Paroreomyza montana*

Ecology and Evolution of Cooperative Breeding in Birds, ed. W. D. Koenig and J. L. Dickinson. Published by Cambridge University Press.
© Cambridge University Press 2004.

(cont.)

Common name	Scientific name	Common name	Scientific name
Mexican jay	*Aphelocoma ultramarina*	Speckled mousebird	*Colius striatus*
Moustached warbler	*Acrocephalus melanopogon*	Splendid fairy-wren	*Malurus splendens*
Noisy miner	*Manorina melanocephala*	Stitchbird	*Notiomystis cincta*
Ornate tinamou	*Nothoprocta ornata*	Stripe-backed wren	*Campylorynchus nuchalis*
Ostrich	*Struthio camelus*	Superb fairy-wren	*Malurus cyaneus*
Ovenbird	*Seiurus aurocapillus*	Taiwan yuhina	*Yuhina brunneiceps*
Pale chanting goshawk	*Melierax canorus*	Tasmanian native hen	*Gallinula mortierii*
Pale-winged trumpeter	*Psophia leucoptera*	Treecreeper	*Certhia familiaris*
Penduline tit	*Remiz pendulinus*	Tree swallow	*Tachycineta bicolor*
Pied kingfisher	*Ceryle rudis*	Tufted titmouse	*Parus bicolor*
Pinyon jay	*Gymnorhinus cyanocephalus*	Unicolored jay	*Aphelocoma unicolor*
		Western bluebird	*Sialia mexicana*
Pukeko (purple swamphen)	*Porphyrio porphyrio*	Western scrub-jay	*Aphelocoma californica*
Pygmy nuthatch	*Sitta pygmaea*	White-backed mousebird	*Colius colius*
Red grouse	*Lagopus scoticus*	White-browed scrubwren	*Sericornis frontalis*
Red-backed mousebird	*Colius castanotus*	White-browed sparrow-weaver	*Plocepasser mahali*
Red-browed buffalo-weaver	*Bubalornis niger*		
		White-fronted bee-eater	*Merops bullockoides*
Red-cockaded woodpecker	*Picoides borealis*	White-headed woodhoopoe	*Phoeniculus bollei*
Red-throated bee-eater	*Merops bullocki*	White-throated magpie-jay	*Calocitta formosa*
Red-winged blackbird	*Agelaius phoeniceus*	White-throated sparrow	*Zonotrichia albicollis*
Rifleman	*Acanthisitta chloris*	White-winged chough	*Corcorax melanorhamphos*
Seychelles warbler	*Acrocephalus sechellensis*	White-winged fairy-wren	*Malurus leucopterus*
Siberian jay	*Perisoreus infaustus*	Wren	*Troglodytes troglodytes*
Slaty-breasted tinamou	*Crypturellus boucardii*	Yellow-billed cuckoo	*Coccyzus americanus*
Smith's longspur	*Calcarius pictus*	Yellow-billed shrike	*Corvinella corvina*
Smooth-billed ani	*Crotophaga ani*	Yellow-faced honeyeater	*Lichenostomus chrysops*
Sociable weaver	*Philetairus socius*	Zebra finch	*Taeniopygia guttata*

MAMMALS

Common name	Scientific name	Common name	Scientific name
African elephant	*Loxodonta africana*	Golden jackal	*Canis aureus*
African lion	*Panthera leo*	Golden lion tamarin	*Leontopithecus rosalia*
African wild dog	*Lycaon pictus*	Gray wolf	*Canis lupus*
Alpine marmot	*Marmota marmota*	Killer whale	*Orcinus orca*
Banded mongoose	*Mungos mungo*	Mashona mole-rat	*Cryptomys darlingi*
Belding's ground squirrel	*Spermophilus beldingi*	Meerkat	*Suricata suricatta*
Black-backed jackal	*Canis mesomelas*	Mongolian gerbil	*Meriones unguiculatus*
Black-tailed prairie dog	*Cynomys ludovicianus*	Muskox	*Ovibos moschatus*
Brown hyena	*Hyaena brunnea*	Naked mole-rat	*Heterocephalus glaber*
Bush baby	*Galago crassicaudatus*	Pilot whale	*Globicephala melas*
Coati	*Nasua narica*	Pine vole	*Microtus pinetorum*
Common marmoset	*Callithrix jacchus*	Prairie vole	*Microtus ochrogaster*
Common mole-rat	*Cryptomys anselli*	Red deer (elk)	*Cervus elaphus*
Cotton-top tamarin	*Saguinus oedipus*	Saddle-back tamarin	*Saguinus fuscicollis*
Damaraland mole-rat	*Cryptomys damarensis*	Spotted hyena	*Crocuta crocuta*
Dingo	*Canis familiaris*	Wied's black tufted-ear marmoset	*Callithrix kuhli*
Dwarf mongoose	*Helogale parvula*		
Ethiopian wolf	*Canis simensis*	Zambian mole-rat	*Cryptomys mechowi*

References

Abbott, D. H. (1984). Behavioral and physiological suppression of fertility in subordinate marmoset monkeys. *Am. J. Primatol.*, **6**, 169–186.

Alexander, R. D. (1974). The evolution of social behavior. *Annu. Rev. Ecol. Syst.*, **5**, 325–383.

Allee, W. C. (1931). *Animal Aggregations: a Study in General Sociology*. Chicago, IL: University of Chicago Press.

Alonso, J. A. and Alonso, J. C. (1993). Age-related differences in time budgets and parental care in wintering common cranes. *Auk*, **110**, 78–88.

Amos, B., Schlotterer, C. and Tautz, D. (1993). Social structure of pilot whales revealed by analytical DNA profiling. *Science*, **260**, 670–672.

Anava, A. (1998). Seasonal water and energy fluxes in Arabian babblers (*Turdoides squamiceps*), a passerine that inhabits extreme deserts. Ph.D. dissertation, Ben-Gurion University of the Negev, Israel.

Anava, A., Kam, M., Shkolnik, A. and Degen, A. A. (2000). Seasonal field metabolic rate and dietary intake in Arabian babblers (*Turdoides squamiceps*) inhabiting extreme deserts. *Funct. Ecol.*, **14**, 607–613.

(2001a). Growth rate and energetics of Arabian babbler (*Turdoides squamiceps*) nestlings. *Auk*, **118**, 519–524.

(2001b). Effect of group size on field metaolic rate of Arabian babblers provisioning nestlings. *Condor*, **103**, 376–380.

(2001c). Does group size affect field metabolic rate of Arabian babbler *(Turdoides squamiceps)* nestlings? *Auk*, **118**, 525–528.

Anderson, A. H. and Anderson, A. (1972). *The Cactus Wren*. Tucson, AZ: University of Arizona Press.

Andersson, M. (1984). Brood parasitism within species. In: *Producers and Scroungers*, ed. C. J. Barnard. London: Croom-Helm. pp. 195–228.

(1994) *Sexual Selection*. Princeton, NJ: Princeton University Press.

(2001). Relatedness and the evolution of conspecific brood parasitism. *Am. Nat.*, **158**, 599–614.

Andersson, M. and Åhlund, M. (2000). Host-parasite relatedness shown by protein fingerprinting in a brood parasitic bird. *Proc. Natl. Acad. Sci. USA*, **97**, 13188–13193.

Ankney, C. D. (1982). Sex ratio varies with egg sequence in lesser snow geese. *Auk*, **99**, 662–666.

Arcese, P. (1989). Intrasexual competition, mating system, and natal dispersal in song sparrows. *Anim. Behav.*, **38**, 958–979.

Armitage, K. B. (1981). Sociality as a life-history tactic of ground-squirrels. *Oecologia*, **48**, 36–49.

(1999). Evolution of sociality in marmots. *J. Mammal.*, **80**, 1–10.

Armstrong, E. A. and Whitehouse, H. L. K. (1977). Behavioral adaptations of the wren (*Troglodytes troglodytes*). *Biol. Rev.*, **52**, 235–294.

Arnold, K. E. and Owens, I. P. F. (1998). Cooperative breeding in birds: a comparative test of the life history hypothesis. *Proc. R. Soc. London Ser. B*, **265**, 739–745.

(1999). Cooperative breeding in birds: the role of ecology. *Behav. Ecol.*, **10**, 465–471.

Arnold, K. E., Griffith, S. C. and Goldizen, A. W. (2001). Sex-biased hatching sequences in the cooperatively breeding noisy miner. *J. Avian Biol.*, **32**, 219–223.

Arnold, W. (1990a). The evolution of marmot sociality. 1. Why disperse late? *Behav. Ecol. Sociobiol.*, **27**, 229–237.

(1990b). The evolution of marmot sociality. 2. Costs and benefits of joint hibernation. *Behav. Ecol. Sociobiol.*, **27**, 239–246.

Asa, C. S. (1997). Hormonal and experiential factors in the expression of social and parental behavior in canids. In: *Cooperative Breeding in Mammals*, ed. N. G. Solomon and J. A. French. Cambridge: Cambridge University Press. pp. 129–149.

Baglione, V., Canestrari, D., Marcos, J. M., Griesser, M. and Ekman, J. (2002a). History, environment and social behaviour: experimentally induced cooperative breeding in the carrion crow. *Proc. R. Soc. London Ser. B*, **269**, 1247–1251.

Baglione, V., Marcos, J. M. and Canestrari, D. (2002b). Cooperatively breeding groups of the carrion crow *Corvus corone corone* in northern Spain. *Auk*, **119**, 790–799.

Bailey, R. E. (1952). The incubation patch of passerine birds. *Condor*, **54**, 121–136.

Baker, A. J. and Woods, F. (1992). Reproduction of the emperor tamarin (*Saguinus imperator*) in captivity, with comparisons to cotton-top and golden lion tamarins. *Am. J. Primatol.*, **26**, 1–10.

Baker, A. J., Dietz, J. M. and Kleiman, D. G. (1993). Behavioural evidence for monopolization of paternity in multi-male groups of golden lion tamarins. *Anim. Behav.*, **46**, 1091–1103.

Balcombe, J. P. (1989). Non-breeder asymmetry in Florida scrub jays. *Evol. Ecol.*, **3**, 77–79.

Balda, R. P. and Bateman, G. C. (1971). Flocking and the annual cycle of the pinyon jay. *Condor*, **73**, 287–302.

Bales, K., O'Herron, M., Baker, A. J. and Dietz, J. M. (2001). Sources of variability in numbers of live births in wild golden lion tamarins (*Leontopithecus rosalia*). *Am. J. Primatol.*, **54**, 211–221.

Ball, G. F. (1983). Functional incubation in male barn swallows. *Auk*, **100**, 998–1000.

(1991). Endocrine mechanisms and the evolution of avian parental care. *Proc. Int. Ornithol. Congr.*, **20**, 984–991.

(1993). The neural integration of environmental information by seasonally breeding birds. *Am. Zool.*, **33**, 185–199.

Balthazart, J. (1983). Hormonal correlates of behavior. In: *Avian Biology*, vol. 7, ed. D. S. Farner and K. C. Parkes. New York, NY: Academic Press. pp. 221–365.

Barkan, C. P. L., Craig, J. L., Strahl, S. D., Stewart, A. M. and Brown, J. L. (1986). Social dominance in communal Mexican jays *Aphelocoma ultramarina*. *Anim. Behav.*, **34**, 175–187.

Barker, F. K., Barrowclough, G. F. and Groth, J. G. (2002). A phylogenetic hypothesis for passerine birds: taxonomic and biogeographic implications of an analysis of nuclear DNA sequence data. *Proc. R. Soc. London Ser. B*, **269**, 295–308.

Barrett, G. W., Ford, H. A. and Recher, H. F. (1994). Conservation of woodland birds in a fragmented rural landscape. *Pacific Cons. Biol.*, **1**, 245–256.

Barrett, S. C. H. and Charlesworth, B. (1991). Effects of a change in the level of inbreeding on the genetic load. *Nature*, **352**, 522–524.

Bartholomew, G. A. and Trost, C. H. (1970). Temperature regulation in the speckled mousebird *(Colius striatus)*. *Condor*, **72**, 141–146.

Bateson, P. P. G. (1982). Preferences for cousins in Japanese quail. *Nature*, **295**, 236–237.

Beauchamp, G. (1999). The evolution of communal roosting in birds: origin and secondary losses. *Behav. Ecol.*, **10**, 675–687.

Bednarz, J. C. (1988). Cooperative hunting in Harris's hawks (*Parabuteo unicinctus*). *Science*, **239**, 1525–1527.

(1995). Harris's hawk (*Parabuteo unicinctus*). In: *The Birds of North America*, ed. A. Poole and F. Gill. Philadelphia, PA and Washington, DC: Academy of Natural Sciences and American Ornithologists' Union.

Bednarz, J. C. and Hayden, T. J. (1991). Skewed brood sex ratio and sex-biased hatching sequence in Harris's hawks. *Am. Nat.*, **137**, 116–132.

Bednekoff, P. (1997). Mutualism among safe, selfish sentinels: a dynamic game. *Am. Nat.*, **150**, 373–392.

Beletsky, L. D. and Orians, G. H. (1987). Territoriality among red-winged blackbirds. II. Removal experiments and site dominance. *Behav. Ecol. Sociobiol.*, **20**, 339–349.

(1989). Familiar neighbors enhance breeding success in birds. *Proc. Natl. Acad. Sci. USA*, **86**, 7933–6.

(1996). *Red-Winged Blackbirds: Decision-Making and Reproductive Success*. Chicago: University of Chicago Press.

Beletsky, L. D., Gori, D. F., Freeman, S. and Wingfield, J. C. (1995). Testosterone and polygyny in birds. *Curr. Ornithol.*, **12**, 1–41.

Bengtsson, B. O. (1978). Avoiding inbreeding: at what cost? *J. Theor. Biol.*, **73**, 439–444.

Bennett, N. C. and Faulkes, C. G. (2000). *African Mole-Rats: Ecology and Eusociality*. Cambridge: Cambridge University Press.

Bennett, N. C., Jarvis, J. U. M., Faulkes, C. G. and Millar, R. P. (1993). LH responses to single doses of exogenous GnRH by freshly captured Damaraland mole-rats, *Cryptomys damarensis*. *J. Repro. Fert.*, **99**, 81–86.

Bennett, N. C., Jarvis, J. U. M., Millar, R. P., Sasano, H. and Ntshinga, K. V. (1994). Reproductive suppression in eusocial *Cryptomys damarensis* colonies: socially-induced infertility in females. *J. Zool.*, **233**, 617–630.

Bennett, N. C., Faulkes, C. G. and Molteno, A. J. (1996). Reproductive suppression in subordinate, non-breeding female Damaraland mole-rats: two components to a lifetime of socially-induced infertility. *Proc. R. Soc. London Ser. B*, **263**, 1599–1603.

Bennett, N. C., Faulkes, C. G. and Spinks, A. C. (1997). LH responses to single doses of exogenous GnRH by social Mashona mole-rats: a continuum of socially induced infertility in the family Bathyergidae. *Proc. R. Soc. London Ser. B*, **264**, 1001–1006.

Bennett, N. C., Molteno, A. J. and Spinks, A. C. (2000). Pituitary sensitivity to exogenous GnRH in giant Zambian mole-rats, *Cryptomys mechowi* (Rodentia: Bathyergidae): support for the 'socially induced infertility continuum'. *J. Zool.*, **252**, 447–452.

Bennett, P. M. and Owens, I. P. F. (2002). *Evolutionary Ecology of Birds*. Oxford: Oxford University Press.

Bensch, S. (1999). Sex allocation in relation to parental quality. *Proc. Int. Ornithol. Congr.*, **22**, 451–466.

Bensch, S., Hasselquist, D., Nielsen, B. and Hansson, B. (1998). Higher fitness for philopatric than for immigrant males in a semi-isolated population of great reed warblers. *Evolution*, **528**, 877–883.

Bertram, B. C. R. (1978). Living in groups: predators and prey. In: *Behavioural Ecology: an Evolutionary Approach*, ed. J. R. Krebs and N. B. Davies. Oxford: Blackwell. pp. 64–96.

(1992). *The Ostrich Communal Nesting System*. Princeton, NJ: Princeton University Press.

Birkhead, T. R. (1991). *The Magpies: the Ecology and Behaviour of Black-billed and Yellow-billed Magpies*. London: Poyser.

Birkhead, T. R., Hatchwell, B. J. and Davies, N. B. (1991). Sperm competition and the reproductive organs of the male and female dunnock *Prunella modularis*. *Ibis*, **133**, 306–311.

Bishop, R. P. and Groves, A. L. (1991). The social structure of Arabian babbler, *Turdoides squamiceps*, roosts. *Anim. Behav.*, **42**, 323–325.

Black, J. M. and Owen, M. (1987). Determinant factors of social rank in goose flocks: acquisition of social rank in young geese. *Behaviour*, **102**, 129–146.

(1989). Parent–offspring relationships in wintering barnacle geese. *Anim. Behav.*, **37**, 187–198.

Blackwell, P. and Bacon, P. J. (1993). A critique of the territory inheritance hypothesis. *Anim. Behav.*, **46**, 821–823.

Blanchard, L. (2000). An investigation of the communal breeding system of the smooth-billed ani (*Crotophaga ani*). M.Sc. dissertation, McMaster University, Canada.

Blumstein, D. T. and Armitage, K. B. (1999). Cooperative breeding in marmots. *Oikos*, **84**, 369–382.

Boal, C. W. and Spaulding, R. L. (2000). Helping at a Cooper's hawk nest. *Wilson Bull.*, **112**, 275–277.

Bock, C. E. (1970). The ecology and behavior of the Lewis woodpecker (*Asyndesmus lewis*). *Univ. Calif. Publ. Zool.*, **92**, 1–91.

Boix-Hinzen, C. and Lovegrove, B. G. (1998). Circadian metabolic and thermoregulatory patterns of red-billed woodhoopoes (*Phoeniculus purpureus*): the influence of huddling. *J. Zool.*, **244**, 33–41.

Boland, C. R. J. and Cockburn, A. (2002). Short sketches from the long history of cooperative breeding in Australian birds. *Emu*, **102**, 9–17.

Boland, C. R. J., Heinsohn, R. G. and Cockburn, A. (1997a). Experimental manipulation of brood reduction and parental care in cooperatively breeding white-winged choughs. *J. Anim. Ecol.*, **66**, 683–691.

(1997b). Deception by helpers in cooperatively breeding white-winged choughs and its experimental manipulation. *Behav. Ecol. Sociobiol.*, **41**, 251–256.

Bortolotti, G. R. (1986). Influence of sibling competition on nestling sex ratios of sexually dimorphic birds. *Am. Nat.*, **127**, 495–507.

Bowen, B. S. (2002). Groove-billed ani (*Crotophaga sulcirostris*). In: *Birds of North America*, ed. A. Poole and F. Gill. Philadelphia, PA and Washington, DC: Academy of Natural Sciences and American Ornithologists' Union.

Bowen, B. S., Koford, R. L. and Vehrencamp, S. L. (1989). Dispersal in communally nesting groove–billed anis (*Crotophaga sulcirostris*). *Condor*, **91**, 52–64.

Bowen, B. S., Koford, R. R. and Vehrencamp, S. L. (1991). Seasonal pattern of reverse mounting in the groove-billed ani (*Crotophaga sulcirostris*). *Condor*, **93**, 159–163.

Bowen, B. S., Koford, R. R. and Brown, J. L. (1995). Genetic evidence for undetected alleles and unexpected parentage in the gray-breasted jay. *Condor*, **97**, 503–511.

Brant, C. L., Schwab, T. M., Vandenbergh, J. G., Schaefer, R. L. and Solomon, N. G. (1998). Behavioural suppression of female pine voles after replacement of the breeding male. *Anim. Behav.*, **55**, 615–627.

Braude, S. (2000). Dispersal and new colony formation in wild naked mole-rats: evidence against inbreeding as the system of mating. *Behav. Ecol.*, **11**, 7–12.

Breininger, D. R. (1999). Florida scrub-jay demography and dispersal in a fragmented landscape. *Auk*, **116**, 520–527.

Breininger, D. R., Larson, V. L., Duncan, B. W., Smith, R. B., Oddy, D. M. and Goodchild, M. (1995). Landscape patterns in Florida scrub-jay habitat preference and demography. *Conserv. Biol.*, **9**, 1142–1153.

Breuner, C. W. and Orchinik, M. (2002). Plasma binding proteins as mediators of corticosteroid action in vertebrates. *J. Endocrinol.*, **175**, 99–112.

Briskie, J. V., Montgomerie, R., Pöldmaa, T. and Boag, P. T. (1998). Paternity and paternal care in the polygynandrous Smith's longspur. *Behav. Ecol. Sociobiol.*, **43**, 181–190.

Brooker, M. G., Rowley, I., Adams, M. and Baverstock, P. R. (1990). Promiscuity: an inbreeding avoidance mechanism in a socially monogamous species? *Behav. Ecol. Sociobiol.*, **26**, 191–200.

Brooks, T. and Thompson, H. S. (2001). Current bird conservation issues in Africa. *Auk*, **118**, 575–582.

Brosset, A. and Fry, C. H. (1988). Order Musophagiformes, Musophagidae, turacos, go-away birds and plantain-eaters. In: *The Birds of Africa*, vol. 3, ed. C. H. Fry, S. Keith and E. K. Urban. New York, NY: Academic Press. pp. 26–57.

Brotherton, P. N. M., Clutton-Brock, T. H., O'Riain, M. J., Gaynor, D., Sharpe, L., Kansky, R. and McIlrath, G. M. (2001). Offspring food allocation by parents and helpers in a cooperative mammal. *Behav. Ecol.*, **12**, 590–599.

Brown, C. R. (1984). Laying eggs in a neighbor's nest: benefit and cost of colonial nesting in swallows. *Science*, **224**, 518–519.

(1988). Enhanced foraging efficiency through information centers: a benefit of coloniality in cliff swallows. *Ecology*, **69**, 602–613.

Brown, C. R. and Foster, G. G. (1992). The thermal and energetic significance of clustering on the speckled mousebird *(Colius striatus)*. *J. Comp. Physiol. B*, **162**, 664–685.

Brown, J. L. (1969). Territorial behavior and population regulation in birds, a review and re-evaluation. *Wilson Bull.*, **81**, 293–329.

(1974). Alternate routes to sociality in jays – with a theory for the evolution of altruism and communal breeding. *Am. Zool.*, **64**, 63–80.

(1978). Avian communal breeding systems. *Annu. Rev. Ecol. Syst.*, **9**, 123–155.

(1980). Fitness in complex avian social systems. In: *Evolution of Social Behavior: Hypotheses and Empirical Tests*, ed. H. Markl. Weinheim: Dahlem Konferenzen Verlag Chemie. pp. 115–128.

(1982). Optimal group size in territorial animals. *J. Theor. Biol.*, **95**, 793–810.

(1985). The evolution of helping behavior – an ontogenetic and comparative perspective. In: *The Evolution of Adaptive Skills: Comparative and Ontogenetic Approaches*, ed. E. S. Gollin. Hillsdale, NJ: Erbaus. pp. 137–171.

(1986). Cooperative breeding and the regulation of numbers. *Proc. Int. Ornithol. Congr.*, **18**, 774–782.

(1987). *Helping and Communal Breeding in Birds*. Princeton, NJ: Princeton University Press.

(1994). Mexican jay *(Aphelocoma ultramarina)*. In: *The Birds of North America*, ed. A. Poole and F. Gill. Philadelphia, PA and Washington, DC: Academy of Natural Sciences and American Ornithologists' Union.

Brown, J. L. and Brown, E. R. (1984). Parental facilitation: parent offspring relations in communally breeding birds. *Behav. Ecol. Sociobiol.*, **14**, 203–209.

(1990). Mexican jays: uncooperative breeding. In: *Cooperative Breeding in Birds*, ed. P. B. Stacey and W. D. Koenig. Cambridge: Cambridge University Press. pp. 267–288.

(1998). Are inbred offspring less fit? Survival in a natural population of Mexican jays. *Behav. Ecol.*, **9**, 60–63.

Brown, J. L. and Vleck, C. M. (1998). Prolactin and helping in birds: has natural selection strengthened helping behavior? *Behav. Ecol.*, **9**, 541–545.

Brown, J. L., Brown, E. R., Brown, S. D. and Dow, D. D. (1982). Helpers: effects of experimental removal on reproductive success. *Science*, **215**, 421–422.

Bruener, C. W. and Orchinik, M. (2001). Seasonal regulation of membrane and intracellular corticosteroid receptors in the house sparrow brain. *J. Neuroendocrinol.*, **13**, 412–420.

(2002). Plasma binding proteins as mediators of corticosteroid action in vertebrates. *J. Endocrinol.*, **175**, 99–112.

Bruning, D. F. (1974). Social structure and reproductive behavior of the greater rhea. *Living Bird*, **13**, 251–294.

Bull, J. J. and Charnov, E. L. (1988). How fundamental are Fisherian sex ratios? *Oxford Surv. Evol. Biol.*, **5**, 96–135.

Bulmer, M. G. and Taylor, P. D. (1980). Dispersal and the sex ratio. *Nature*, **284**, 448–449.

Buntin, J. D. (1996). Neural and hormonal control of parental behavior in birds. *Adv. Study Behav.*, **25**, 161–213.

Burda, H. (1995). Individual recognition and incest avoidance in eusocial common mole-rats rather than reproductive suppression by parents. *Experientia*, **51**, 411–413.

Burke, T., Davies, N. B., Bruford, M. W. and Hatchwell, B. J. (1989). Parental care and mating behaviour of polyandrous dunnocks *Prunella modularis* related to paternity by DNA fingerprinting. *Nature*, **338**, 249–251.

Burley, N. (1981). The evolution of sexual indistinguishability. In: *Natural Selection and Social Behavior: Recent Research and New Theory*, ed. R. D. Alexander and D. W. Tinkle. New York: Chiron Press. pp. 121–137.

Burney, D. A., James, H. F., Burney, L. P., Olson, S. L., Kikuchi, W., Wagner, W. L., Burney, M., McCloskey, D., Kikuchi, D., Grady, F., Gage, R. I. and Nishek, R. (2001). Holocene lake sediments in the Maha'ulepu caves of Kaua'i: evidence for a diverse biotic assemblage from the Hawaiian lowlands and its transformation since human arrival. *Ecol. Monogr.*, **71**, 615–642.

Burt, D. B. (1996). Phylogenetic and ecological aspects of cooperative breeding in the bee-eaters (Aves: Meropidae). Ph.D. dissertation, University of Arizona, Tucson, AZ.

  (2001). Evolutionary stasis, constraint and other terminology describing evolutionary patterns. *Biol. J. Linn. Soc.*, **72**, 509–517.

  (2002). Social and breeding biology of bee-eaters in Thailand. *Wilson Bull.*, **114**, 275–279.

Burt, D. B. and Peterson, A. T. (1993). Biology of cooperative-breeding scrub jays (*Aphelocoma coerulescens*) of Oaxaca, Mexico. *Auk*, **110**, 207–214.

Caffrey, C. (2000). Correlates of reproductive success in cooperatively breeding western American crows: if helpers help, it's not by much. *Condor*, **102**, 333–341.

Cant, M. A. (1998). A model for the evolution of reproductive skew without reproductive suppression. *Anim. Behav.*, **55**, 163–169.

Cant, M. A. and Field, J. (2001). Helping effort and future fitness in cooperative animal societies. *Proc. R. Soc. London Ser. B*, **268**, 1959–1964.

Cant, M. A. and Johnstone, R. A. (1999). Costly young and reproductive skew in animal societies. *Behav. Ecol.*, **10**, 178–184.

Cant, M. A. and Reeve, H. K. (2002). Female control of the distribution of paternity in cooperative breeders. *Am. Nat.*, **160**, 602–611.

Cariello, M. O., Schwabl, H. G., Lee, R. W. and Macedo, R. H. F. (2002). Individual female clutch identification through yolk protein electrophoresis in the communally breeding guira cuckoo (*Guira guira*). *Mol. Ecol.*, **11**, 2417–2424.

Carlson, A. A., Young, A. J., Bennett, N. C., Russell, A. F., McNeilly, A. S. and Clutton-Brock, T. H. (in press). Hormonal correlates of dominance in cooperative meerkats (*Suricata suricatta*). *Hormones and Behavior*, in press.

Carmen, W. J. (2004). Behavioral ecology of the California scrub-jay (*Aphelocoma californica*): a noncooperative breeder with close cooperative relatives. *Stud. Avian Biol.*, in press.

Carrick, R. (1963). Ecological significance of territory in the Australian magpie, *Gymnorhina tibicen*. *Proc. Int. Ornithol. Congr.*, **13**, 740–753.

  (1972). Population ecology of the black-backed magpie, royal penguin and silver gull. *US Dept. Int. Wildl. Res. Rep.*, **2**, 41–99.

Carter, C. S. and Roberts, R. L. (1997). The psychobiological basis of cooperative breeding in rodents. In: *Cooperative Breeding in Mammals*, ed. N. G. Solomon and J. A. French. Cambridge: Cambridge University Press. pp. 231–266.

Castro, I., Minot, E. O., Fordham, R. A. and Birkhead, T. R. (1996). Polygynandry, face-to-face copulation and sperm competition in the hihi *Notiomystis cincta* (Aves: Meliphagidae). *Ibis*, **138**, 765–771.

Caughley, G. (1994). Directions in conservation biology. *J. Anim. Ecol.*, **63**, 215–244.

Chao, L. (1997). Evolution of polyandry in a communal breeding system. *Behav. Ecol.*, **8**, 668–674.

Chaplin, S. B. (1982). The energetic significance of huddling behavior in common bushtits (*Psaltriparus minimus*). *Auk*, **99**, 424–430.

Charnov, E. L. (1982). *The Theory of Sex Allocation*. Princeton, NJ: Princeton University Press.

Cheeseman, C. L., Mallinson, P. J., Ryan, J. and Wilesmith, J. W. (1993). Recolonisation by badgers in Gloucestershire. In: *The Badger*, ed. T. J. Hayden. Dublin: Royal Irish Academy. pp. 78–93.

Chen, J. and Franklin, J. F. (1990). Microclimate pattern and basic biological responses at the clearcut edges of old-growth Douglas fir stands. *Northwest Environ. J.*, **6**, 424–425.

Cheney, D. L. and Seyfarth, R. M. (1985). Vervet monkey alarm calls: manipulation through shared information? *Behaviour*, **94**, 150–166.

Cisek, D. (2000). New colony formation in the "highly inbred" eusocial naked mole-rat: outbreeding is preferred. *Behav. Ecol.*, **11**, 1–6.

Clark, A. B. (1978). Sex ratio and local resource competition in a prosimian primate. *Science*, **201**, 163–165.

Clark, M. M. and Galef, B. G. (2000). Why some male Mongolian gerbils may help at the nest: testosterone, asexuality and alloparenting. *Anim. Behav.*, **59**, 801–806.

Clarke, F. M. and Faulkes, C. G. (1999). Kin discrimination and female mate choice in the naked mole-rat, *Heterocephalus glaber*. *Proc. R. Soc. London Ser. B*, **266**, 1995–2002.

Clarke, M. F. (1988). The reproductive behaviour of the bell miner *Manorina melanophrys*. *Emu*, **88**, 88–100.

(1989). The pattern of helping in the bell miner (*Manorina melanophrys*). *Ethology*, **80**, 292–306.

(1995). Co-operative breeding in Australasian birds: a review of hypotheses and evidence. *Corella*, **19**, 73–90.

Clarke, M. F and Fitz-Gerald, G. F. (1994). Spatial organisation of the cooperatively breeding bell miner *Manorina melanophrys*. *Emu*, **94**, 96–105.

Clarke, M. F., Jones, D. A., Ewne, J. G., Robertson, R. J., Griffiths, R., Painter, J., Boag, P. T. and Crozier, R. (2002). Male-biased sex ratios in broods of the cooperatively breeding bell miner *Manorina melanophrys*. *J. Avian Biol.*, **33**, 71–76.

Clotfelter, E. D. (1996). Mechanisms of facultative sex-ratio variation in zebra finches (*Taeniopygia guttata*). *Auk*, **113**, 441–449.

Clutton-Brock, T. H. (1986). Sex ratio variation in birds. *Ibis*, **128**, 317–329.

ed. (1988). *Reproductive Success*. Chicago, IL: University of Chicago Press.

(1989a). Mammalian mating systems. *Proc. R. Soc. London Ser. B*, **236**, 339–372.

(1989b). Female transfer and inbreeding avoidance in social mammals. *Nature*, **337**, 70–72.

(1991). *The Evolution of Parental Care*. Princeton, NJ: Princeton University Press.

(1998). Reproductive skew, concessions and limited control. *Trends Ecol. Evol.*, **13**, 288–292.

(2002). Breeding together: kin selection and mutualism in cooperative vertebrates. *Science*, **296**, 69–72.

Clutton-Brock, T. H. and Harvey, P. H. (1977). Primate ecology and social organization. *J. Zool.*, **183**, 1–39.

(1978). Mammals, resources and reproductive strategies. *Nature*, **273**, 191–195.

Clutton-Brock, T. H., Albon, S. D. and Guinness, F. E. (1981). Parental investment in male and female offspring in polygynous mammals. *Nature*, **289**, 487–489.

(1984). Maternal dominance, breeding success, and birth sex ratios in red deer. *Nature*, **308**, 358–360.

(1989). Fitness costs of gestation and lactation in wild mammals. *Nature*, **337**, 260–262.

Clutton-Brock, T. H., Gaynor, D., Kansky, R., MacColl, A. D. C., McIlrath, G., Chadwick, P., Brotherton, P. N. M., O'Riain, M. J., Manser, M. and Skinner, J. D. (1998). Costs of cooperative behaviour in suricates (*Suricata suricatta*). *Proc. R. Soc. London Ser. B*, **265**, 185–190.

Clutton-Brock, T. H., Gaynor, D., McIlrath, G. M., MacColl, A. D. C., Kansky, R., Chadwick, P., Manser, M., Skinner, J. D. and Brotherton, P. N. M. (1999a). Predation, group size and mortality in a cooperative mongoose, *Suricata suricatta*. *J. Anim. Ecol.*, **68**, 672–683.

Clutton-Brock, T. H., O'Riain, M. J., Brotherton, P. N. M., Gaynor, D., Kansky, R., Griffin, A. S. and Manser, M. (1999b). Selfish sentinels in cooperative mammals. *Science*, **284**, 1640–1644.

Clutton-Brock, T. H., Brotherton, P. N. M., O'Riain, M. J., Griffin, A. S., Gaynor, D., Sharpe, L., Kansky, R., Manser, M. B. and McIlrath, G. M. (2000). Individual contributions to babysitting in a cooperative mongoose, *Suricata suricatta*. *Proc. R. Soc. London Ser. B*, **267**, 301–305.

Clutton-Brock, T. H., Brotherton, P. N. M., O'Riain, M. J., Griffin, A. S., Gaynor, D., Kansky, R., Sharpe, L. and McIlrath, G. M. (2001a). Contributions to cooperative rearing in meerkats. *Anim. Behav.*, **61**, 705–710.

Clutton-Brock, T. H., Brotherton, P. N. M., Russell, A. F., O'Riain, M. J., Gaynor, D., Kansky, R., Griffin, A., Manser, M., Sharpe, L., McIlrath, G. M., Small, T., Moss, A. and Monfort, S. (2001b). Cooperation, control, and concession in meerkat groups. *Science*, **291**, 478–481.

Clutton-Brock, T. H., Russell, A. F., Sharpe, L. L., Brotherton, P. N. M., McIlrath, G. M., White, S. and Cameron, E. Z. (2001c). Effects of helpers on juvenile development and survival in meerkats. *Science*, **293**, 2446–2449.

Clutton-Brock, T. H., Russell, A. F., Sharpe, L. L., Young, A. J., Balmforth, Z. and McIlrath, G. M. (2002). Evolution and development of sex differences in cooperative behavior in meerkats. *Science*, **297**, 253–256.

Clutton-Brock, T. H., Russell, A. F. and Sharpe, L. L. (2003). Meerkat helpers do not specialize on particular activities. *Anim. Behav.*, **66**, 531–540.

(2004). Behavioural tactics of breeders in cooperative meerkats. *Anim. Behav.*, in press.

Cockburn, A. (1990). Sex ratio variation in marsupials. *Aust. J. Zool.*, **37**, 467–479.

(1996). Why do so many Australian birds cooperate: social evolution in the Corvida? In: *Frontiers of Population Ecology*, ed. R. B. Floyd, A. W. Sheppard and P. J. De Barro. East Melbourne: CSIRO. pp. 451–472.

(1998). Evolution of helping behavior in cooperatively breeding birds. *Annu. Rev. Ecol. Syst.*, **29**, 141–177.

Cockburn, A., Legge, S. and Double, M. C. (2002). Sex ratios in birds and mammals: can the hypotheses be disentangled? In: *The Sex Ratio Handbook*, ed. I. C. W. Hardy. Cambridge: Cambridge University Press. pp. 266–286.

Cockburn, A., Osmond, H. L., Mulder, R. A., Green, D. J. and Double, M. C. (2003). Divorce, dispersal, density-dependence and incest avoidance in the cooperatively breeding superb fairy-wren *Malurus cyaneus*. *J. Anim. Ecol.*, **72**, 189–202.

Coddington, C. L. and Cockburn, A. (1995). The mating system of free-living emus. *Aust. J. Zool.*, **43**, 365–372.

Codenotti, T. L. and Alvarez, F. (1997). Cooperative breeding between males in the greater rhea *Rhea americana*. *Ibis*, **139**, 568–571.

Conner, R. N. and Rudolph, D. C. (1991). Forest habitat loss, fragmentation, and red-cockaded woodpecker populations. *Wilson Bull.*, **103**, 446–457.

Conner, R. N., Rudolph, D. C. and Walters, J. R. (2001). *The Red-Cockaded Woodpecker: Surviving in a Fire-Maintained Ecosystem*. Austin: University of Texas Press.

Connor, R. C. (1995). Altruism among non-relatives: alternatives to the 'prisoner's dilemma'. *Trends Ecol. Evol.*, **10**, 84–86.

Conrad, K. F., Clarke, M. F., Robertson, R. J. and Boag, P. T. (1998). Paternity and the relatedness of helpers in the cooperatively breeding bell miner. *Condor*, **100**, 343–349.

Cooney, R. and Bennett, N. C. (2000). Inbreeding avoidance and reproductive skew in a cooperative mammal. *Proc. R. Soc. London Ser. B*, **267**, 801–806.

Cooper, C. B. (2000). Behavioral ecology and conservation of the Australian brown treecreeper (*Climacteris picumnus*). Ph.D. dissertation, Virginia Polytechnic Institute and State University, Blacksburg, VA.

Cooper, C. B. and Walters, J. R. (2002a). Experimental evidence of disrupted dispersal causing decline of an Australian passerine in fragmented habitat. *Conserv. Biol.*, **16**, 471–478.

(2002b). Independent effects of woodland loss and fragmentation on brown treecreeper distribution. *Biol. Conserv.*, **105**, 1–10.

Cooper, C. B., Walters, J. R. and Ford, H. A. (2002a). Effects of remnant size and connectivity on the response of brown treecreepers to habitat fragmentation. *Emu*, **102**, 249–256.

Cooper, C. B., Walters, J. R. and Priddy, J. A. (2002b). Landscape patterns and dispersal success: simulated population dynamics in the brown treecreeper. *Ecol. Appl.*, **12**, 1576–1587.

Copeyon, C. K., Walters, J. R. and Carter, J. H. I. (1991). Induction of red-cockaded woodpecker group formation by artificial cavity construction. *J. Wildl. Manage.*, **55**, 549–556.

Coyne, J. A. and Price, T. D. (2000). Little evidence for sympatric speciation in island birds. *Evolution*, **54**, 2166–2171.

Craig, J. L. (1979). Habitat variation in the social organization of a communal gallinule, the pukeko, *Porphyrio porphyrio melanotus*. *Behav. Ecol. Sociobiol.*, **5**, 331–358.

(1980). Pair and group breeding behaviour of a communal gallinule, the pukeko *Porphyrio porphyrio*. *Anim. Behav.*, **32**, 23–32.

Craig, J. L. and Jamieson, I. G. (1988). Incestuous mating in a communal bird: a family affair. *Am. Nat.*, **131**, 58–70.

(1990). Pukeko: different approaches and some different answers. In: *Cooperative Breeding in Birds*, ed. P. B. Stacey and W. D. Koenig. Cambridge: Cambridge University Press. pp. 385–412.

Cramp, S. and Perrins, C. M. (1993). *The Birds of the Western Palearctic, Vol. 7: Flycatchers to Shrikes*. Oxford: Oxford University Press.

Creel, S. R. (2001). Social dominance and stress hormones. *Trends Ecol. Evol.*, **16**, 491–497.

Creel, S. R. and Creel, N. M. (1991). Energetics, reproductive suppression and obligate communal breeding in carnivores. *Behav. Ecol. Sociobiol.*, **28**, 263–270.

(1995). Communal hunting and pack size in African wild dogs, *Lycaon pictus*. *Anim. Behav.*, **50**, 1325–1339.

(2001). *The African Wild Dog*. Princeton, NJ: Princeton University Press.

Creel, S. R. and Macdonald, D. (1995). Sociality, group size, and reproductive suppression among carnivores. *Adv. Study Behav.*, **24**, 203–257.

Creel, S. R. and Waser, P. M. (1994). Inclusive fitness and reproductive strategies in dwarf mongooses. *Behav. Ecol. Sociobiol.*, **5**, 339–348.

(1997). Variation in reproductive suppression among dwarf mongooses: interplay between mechanisms and evaluation. In: *Cooperative Breeding in Mammals*, ed. N. G. Solomon and J. A. French. Cambridge: Cambridge University Press. pp. 150–170.

Creel, S. R., Monfort, S. L., Wildt, D. E. and Waser, P. M. (1991). Spontaneous lactation is an adaptive result of pseudopregnancy. *Nature*, **351**, 660–662.

Creel, S. R., Creel, N. M., Wildt, D. E. and Monfort, S. L. (1992). Behavioural and endocrine mechanisms of reproductive suppression in Serengeti dwarf mongooses. *Anim. Behav.*, **43**, 231–245.

Creel, S. R., Wildt, D. E. and Monfort, S. L. (1993). Aggression, reproduction, and androgens in wild dwarf mongooses: a test of the challenge hypothesis. *Am. Nat.*, **141**, 816–825.

Crespi, B. J. and Ragsdale, J. E. (2000). A skew model for the evolution of sociality via manipulation: why it is better to be feared than loved. *Proc. R. Soc. London Ser. B*, **267**, 821–828.

Crick, H. Q. P. (1992). Load-lightening in cooperatively breeding birds and the cost of reproduction. *Ibis*, **134**, 56–61.

Crick, H. Q. P. and Fry, C. H. (1986). Effects of helpers on parental condition in red-throated bee-eaters (*Merops bullocki*). *J. Anim. Ecol.*, **55**, 893–905.

Crome, F. H. J. (1976). Some observations on the biology of the cassowary in northern Queensland. *Emu*, **76**, 8–14.

Curry, R. L. (1988a). Influence of kinship on helping behavior in Galápagos mockingbirds. *Behav. Ecol. Sociobiol.*, **22**, 141–152.

(1988b). Group structure, within-group conflict and reproductive tactics in cooperatively breeding Galápagos mockingbirds, *Nesomimus parvulus*. *Anim. Behav.*, **36**, 1708–1728.

Curry, R. L. and Grant, P. R. (1989). Demography of the cooperatively breeding Galápagos mockingbird, *Nesomimus parvulus*, in a climatically variable environment. *J. Anim. Ecol.*, **58**, 441–464.

(1990). Galápagos mockingbirds: territorial cooperative breeding in a climatically variable environment. In: *Cooperative Breeding in Birds*, ed. P. B. Stacey and W. D. Koenig. Cambridge: Cambridge University Press. pp. 289–331.

Daniels, S. J. (1997). *Female Dispersal and Inbreeding in the Red-Cockaded Woodpecker*. M.S. dissertation, Virginia Polytechnic Institute and State University, Blacksburg, VA.

Daniels, S. J. and Walters, J. R. (1999). Inbreeding depression and its effects on natal dispersal in red-cockaded woodpeckers. *Proc. Int. Ornithol. Congr.*, **22**, 2492–2498.

(2000a). Inbreeding depression and its effects on natal dispersal in red-cockaded woodpeckers. *Condor*, **103**, 482–491.

(2000b). Between-year breeding dispersal in red-cockaded woodpeckers: multiple causes and estimated cost. *Ecology*, **81**, 2473–2484.

Daniels, S. J., Priddy, J. A. and Walters, J. R. (2000). Inbreeding in small populations of red-cockaded woodpeckers: insights from a spatially explicit individual-based model. In: *Genetics, Demography, and Viability of Fragmented Populations*, ed. A. G. Young and G. M. Clarke. Cambridge: Cambridge University Press. pp. 129–147.

Darwin, C. (1871). *The Descent of Man, and Selection in Relation to Sex*. London: John Murray.

Davies, N. B. (1983). Polyandry, cloaca-pecking and sperm competition in dunnocks. *Nature*, **302**, 334–336.

(1985). Cooperation and conflict among dunnocks, *Prunella modularis*, in a variable mating system. *Anim. Behav.*, **33**, 628–648.

(1989). Sexual conflict and the polygamy threshold. *Anim. Behav.*, **38**, 226–234.

(1990). Dunnocks: cooperation and conflict among males and females in a variable mating system. In: *Cooperative Breeding in Birds*, ed. P. B. Stacey and W. D. Koenig. Cambridge: Cambridge University Press. pp. 455–485.

(1992). *Dunnock Behaviour and Social Evolution*. Oxford: Oxford University Press.

Davies, N. B. and Hatchwell, B. J. (1992). The value of male parental care and its influence on reproductive allocation by male and female dunnocks. *J. Anim. Ecol.*, **61**, 259–272.

Davies, N. B. and Lundberg, A. (1984). Food distribution and a variable mating system in the dunnock, *Prunella modularis*. *J. Anim. Ecol.*, **53**, 895–912.

Davies, N. B., Hatchwell, B. J., Robson, T. and Burke, T. (1992). Paternity and parental effort in dunnocks *Prunella modularis*: how good are chick-feeding rules? *Anim. Behav.*, **43**, 729–745.

Davies, N. B., Hartley, I. R., Hatchwell, B. J., Desrochers, A., Skeer, J. and Nebel, D. (1995). The polygynandrous mating system of the alpine accentor, *Prunella collaris*. 1. Ecological causes and reproductive conflicts. *Anim. Behav.*, **49**, 769–788.

Davies, N. B., Hartley, I. R., Hatchwell, B. J. and Langmore, N. E. (1996). Female control of copulations to maximize male help: a comparison of polygynandrous alpine accentors, *Prunella collaris*, and dunnocks, *P. modularis*. *Anim. Behav.*, **51**, 27–47.

Davis, D. E. (1940). Social nesting habits of the smooth-billed ani. *Auk*, **57**, 197–218.

(1942). The phylogeny of social nesting habits in the Crotophaginae. *Q. Rev. Biol.*, **17**, 115–134.

Dawkins, R. (1979). Twelve misunderstandings of kin selection. *Z. Tierpsychol.*, **51**, 184–200.

Dawson, J. W. and Mannan, R. W. (1991). Dominance hierarchies and helper contributions in Harris' hawks. *Auk*, **108**, 649–660.

Dean, W. R. J. and Williams, J. B. (1999). Sunning behaviour and its possible influence on digestion in the white-backed mousebird *Colius colius*. *Ostrich*, **70**, 239–241.

Decoux, J. P. (1988a). Order Coliiformes, Coliidae, mousebirds or colies. In: *The Birds of Africa*, vol. 3, ed. C. H. Fry, S. Keith and E. K. Urban. New York, NY: Academic Press. pp. 243–254.

(1988b). Régime comportement alimentaire et régulation écologique du métabolisme chez *Colius striatus*. *Terre et la Vie*, **30**, 395–420.

DeGange, A. R., Fitzpatrick, J. W., Layne, J. N. and Woolfenden, G. E. (1989). Acorn harvesting by Florida scrub jays. *Ecology*, **70**, 348–356.

del Hoya, J., Elliott, A., Sargatal, J. and Cabot, J. (1996). *Handbook of the Birds of the World, vol. 3: Hoatzin to Auks*. Barcelona: Lynx Edicions.

DeLay, L. S., Faaborg, J., Naranjo, J., Paz, S. M., DeVries, T. and Parker, P. G. (1996). Paternal care in the cooperatively polyandrous Galápagos hawk. *Condor*, **98**, 300–311.

Dement, M. W. and van Soest, P. J. (1983). *Body Size, Digestive Capacity and Feeding Strategies of Herbivores*. Morilton, AR: Winrock International.

Dhondt, A. A. and Hochachka, W. M. (2001). Adaptive sex ratios and parent–offspring conflict. *Trends Ecol. Evol.*, **16**, 61–62.

Dickinson, J. L. (2001). Extrapair copulations in western bluebirds (*Sialia mexicana*): female receptivity favors older males. *Behav. Ecol. Sociobiol.*, **50**, 423–429.

(2004). Facultative sex ratio adjustment by western bluebird mothers with stay-at-home helpers at the nest. *Anim Behav.*, in press.

Dickinson, J. L. and Akre, J. J. (1998). Extrapair paternity, inclusive fitness, and within-group benefits of helping in western bluebirds. *Mol. Ecol.*, **7**, 95–105.

Dickinson, J. L. and Weathers, W. W. (1999). Replacement males in the western bluebird: opportunity for parentage, chick-feeding rules, and the importance of male parental care. *Behav. Ecol. Sociobiol.*, **45**, 201–209.

Dickinson, J. L., Haydock, J., Koenig, W. D., Stanback, M. T. and Pitelka, F. A. (1995). Genetic monogamy in single-male groups of acorn woodpeckers, *Melanerpes formicivorus*. *Mol. Ecol.*, **4**, 765–769.

Dickinson, J. L., Koenig, W. D. and Pitelka, F. A. (1996). Fitness consequences of helping behavior in the western bluebird. *Behav. Ecol.*, **7**, 168–177.

Dietz, J. M. and Baker, A. J. (1993). Polygyny and female reproductive success in golden lion tamarins, *Leontopithecus rosalia*. *Anim. Behav.*, **46**, 1067–1078.

Dijkstra, C., Daan, S. and Pen, I. (1998). Fledgling sex ratios in relation to brood size in size-dimorphic altricial birds. *Behav. Ecol.*, **9**, 287–296.

Dixon, A. F. (1992). Prolactin and parental behaviour in a male new world primate. *Nature*, **299**, 551–553.

Dobson, F. S., Chesser, R. K., Hoogland, J. L., Sugg, D. W. and Foltz, D. W. (1997). Do black-tailed prairie dogs minimize inbreeding? *Evol. Ecol.*, **51**, 970–978.

Doerr, V. A. J. and Doerr, E. D. (2001). Brown treecreeper. In: *Handbook of Australian, New Zealand and Antarctic Birds*, vol. 5, ed. P. J. Higgins, J. M. Peter and W. K. Steele. Oxford: Oxford University Press.

Doncaster, C. P. and Woodroffe, R. (1993). Den site can determine shape and size of badger territories: implications for group-living. *Oikos*, **66**, 88–93.

Double, M. C. and Cockburn, A. (2000). Pre-dawn infidelity: females control extra-pair mating in superb fairy-wrens. *Proc. R. Soc. London Ser. B*, **267**, 465–470.

(2003). Subordinate superb fairy-wrens (*Malurus cyaneus*) parasitize the reproductive success of attractive dominant males. *Proc. R. Soc. London Ser. B*, **270**, 379–384.

Dow, D. D. (1980). Communally breeding Australian birds with an analysis of distributional and environmental factors. *Emu*, **80**, 121–140.

Dow, D. D. and Whitmore, M. J. (1990). Noisy miners: variations on the theme of communality. In: *Cooperative Breeding in Birds*, ed. P. B. Stacey and W. D. Koenig. Cambridge: Cambridge University Press. pp. 561–592.

Du Plessis, M. A. (1989). Behavioural ecology of the red-billed woodhoopoe *Phoeniculus purpureus* in South

Africa. Ph.D. dissertation, University of Cape Town, South Africa.

(1992). Obligate cavity-roosting as a constraint on dispersal of green (red-billed) woodhoopoes: consequences for philopatry and the likelihood of inbreeding. *Oecologia*, **90**, 205–11.

(1993). Helping behaviour in cooperatively-breeding green woodhoopoes: selected or unselected trait? *Behaviour*, **127**, 49–65.

Du Plessis, M. A. and Williams, J. B. (1994). Communal cavity roosting in green woodhoopoes: consequences for energy expenditure and the seasonal pattern of mortality. *Auk*, **111**, 292–299.

Du Plessis, M. A., Weathers, W. W. and Koenig, W. D. (1994). Energetic benefits of communal roosting by acorn woodpeckers during the nonbreeding season. *Condor*, **96**, 631–637.

Du Plessis, M. A., Siegfried, W. R. and Armstrong, A. J. (1995). Ecological and life-history correlates of cooperative breeding in South African birds. *Oecologia*, **102**, 180–188.

Dunford, C. (1977). Kin selection for ground squirrel alarm calls. *Am. Nat.*, **111**, 782–785.

Dunn, P. O. and Cockburn, A. (1996). Evolution of male parental care in a bird with almost complete cuckoldry. *Evolution*, **50**, 2542–2548.

(1999). Extrapair mate choice and honest signaling in cooperatively breeding superb fairy-wrens. *Evolution*, **53**, 938–946.

Dunn, P. O., Cockburn, A. and Mulder, R. A. (1995). Fairy-wren helpers often care for young to which they are unrelated. *Proc. R. Soc. London Ser. B*, **259**, 339–343.

Dyer, M. (1983). Effect of nest helpers on growth of red-throated bee-eaters. *Ostrich*, **54**, 43–46.

Eden, S. F. (1987). When do helpers help? Food availability and helping in the moorhen, *Gallinula chloropus*. *Behav. Ecol. Sociobiol.*, **21**, 191–195.

Edwards, S. V. and Boles, W. E. (2002). Out of Gondwana: the origin of passerine birds. *Trends Ecol. Evol.*, **17**, 347–349.

Edwards, S. V. and Naeem, S. (1993). The phylogenetic component of cooperative breeding in perching birds. *Am. Nat.*, **141**, 754–789.

(1994). Homology and comparative methods in the study of avian cooperative breeding. *Am. Nat.*, **143**, 723–733.

Edwards, T. C., Jr. (1986). Ecological distribution of the grey-breasted jay. *Condor*, **88**, 456–460.

Ekman, J. and Griesser, M. (2002). Why offspring delay dispersal: experimental evidence for a role of parental tolerance. *Proc. R. Soc. London Ser. B*, **269**, 1709–1714.

Ekman, J. and Hake, M. K. (1990). Monitoring starvation risk: adjustments of greenfinches' body reserves during periods of unpredictable foraging success. *Behav. Ecol.*, **1**, 62–67.

Ekman, J. and Lilliendahl, K. (1993). Using priority to food access: fattening strategies in dominance-structured willow tit (*Parus montanus*) flocks. *Behav. Ecol.*, **4**, 232–238.

Ekman, J. and Rosander, B. (1992). Survival enhancement through food sharing: a means for parental control of natal dispersal. *Theor. Pop. Biol.*, **42**, 117–129.

Ekman, J., Sklepkovych, B. and Tegelström, H. (1994). Offspring retention in the Siberian jay (*Perisoreus infaustus*): the prolonged brood care hypothesis. *Behav. Ecol.*, **5**, 245–253.

Ekman, J., Bylin, A. and Tegelström, H. (1999). Increased lifetime reproductive success for Siberian jay *Perisoreus infaustus* males with delayed dispersal. *Proc. R. Soc. London Ser. B*, **266**, 911–915.

(2000). Parental nepotism enhances survival of retained offspring in the Siberian jay. *Behav. Ecol.*, **11**, 416–420.

Ekman, J., Baglione, V., Eggers, S. and Griesser, M. (2001a). Delayed dispersal: living under the reign of nepotistic parents. *Auk*, **118**, 1–10.

Ekman, J., Eggers, S., Griesser, M. and Tegelstrom, H. (2001b). Queuing for preferred territories: delayed dispersal of Siberian jays. *J. Anim. Ecol.*, **70**, 317–324.

Ekman, J., Eggers, S. and Griesser, M. (2002). Fighting to stay: the role of sibling rivalry for delayed dispersal. *Anim. Behav.*, **64**, 453–459.

Eltzroth, E. K. and Robertson, S. R. (1984). Violet-green swallows help western bluebirds at the nest. *J. Field Ornithol.*, **55**, 259–261.

Emlen, S. T. (1982a). The evolution of helping. I. An ecological constraints model. *Am. Nat.*, **119**, 29–39.

(1982b). The evolution of helping. II. The role of behavioral conflict. *Am. Nat.*, **119**, 40–53.

(1984). Cooperative breeding in birds and mammals. In: *Behavioural Ecology: an Evolutionary Approach*, 2nd edn., ed. J. R. Krebs and N. B. Davies. Oxford: Blackwell. pp. 305–339.

(1990). White-fronted bee-eaters: helping in a colonially nesting species. In: *Cooperative Breeding in Birds*, ed. P. B. Stacey and W. D. Koenig. Cambridge: Cambridge University Press. pp. 489–526.

(1991). Evolution of cooperative breeding in birds and mammals. In: *Behavioural Ecology: an Evolutionary Approach*, 3rd edn., ed. J. R. Krebs and N. B. Davies. Oxford: Blackwell. pp. 301–337.

(1995). An evolutionary theory of the family. *Proc. Natl. Acad. Sci. USA*, **92**, 8092–8099.

(1996). Reproductive sharing in different types of kin associations. *Am. Nat.*, **148**, 756–763.

(1997a). Predicting family dynamics in social vertebrates. In *Behavioural Ecology: an Evolutionary Approach*, 4th edn., ed. J. R. Krebs and N. B. Davies. Oxford: Blackwell. pp. 228–253.

(1997b). When mothers prefer daughters over sons. *Trends Ecol. Evol.*, **12**, 291–292.

(1999). Reproductive skew in cooperatively breeding birds: an overview of the issues. *Proc. Int. Ornithol. Congr.*, **22**, 2922–2931.

Emlen, S. T. and Vehrencamp, S. L. (1983). Cooperative breeding strategies among birds. In: *Perspectives in Ornithology*, ed. A. Brush. Cambridge: Cambridge University Press. pp. 93–120.

(1985). Cooperative breeding strategies among birds. In: *Experimental Behavioral Ecology*, ed. B. Hölldobler and M. Lindauer. Stuttgart: Gustav Fischer. pp. 359–374.

Emlen, S. T. and Wrege, P. H. (1986). Forced copulations and intraspecific parasitism: two costs of social living in the white-fronted bee-eater. *Ethology*, **71**, 2–29.

(1988). The role of kinship in helping decisions among white-fronted bee-eaters. *Behav. Ecol. Sociobiol.*, **23**, 305–316.

(1989). A test of alternate hypotheses for helping behavior in white-fronted bee-eaters of Kenya. *Behav. Ecol. Sociobiol.*, **25**, 303–320.

(1991). Breeding biology of white-fronted bee-eaters at Nakuru: the influence of helpers on breeder fitness. *J. Anim. Ecol.*, **60**, 309–326.

(1992). Parent–offspring conflict and the recruitment of helpers among bee-eaters. *Nature*, **356**, 331–333.

Emlen, S. T., Emlen, J. M. and Levin, S. A. (1986). Sex-ratio selection in species with helpers-at-the-nest. *Am. Nat.*, **127**, 1–8.

Emlen, S. T., Reeve, H. K., Sherman, P. W. and Wrege, P. H. (1991). Adaptive versus non-adaptive explanations of behavior: the case of alloparental helping. *Am. Nat.*, **138**, 259–270.

Emlen, S. T., Reeve, H. K. and Keller, L. (1998). Reproductive skew: disentangling concessions from control. *Trends Ecol. Evol.*, **13**, 458–459.

Engh, A. L., Esch, K., Smale, L. and Holekamp, K. E. (2000). Mechanisms of maternal rank "inheritance" in the spotted hyaena, *Crocuta crocuta*. *Anim. Behav.*, **60**, 323–332.

Engh, A. L., Funk, S. M., Van Horn, R. C., Scribner, K. T., Bruford, M. W., Libants, S., Szykman, M., Smale, L. and Holekamp, K. E. (2002). Reproductive skew among males in a female-dominated mammalian society. *Behav. Ecol.*, **13**, 193–200.

Ens, B. J., Safriel, U. N. and Harris, M. P. (1993). Divorce in the long-lived and monogamous oystercatcher, *Haematopus estralegus*: incompatibility or choosing the better option? *Anim. Behav.*, **45**, 1199–1217.

Ericson, P. G. P., Christidis, L., Cooper, A., Irestedt, M., Jackson, J., Johansson, U. S. and Norman, J. A. (2002). A Gondwanan origin of passerine birds supported by DNA sequences of the endemic New Zealand wrens. *Proc. R. Soc. London Ser. B*, **269**, 235–241.

Evans, P. G. H. (1988). Intraspecific nest parasitism in the European starling *Sturnus vulgaris*. *Anim. Behav.*, **36**, 1282–1294.

Ewen, J. G. (2001). Primary sex ratio variation in the Meliphagidae (honeyeaters). Ph.D. dissertation, La Trobe University, Melbourne, Australia.

Ewen, J. G. and Armstrong, D. P. (2000). Male provisioning is negatively correlated with attempted extrapair copulation frequency in the stitchbird (or hihi). *Anim. Behav.*, **60**, 429–433.

(2002). Unusual sexual behaviour in the stitchbird (or hihi) *Notiomystis cincta*. *Ibis*, **144**, 530–531.

Ewen, J. G., Armstrong, D. P. and Lambert, D. M. (1999). Floater males gain reproductive success through extrapair fertilizations in the stitchbird. *Anim. Behav.*, **58**, 321–328.

Ewen, J. G., Clarke, R. H., Moysey, E., Boulton, R. L., Crozier, R. H. and Clarke, M. F. (2001). Primary sex ratio bias in an endangered cooperatively breeding bird, the black-eared miner, and its complications for conservation. *Biol. Conserv.*, **101**, 137–145.

Ewen, J. G., Crozier, R. H., Cassey, P., Ward-Smith, T., Painter, J. N., Robertson, R. J., Jones, D. A. and Clarke, M. F. (2003). Facultative control of offspring sex in the cooperatively breeding bell miner, *Manorina melanophrys*. *Behav. Ecol.*, **14**, 157–164.

Faaborg, J. and Bednarz, J. C. (1990). Galápagos and Harris' hawks: divergent causes of sociality in two raptors. In: *Cooperative Breeding in Birds*, ed. P. B. Stacey and W. D. Koenig. Cambridge: Cambridge University Press. pp. 359–383.

Faaborg, J. and Patterson, C. B. (1981). The characteristics and occurrence of cooperative polyandry. *Ibis*, **123**, 477–484.

Faaborg, J., Parker, P. G., DeLay, L., DeVries, T., Bednarz, J. C., Paz, S. M., Naranjo, J. and Waite, T. A. (1995). Confirmation of cooperative polyandry in the Galápagos hawk (*Buteo galapagoensis*). *Behav. Ecol. Sociobiol.*, **36**, 83–90.

Faaborg, J., Thompson, F. R., III, Robinson, S. R., Donovan, T. M., Whitehead, D. R. and Brawn, J. D. (1998). Understanding fragmented Midwestern landscapes: the future. In: *Avian Conservation: Research and Management*, ed. J. M. Marzluff and R. Sallabanks. Washington, DC: Island Press. pp. 193–207.

Fahrig, L. (2001). How much habitat is enough? *Biol. Conserv.*, **100**, 65–74.

Farley, G. H. (1995). Thermal, social and distributional consequences of nighttime cavity roosting in *Campylorhynchus* wrens. Ph.D. dissertation, University of New Mexico, Albuquerque, NM.

Faulkes, C. G. and Abbott, D. H. (1991). Social control of reproduction in both breeding and non-breeding male naked mole-rats, *Heterocephalus glaber*. *J. Repro. Fert.*, **93**, 427–435.

(1997). The physiology of a reproductive dictatorship: regulation of male and female reproduction by a single breeding female in colonies of naked mole-rats. In: *Cooperative Breeding in Mammals*, ed. N. G. Solomon and J. A. French. Cambridge: Cambridge University Press. pp. 302–334.

Faulkes, C. G. and Bennett, N. C. (2001). Family values: group dynamics and social control of reproduction in African mole-rats. *Trends Ecol. Evol.*, **16**, 184–190.

Faulkes, C. G., Abbott, D. H., Jarvis, J. U. M. and Sherriff, F. E. (1990a). LH responses of female naked mole-rats, *Heterocephalus glaber*, to single and multiple doses of exogenous GnRH. *J. Repro. Fert.*, **89**, 317–323.

Faulkes, C. G., Abbott, D. H. and Mellor, A. L. (1990b). Investigation of genetic diversity in wild colonies of naked mole-rats (*Heterocephalus glaber*) by DNA fingerprinting. *J. Zool.*, **221**, 87–97.

Faulkes, C. G., Abbott, D. H. and Jarvis, J. U. M. (1991). Social suppression of reproduction in male naked mole-rats, *Heterocephalus glaber*. *J. Repro. Fert.*, **91**, 593–604.

Faulkes, C. G., Trowell, S. N., Jarvis, J. U. M. and Bennett, N. C. (1994). Investigation of numbers and motility of spermatozoa in reproductively active and socially suppressed males of two eusocial African mole-rats, the naked mole-rat (*Heterocephalus glaber*) and the Damaraland mole-rat (*Cryptomys damarensis*). *J. Repro. Fert.*, **100**, 411–416.

Faulkes, C. G., Bennett, N. C., Bruford, M. W., O'Brien, H. P., Aguilar, G. H. and Jarvis, J. U. M. (1997). Ecological constraints drive social evolution in the African mole-rats. *Proc. R. Soc. London Ser. B*, **264**, 1619–1627.

Feduccia, A. (1996). *The Origin and Evolution of Birds*. New Haven, CT: Yale University Press.

Fernández, G. J. and Reboreda, J. C. (1998). Effects of clutch size and timing of breeding on reproductive success of greater rheas. *Auk*, **115**, 340–348.

Fessl, B., Kleindorfer, S., Hoi, H. and Lorenz, K. (1996). Extra male parental behaviour: evidence for an alternative mating strategy in the moustached warbler *Acrocephalus melanopogon*. *J. Avian Biol.*, **27**, 88–91.

Festa-Bianchet, M. (1996). Offspring sex ratio studies of mammals: does publication depend on the quality of the research or the direction of the results? *Ecoscience*, **3**, 42–44.

Ficken, M. S., Weise, C. M. and Popp, J. W. (1990). Dominance rank and resources access in winter flocks of black-capped chickadees. *Wilson Bull.*, **102**, 623–633.

Firestone, K. B., Thompson, K. V. and Carter, C. S. (1991). Female–female interactions and social stress in prairie voles. *Behav. Neural Biol.*, **55**, 31–41.

Fisher, R. A. (1930). *The Genetical Theory of Natural Selection*. Oxford: Clarendon Press.

Ford, H. A., Bell, H., Nias, R. and Noske, R. (1988). The relationship between ecology and the incidence of cooperative breeding in Australian birds. *Behav. Ecol. Sociobiol.*, **22**, 239–250.

Ford, H. A., Barrett, G. W., Saunders, D. A. and Recher, H. F. (2001). Why have birds in the woodlands of southern Australia declined? *Biol. Conserv.*, **97**, 71–88.

Forsgren, E., Karlsson, A. and Kvarnemo, C. (1996). Female sand gobies gain direct benefits by choosing males with eggs in their nests. *Behav. Ecol. Sociobiol.*, **39**, 91–96.

Fraga, R. M. (1979). Helpers at the nest in passerines from Buenos Aires province, Argentina. *Auk*, **96**, 606–608.

Frank, S. A. (1987). Individual and population sex allocation patterns. *Theor. Pop. Biol.*, **31**, 47–74.

(1990). Sex allocation theory for birds and mammals. *Annu. Rev. Ecol. Syst.*, **21**, 13–56.

Franklin, J. F. and Forman, R. T. T. (1987). Creating landscape patterns by forest cutting: ecological consequences and principles. *Landscape Ecol.*, **1**, 5–18.

French, J. A. (1997). Proximate regulation of singular breeding in callitrichid primates. In: *Cooperative Breeding in Mammals*, ed. N. G. Solomon and J. A. French. Cambridge: Cambridge University Press. pp. 34–75.

French, J. A., Abbott, D. H. and Snowdon, C. T. (1984). The effects of social environment on estrogen excretion, scent marking, and sociosexual behavior in tamarins (*Saguinus oedipus*). *Am. J. Primatol.*, **6**, 155–167.

French, J. A., Inglett, B. J. and Dethlefs, T. M. (1989). The reproductive status of nonbreeding group members in captive golden lion tamarin social groups. *Am. J. Primatol.*, **18**, 73–86.

French, J. A., Degraw, W. A., Hendricks, S. E., Wegner, F. and Bridson, W. E. (1992). Urinary and plasma gonadotropin concentrations in golden lion tamarins (*Leontopithecus r. rosalia*). *Am. J. Primatol.*, **26**, 53–59.

Frith, C. B., Frith, D. W. and Jansen, A. (1997). The nesting biology of the chowchilla, *Orthonyx spaldingii* (Orthonychidae). *Emu*, **97**, 18–30.

Frith, H. J. and Davies, S. J. J. F. (1961). Ecology of the magpie goose *Anseranas semipalmata*. *CSIRO Wildl. Res.*, **113**, 555–557.

Fry, C. H. (1977). The evolutionary significance of cooperative breeding in birds. In: *Evolutionary Ecology*, ed. B. Stonehouse and C. M. Perrins. Baltimore, MD: University Park Press. pp. 127–136.

Fry, C. H. (1980). Survival and longevity among tropical land birds. *Proc. Pan-African Ornithol. Congr.*, **4**, 333–343.

Fry, C. H., Keith, S. and Urban, E. K. (1988). *The Birds of Africa*, vol. 3. New York, NY: Academic Press.

Garland, T., Jr., Midford, P. E. and Ives, A. R. (1999). An introduction to phylogenetically based statistical methods, with a new method for confidence intervals on ancestral values. *Am. Zool.*, **39**, 374–388.

Gaston, A. J. (1973). The ecology and behaviour of the long-tailed tit. *Ibis*, **115**, 330–351.

Gaston, A. J. (1977). Social behaviour within groups of jungle babblers (*Turdoides striatus*). *Anim. Behav.*, **25**, 828–848.

Gaston, A. J. (1978). The evolution of group territorial behavior and cooperative breeding. *Am. Nat.*, **112**, 1091–1100.

Gayou, D. C. (1986). The social system of the Texas green jay. *Auk*, **103**, 540–547.

Getz, L. L., McGuire, B., Pizzuto, T., Hofmann, J. E. and Frase, B. (1993). Social organization of the prairie vole (*Microtus ochrogaster*). *J. Mammal.*, **74**, 44–58.

Getz, L. L., McGuire, B., Hofmann, J. E., Pizzuto, T. and Frase, B. (1994). Natal dispersal and philopatry in prairie voles (*Microtus ochrogaster*): settlement, survival, and potential reproductive success. *Ethol. Ecol. Evol.*, **6**, 267–284.

Gibbons, D. W. (1986). Brood parasitism and cooperative breeding in the moorhen, *Gallinula chloropus*. *Behav. Ecol. Sociobiol.*, **19**, 221–232.

Gibbs, H. L., Goldizen, A. W., Bullough, C. and Goldizen, A. R. (1994). Parentage analysis of multi-male social groups of Tasmanian native hens (*Tribonyx mortierii*): genetic evidence for monogamy and polyandry. *Behav. Ecol. Sociobiol.*, **35**, 363–371.

Gilchrist, J. S. (2001). Reproduction and pup care in the communal breeding banded mongoose. Ph.D. dissertation, University of Cambridge.

Gill, F. B. (1995). *Ornithology*, 2nd edn. New York: W. H. Freeman.

Ginther, A. J., Ziegler, T. E. and Snowdon, C. T. (2001). Reproductive biology of captive male cottontop tamarin monkeys as a function of social environment. *Anim. Behav.*, **61**, 65–78.

Ginther, A. J., Carlson, A. A., Ziegler, T. E. and Snowdon, C. T. (2002). Neonatal and pubertal development in males of a cooperatively breeding primate, the cotton-top tamarin (*Saguinus oedipus oedipus*). *Biol. Repro.*, **66**, 282–290.

Gittleman, J. L. (1989). Carnivore group living: comparative trends. In: *Carnivore Behavior, Ecology and Evolution*, ed. J. L. Gittleman. Ithaca, NY: Cornell University Press. pp. 183–207.

Goldizen, A. W. (1987). Facultative polyandry and the role of infant-carrying in wild saddle-back tamarins (*Saguinus fuscicollis*). *Behav. Ecol. Sociobiol.*, **20**, 99–109.

Goldizen, A. W. and Terborgh, J. (1989). Demography and dispersal patterns of a tamarin population: possible causes of delayed breeding. *Am. Nat.*, **134**, 208–224.

Goldizen, A. W., Putland, D. A. and Goldizen, A. R. (1998). Variable mating patterns in Tasmanian native hens (*Gallinula mortierii*): correlates of reproductive success. *J. Anim. Ecol.*, **67**, 307–317.

Goldizen, A. W., Buchan, J. C., Putland, D. A., Goldizen, A. R. and Krebs, E. A. (2000). Patterns of mate-sharing in a population of Tasmanian native hens *Gallinula mortierii*. *Ibis*, **142**, 40–47.

Gompper, M. E., Gittleman, J. L. and Wayne, R. K. (1997). Genetic relatedness, coalitions and social behaviour of white-nosed coatis, *Nasua nirica*. *Anim. Behav.*, **53**, 781–797.

Gorman, M. L., Mills, M. G., Raath, J. P. and Speakman, J. R. (1998). High hunting costs make African wild dogs vulnerable to kleptoparasitism by hyaenas. *Nature*, **391**, 479–481.

Gould, S. J. and Vrba, E. S. (1982). Exaptation – a missing term in the science of form. *Paleobiology*, **8**, 4–15.

Gowaty, P. A. (1993). Differential dispersal, local resource competition, and sex ratio variation in birds. *Am. Nat.*, **141**, 263–280.

  (1996a). Field studies of parental care in birds: new data focus questions on variation among females. *Adv. Study Behav.*, **26**, 477–531.

  (1996b). Multiple mating by females selects for males that stay: another hypothesis for social monogamy in passerine birds. *Anim. Behav.*, **51**, 482–484.

Gowaty, P. A. and Lennartz, M. R. (1985). Sex ratios of nestling and fledgling red-cockaded woodpeckers (*Picoides borealis*) favor males. *Am. Nat.*, **126**, 347–353.

Gowaty, P. A. and Plissner, J. H. (1998). Eastern bluebird (*Sialia sialis*). In: *The Birds of North America*, ed. A. Poole and F. Gill. Philadelphia, PA and Washington, DC: Academy of Natural Sciences and American Ornithologists' Union.

Grant, B. R. and Grant, P. R. (1996). Cultural inheritance of song and its role in the evolution of Darwin's finches. *Evolution*, **50**, 2471–2487.

Greeff, J. M. and Bennett, N. C. (2000). Causes and consequences of incest avoidance in the cooperatively breeding mole-rat, *Cryptomys darlingi* (Bathyergidae). *Ecol. Lett.*, **3**, 318–328.

Green, D. J. and Cockburn, A. (2001). Post-fledging care, philopatry and recruitment in brown thornbills. *J. Anim. Ecol.*, **70**, 505–514.

Green, D. J., Cockburn, A., Hall, M. L., Osmond, H. L. and Dunn, P. O. (1995). Increased opportunities for cuckoldry may be why dominant male fairy-wrens tolerate helpers. *Proc. R. Soc. London Ser. B*, **262**, 297–303.

Green, D. J., Osmond, H. L., Double, M. C. and Cockburn, A. (2000). Display rate by male fairy-wrens (*Malurus cyaneus*) during the fertile period of females has little influence on extra-pair mate choice. *Behav. Ecol. Sociobiol.*, **48**, 438–446.

Greenwood, P. J. (1980). Mating systems, philopatry and dispersal in birds and mammals. *Anim. Behav.*, **28**, 1140–1162.

Griesser, M. (2003). Nepotistic vigilance behavior in Siberian jay parents. *Behav. Ecol.*, **14**, 246–250.

Griesser, M. and Ekman, J. (2004). Nepotistic alarm calling in the Siberian jay (*Perisoreus infaustus*). *Anim. Behav.*, in press.

Griffith, S. C. (2000). High fidelity on islands: a comparative study of extrapair paternity in passerine birds. *Behav. Ecol.*, **11**, 265–273.

Griffiths, R., Daan, S. and Dijkstra, C. (1996). Sex identification in birds using two CHD genes. *Proc. R. Soc. London Ser. B*, **263**, 1251–1256.

Griffiths, R., Double, M. C., Orr, K. and Dawson, R. J. G. (1998). A DNA test to sex most birds. *Mol. Ecol.*, **7**, 1071–1075.

Grimes, L. G. (1976). The occurrence of cooperative breeding behaviour in African birds. *Ostrich*, **47**, 1–15.

  (1980). Observations of group behaviour and breeding biology of the yellow-billed shrike *Corvinella corvina*. *Ibis*, **122**, 166–192.

Grindstaff, J. L., Buerkle, C. A., Casto, J. M., Nolan, V., Jr. and Ketterson, D. (2001). Offspring sex ratio is unrelated to male attractiveness in dark-eyed juncos (*Junco hyemalis*). *Behav. Ecol. Sociobiol.*, **50**, 312–316.

Hager, R. (2003). Reproductive skew models applied to primates. In: *Sexual Selection and Reproductive Competition in Primates: New Perspectives and Directions*, ed. C. B. Jones. Norman, OK: American Society of Primatologists. pp. 65–101.

Haig, D. (2000). Genomic imprinting, sex-biased dispersal, and social behavior. *Ann. NY Acad. Sci.*, **907**, 149–163.

Haig, S. M., Walters, J. R. and Plissner, J. H. (1994). Genetic evidence for monogamy in the cooperatively breeding red-cockaded woodpecker. *Behav. Ecol. Sociobiol.*, **34**, 295–303.

Hailman, J. P., McGowan, K. J. and Woolfenden, G. E. (1994). Role of helpers in the sentinel behaviour of the Florida scrub jay (*Aphelocoma c. coerulescens*). *Ethology*, **97**, 119–140.

Hamilton, W. D. (1963). The evolution of altruistic behavior. *Am. Nat.*, **97**, 354–356.

  (1964). The genetical evolution of social behaviour. I, II. *J. Theor. Biol.*, **7**, 1–52.

  (1967). Extraordinary sex ratios. *Science*, **156**, 477–488.

  (1971). Geometry for the selfish herd. *J. Theor. Biol.*, **31**, 295–311.

Hammel, E. A., McDaniel, C. K. and Wachter, K. W. (1979). Demographic consequences of incest tabus: a microsimulation analysis. *Science*, **205**, 972–977.

Handford, P. and Mares, M. A. (1985). The mating systems of ratites and tinamous: an evolutionary perspective. *J. Linn. Soc.*, **25**, 77–104.

Hannon, S. J., Mumme, R. L., Koenig, W. D. and Pitelka, F. A. (1985). Replacement of breeders and within-group conflict in the cooperatively breeding acorn woodpecker. *Behav. Ecol. Sociobiol.*, **17**, 303–312.

Hardy, I. C. W. (1997). Possible factors influencing vertebrate sex ratios: an introductory overview. *Appl. Anim. Behav. Sci.*, **51**, 217–241.

Harrison, C. J. O. (1969). Helpers at the nest in Australian passerine birds. *Emu*, **69**, 30–40.

Harshman, J. (1994). Reweaving the tapestry: what can we learn from Sibley and Ahlquist (1990)? *Auk*, **111**, 377–388.

Hartley, I. R. and Davies, N. B. (1994). Limits to cooperative polyandry in birds. *Proc. R. Soc. London Ser. B*, **257**, 67–73.

Hartley, I. R., Davies, N. B., Hatchwell, B. J., Desrochers, A., Nebel, D. and Burke, T. (1995). The polygynandrous mating system of the alpine accentor, *Prunella collaris*. 2. Multiple paternity and parental effort. *Anim. Behav.*, **49**, 789–803.

Hartley, I. R., Griffith, S. C., Wilson, K., Shepherd, M. and Burke, T. (1999). Nestling sex ratios in the polygynously breeding corn bunting, *Miliaria calancra*. *J. Avian Biol.*, **30**, 7–14.

Hasselquist, D. and Kempenaers, B. (2002). Parental care and adaptive brood sex ratio manipulation in birds. *Phil. Trans. R. Soc. London B*, **357**, 363–372.

Hasselquist, D. and Sherman, P. W. (2001). Social mating systems and extrapair fertilizations in passerine birds. *Behav. Ecol.*, **12**, 457–466.

Hatchwell, B. J. (1999). Investment strategies of breeders in avian cooperative breeding systems. *Am. Nat.*, **154**, 205–219.

Hatchwell, B. J. and Davies, N. B. (1992). An experimental study of mating competition in monogamous and polyandrous dunnocks, *Prunella modularis*. II. Influence of removal and replacement experiments on mating systems. *Anim. Behav.*, **43**, 611–622.

Hatchwell, B. J. and Komdeur, J. (2000). Ecological constraints, life history traits and the evolution of cooperative breeding. *Anim. Behav.*, **59**, 1079–1086.

Hatchwell, B. J. and Russell, A. F. (1996). Provisioning rules in cooperatively breeding long-tailed tits *Aegithalos caudatus*: an experimental study. *Proc. R. Soc. London Ser. B*, **263**, 83–88.

Hatchwell, B. J., Russell, A. F., Fowlie, M. K. and Ross, D. J. (1999). Reproductive success and nest-site selection in a cooperative breeder: effect of experience and a direct benefit of helping. *Auk*, **116**, 355–363.

Hatchwell, B. J., Russell, A. F., Ross, D. J. and Fowlie, M. K. (2000). Divorce in cooperatively breeding long-tailed tits: a consequence of inbreeding avoidance? *Proc. R. Soc. London Ser. B*, **267**, 813–819.

Hatchwell, B. J., Anderson, C., Ross, D. J., Fowlie, M. K. and Blackwell, P. G. (2001a). Social organization of cooperatively breeding long-tailed tits: kinship and spatial dynamics. *J. Anim. Ecol.*, **70**, 820–830.

Hatchwell, B. J., Ross, D. J., Fowlie, M. K. and McGowan, A. (2001b). Kin discrimination in cooperatively breeding long-tailed tits. *Proc. R. Soc. London Ser. B*, **268**, 885–890.

Hatchwell, B. J., Ross, D. J., Chaline, N., Fowlie, M. K. and Burke, T. A. (2002). Parentage in the cooperative breeding system of long-tailed tits, *Aegithalos caudatus*. *Anim. Behav.*, **64**, 55–63.

Hatchwell, B. J., Russell, A. F., MacColl, A. D. C., Ross, D. J., Fowlie, M. K. and McGowan, A. (2003). Helpers increase long-term but not short-term productivity in cooperatively breeding long-tailed tits. *Behav. Ecol.*, in press.

Haydock, J. (1993). Cooperative breeding in bicolored wrens, *Campylorhynchus griseus*. Ph.D. dissertation, Purdue University, West Lafayette, IN.

Haydock, J. and Koenig, W. D. (2002). Reproductive skew in the polygynandrous acorn woodpecker. *Proc. Natl. Acad. Sci. USA*, **99**, 7178–7183.

(2003). Patterns of reproductive skew in the polygynandrous acorn woodpecker. *Am. Nat.*, **162**, 277–289.

Haydock, J., Parker, P. G. and Rabenold, K. N. (1996). Extra-pair paternity uncommon in the cooperatively breeding bicolored wren. *Behav. Ecol. Sociobiol.*, **38**, 1–16.

Haydock, J., Koenig, W. D. and Stanback, M. T. (2001). Shared parentage and incest avoidance in the cooperatively breeding acorn woodpecker. *Mol. Ecol.*, **10**, 1515–1525.

Hedrick, P. W. (1994). Purging inbreeding depression and the probability of extinction. *Heredity*, **73**, 363–372.

Heer, L. (1996). Cooperative breeding by alpine accentors *Prunella collaris*: polygynandry, territoriality and multiple paternity. *J. Ornithol.*, **137**, 35–51.

Hegner, R. E. and Emlen, S. T. (1987). Territorial organization of the white-fronted bee-eater in Kenya. *Ethology*, **76**, 189–222.

Heinsohn, R. G. (1991a). Kidnapping and reciprocity in cooperatively breeding white-winged choughs. *Anim. Behav.*, **41**, 1097–1100.

(1991b). Slow learning of foraging skills and extended parental care in cooperatively breeding white-winged choughs. *Am. Nat.*, **137**, 864–881.

(1991c). Evolution of obligate cooperative breeding in white-winged choughs: a statistical approach. *Proc. Int. Ornithol. Congr.*, **20**, 1309–1316.

(1992). Cooperative enhancement of reproductive success in white-winged choughs. *Evol. Ecol.*, **6**, 97–114.

(1995). Hatching asynchrony and brood reduction in cooperatively breeding white-winged choughs *Corcorax melanorhamphos*. *Emu*, **95**, 252–258.

Heinsohn, R. G. and Cockburn, A. (1994). Helping is costly to young birds in cooperatively breeding white-winged choughs. *Proc. R. Soc. London Ser. B*, **256**, 293–298.

Heinsohn, R. G. and Legge, S. (1999). The cost of helping. *Trends Ecol. Evol.*, **14**, 53–57.

(2003). Breeding biology of the reverse-dichromatic, cooperative parrot, *Eclectus roratus*. *J. Zool.*, **259**, 197–208.

Heinsohn, R. G., Cockburn, A. and Cunningham, R. B. (1988). Foraging, delayed maturity and cooperative breeding in white-winged choughs (*Corcorax melanorhamphos*). *Ethology*, **77**, 177–186.

Heinsohn, R. G., Cockburn, A. and Mulder, R. A. (1990). Avian cooperative breeding: old hypotheses and new directions. *Trends Ecol. Evol.*, **5**, 403–407.

Heinsohn, R. G., Legge, S. and Barry, S. (1997). Extreme bias in sex allocation in eclectus parrots. *Proc. R. Soc. London Ser. B*, **264**, 1325–1329.

Heinsohn, R. G., Legge, S. and Dunn, P. O. (1999). Extreme reproductive skew in cooperatively-breeding birds: tests of theory in white-winged choughs. *Proc. Int. Ornithol. Congr.*, **22**, 2858–2878.

Heinsohn, R. G., Dunn, P. O., Legge, S. and Double, M. C. (2000). Coalitions of relatives and reproductive skew in cooperatively breeding white-winged choughs. *Proc. R. Soc. London Ser. B*, **267**, 243–249.

Heistermann, M., Kleis, E., Prove, E. and Wolters, H. J. (1989). Fertility status, dominance, and scent marking behavior of family-housed female cotton-top tamarins (*Saguinus oedipus*) in absence of their mothers. *Am. J. Primatol.*, **18**, 177–189.

Helsper, J. P. F. G., van Loon, Y. P. J. and Kwakkel, R. P. (1996). Growth of broiler chicks fed diets containing tannin-free and tannin-containing near-isogenic lines of faba bean *(Vicia faba)*. *J. Agron. Food Sci.*, **44**, 1070–1075.

Heppell, S. S., Walters, J. R. and Crowder, L. B. (1994). Evaluating management alternatives for red-cockaded woodpeckers: a modeling approach. *J. Wildl. Manage.*, **58**, 479–487.

Herbst, M. and Bennett, N. C. (2001). Recrudescence of sexual activity in a colony of the Mashona mole-rat (*Cryptomys darlingi*): an apparent case of incest avoidance. *J. Zool.*, **254**, 163–175.

Higgins, P. J., Peter, J. M. and Steele, W. K. (2001). *Handbook of Australian, New Zealand and Antarctic Birds, vol. 5. Tyrant-Flycatchers to Chats*. Oxford: Oxford University Press.

Hillis, D. M., Huelsenbeck, J. P. and Cunningham, C. W. (1994). Application and accuracy of molecular phylogenies. *Science*, **264**, 671–677.

Hodge, S. J. (2003). Evolution of cooperation in the communal breeding banded mongoose. Ph.D. dissertation, University of Cambridge.

Hodges, K. E., Mech, S. and Wolff, J. O. (2002). Sex and the single vole: effects of social grouping on prairie vole reproductive success. *Ethology*, **108**, 871–884.

Holmes, W. G. and Sherman, P. W. (1983). Kin recognition in animals. *Am. Sci.*, **71**, 46–55.

Hooge, P. N. (1995). Dispersal dynamics of the cooperatively breeding acorn woodpecker. Ph.D. dissertation, University of California, Berkeley, CA.

Hoogland, J. L. (1983). Nepotism and alarm calling in the black-tailed prairie dog (*Cynomys ludovicianus*). *Anim. Behav.*, **31**, 472–479.

(1995). *The Black-Tailed Prairie Dog*. Chicago, IL: University of Chicago Press.

Horn, P. L., Rafalski, J. A. and Whitehead, P. J. (1996). Molecular genetic (RAPD) analysis of breeding magpie geese. *Auk*, **113**, 552–557.

Hubrecht, R. C. (1989). The fertility of daughters in common marmoset (*Callithrix jacchus jacchus*) family groups. *Primates*, **30**, 423–432.

Hurxthal, L. M. (1979). Breeding behaviour of the ostrich, *Struthio camelus massaicus*, in Nairobi National Park. Ph.D. dissertation, University of Nairobi, Kenya.

Innes, K. E. and Johnston, R. E. (1996). Cooperative breeding in the white-throated magpie-jay. How do auxiliaries influence nesting success? *Anim. Behav.*, **51**, 519–533.

IUCN (2000). *Red List of Threatened Species*. Cambridge: IUCN, the World Conservation Union. www.redlist.org.

Jakobsson, S. (1988). Territorial fidelity of willow warbler (*Phylloscopus trochilus*) males and success in competition over territories. *Behav. Ecol. Sociobiol.*, **22**, 79–84.

James, P. C. and Oliphant, L. W. (1986). Extra birds and helpers at the nest of Richardson's merlin. *Condor*, **88**, 533–534.

James, W. H. (1993). Continuing confusion. *Nature*, **365**, 8.

Jamieson, I. G. (1989). Behavioral heterochrony and the evolution of birds' helping at the nest: an unselected consequence of communal breeding? *Am. Nat.*, **133**, 394–406.

(1991). The unselected hypothesis for the evolution of helping behavior: too much or too little emphasis on natural selection? *Am. Nat.*, **138**, 271–282.

(1997). Testing reproductive skew models in a communally breeding bird, the pukeko, *Porphyrio porphyrio*. *Proc. R. Soc. London Ser. B*, **264**, 335–340.

(1999). Reproductive skew models and inter-species variation in adjustment of individual clutch sizes in joint-nesting birds. *Proc. Int. Ornithol. Congr.*, **22**, 2894–2909.

Jamieson, I. G. and Craig, J. L. (1987a). Critique of helping behaviour in birds: a departure from functional explanations. In: *Perspectives in Ethology*, vol. 7, ed. P. P. G. Bateson and P. Klopfer. New York, NY: Plenum. pp. 79–98.

(1987b). Dominance and mating in a communal polygynandrous bird: cooperation or indifference towards mating competitors? *Ethology*, **75**, 317–327.

(1987c). Male–male and female–female courtship and copulation behaviour in a communally breeding bird. *Anim. Behav.*, **35**, 1251–1253.

Jamieson, I. G., Quinn, J. S., Rose, P. A. and White, B. N. (1994). Shared paternity among non-relatives is a result of an egalitarian mating system in a communally breeding bird, the pukeko. *Proc. R. Soc. London Ser. B*, **257**, 271–277.

Jansen, A. (1999). Home ranges and group-territoriality in chowchillas *Orthonyx spaldingii*. *Emu*, **99**, 280–290.

Jarman, P. J. (1974). The social organisation of antelopes in relation to their ecology. *Behaviour*, **48**, 215–267.

Jarvis, J. U. M. and Bennett, N. C. (1993). Eusociality has evolved independently in two genera of bathyergid mole-rats – but occurs in no other subterranean mammal. *Behav. Ecol. Sociobiol.*, **33**, 353–360.

Jarvis, J. U. M., Oriain, M. J., Bennett, N. C. and Sherman, P. W. (1994). Mammalian eusociality: a family affair. *Trends Ecol. Evol.*, **9**, 47–51.

Jarvis, J. U. M., Bennett, N. C. and Spinks, A. C. (1998). Food availability and foraging by wild colonies of Damaraland mole-rats (*Cryptomys damarensis*): implications for sociality. *Oecologia*, **113**, 290–298.

Jennions, M. D. and Macdonald, D. W. (1994). Cooperative breeding in mammals. *Trends Ecol. Evol.*, **9**, 89–93.

Jennions, M. D. and Petrie, M. (2000). Why do females mate multiply? A review of the genetic benefits. *Biol. Rev.*, **75**, 21–64.

Johnson, K. J. (2001). Taxon sampling and the phylogenetic position of passeriformes: evidence from 916 avian cytochrome b sequences. *Syst. Biol.*, **50**, 128–136.

Johnson, M. L. and Gaines, M. S. (1990). Evolution of dispersal: theoretical models and empirical tests using birds and mammals. *Annu. Rev. Ecol. Syst.*, **21**, 449–480.

Johnstone, R. A. (2000). Models of reproductive skew: a review and synthesis. *Ethology*, **106**, 5–26.

Johnstone, R. A. and Cant, M. A. (1999). Reproductive skew and the threat of eviction: a new perspective. *Proc. R. Soc. London Ser. B*, **266**, 275–279.

Johnstone, R. A., Woodroffe, R., Cant, M. A. and Wright, J. (1999). Reproductive skew in multimember groups. *Am. Nat.*, **153**, 315–331.

Jones, C. S., Lessells, C. M. and Krebs, J. R. (1991). Helpers-at-the-nest in European bee-eaters (*Merops apiaster*): a genetic analysis. In: *DNA Fingerprinting Approaches and Applications*, ed. T. Burke, G. Dolf, A. J. Jeffreys and R. Wolff. Basel: Birkhauser. pp. 169–192.

Jones, D. A. (1998). Parentage, mate removal experiments and sex allocation in the co-operatively breeding bell miner, *Manorina melanophrys*. M.Sc. dissertation, Queen's University, Kingston, ON, Canada.

Jones, R. E. (1969). Hormonal control of incubation patch development in the California quail, *Lophortyx californicus*. *Gen. Comp. Endocrinol.*, **13**, 1–13.

(1971). The incubation patch of birds. *Biol. Rev.*, **46**, 315–339.

Joste, N. E., Koenig, W. D., Mumme, R. L. and Pitelka, F. A. (1982). Intra-group dynamics of a cooperative breeder: an analysis of reproductive roles in the acorn woodpecker. *Behav. Ecol. Sociobiol.*, **11**, 195–201.

Kattan, G. (1988). Food habits and social organization of acorn woodpeckers in Colombia. *Condor*, **90**, 100–106.

Keane, B., Waser, P. M., Creel, S. R., Creel, N. M., Elliott, L. F. and Minchella, D. J. (1994). Subordinate reproduction in dwarf mongooses. *Anim. Behav.*, **47**, 65–75.

Keane, B., Creel, S. R. and Waser, P. M. (1996). No evidence of inbreeding avoidance or inbreeding depression in a social carnivore. *Behav. Ecol.*, **7**, 480–489.

Keller, L. (1997). Indiscriminate altruism: unduly nice parents and siblings. *Trends Ecol. Evol.*, **12**, 99–103.

Keller, L. and Chapuisat, M. (1999). Cooperation among selfish individuals in insect societies. *BioScience*, **49**, 899–909.

Keller, L. and Reeve, H. K. (1994). Partitioning of reproduction in animal societies. *Trends Ecol. Evol.*, **9**, 98–102.

Keller, L. F. (1998). Inbreeding and its fitness effects in an insular population of song sparrows (*Melospiza melodia*). *Evolution*, **52**, 240–250.

Keller, L. F. and Arcese, P. (1998). No evidence for inbreeding avoidance in a natural population of song sparrows (*Melospiza melodia*). *Am. Nat.*, **152**, 380–392.

Kemp, A. C. (2001). Family Bucerotidae (Hornbills). In: *Handbook of the Birds of the World*, vol. 6, ed. J. del Hey, A. Elliott and J. Sargatal. Barcelona: Lynx Edicions. pp. 436–523.

Kempenaers, B., Congdon, B., Boag, P. and Robertson, R. J. (1999). Extrapair paternity and egg hatchability in tree swallows: evidence for the genetic compatibility hypothesis? *Behav. Ecol.*, **10**, 304–311.

Ketterson, E. D. and Nolan, V. (1994). Male parental behavior in birds. *Annu. Rev. Ecol. Syst.*, **25**, 601–628.

Khan, M. Z. and Walters, J. R. (1997). Is helping a beneficial learning experience for red-cockaded woodpecker (*Picoides borealis*) helpers? *Behav. Ecol. Sociobiol.*, **41**, 69–73.

(2002). Effects of helpers on breeder survival in the red-cockaded woodpecker (*Picoides borealis*). *Behav. Ecol. Sociobiol.*, **52**, 336–344.

Khan, M. Z., McNabb, F. M. A., Walters, J. R. and Sharp, P. J. (2001). Patterns of testosterone and prolactin concentrations and reproductive behavior of helpers and breeders in the cooperatively breeding red-cockaded woodpecker (*Picoides borealis*). *Horm. Behav.*, **40**, 1–13.

Kilner, R. (1998). Primary and secondary sex ratio manipulation by zebra finches. *Anim. Behav.*, **56**, 155–164.

Kimwele, C. N. and Graves, J. A. (2003). A molecular genetic analysis of the communal nesting of the ostrich (*Struthio camelus*). *Mol. Ecol.*, **12**, 229–236.

King, D. I., Champlin, T. B. and Champlin, P. J. (2000). An observation of cooperative breeding in the ovenbird. *Wilson Bull.*, **112**, 287–288.

Knapp, R. A. and Kovach, J. T. (1991). Courtship as an honest indicator of male parental quality in the bicolor damselfish, *Stegaster partitus*. *Behav. Ecol.*, **2**, 295–300.

Koenig, W. D. (1981). Space competition in the acorn woodpecker: power struggles in a cooperative breeder. *Anim. Behav.*, **29**, 396–409.

(1990). Opportunity of parentage and nest destruction in polygynandrous acorn woodpeckers, *Melanerpes formicivorus*. *Behav. Ecol.*, **1**, 55–61.

(1991). The effects of tannins and lipids on digestion of acorns by acorn woodpeckers. *Auk*, **108**, 79–88.

Koenig, W. D. and Benedict, L. S. (2002). Size, insect parasitism and energetic value of acorns stored by acorn woodpeckers. *Condor*, **104**, 539–547.

Koenig, W. D. and Dickinson, J. L. (1996). Nestling sex-ratio variation in western bluebirds. *Auk*, **113**, 902–910.

Koenig, W. D. and Heck, M. K. (1988). Ability of two species of oak woodland birds to subsist on acorns. *Condor*, **90**, 705–708.

Koenig, W. D. and Mumme, R. L. (1987). *Population Ecology of the Cooperatively Breeding Acorn Woodpecker*. Princeton, NJ: Princeton University Press.

(1990). Levels of analysis and the functional significance of helping behavior. In: *Interpretation and Explanation in the Study of Animal Behavior. Vol. 2: Explanation, Evolution, and Adaptation*, ed. M. Bekoff and D. Jamieson. Boulder, CO: Westview Press. pp. 268–303.

Koenig, W. D. and Pitelka, F. A. (1979). Relatedness and inbreeding avoidance: counterploys in the communally nesting acorn woodpecker. *Science*, **206**, 1103–1105.

(1981). Ecological factors and kin selection in the evolution of cooperative breeding in birds. In: *Natural Selection and Social Behavior: Recent Research and New Theory*, ed. R. D. Alexander and D. W. Tinkle. New York, NY: Chiron Press. pp. 261–280.

Koenig, W. D. and Stacey, P. B. (1990). Acorn woodpeckers: group living and food storage under contrasting ecological conditions. In: *Cooperative Breeding in Birds*, ed. P. B. Stacey and W. D. Koenig. Cambridge: Cambridge University Press. pp. 413–454.

Koenig, W. D. and Walters, J. R. (1999). Sex-ratio selection in species with helpers at the nest: the repayment model revisited. *Am. Nat.*, **153**, 124–130.

Koenig, W. D., Mumme, R. L. and Pitelka, F. A. (1983). Female roles in cooperatively breeding acorn woodpeckers. In: *Social Behavior of Female Vertebrates*, ed. S. K. Wasser. New York, NY: Academic Press. pp. 235–261.

  (1984). The breeding system of the acorn woodpecker in central coastal California. *Z. Tierpsychol.*, **65**, 289–308.

Koenig, W. D., Pitelka, F. A., Carmen, W. J., Mumme, R. L. and Stanback, M. T. (1992). The evolution of delayed dispersal in cooperative breeders. *Q. Rev. Biol.*, **67**, 111–150.

Koenig, W. D., Mumme, R. L., Stanback, M. T. and Pitelka, F. A. (1995). Patterns and consequences of egg destruction among joint-nesting acorn woodpeckers. *Anim. Behav.*, **50**, 607–621.

Koenig, W. D., Van Vuren, D. and Hooge, P. N. (1996). Detectability, philopatry, and the distribution of dispersal distances in vertebrates. *Trends Ecol. Evol.*, **11**, 514–517.

Koenig, W. D., Haydock, J. and Stanback, M. T. (1998). Reproductive roles in the cooperatively breeding acorn woodpecker: incest avoidance versus reproductive competition. *Am. Nat.*, **151**, 243–255.

Koenig, W. D., Stanback, M. T. and Haydock, J. (1999). Demographic consequences of incest avoidance in the cooperatively breeding acorn woodpecker. *Anim. Behav.*, **57**, 1287–1293.

Koenig, W. D., Hooge, P. N., Haydock, J. and Stanback, M. T. (2000). Natal dispersal in the cooperatively breeding acorn woodpecker. *Condor*, **102**, 492–502.

Koenig, W. D., Stanback, M. T., Haydock, J. and Kraaijeveld-Smit, F. (2001). Nestling sex ratio variation in the cooperatively breeding acorn woodpecker (*Melanerpes formicivorus*). *Behav. Ecol. Sociobiol.*, **49**, 357–365.

Koford, R. R., Bowen, B. S. and Vehrencamp, S. L. (1986). Habitat saturation in groove–billed anis (*Crotophaga sulcirostris*). *Am. Nat.*, **127**, 317–337.

  (1990). Groove-billed anis: joint nesting in a tropical cuckoo. In: *Cooperative Breeding in Birds*, ed. P. B. Stacey and W. D. Koenig. Cambridge: Cambridge University Press. pp. 333–356.

Kokko, H. (1999). Cuckoldry and the stability of biparental care. *Ecol. Lett.*, **2**, 247–255.

  (2003). Are reproductive skew models evolutionarily stable? *Proc. R. Soc. London Ser. B*, **270**, 265–270.

Kokko, H. and Ekman, J. (2002). Delayed dispersal as a route to breeding: territorial inheritance, safe havens, and ecological constraints. *Am. Nat.*, **160**, 468–484.

Kokko, H. and Johnstone, R. A. (1999). Social queuing in animal societies: a dynamic model of reproductive skew. *Proc. R. Soc. London Ser. B*, **266**, 571–578.

Kokko, H. and Lundberg, P. (2001). Dispersal, migration, and offspring retention in saturated habitats. *Am. Nat.*, **157**, 188–202.

Kokko, H., Johnstone, R. A. and Clutton-Brock, T. H. (2001). The evolution of cooperative breeding through group augmentation. *Proc. R. Soc. London Ser. B*, **268**, 187–196.

Kokko, H., Johnstone, R. A. and Wright, J. (2002). The evolution of parental and alloparental effort in cooperatively breeding groups: when should helpers pay to stay? *Behav. Ecol.*, **13**, 291–300.

Kölliker, M., Heeb, P., Werner, I., Mateman, A. C., Lessells, C. M. and Richner, H. (1999). Offspring sex ratio is related to male body size in the great tit (*Parus major*). *Behav. Ecol.*, **10**, 68–72.

Komdeur, J. (1992). Importance of habitat saturation and territory quality for evolution of cooperative breeding in the Seychelles warbler. *Nature*, **358**, 493–495.

  (1994a). The effect of kinship on helping in the cooperative breeding Seychelles warbler (*Acrocephalus sechellensis*). *Proc. R. Soc. London Ser. B*, **256**, 47–52.

  (1994b). Experimental evidence for helping and hindering by previous offspring in the cooperative-breeding Seychelles warbler *Acrocephalus sechellensis*. *Behav. Ecol. Sociobiol.*, **34**, 175–186.

  (1996). Influence of helping and breeding experience on reproductive performance in the Seychelles warbler: a translocation experiment. *Behav. Ecol.*, **7**, 326–333.

  (1998). Long-term fitness benefits of egg sex modification by the Seychelles warbler. *Ecol. Lett.*, **1**, 56–62.

  (2001). Mate guarding in the Seychelles warbler is energetically costly and adjusted to paternity risk. *Proc. R. Soc. London Ser. B*, **268**, 2103–2111.

Komdeur, J. and Edelaar, P. (2001a). Evidence that helping at the nest does not result in territory inheritance in the Seychelles warbler. *Proc. R. Soc. London Ser. B*, **268**, 2007–2012.

(2001b). Male Seychelles warblers use territory budding to maximize lifetime fitness in a saturated environment. *Behav. Ecol.*, **12**, 706–715.

Komdeur, J. and Hatchwell, B. J. (1999). Kin recognition: function and mechanism in avian societies. *Trends Ecol. Evol.*, **14**, 237–241.

Komdeur, J. and Pen, I. (2002). Adaptive sex allocation in birds: the complexities of linking theory and practice. *Phil. Trans. R. Soc. London B*, **357**, 373–380.

Komdeur, J., Huffstadt, A., Prast, W., Castle, G., Mileto, R. and Wattel, J. (1995). Transfer experiments of Seychelles warblers to new islands: changes in dispersal and helping behaviour. *Anim. Behav.*, **49**, 695–708.

Komdeur, J., Daan, S., Tinbergen, J. and Mateman, A. C. (1997). Extreme adaptive modification in sex ratio of Seychelles warbler's eggs. *Nature*, **385**, 522–525.

Komdeur, J., Magrath, M. J. L. and Krackow, S. (2002). Pre-ovulation control of hatchling sex ratio in the Seychelles warbler. *Proc. R. Soc. London Ser. B*, **269**, 1067–1072.

Komdeur, J., Piersma, T., Kraaijeveld, K., Kraaijeveld-Smit, F. and Richardson, D. S. (2004). Why Seychelles warblers fail to recolonize nearby islands: unwilling or unable to fly there? *Ibis*, in press.

Kossenko, S. M. and Fry, C. H. (1998). Competition and coexistence of the European bee-eater *Merops apiaster* and the blue-cheeked bee-eater *Merops persicus* in Asia. *Ibis*, **140**, 2–13.

Köster, F. (1971). Zum Nistverhalten des Ani, *Crotophaga ani. Bonn. Zool. Beitr.*, **22**, 4–27.

Kraaijeveld, K. and Dickinson, J. L. (2001). Family-based winter territoriality in western bluebirds: the structure and dynamics of winter groups. *Anim. Behav.*, **61**, 109–117.

Krackow, S. (1995). Potential mechanisms for sex ratio adjustment in mammals and birds. *Biol. Rev.*, **70**, 225–241.

(1999). Avian sex ratio distortions: the myth of maternal control. *Proc. Int. Ornithol. Congr.*, **22**, 425–433.

Krebs, J. R. (1982). Territorial defense in the great tit (*Parus major*): do residents always win? *Behav. Ecol. Sociobiol.*, **11**, 185–194.

Krebs, J. R. and Davies, N. B., eds. (1993). *An Introduction to Behavioural Ecology*, 3rd edn. Oxford: Blackwell.

Kruuk, H. (1975). Functional aspects of social hunting in carnivores. In: *Function and Evolution in Behavior*, ed. G. Baerends, C. Beer and A. Manning. Oxford: Clarendon Press. pp. 119–141.

Lacey, E. A. and Sherman, P. W. (1997). Cooperative breeding in naked mole-rats: implications for vertebrate and invertebrate sociality. In: *Cooperative Breeding in Mammals*, ed. N. G. Solomon and J. A. French. Cambridge: Cambridge University Press. pp. 267–301.

Lack, D. (1968). *Ecological Adaptations for Breeding in Birds*. London: Chapman and Hall.

Lacy, R. C. and Ballou, J. D. (1998). Effectiveness of selection in reducing the genetic load in populations of *Peromyscus polionotus* during generations of inbreeding. *Evol. Ecol.*, **52**, 900–909.

Lambert, D. M., Millar, C. D., Jack, K., Anderson, S. and Craig, J. L. (1994). Single- and multilocus DNA fingerprinting of communally breeding pukeko: do copulations or dominance ensure reproductive success? *Proc. Natl. Acad. Sci. USA*, **91**, 9641–9645.

Lancaster, D. A. (1964a). Biology of the brushland tinamou, *Nothoprocta cinerascens. Bull. Am. Mus. Nat. Hist.*, **127**, 269–314.

(1964b). Life history of the Boucard tinamou in British Honduras: II. breeding biology. *Condor*, **66**, 253–276.

Langen, T. A. (1996a). The mating system of the white-throated magpie-jay *Calocitta formosa* and Greenwood's hypothesis for sex-biased dispersal. *Ibis*, **138**, 506–513.

(1996b). Skill acquisition and the timing of natal dispersal in the white-throated magpie-jay, *Calocitta formosa. Anim. Behav.*, **51**, 575–588.

(2000). Prolonged offspring dependence and cooperative breeding in birds. *Behav. Ecol.*, **11**, 367–377.

Langen, T. A. and Vehrencamp, S. L. (1999). How white-throated magpie-jay helpers contribute during breeding. *Auk*, **116**, 131–140.

Langham, G. M., Hite, J. M. and DaCosta, J. M. (2003). Sex-biased territoriality, movement patterns, and helping behavior in rufous-tailed jacamars. *Auk*, **120** (2, suppl.), 34AA.

Legge, S. (2000a). The effect of helpers on reproductive success in the laughing kookaburra. *J. Anim. Ecol.*, **69**, 714–724.

(2000b). Helper contributions in the cooperatively breeding laughing kookaburra: feeding young is no laughing matter. *Anim. Behav.*, **59**, 1009–1018.

Legge, S. and Cockburn, A. (2000). Social and mating system of cooperatively breeding laughing kookaburras (*Dacelo novaeguineae*). *Behav. Ecol. Sociobiol.*, **47**, 220–229.

Legge, S., Heinsohn, R. G., Double, M. C., Griffiths, R. and Cockburn, A. (2001). Complex sex allocation in the laughing kookaburra. *Behav. Ecol.*, **12**, 524–533.

Leimar, O. (1996). Life-history analysis of the Trivers and Willard sex-ratio problem. *Behav. Ecol.*, **7**, 316–325.

Leonard, M. L., Horn, A. G. and Eden, S. F. (1989). Does juvenile helping enhance breeder reproductive success? A removal experiment on moorhens. *Behav. Ecol. Sociobiol.*, **25**, 357–361.

Lessells, C. M. (1990). Helping at the nest in European bee-eaters: who helps and why? In: *Population Biology of Passerine Birds, an Integrated Approach*, ed. J. Blondel, A. Gosler, J. D. Lebreton and R. McCleery. Berlin: Springer-Verlag. pp. 357–368.

  (1991). The evolution of life histories. In: *Behavioural Ecology: an Evolutionary Approach*, 3rd edn., ed. J. R. Krebs and N. B. Davies. Oxford: Blackwell. pp. 32–68.

Lessells, C. M. and Avery, M. I. (1987). Sex-ratio selection in species with helpers at the nest: some extensions of the repayment model. *Am. Nat.*, **129**, 610–620.

Lessells, C. M. and Mateman, A. C. (1996). Molecular sexing of birds. *Nature*, **383**, 761–762.

  (1998). Sexing birds using random amplified polymorphic DNA (RAPD) markers. *Mol. Ecol.*, **7**, 187–195.

Lessells, C. M. and Quinn, J. S. (1999). Primary sex ratios: variation, causes and consequences. *Proc. Int. Ornithol. Congr.*, **22**, 422–424.

Lessells, C. M., Avery, M. I. and Krebs, J. R. (1994). Nonrandom dispersal of kin: why do European bee-eater (*Merops apiaster*) brothers nest close together? *Behav. Ecol.*, **5**, 105–113.

Lessells, C. M., Mateman, A. C. and Visser, J. (1996). Great tit hatchling sex ratios. *J. Avian Biol.*, **27**, 135–142.

Letcher, B. H., Priddy, J. A., Walters, J. R. and Crowder, L. B. (1998). An individual-based, spatially-explicit simulation model of the population dynamics of the endangered red-cockaded woodpecker. *Biol. Conserv.*, **86**, 1–14.

Lewis, D. M. (1982). Cooperative breeding in a population of white-browed sparrow-weavers. *Ibis*, **124**, 511–522.

Lewis, S. E. and Pusey, A. E. (1997). Factors influencing the occurrence of communal care in plural breeding mammals. In: *Cooperative Breeding in Mammals*, ed. N. G. Solomon and J. A. French. Cambridge: Cambridge University Press. pp. 335–363.

Li, S.-H. and Brown, J. L. (2000). High frequency of extrapair fertilization in a plural breeding bird, the Mexican jay, revealed by DNA microsatellites. *Anim. Behav.*, **60**, 867–877.

  (2002). Reduction of maternal care: a new benefit of multiple mating? *Behav. Ecol.*, **13**, 87–93.

Lifson, N. and McClintock, R. (1966). Theory of use of the turnover rates of body water for measuring energy and material balance. *J. Theor. Biol.*, **180**, 803–811.

Ligon, J. D. (1970). Behavior and breeding biology of the red-cockaded woodpecker. *Auk*, 87, 255–278.

  (1978). Reproductive interdependence of pinyon jays and pinyon pines. *Ecol. Monogr.*, **48**, 111–126.

  (1981). Demographic patterns and communal breeding in the green woodhoopoe (*Phoeniculus purpureus*). In: *Natural Selection and Social Behavior: Recent Research and New Theory*, ed. R. D. Alexander and D. W. Tinkle. New York, NY: Chiron Press. pp. 231–243.

  (1985). [Book review] The Florida scrub jay: demography of a cooperative-breeding bird. *Science*, **227**, 1573–1574.

  (1993). The role of phylogenetic history in the evolution of contemporary avian mating and parental care systems. *Curr. Ornithol.*, **10**, 1–46.

  (1999). *The Evolution of Avian Breeding Systems*. Oxford: Oxford University Press.

  (2001). Family Phoeniculidae (Woodhoopoes). In: *Handbook of the Birds of the World*, vol. 6, ed. J. del Hoyo, A. Elliott and J. Sargatal. Barcelona: Lynx Edicions. pp. 412–435.

Ligon, J. D. and Davidson, N. K. (1988). Order Coraciiformes, Phoeniculidae, Woodhoopoes. In: *The Birds of Africa*, vol. 3, ed. E. K. Urban, C. H. Fry and S. Keith. New York, NY: Academic Press. pp. 356–370.

Ligon, J. D. and Ligon, S. H. (1978a). Communal breeding in green woodhoopoes as a case for reciprocity. *Nature*, **276**, 496–498.

  (1978b). The communal social system of the green woodhoopoe in Kenya. *Living Bird*, **17**, 159–197.

  (1983). Reciprocity in the green woodhoopoe (*Phoeniculus purpureus*). *Anim. Behav.*, **31**, 480–489.

  (1988). Territory quality: key determinant of fitness in the group-living green woodhoopoe. In: *The Ecology of Social Behavior*, ed. C. Slobodchikoff. New York, NY: Academic Press. pp. 229–254.

  (1990a). Female-biased sex ratio at hatching in the green woodhoopoe. *Auk*, **107**, 765–771.

  (1990b). Green woodhoopoes: life history traits and sociality. In: *Cooperative Breeding in Birds*, ed. P. B. Stacey and W. D. Koenig. Cambridge: Cambridge University Press. pp. 31–66.

Ligon, J. D. and Stacey, P. B. (1989). On the significance of helping behavior in birds. *Auk*, **106**, 700–705.

(1991). The origin and maintenance of helping behavior in birds. *Am. Nat.*, **138**, 254–258.

(1996). Land use, lag times and the detection of demographic change: the case of the acorn woodpecker. *Conserv. Biol.*, **10**, 840–846.

Ligon, J. D., Carey, C. and Ligon, S. H. (1988). Cavity roosting, philopatry and cooperative breeding in the green woodhoopoe may reflect a physiological trait. *Auk*, **105**, 123–127.

Ligon, J. D., Ligon, S. H. and Ford, H. A. (1991). An experimental study of the basis of male philopatry in the cooperatively breeding superb fairy-wren *Malurus cyaneus*. *Ethology*, **87**, 134–148.

Lindström, E. (1986). Territory inheritance and the evolution of group-living in carnivores. *Anim. Behav.*, **34**, 1825–1835.

Loflin, R. K. (1983). Communal behaviors of the smooth-billed ani (*Crotophaga ani*). Ph.D. dissertation, University of Miami, Coral Gables, FL.

Lucas, J. R., Creel, S. R. and Waser, P. M. (1997). Dynamic optimization and cooperative breeding: an evaluation of future fitness effects. In: *Cooperative Breeding in Mammals*, ed. N. G. Solomon and J. A. French. Cambridge: Cambridge University Press. pp. 171–198.

Lundy, K. J., Parker, P. G. and Zahavi, A. (1998). Reproduction by subordinates in cooperatively breeding Arabian babblers is uncommon but predictable. *Behav. Ecol. Sociobiol.*, **43**, 173–180.

Lyon, B. E. (1993a). Conspecific brood parasitism as a flexible female reproductive tactic in American coots. *Anim. Behav.*, **46**, 911–928.

(1993b). Tactics of parasitic American coots: host choice and the pattern of egg dispersion among host nests. *Behav. Ecol. Sociobiol.*, **33**, 87–100.

MacColl, A. and Hatchwell, B. J. (2002). Temporal variation in fitness payoffs promotes cooperative breeding in long-tailed tits *Aegithalos caudatus*. *Am. Nat.*, **160**, 186–194.

Macedo, R. H. F. (1992). Reproductive patterns and social organization of the communal guira cuckoo (*Guira guira*) in central Brazil. *Auk*, **109**, 786–799.

Macedo, R. H. F. and Bianchi, C. A. (1997). Communal breeding in tropical guira cuckoos *Guira guira*: sociality in the absence of a saturated habitat. *J. Avian Biol.*, **28**, 207–215.

Macedo, R. H. F. and Melo, C. (1999). Confirmation of infanticide in the communally breeding guira cuckoo. *Auk*, **116**, 847–851.

Macedo, R. H. F., Cariello, M. and Muniz, L. (2001). Context and frequency of infanticide in communally breeding guira cuckoos. *Condor*, **103**, 170–175.

Macgregor, N. A. and Cockburn, A. (2002). Sex differences in parental response to begging nestlings in superb fairy-wrens. *Anim. Behav.*, **63**, 923–932.

Macnamee, M. C., Sharp, P. J., Lea, R. W., Sterling, R. J. and Harvey, S. (1986). Evidence that vasoactive intestinal polypeptide is a physiological prolactin-releasing factor in the bantam hen. *Gen. Comp. Endocrinol.*, **62**, 470–478.

MacRoberts, M. H. and MacRoberts, B. R. (1976). Social organization and behavior of the acorn woodpecker in central coastal California. *Ornithol. Monogr.*, **21**, 1–115.

Maddison, D. R. and Maddison, W. P. (2000). *MacClade 4: Analysis of Phylogeny and Character Evolution*. Sunderland, MA: Sinauer Associates.

Maddison, W. P. (1990). A method for testing the correlated evolution of two binary characters: are gains or losses concentrated on certain branches of a phylogenetic tree? *Evolution*, **44**, 539–557.

(1995). Calculating the probability distribution of ancestral states reconstructed by parsimony on phylogenetic trees. *Syst. Biol.*, **44**, 474–481.

Magrath, R. D. (1999). Problems of distinguishing among models of reproductive skew within populations of cooperatively-breeding birds. *Proc. Int. Ornithol. Congr.*, **22**, 2879–2893.

(2001). Group breeding dramatically increases reproductive success of yearling but not older female scrubwrens: a model for cooperatively breeding birds? *J. Anim. Ecol.*, **70**, 370–385.

Magrath, R. D. and Heinsohn, R. G. (2000). Reproductive skew in birds: models, problems and prospects. *J. Avian Biol.*, **31**, 247–258.

Magrath, R. D. and Whittingham, L. A. (1997). Subordinate males are more likely to help if unrelated to the breeding female in cooperatively breeding white-browed scrubwrens. *Behav. Ecol. Sociobiol.*, **41**, 185–192.

Magrath, R. D. and Yezerinac, S. M. (1997). Facultative helping does not influence reproductive success or survival in cooperatively breeding white-browed scrubwrens. *J. Anim. Ecol.*, **66**, 658–670.

Malan, G. (1998). Solitary and social hunting in pale chanting goshawk *(Melierax canorus)* families: why use both strategies? *J. Raptor Res.*, **32**, 195–201.

Malan, G., Crowe, T. M., Biggs, R. and Herholdt, J. J. (1997). The social system of the pale chanting goshawk *Melierax canorus*, monogamy vs. polyandry and delayed dispersal. *Ibis*, **139**, 313–321.

Maney, D. L., Hahn, T. P., Schoech, S. J., Sharp, P. J., Morton, M. L. and Wingfield, J. C. (1999). Effects of ambient temperature on photo-induced prolactin secretion in three subspecies of white-crowned sparrow, *Zonotrichia leucophrys*. *Gen. Comp. Endocrinol.*, **113**, 445–456.

Manser, M. B. (1999). Response of foraging group members to sentinel calls in suricates *Suricata suricatta*. *Proc. R. Soc. London Ser. B*, **266**, 1013–1019.

Marchant, S. and Higgins, P. J. (1990). *Handbook of Australian, New Zealand and Antarctic Birds*. Oxford: Oxford University Press.

Martín-Vivaldi, M., Martínez, J. G., Palomino, J. J. and Soler, M. (2002). Extrapair paternity in the hoopoe *Upupa epops*: an exploration of the influence of interactions between breeding pairs, non-pair males and strophe length. *Ibis*, **144**, 236–247.

Marzluff, J. M. and Balda, R. P. (1990). Pinyon jays: making the best of a bad situation by helping. In: *Cooperative Breeding in Birds*, ed. P. B. Stacey and W. D. Koenig. Cambridge: Cambridge University Press. pp. 199–237.

Marzluff, J. M., Woolfenden, G. E., Fitzpatrick, J. W. and Balda, R. P. (1996). Breeding partnerships of two New World jays. In: *Partnerships in Birds: the Study of Monogamy*, ed. J. M. Black. Oxford: Oxford University Press. pp. 138–161.

Maswanganye, K. A., Bennett, N. C., Brinders, J. and Cooney, R. (1999). Oligospermia and azoospermia in non-reproductive male Damaraland mole-rats *Cryptomys damarensis* (Rodentia: Bathyergidae). *J. Zool.*, **248**, 411–418.

Mateo, J. M. (2002). Kin-recognition abilities and nepotism as a function of sociality. *Proc. R. Soc. London Ser. B*, **269**, 721–727.

Matthysen, E. (1999). Nuthatches (*Sitta europaea*: Aves) in forest fragments: demography of a patchy population. *Oecologia*, **119**, 501–509.

Maynard Smith, J. (1979). Game theory and the evolution of behaviour. *Proc. R. Soc. London Ser. B*, **205**, 475–488.

Maynard Smith, J. and Parker, G. A. (1976). The logic of asymmetric contests. *Anim. Behav.*, **24**, 159–175.

Mayr, G. (2000). Tiny hoopoe-like birds from the middle Eocene of Messel (Germany). *Auk*, **117**, 964–970.

(2001). A new specimen of the tiny middle Eocene bird *Gracilitarsus mirabilis* (New Family: Gracilitarsidae). *Condor*, **103**, 78–84.

Mays, N. A., Vleck, C. M. and Dawson, J. (1991). Plasma luteinizing hormone, steroid hormones, behavioral role, and nest stage in cooperatively breeding Harris' hawks (*Parabuteo unicinctus*). *Auk*, **108**, 619–637.

McGowan, K. J. and Woolfenden, G. E. (1989). A sentinel system in the Florida scrub jay. *Anim. Behav.*, **37**, 1000–1006.

McGuire, B., Getz, L. L., Hofmann, J. E., Pizzuto, T. and Frase, B. (1993). Natal dispersal and philopatry in prairie voles (*Microtus ochrogaster*) in relation to population density, season, and natal social environment. *Behav. Ecol. Sociobiol.*, **32**, 293–302.

McKechnie, A. E. and Lovegrove, B. G. (2001a). Thermoregulation and the energetic significance of clustering behavior in the white-backed mousebird (*Colius colius*). *Physiol. Biochem. Zool.*, **74**, 238–249.

(2001b). Heterothermic responses in the speckled mousebird (*Colius striatus*). *J. Comp. Physiol. B*, **171**, 507–518.

McLennan, D. A. and Brooks, D. R. (1993). The phylogenetic component of cooperative breeding in perching birds: a commentary. *Am. Nat.*, **141**, 790–795.

McNamara, J. C. and Houston, A. I. (1990). The value of fat reserves and the trade-off between starvation and predation. *Acta Biotheor.*, **38**, 37–61.

McRae, S. B. (1995). Temporal variation in responses to intraspecific brood parasitism in the moorhen. *Anim. Behav.*, **49**, 1073–1088.

(1996a). Brood parasitism in the moorhen: brief encounters between parasites and hosts and the significance of an evening laying hour. *J. Avian Biol.*, **27**, 311–320.

(1996b). Family values: costs and benefits of communal nesting in the moorhen. *Anim. Behav.*, **52**, 225–245.

(1997). A rise in nest predation enhances the frequency of intraspecific brood parasitism in a moorhen population. *J. Anim. Ecol.*, **66**, 143–153.

McRae, S. B. and Amos, W. (1999). Can incest within cooperative breeding groups be detected using DNA fingerprinting? *Behav. Ecol. Sociobiol.*, **47**, 104–107.

McRae, S. B. and Burke, T. (1996). Intraspecific brood parasitism in the moorhen: parentage and parasite-host relationships determined by DNA fingerprinting. *Behav. Ecol. Sociobiol.*, **38**, 115–129.

Meng, H. (1951). The Cooper's hawk. Ph.D. dissertation, Cornell University, Ithaca, NY.

Messier, F. and Barrette, C. (1982). The social system of the coyote (*Canis latrans*) in a forested habitat. *Can. J. Zool.*, **60**, 1743–1753.

Millar, C. D., Anthony, I., Lambert, D. M., Stapleton, P. M., Bergmann, C. C., Bellamy, A. R. and Young, E. C. (1994). Patterns of reproductive success determined by DNA fingerprinting in a communally breeding oceanic bird. *Biol. J. Linn. Soc.*, **52**, 31–48.

Miller, A. H. (1964). Social parasites among birds. *Sci. Monthly*, **42**, 238–246.

Mindell, D. P., Sorenson, M. D., Dimcheff, D. E., Hasegawa, M., Ast, J. C. and Yuri, T. (1999). Interordinal relationships of birds and other reptiles based on whole mitochondrial genomes. *Syst. Biol.*, **48**, 138–152.

Moehlman, P. D. and Hofer, H. (1997). Cooperative breeding, reproductive suppression, and body mass in canids. In: *Cooperative Breeding in Mammals*, ed. N. G. Solomon and J. A. French. Cambridge: Cambridge University Press. pp. 76–128.

Møller, A. P. (1987). Intraspecific nest parasitism and anti-parasite behaviour in swallows, *Hirundo rustica*. *Anim. Behav.*, **35**, 247–254.

(1992). Frequency of female copulations with multiple males and sexual selection. *Am. Nat.*, **139**, 1089–1101.

Moore, F. L. and Zoeller, R. T. (1985). Stress-induced inhibition of reproduction: evidence of suppressed secretion of LH-RH in an amphibian. *Gen. Comp. Endocrinol.*, **60**, 252–258.

Moore, J. and Ali, R. (1984). Are dispersal and inbreeding avoidance related? *Anim. Behav.*, **34**, 94–112.

Moore, M. C., Thompson, C. W. and Marler, C. A. (1991). Reciprocal changes in corticosterone and testosterone levels following acute and chronic handling stress in the tree lizard, *Urosaurus ornatus*. *Gen. Comp. Endocrinol.*, **81**, 217–226.

Mota, M. T. and Sousa, M. B. C. (2000). Prolactin levels of fathers and helpers related to alloparental care in common marmosets, *Callithrix jacchus*. *Folia Primatol.*, **71**, 22–26.

Mulder, R. A. (1995). Natal and breeding dispersal in a co-operative, extra-group-mating bird. *J. Avian Biol.*, **26**, 234–240.

(1997). Extra-group courtship displays and other reproductive tactics of superb fairy-wrens. *Aust. J. Zool.*, **45**, 131–143.

Mulder, R. A. and Cockburn, A. (1993). Sperm competition and the reproductive anatomy of male superb fairy-wrens. *Auk*, **110**, 588–593.

Mulder, R. A. and Langmore, N. E. (1993). Dominant males punish helpers for temporary defection in superb fairy-wrens. *Anim. Behav.*, **45**, 830–833.

Mulder, R. A., Dunn, P. O., Cockburn, A., Lazenby-Cohen, K. A. and Howell, M. J. (1994). Helpers liberate female fairy-wrens from constraints on extra-pair mate choice. *Proc. R. Soc. London Ser. B*, **255**, 223–229.

Mumme, R. L. (1992a). Delayed dispersal and cooperative breeding in the Seychelles warbler. *Trends Ecol. Evol.*, **7**, 330–331.

(1992b). Do helpers increase reproductive success? An experimental analysis in the Florida scrub jay. *Behav. Ecol. Sociobiol.*, **31**, 319–328.

(1997). A bird's-eye view of mammalian cooperative breeding. In: *Cooperative Breeding in Mammals*, ed. N. G. Solomon and J. A. French. Cambridge: Cambridge University Press. pp. 364–388.

Mumme, R. L. and de Queiroz, A. (1985). Individual contributions to cooperative behaviour in the acorn woodpecker: effects of reproductive status, sex, and group size. *Behaviour*, **95**, 290–313.

Mumme, R. L., Koenig, W. D. and Pitelka, F. A. (1983a). Reproductive competition in the communal acorn woodpecker: sisters destroy each other's eggs. *Nature*, **305**, 583–584.

(1983b). Mate guarding in the acorn woodpecker: within-group reproductive competition in a cooperative breeder. *Anim. Behav.*, **31**, 1094–1106.

Mumme, R. L., Koenig, W. D. and Ratnieks, F. L. W. (1989). Helping behaviour, reproductive value, and the future component of indirect fitness. *Anim. Behav.*, **38**, 331–343.

Mumme, R. L., Koenig, W. D. and Pitelka, F. A. (1990). Individual contributions to cooperative nest care in the acorn woodpecker. *Condor*, **92**, 360–368.

Myers, J. H. (1978). Sex ratio adjustment under food stress: maximization of quality or numbers of offspring. *Am. Nat.*, **112**, 381–388.

Nagy, K. A., Girard, I. A. and Brown, T. K. (1999). Energetics of free-ranging mammals, reptiles and birds. *Annu. Rev. Nutrition*, **19**, 247–277.

Nakamura, M. (1990). Cloacal protuberance and copulatory behavior of the alpine accentor (*Prunella collaris*). *Auk*, **107**, 284–295.

(1998a). Multiple mating and cooperative breeding in polygynandrous alpine accentors. I. Competition among females. *Anim. Behav.*, **55**, 257–273.

(1998b). Multiple mating and cooperative breeding in polygynandrous alpine accentors. II. Male mating tactics. *Anim. Behav.*, **55**, 275–287.

Nakamura, M., Yamagishi, S. and Nishiumi, I. (2001). Cooperative breeding of the white-headed vanga *Leptopterus viridis*, and endemic species in Madagascar. *J. Yamashina Inst. Ornithol.*, **33**, 1–14.

Negro, J. J., Villarroel, M., Tella, J. L., Kuhnlein, U., Hiraldo, F., Donazar, J. A. and Bird, D. M. (1996). DNA fingerprinting reveals a low incidence of extra-pair fertilizations in the lesser kestrel. *Anim. Behav.*, **51**, 935–943.

Newton, I., ed. (1989). *Lifetime Reproduction in Birds*. New York, NY: Academic Press.

Nilsson, J.-A. (1989). Causes and consequences of natal dispersal in the marsh tit, *Parus palustris*. *J. Anim. Ecol.*, **58**, 619–636.

Nocedal, J. and Ficken, M. S. (1998). Bridled titmouse, *Baeolophus wollweberi* (Passeriformes; Paridae). In *Check-List of North American Birds*, 7th edn. Washington, DC: American Ornithologists' Union. p. 466.

Nolan, J. V. and Thompson, C. F. (1975). The occurrence and significance of anomalous reproductive activities in two North American non-parasitic cuckoos *Coccyzus* spp. *Ibis*, **117**, 496–503.

Noske, R. A. (1991). A demographic comparison of cooperatively breeding and non-cooperative treecreepers (Climacteridae). *Emu*, **91**, 73–86.

Oddie, K. R. (1998). Sex discrimination before birth. *Trends Ecol. Evol.*, **13**, 130–131.

(2000). Size matters: competition between male and female great tit offspring. *J. Anim. Ecol.*, **69**, 903–912.

O'Riain, M. J. and Braude, S. (2001). Inbreeding versus outbreeding in captive and wild populations of naked mole-rats. In: *Dispersal*, ed. J. Clobert, E. Danchin, A. A. Dhondt and J. D. Nichols. Oxford: Oxford University Press. pp. 143–154.

O'Riain, M. J., Jarvis, J. U. M. and Faulkes, C. G. (1996). A dispersive morph in the naked mole-rat. *Nature*, **380**, 619–621.

O'Riain, M. J., Bennett, N. C., Brotherton, P. N. M., McIlrath, G. M. and Clutton-Brock, T. H. (2000a). Reproductive suppression and inbreeding avoidance in wild populations of co-operatively breeding meerkats (*Suricata suricatta*). *Behav. Ecol. Sociobiol.*, **48**, 471–477.

O'Riain, M. J., Jarvis, J. U. M., Alexander, R., Buffenstein, R. and Peeters, C. (2000b). Morphological castes in a vertebrate. *Proc. Natl. Acad. Sci. USA*, **97**, 13194–13197.

Ostlund, S. and Ahnesjo, I. (1998). Female fifteen-spined sticklebacks prefer better fathers. *Anim. Behav.*, **56**, 1177–1183.

Otter, K., Ramsay, S. M. and L., R. (1999). Enhanced reproductive success of female black-capped chickadees mated to high-ranking males. *Auk*, **116**, 345–354.

Owens, D. D. and Owens, M. J. (1984). Helping behaviour in brown hyenas. *Nature*, **308**, 843–845.

Owens, I. P. F. and Bennett, P. M. (1995). Ancient ecological diversification explains life-history variation among living birds. *Proc. R. Soc. London Ser. B*, **261**, 227–232.

Packer, C. and Pusey, A. E. (1982). Cooperation and competition within coalitions of male lions: kinship or game theory? *Nature*, **55**, 163–169.

Packer, C. and Ruttan, L. (1988). The evolution of cooperative hunting. *Am. Nat.*, **132**, 159–198.

Packer, C., Scheel, D. and Pusey, A. E. (1990). Why lions form groups: food is not enough. *Am. Nat.*, **136**, 1–19.

Packer, C., Gilbert, D. A., Pusey, A. E. and O'Brian, S. J. (1991). A molecular genetic analysis of kinship and cooperation in African lions. *Nature*, **351**, 562–565.

Packer, C., Pusey, A. E. and Eberly, L. E. (2001). Egalitarianism in female African lions. *Science*, **293**, 690–693.

Pagliani, A. C., Lee, P. L. M. and Bradbury, R. B. (1999). Molecular determination of sex ratio in yellowhammer *Emberiza citrinella* offspring. *J. Avian Biol.*, **30**, 239–244.

Painter, J. N., Crozier, R. H., Poiani, A., Robertson, R. J. and Clarke, M. F. (2000). Complex social organization reflects genetic structure and relatedness in the cooperatively breeding bell miner, *Manorina melanophrys*. *Mol. Ecol.*, **9**, 1339–1347.

Palmer, A. R. (2000). Quasireplication and the contract of error: lessons from sex ratios, heritabilities and fluctuating asymmetry. *Annu. Rev. Ecol. Syst.*, **31**, 441–480.

Parker, G. A. (1974). Assessment strategy and the evolution of animal conflicts. *J. Theor. Biol.*, **47**, 223–243.

Parker, P. G., Jones, T. C., Haydock, J., Dickinson, J. L. and Worden, B. D. (1999). Multilocus minisatellite DNA fingerprinting and cooperative breeding. *Behav. Ecol. Sociobiol.*, **47**, 108–111.

Pärt, T. (1996). Problems with testing inbreeding avoidance: the case of the collared flycatcher. *Evolution*, **50**, 1625–1630.

Partridge, L. (1989). Lifetime reproductive success and life-history evolution. In *Lifetime Reproduction in Birds*, ed. I. Newton. New York, NY: Academic Press. pp. 421–440.

Pasinelli, G. and Walters, J. R. (2002). Social and environmental factors affect natal dispersal and philopatry of male red-cockaded woodpeckers. *Ecology*, **83**, 2229–2239.

Payne, R. B. (1991). Natal dispersal and population structure in a migratory songbird, the indigo bunting. *Evolution*, **45**, 49–62.

Payne, R. B., Payne, L. L. and Rowley, I. (1985). Splendid wren *Malurus splendens* response to cuckoos: an experimental test of social organization in a communal bird. *Behaviour*, **94**, 108–127.

Pearson, A. K. and Pearson, O. P. (1955). Natural history and breeding behavior of the tinamou, *Nothoprocta ornata*. *Auk*, **72**, 113–127.

Pen, I. (2000). Sex allocation in a life history context. Ph.D. dissertation, University of Groningen, The Netherlands.

Pen, I. and Weissing, F. J. (2000). Sex-ratio optimization with helpers at the nest. *Proc. R. Soc. London Ser. B*, **267**, 539–543.

  (2002). Optimal sex allocation: steps towards a mechanistic theory. In: *The Sex Ratio Handbook*, ed. I. C. W. Hardy. Cambridge: Cambridge University Press. pp. 26–45.

Pen, I., Weissing, F. J. and Daan, S. (1999). Seasonal sex ratio trend in the European kestrel: an ESS analysis. *Am. Nat.*, **153**, 384–397.

Peters, A. (2000). Testosterone treatment is immunosuppressive in superb fairy-wrens, yet free-living males with high testosterone are more immunocompetent. *Proc. R. Soc. London Ser. B*, **267**, 883–889.

  (2003). To court or not to court: testosterone and the trade-off between mating and paternal effort in extra-pair mating superb fairy-wrens. *Anim. Behav.*, **64**, 103–112.

Peters, A., Astheimer, L. B., Boland, C. R. J. and Cockburn, A. (2000). Testosterone is involved in acquisition and maintenance of sexually selected male plumage in superb fairy-wrens, *Malurus cyaneus*. *Behav. Ecol. Sociobiol.*, **47**, 438–445.

Peters, A., Astheimer, L. B. and Cockburn, A. (2001). The annual testosterone profile in cooperatively breeding superb fairy-wrens, *Malurus cyaneus*, reflects their extreme infidelity. *Behav. Ecol. Sociobiol.*, **50**, 519–527.

Peters, A., Cockburn, A. and Cunningham, R. (2002). Testosterone treatment suppresses paternal care in superb fairy-wrens, *Malurus cyaneus*, despite their concurrent investment in courtship. *Behav. Ecol. Sociobiol.*, **51**, 538–547.

Peterson, A. T. and Burt, D. B. (1992). Phylogenetic history of social evolution and habitat use in the *Aphelocoma* jays. *Anim. Behav.*, **44**, 859–866.

Petrie, M. (1983). Female moorhens compete for small fat males. *Science*, **220**, 413–415.

  (1984). Territory size in the moorhen (*Gallinula chloropus*): an outcome of RHP asymmetry between neighbours. *Anim. Behav.*, **32**, 861–870.

Petrie, M. and Kempenaers, B. (1998). Extra-pair paternity in birds: explaining variation between species and populations. *Trends Ecol. Evol.*, **13**, 52–58.

Petrie, M. and Lipsitch, M. (1994). Avian polygyny is most likely in populations with high variability in heritable male fitness. *Proc. R. Soc. London Ser. B*, **256**, 275–280.

Petrie, M. and Møller, A. P. (1991). Laying eggs in others' nests: intraspecific brood parasitism in birds. *Trends Ecol. Evol.*, **6**, 315–320.

Petrie, M., Doums, C. and Møller, A. P. (1998). The degree of extra-pair paternity increases with genetic variability. *Proc. Natl. Acad. Sci. USA*, **95**, 9390–9395.

Petrie, M., Krupa, A. and Burke, T. (1999). Peacocks lek with relatives even in the absence of social and environmental cues. *Nature*, **401**, 155–157.

Pettifor, R. A., Perrins, C. M. and McCleery, R. H. (1988). Individual optimization of clutch size in great tits. *Nature*, **336**, 160–162.

Piper, W. H. (1994). Courtship, copulation, nesting behavior and brood parasitism in the Venezuelan stripe-backed wren. *Condor*, **96**, 654–671.

Piper, W. H. and Slater, G. (1993). Polyandry and incest avoidance in the cooperative stripe-backed wren of Venezuela. *Behaviour*, **124**, 227–247.

Piper, W. H., Parker, P. G. and Rabenold, K. N. (1995). Facultative dispersal by juvenile males in the cooperative stripe-backed wren. *Behav. Ecol.*, **6**, 337–342.

Place, N. J., Holekamp, K. E., Sisk, C. L., Weldele, M. L., Coscia, E. M., Drea, C. M. and Glickman, S. E. (2002). Effects of prenatal treatment with antiandrogens on luteinizing hormone secretion and sex steroid concentrations in adult spotted hyenas, *Crocuta crocuta*. *Biol. Repro.*, **67**, 1405–1413.

Poiani, A. and Fletcher, T. (1994). Plasma levels of androgens and gonadal development of breeders and helpers in the bell miner (*Manorina melanophrys*). *Behav. Ecol. Sociobiol.*, **34**, 31–41.

Poiani, A. and Jermin, L. S. (1994). A comparative analysis of some life-history traits between cooperatively and non-cooperatively breeding Australian passerines. *Evol. Ecol.*, **8**, 471–488.

Põldmaa, T., Montgomerie, R. and Boag, P. (1995). Mating system of the cooperatively breeding noisy miner *Manorina melanocephala*, as revealed by DNA profiling. *Behav. Ecol. Sociobiol.*, **37**, 137–143.

Pottinger, T. G. (1999). The impact of stress on animal reproductive activities. In: *Stress Physiology in Animals*, ed. P. H. M. Balm. Sheffield: Sheffield Academic Press. pp. 130–177.

Powell, R. A. and Fried, J. J. (1992). Helping by juvenile pine voles (*Microtus pinetorum*), growth and survival of younger siblings, and the evolution of pine vole sociality. *Behav. Ecol.*, **3**, 325–333.

Pravosudova, E. V. (1999). Forest fragmentation and the social and genetic structure of a permanent-resident bird. M.S. dissertation, Ohio State University, Columbus, OH.

Pravosudova, E. V. and Grubb, T. C. (2001). An experimental test of the prolonged brood care model in the tufted titmouse (*Baeolophus bicolor*). *Behav. Ecol.*, **11**, 309–314.

Price, E. C. (1992). The costs of infant carrying in captive cotton-top tamarins. *Am. J. Primatol.*, **26**, 23–33.

Prinzinger, R. (1988). Energy metabolism, body temperature and breathing parameters in non-torpid blue-naped mousebirds *(Urocolius macrourus)*. *J. Comp. Physiol. B*, **157**, 801–806.

Prinzinger, R., Goppel, R., Lorenz, A. and Kulzer, E. (1981). Body temperature and metabolism in the red-backed mousebird *(Colius castanotus)* during fasting and torpor. *Comp. Biochem. Physiol. A*, **69**, 689–692.

Pruett-Jones, S. G. and Lewis, M. J. (1990). Sex ratio and habitat limitation promote delayed dispersal in superb fairy-wrens. *Nature*, **348**, 541–542.

Pulliam, H. R. (1973). On the advantages of group living. *J. Theor. Biol.*, **38**, 419–422.

Pusey, A. E. and Packer, C. (1994). Non-offspring nursing in social carnivores: minimizing the costs. *Behav. Ecol.*, **5**, 362–374.

Pusey, A. E. and Wolf, M. (1996). Inbreeding avoidance in animals. *Trends Ecol. Evol.*, **11**, 201–206.

Putland, D. (2001). Has sexual selection been overlooked in the study of avian helping behaviour? *Anim. Behav.*, **62**, 811–814.

Questiau, S., Escaravage, N., Eybert, M. C. and Taberlet, P. (2000). Nestling sex ratios in a population on bluethroats *Luscinia svecica* inferred from AFLPTM analysis. *J. Avian Biol.*, **31**, 8–14.

Quinn, J. S. and Startek-Foote, J. (2000). Smooth-billed ani (*Crotophaga ani*). In: *Birds of North America*, ed. A. Poole and F. Gill. Philadelphia, PA and Washington, DC: Academy of Natural Sciences and American Ornithologists' Union.

Quinn, J. S., Macedo, R. and White, B. N. (1994). Genetic relatedness of communally breeding guira cuckoos. *Anim. Behav.*, **47**, 515–529.

Quinn, J. S., Woolfenden, G. E., Fitzpatrick, J. W. and White, B. N. (1999). Multi-locus DNA fingerprinting supports genetic monogamy in Florida scrub-jays. *Behav. Ecol. Sociobiol.*, **45**, 1–10.

Rabenold, K. N. (1984). Cooperative enhancement of reproductive success in tropical wren societies. *Ecology*, **65**, 871–885.

Rabenold, K. N. (1990). *Campylorhynchus* wrens: the ecology of delayed dispersal and cooperation in the Venezuelan savanna. In: *Cooperative Breeding in Birds*, ed. P. B. Stacey and W. D. Koenig. Cambridge: Cambridge University Press. pp. 157–196.

Rabenold, P. P., Rabenold, K. N., Piper, W. H., Haydock, J. and Zack, S. N. (1990). Shared paternity revealed by genetic analysis in cooperatively breeding tropical wrens. *Nature*, **348**, 538–540.

Rabinowitz, D., Cairns, S. and Dillon, T. (1986). Seven forms of rarity and their frequency in the flora of the British Isles. In: *Conservation Biology: the Science of Scarcity and Diversity*, ed. M. E. Soulé. Sunderland, MA: Sinauer Associates. pp. 182–204.

Radford, A. N. and Blakey, J. K. (2000). Is variation in brood sex ratios adaptive in the great tit (*Parus major*)? *Behav. Ecol.*, **11**, 294–298.

Ragsdale, J. E. (1999). Reproductive skew theory extended: the effect of resource inheritance on social organization. *Evol. Ecol. Res.*, **1**, 859–874.

Ralls, K., Harvey, P. H. and Lyles, A. M. (1986). Inbreeding in natural populations of birds and mammals. In: *Conservation Biology: the Science of Scarcity and Diversity*, ed. M. E. Soulé. Sunderland, MA: Sinauer Associates. pp. 35–56.

Rasa, O. A. E. (1986). Coordinated vigilence in dwarf mongoose family groups: the "watchman's song" hypothesis and the cost of guarding. *Z. Tierpsychol.*, **71**, 340–344.

Recher, H. F. and Lim, L. (1990). A review of current ideas of the extinction, conservation and management of Australia's terrestrial vertebrate fauna. *Proc. Ecol. Soc. Aust.*, **16**, 287–301.

Reed, J. M., Boulinier, T., Danchin, E. and Oring, L. W. (1999). Informed dispersal: prospecting by birds for breeding sites. *Curr. Ornithol.*, **15**, 189–259.

Reeve, H. K. (1998). Game theory, reproductive skew, and nepotism. In: *Game Theory and Animal Behavior*, ed. L. Dugatkin and H. K. Reeve. Oxford: Oxford University Press. pp. 118–145.

(2000). A transactional theory of within-group conflict. *Am. Nat.*, **155**, 365–382.

Reeve, H. K. and Emlen, S. T. (2000). Reproductive skew and group size: an N-person staying incentive model. *Behav. Ecol.*, **11**, 640–647.

Reeve, H. K. and Keller, L. (1995). Partitioning of reproduction in mother–daughter versus sibling associations: a test of optimal skew theory. *Am. Nat.*, **145**, 119–132.

(1996). Relatedness asymmetry and reproductive sharing in animal societies. *Am. Nat.*, **148**, 764–769.

(1997). Reproductive bribing and policing as evolutionary mechanisms for the suppression of within-group selfishness. *Am. Nat.*, **150**, S42-S58.

(2001). Tests of reproductive-skew models in social insects. *Annu. Rev. Entomol.*, **46**, 347–385.

Reeve, H. K. and Ratnieks, F. L. W. (1993). Queen–queen conflicts in polygynous societies: mutual tolerance and reproductive skew. In: *Queen Number and Sociality in Insects*, ed. L. Keller. Oxford: Oxford University Press. pp. 45–85.

Reeve, H. K. and Sherman, P. W. (1993). Adaptation and the goals of evolutionary research. *Q. Rev. Biol.*, **68**, 1–31.

Reeve, H. K., Westneat, D. F., Noon, W. A., Sherman, P. W. and Aquadro, C. F. (1990). DNA "fingerprinting" reveals high levels of inbreeding in colonies of the eusocial naked mole-rat. *Proc. Natl. Acad. Sci. USA*, **87**, 2496–2500.

Reeve, H. K., Emlen, S. T. and Keller, L. (1998). Reproductive sharing in animal societies: reproductive incentives or incomplete control by dominant breeders? *Behav. Ecol.*, **9**, 267–278.

Reid, B. and Williams, G. R. (1975). The kiwi. In: *Biogeography and Ecology in New Zealand*, ed. G. Kuschel. The Hague: W. Junk. pp. 301–220.

Restrepo, C. and Mondragón, M. L. (1998). Cooperative breeding in the frugivorous toucan barbet (*Semnornis ramphastinus*). *Auk*, **115**, 4–15.

Reyer, H.-U. (1980). Flexible helper structure as an ecological adaptation in the pied kingfisher (*Ceryle rudis rudis* L.). *Behav. Ecol. Sociobiol.*, **6**, 219–227.

(1984). Investment and relatedness: a cost/benefit analysis of breeding and helping in the pied kingfisher (*Ceryle rudis*). *Anim. Behav.*, **32**, 1163–1178.

(1990). Pied kingfishers: ecological causes and reproductive consequences of cooperative breeding. In: *Cooperative Breeding in Birds*, ed. P. B. Stacey and W. D. Koenig. Cambridge: Cambridge University Press. pp. 527–557.

Reyer, H.-U. and Westerterp, K. (1985). Parental energy expenditure: a proximate cause of helper recruitment in the pied kingfisher (*Ceryle rudis*). *Behav. Ecol. Sociobiol.*, **17**, 363–369.

Reyer, H.-U., Dittami, J. P. and Hall, M. R. (1986). Avian helpers at the nest: are they psychologically castrated? *Ethology*, **71**, 216–228.

Richardson, D. S., Jury, F. L., Blaakmeer, K., Komdeur, J. and Burke, T. (2001). Parentage assignment and extra-group paternity in a cooperative breeder: the Seychelles warbler (*Acrocephalus sechellensis*). *Mol. Ecol.*, **10**, 2263–2273.

Richardson, D. S., Burke, T. and Komdeur, J. (2002). Direct benefits and the evolution of female-biased cooperative breeding in Seychelles warblers. *Evolution*, **56**, 2313–2321.

Richardson, D. S., Komdeur, J. and Burke, T. (2003). Avian behaviour: altruism and infidelity among warblers. *Nature*, **422**, 580.

Richner, H. (1990). Helpers-at-the-nest in carrion crows *Corvus corone corone*. *Ibis*, **132**, 105–108.

Ricklefs, R. E. (1968). Patterns of growth in birds. *Ibis*, **110**, 419–451.

(1976). Growth rates of birds in the humid New World tropics. *Ibis*, **118**, 179–207.

Ricklefs, R. E. and Starck, J. M. (1998). The evolution of the development mode in birds. In: *Avian Growth and Development: Evolution Within the Altricial–Precocial Spectrum*, ed. J. M. Starck and R. E. Ricklefs. Oxford: Oxford University Press. pp. 366–380.

Ridley, A. R. (2003). The causes and consequences of helping behaviour in the cooperatively breeding Arabian babbler. Ph.D. dissertation, University of Cambridge.

Ridpath, M. G. (1972). The Tasmanian native hen, *Tribonyx mortierii*. I-III. *CSIRO Wildl. Res.*, **17**, 1–118.

Roberts, R. L., Zullo, A., Gustafson, E. A. and Carter, C. S. (1996). Perinatal steroid treatments alter alloparental and affiliative behavior in prairie voles. *Horm. Behav.*, **30**, 576–582.

Robinson, D. and Traill, B. J., Jr. (1996). *Conserving Woodland Birds in the Wheat and Sheep Belts of Southern Australia*. Royal Australasian Ornithologist's Union Conservation Statement No. 10.

Robinson, S. K., Thompson, F. R. I., Donovan, T. M., Whitehead, D. R. and Faaborg, J. (1995). Regional forest fragmentation and the nesting success of migratory birds. *Science*, **267**, 1987–1990.

Robinson, T. R. (2000). Factors affecting natal dispersal by song wrens (*Cyphorhinus phaeocephalus*): ecological constraints and demography. Ph.D. dissertation, University of Illinois, Urbana, IL.

Rood, J. P. (1986). Ecology and social evolution in the mongooses. In: *Ecological Aspects of Social Evolution*, ed. D. I. Rubenstein and R. W. Wrangham. Princeton, NJ: Princeton University Press. pp. 131–152.

(1990). Group size, survival, reproduction, and routes to breeding in dwarf mongooses. *Anim. Behav.*, **39**, 566–572.

Rosenfield, R. N., Bielefeldt, J., Anderson, J. L. and Beckmann, D. J. (1985). Sex ratios in broods of Cooper's hawk. *Wilson Bull.*, **97**, 113–115.

Rosenfield, R. N., Bielefeldt, J. and Vos, S. M. (1996). Skewed sex ratios in Cooper's hawk offspring. *Auk*, **113**, 957–960.

Rowan, M. K. (1967). A study of the colies of southern Africa. *Ostrich*, **38**, 63–115.

Rowley, I. (1965). The life history of the superb blue wren, *Malurus cyaneus*. *Emu*, **64**, 251–297.

(1968). Communal species of Australian birds. *Bonn. Zool. Beitr.*, **19**, 362–368.

(1976). Co-operative breeding in Australian birds. *Proc. Int. Ornithol. Congr.*, **16**, 657–666.

(1978). Communal activities among white-winged choughs *Corcorax melanorhamphos*. *Ibis*, **120**, 178–197.

Rowley, I. and Russell, E. M. (1990). Splendid fairy-wrens: demonstrating the importance of longevity. In: *Cooperative Breeding in Birds*, ed. P. B. Stacey and W. D. Koenig. Cambridge: Cambridge University Press. pp. 1–30.

(1995). The breeding biology of the white-winged fairy-wren *Malurus leucopterus leuconotus* in a Western Australian coastal heathland. *Emu*, **95**, 175–184.

(1997). *Fairy-Wrens and Grasswrens*. Oxford: Oxford University Press.

Rowley, I., Russell, E. M. and Brooker, M. G. (1986). Inbreeding: benefits may outweigh costs. *Anim. Behav.*, **34**, 939–941.

Rowley, I., Russell, E. M., Payne, R. B. and Payne, L. L. (1989). Plural breeding in the splendid fairy-wren, *Malurus splendens* (Aves: Maluridae), a cooperative breeder. *Ethology*, **83**, 229–247.

Rowley, I., Russell, E. M. and Brooker, M. G. (1993). Inbreeding in birds. In: *The Natural History of Inbreeding and Outbreeding*, ed. N. W. Thornhill. Chicago, IL: University of Chicago Press, pp. 304–328.

Russell, A. F. (1999). Ecological constraints and the cooperative breeding systems of the long-tailed tit *Aegithalos caudatus*. Ph.D. dissertation, University of Sheffield.

(2001). Dispersal costs set the scene for helping in an atypical avian cooperative breeder. *Proc. R. Soc. London Ser. B*, **268**, 95–99.

Russell, A. F. and Hatchwell, B. J. (2001). Experimental evidence for kin-biased helping in a cooperatively breeding vertebrate. *Proc. R. Soc. London Ser. B*, **268**, 2169–2174.

Russell, A. F., Clutton-Brock, T. H., Brotherton, P. N. M., Sharpe, L. L., McIlrath, G. M., Dalerum, F. D., Cameron, E. Z. and Barnard, J. A. (2002). Factors affecting pup growth and survival in co-operatively breeding meerkats *Suricata suricatta*. *J. Anim. Ecol.*, **71**, 700–709.

Russell, A. F., Brotherton, P. N. M., McIlrath, G. M., Sharpe, L. L. and Clutton-Brock, T. H. (2003a). Breeding success in cooperative meerkats: effects of helper number and maternal state. *Behav. Ecol.*, **14**, 486–492.

Russell, A. F., Sharpe, L. L., Brotherton, P. N. M. and Clutton-Brock, T. H. (2003b). Cost minimization by helpers in cooperative vertebrates. *Proc. Natl. Acad. Sci. USA*, **100**, 3333–3338.

Russell, E. M. (1989). Co-operative breeding: a Gondwanan perspective. *Emu*, **89**, 61–62.

(2000). Avian life histories: is extended parental care the southern secret? *Emu*, **100**, 377–399.

Russell, E. M. and Rowley, I. (1988). Helper contributions to reproductive success in the splendid fairy-wren, *Malurus splendens*. *Behav. Ecol. Sociobiol.*, **22**, 131–140.

(1993a). Demography of the cooperatively breeding splendid fairy-wren, *Malurus splendens* (Maluridae). *Aust. J. Zool.*, **41**, 475–505.

(1993b). Philopatry or dispersal: competition for territory vacancies in the splendid fairy-wren, *Malurus splendens*. *Anim. Behav.*, **45**, 519–539.

(2000). Demography and social organisation of the red-winged fairy-wren, *Malurus elegans*. *Aust. J. Zool.*, **48**, 161–200.

Russo, C. A. M., Takezaki, N. and Nei, M. (1996). Efficiencies of different genes and different tree-building methods in recovering a known vertebrate phylogeny. *Mol. Biol. Evol.*, **13**, 525–536.

Rylands, A. B. (1996). Habitat and the evolution of social and reproductive behavior in callitrichidae. *Am. J. Primatol.*, **38**, 5–18.

Saether, B.-E. (1990). Age-specific variation in reproductive performance in birds. *Curr. Ornithol.*, **7**, 251–283.

Salo, A. L. and French, J. A. (1989). Early experience, reproductive success, and development of parental behaviour in Mongolian gerbils. *Anim. Behav.*, **38**, 693–702.

Sanchez, S., Pelaez, F., Gil-Burmann, C. and Kaumanns, W. (1999). Costs of infant-carrying in the cotton-top tamarin (*Saguinus oedipus*). *Am. J. Primatol.*, **48**, 99–111.

Sanderson, M. J. and Kim, J. (2000). Parametric phylogenetics? *Syst. Biol.*, **49**, 817–829.

Sapolsky, R. M. (1992). Neuroendocrinology of the stress-response. In: *Behavioral Endocrinology*, ed. J. B. Becker, S. M. Breedlove and D. Crews. Cambridge, MA: MIT Press. pp. 287–324.

Scantlebury, M., Russell, A. F., McIlrath, G. M., Speakman, J. R. and Clutton-Brock, T. H. (2002). The energetics of lactation in cooperatively breeding meerkats *Suricata suricatta*. *Proc. R. Soc. London Ser. B*, **269**, 2147–2153.

Schaffner, C. M. and French, J. A. (1997). Group size and aggression: "recruitment incentives" in a cooperatively breeding primate. *Anim. Behav.*, **54**, 171–180.

Schiegg, K., Walters, J. R. and Priddy, J. A. (2002). The consequences of disrupted dispersal in fragmented red-cockaded woodpecker populations. *J. Anim. Ecol.*, **71**, 710–721.

Schleicher, B., Hoi, H., Valera, F. and Hoi-Leitner, M. (1997). The importance of different paternity guards in the polygynandrous penduline tit (*Remiz pendulinus*). *Behaviour*, **134**, 941–959.

Schlinger, B. A., Lane, N. I., Grisham, W. and Thompson, L. (1999). Androgen synthesis in a songbird: a study of cyp17 (17 alpha-hydroxylase/C17, 20-lyase) activity in the zebra finch. *Gen. Comp. Endocrinol.*, **113**, 46–58.

Schmidt, L. G., Bradshaw, S. D. and Follett, B. K. (1991). Plasma levels of luteinizing hormone and androgens in relation to age and breeding status among cooperatively breeding Australian magpies (*Gymnorhina tibicen* Latham). *Gen. Comp. Endocrinol.*, **83**, 48–55.

Schoech, S. J. (1996). The effect of supplemental food on body condition and the timing of reproduction in a cooperative breeder, the Florida scrub jay. *Condor*, **98**, 234–244.

Schoech, S. J., Mumme, R. L. and Moore, M. C. (1991). Reproductive endocrinology and mechanisms of breeding inhibition in cooperatively breeding Florida scrub jays (*Aphelocoma c. coerulescens*). *Condor*, **93**, 354–364.

Schoech, S. J., Mumme, R. L. and Wingfield, J. C. (1996a). Delayed breeding in the cooperatively breeding Florida scrub-jay (*Aphelocoma coerulescens*): inhibition or the absence of stimulation? *Behav. Ecol. Sociobiol.*, **39**, 77–90.

(1996b). Prolactin and helping behaviour in the cooperatively breeding Florida scrub-jay (*Aphelocoma coerulescens*). *Anim. Behav.*, **52**, 445–456.

(1997). Corticosterone, reproductive status, and body mass in a cooperative breeder, the Florida scrub-jay (*Aphelocoma coerulescens*). *Physiol. Zool.*, **70**, 68–73.

Schoech, S. J., Ketterson, E. D., Nolan Jr., V., Sharp, P. J. and Buntin, J. D. (1998). The effect of exogenous testosterone on parental behavior, plasma prolactin and prolactin binding sites in dark-eyed juncos. *Horm. Behav.*, **34**, 1–10.

Schradin, C. and Anzenberger, G. (2001). Costs of infant carrying in common marmosets, *Callithrix jacchus*: an experimental analysis. *Anim. Behav.*, **62**, 289–295.

Schultz, T. R., Cocroft, R. B. and Churchill, G. A. (1996). The reconstruction of ancestral character states. *Evolution*, **50**, 504–511.

Scott, D. K. (1980). Functional aspects of prolonged parental care in Bewick's swans. *Anim. Behav.*, **28**, 938–952.

Selander, R. K. (1964). Speciation in wrens of the genus *Campylorynchus*. *Univ. Calif. Publ. Zool.*, **74**, 1–305.

Selye, H. (1978). *The Stress of My Life*. New York, NY: Van Nostrand-Reinhold.

Shaffer, M. L. (1981). Minimum population sizes for species conservation. *BioScience*, **31**, 131–134.

(1987). Minimum viable populations: coping with uncertainty. In: *Viable Populations for Conservation*, ed. M. E. Soulé. Cambridge: Cambridge University Press. pp. 69–86.

Sheldon, B. C. (1998). Recent studies of avian sex ratios. *Heredity*, **80**, 397–402.

Shellman-Reeve, J. S. and Reeve, H. K. (2000). Extra-pair paternity as the result of reproductive transactions between paired mates. *Proc. R. Soc. London Ser. B*, **267**, 2543–2546.

Sherley, G. H. (1989). Benefits of courtship-feeding for rifleman (*Acanthisitta chloris*) parents. *Behaviour*, **109**, 309–318.

(1990). Co-operative breeding in rifleman (*Acanthisitta chloris*) benefits to parents, offspring and helpers. *Behaviour*, **112**, 1–22.

Sherman, P. T. (1995a). Breeding biology of white-winged trumpeters (*Psophia leucoptera*) in Peru. *Auk*, **112**, 285–295.

(1995b). Social organization of cooperatively polyandrous white-winged trumpeters (*Psophia leucoptera*). *Auk*, **112**, 296–309.

Sherman, P. W. (1977). Nepotism and the evolution of alarm calls. *Science*, **197**, 1246–1253.

(1985). Alarm calls of Belding's ground squirrels to aerial predators: nepotism or self-preservation? *Behav. Ecol. Sociobiol.*, **17**, 313–323.

(1988). The levels of analysis. *Anim. Behav.*, **36**, 616–619.

Sherman, P. W., Jarvis, J. U. M. and Alexander, R. D., eds. (1991). *The Biology of the Naked Mole-Rat*. Princeton, NJ: Princeton University Press.

Sherman, P. W., Braude, S. and Jarvis, J. U. M. (1999). Litter sizes and mammary numbers of naked mole-rats: breaking the one-half rule. *J. Mammal.*, **80**, 720–733.

Sherry, D. F. (1985). Food storage by birds and mammals. *Adv. Study Behav.*, **15**, 153–188.

Shields, W. M. (1982). *Philopatry, Inbreeding and the Evolution of Sex*. Albany, NY: State University of New York Press.

(1987). Dispersal and mating systems: investigating their causal connections. In: *Mammalian Dispersal Patterns: the Effects of Social Structure on Population Genetics*, ed. B. D. Chepko-Sade and Z. T. Halpin. Chicago: University of Chicago Press. pp. 3–24.

Short, L. L. and Horne, J. F. M. (1988). Order Piciformes, family Capitonidae. In: *The Birds of Africa*, vol. 3, ed. E. K. Urban, C. H. Fry and S. Keith. New York, NY: Academic Press. pp. 413–486.

Shy, M. M. (1982). Interspecific feeding among birds: a review. *J. Field Ornithol.*, **53**, 370–393.

Sibley, C. G. and Ahlquist, J. E. (1985). The phylogeny and classification of the Australo–Papuan passerine birds. *Emu*, **85**, 1–14.

(1990). *Phylogeny and Classification of Birds: a Study in Molecular Evolution*. New Haven, CT: Yale University Press.

Sibley, C. G. and Monroe, B. L., Jr. (1990). *Distribution and Taxonomy of Birds of the World*. New Haven, CT: Yale University Press.

(1993). *A Supplement to Distribution and Taxonomy of Birds of the World*. New Haven, CT: Yale University Press.

Sick, H. (1964). Tinamou. In: *A New Dictionary of Birds*, ed. A. L. Thomson. New York, NY: McGraw-Hill. pp. 821–822.

(1993). *Birds in Brazil*. Princeton, NJ: Princeton University Press.

Sillero-Zubiri, C., Gottelli, D. and MacDonald, D. W. (1996). Male philopatry, extra-pack copulations and inbreeding avoidance in Ethiopian wolves (*Canis simensis*). *Behav. Ecol. Sociobiol.*, **38**, 331–340.

Siriwardena, G. M., Baillie, S. R. and Wilson, J. D. (1998). Variation in the survival rates of some British passerines with respect to their population trends on farmland. *Bird Study*, **45**, 276–292.

Skutch, A. F. (1935). Helpers at the nest. *Auk*, **52**, 257–273.

(1957). The incubation patterns of birds. *Ibis*, **99**, 69–93.

(1959). Life history of the groove-billed ani. *Auk*, **76**, 281–317.

(1961). Helpers among birds. *Condor*, **63**, 198–226.

(1976). *Parent Birds and Their Young*. Austin, TX: University of Texas Press.

(1999). *Helpers at Birds' Nests*. Iowa City, IA: University of Iowa Press.

Smith, J. (1992). Cooperative breeding in the chestnut-crowned babbler *Pomatostomus ruficeps*. *Aust. Birds*, **25**, 64–66.

Smith, J. N. M. (1990). Summary. In: *Cooperative Breeding in Birds*, ed. P. B. Stacey and W. D. Koenig. Cambridge: Cambridge University Press. pp. 593–611.

Smith, S. M. (1978). The "underworld" in a territorial sparrow: adaptive strategy for floaters. *Am. Nat.*, **112**, 571–582.

Solomon, N. G. (1991). Current indirect fitness benefits associated with philopatry in juvenile prairie voles. *Behav. Ecol. Sociobiol.*, **29**, 277–282.

(1993). Body size and social preferences of male and female prairie voles, *Microtus ochrogaster*. *Anim. Behav.*, **45**, 1031–1033.

(1994). Effect of the preweaning environment on subsequent reproduction in prairie voles, *Microtus ochrogaster*. *Anim. Behav.*, **48**, 331–341.

Solomon, N. G. and French, J. A., eds. (1997). *Cooperative Breeding in Mammals*. Cambridge: Cambridge University Press.

Soltis, J. and McElreath, R. (2001). Can females gain extra paternal investment by mating with multiple males? A game theoretic approach. *Am. Nat.*, **158**, 519–529.

Soma, K. K., Sullivan, K. A., Tramontin, A. D., Saldanha, C. J., Schlinger, B. A. and Wingfield, J. C. (2000). Acute and chronic effects of an aromatase inhibitor on territorial aggression in breeding and nonbreeding male song sparrows. *J. Comp. Physiol. A*, **186**, 759–769.

Specht, R. L. (1981). Major vegetation formations in Australia. In: *Ecological Biogeography of Australia*, ed. A. Keast. The Hague: Dr. W. Junk. pp. 163–297.

Spinks, A. C., Bennett, N. C., Faulkes, C. G. and Jarvis, J. U. M. (2000a). Circulating LH levels and the response to exogenous GnRH in the common mole-rat: implications for reproductive regulation in this social, seasonal breeding species. *Horm. Behav.*, **37**, 221–228.

Spinks, A. C., Jarvis, J. U. M. and Bennett, N. C. (2000b). Comparative patterns of philopatry and dispersal in two common mole-rat populations: implications for the evolution of mole-rat sociality. *J. Anim. Ecol.*, **69**, 224–234.

Stacey, P. B. (1979a). Habitat saturation and communal breeding in the acorn woodpecker. *Anim. Behav.*, **27**, 1153–1166.

(1979b). Kinship, promiscuity, and communal breeding in the acorn woodpecker. *Behav. Ecol. Sociobiol.*, **6**, 53–66.

Stacey, P. B. and Koenig, W. D. (1984). Cooperative breeding in the acorn woodpecker. *Sci. Am.*, **251**(2), 114–121.

eds. (1990a). *Cooperative Breeding in Birds: Long-Term Studies of Ecology and Behavior*. Cambridge: Cambridge University Press.

(1990b). Introduction. In: *Cooperative Breeding in Birds*, ed. P. B. Stacey and W. D. Koenig. Cambridge: Cambridge University Press. pp. ix–xviii.

Stacey, P. B. and Ligon, J. D. (1987). Territory quality and dispersal options in the acorn woodpecker, and a challenge to the habitat saturation model of cooperative breeding. *Am. Nat.*, **130**, 654–676.

(1991). The benefits-of-philopatry hypothesis for the evolution of cooperative breeding: variation in territory quality and group size effects. *Am. Nat.*, **137**, 831–846.

Stacey, P. B. and Taper, M. (1992). Environmental variation and the persistence of small populations. *Ecol. Appl.*, **2**, 18–29.

Stallcup, J. A. and Woolfenden, G. E. (1978). Family status and contributions to breeding by Florida scrub jays. *Anim. Behav.*, **26**, 1144–1156.

Stamps, J. A. (1990). When should avian parents differentially provision sons and daughters? *Am. Nat.*, **135**, 671–685.

Stanback, M. T. (1994). Dominance within broods of the cooperatively breeding acorn woodpecker. *Anim. Behav.*, **47**, 1121–1126.

Stanback, M. T., Richardson, D. S., Boix-Hinzen, C. and Mendelsohn, J. (2002). Genetic monogamy in Monteiro's hornbill, *Tockus monteiri*. *Anim. Behav.*, **63**, 787–793.

Starck, J. M. and Ricklefs, R. E. (1998). Patterns of development: the altricial-precocial spectrum. In: *Avian Growth and Development: Evolution Within the Altricial–Precocial Spectrum*, ed. J. M. Starck and R. E. Ricklefs. Oxford: Oxford University Press. pp. 3–30.

Steadman, D. W. (1995). Prehistoric extinctions of Pacific island birds: biodiversity meets zooarchaeology. *Science*, **267**, 1123–1131.

Stearns, S. C. (1992). *The Evolution of Life Histories*. Oxford: Oxford University Press.

Stevens, B. A. and Sutherland, W. J. (1999). Consequences of the Allee effect for behaviour, ecology, and conservation. *Trends Ecol. Evol.*, **14**, 405–410.

Stevens, E. E. and Wiley, R. H. (1995). Genetic consequences of restricted dispersal and incest avoidance in a cooperatively breeding wren. *J. Theor. Biol.*, **175**, 423–436.

Stiles, F. G. and Skutch, A. F. (1989). *A Guide to the Birds of Costa Rica*. London: Christopher Helm.

Stiles, F. G. and Wolf, L. (1979). Ecology and evolution of a lek mating system in the long-tailed hermit hummingbird. *Ornithol. Monogr.*, **27**, 1–78.

Stith, B. M., Fitzpatrick, J. W., Woolfenden, G. E. and Pranty, B. (1996). Classification and conservation of metapopulations: a case study of the Florida scrub-jay. In: *Metapopulations and Wildlife Conservation*, ed. D. R. McCullough. Washington, DC: Island Press. pp. 187–215.

Stouffer, P. and Bierregaard, R. O. J. (1995). Use of Amazonian forest fragments by understory insectivorous birds. *Ecology*, **76**, 2429–2445.

Strahl, S. D. (1985). The behavior and socio-ecology of the hoatzin, *Opisthocomus hoazin*, in the llanos of Venezuela.

Ph.D. dissertation, State University of New York, Albany, NY.

(1988). The social organization and behaviour of the hoatzin *Opisthocomus hoazin* in central Venezuela. *Ibis*, **130**, 483–502.

Strahl, S. D. and Schmitz, A. (1990). Hoatzins: cooperative breeding in a folivorous neotropical bird. In: *Cooperative Breeding in Birds*, ed. P. B. Stacey and W. D. Koenig. Cambridge: Cambridge University Press. pp. 131–155.

Strickland, D. (1991). Juvenile dispersal in gray jays: dominant brood member expels siblings from natal territory. *Can. J. Zool.*, **69**, 2935–2945.

Strickland, D. and Waite, T. A. (2001). Does initial suppression of allofeeding in small jays help to conceal their nests? *Can. J. Zool.*, **79**, 2128–2146.

Sydeman, W. J. (1989). Effects of helpers on nestling care and breeder survival in pygmy nuthatches. *Condor*, **91**, 147–155.

Taborsky, B. and Taborsky, M. (1991). Social organization of North Island brown kiwi: long-term pairs and three types of male spacing behaviour. *Ethology*, **89**, 47–62.

(1992). Spatial organization of the North Island brown kiwi *Apteryx australis mantelli*: sex, pairing status and territoriality. *Ibis*, **134**, 1–10.

Taborsky, M. (1984). Broodcare helpers in the cichlid fish *Lamprologus brichardi*: their costs and benefits. *Anim. Behav.*, **32**, 1236–1252.

Tardif, S. D. (1997). The bioenergetics of parental behavior and the evolution of alloparental care in marmosets and tarmarins. In: *Cooperative Breeding in Mammals*, ed. N. G. Solomon and J. A. French. Cambridge: Cambridge University Press. pp. 11–33.

Tardif, S. D. and Bales, K. (1997). Is infant-carrying a courtship strategy in callitrichid primates? *Anim. Behav.*, **53**, 1001–1007.

Tardif, S. D. and Harrison, M. L. (1990). Estimates of the energetic cost of infant transport in tamarins. *Am. J. Phys. Anthropol.*, **81**, 306.

Tarof, S. A. and Stutchbury, B. J. (1996). A case of cooperative breeding in the hooded warbler. *Wilson Bull.*, **108**, 382–384.

Tatner, P. (1990). Energetic demands during brood rearing in the wheatear *Oenanthe oenanthe*. *Ibis*, **132**, 423–435.

Taylor, B. and van Perlo, B. (1998). *Rails: a Guide to the Rails, Gallinules and Coots of the World*. Sussex: Pica Press.

Thaxton, J. E. and Hingtgen, T. M. (1996). Effects of suburbanization and habitat fragmentation on Florida scrub-jay dispersal. *Florida Field Nat.*, **24**, 25–60.

Thorneycroft, H. B. (1976). A cytogenetic study of the white-throated sparrow *Zonotrichia albicollis* (Gmelin). *Evolution*, **29**, 611–621.

Tobias, J. A., Züchner, T. and de Melo-Júnior, T. A. (2002). Family Galbulidae (Jacamars). In: *Handbook of the Birds of the World*, vol. 7, ed. J. del Hoyo, A. Elliott and J. Sargatal. Barcelona: Lynx Edicions. pp. 74–101.

Torres, R. and Drummond, H. (1999). Variably male-biased sex ratio in a marine bird with females larger than males. *Oecologia*, **118**, 16–22.

Trainor, B. C. and Marler, C. A. (2001). Testosterone, paternal behavior, and aggression in the monogamous California mouse (*Peromyscus californicus*). *Horm. Behav.*, **40**, 32–42.

(2002). Testosterone promotes paternal behaviour in a monogamous mammal via conversion to oestrogen. *Proc. R. Soc. London Ser. B*, **269**, 823–829.

Treganza, T. and Wedell, N. (2000). Genetic compatibility, mate choice and patterns of parentage. *Mol. Ecol.*, **9**, 1013–1027.

Treisman, M. (1975). Predation and the evolution of gregariousness. I. Models for concealment and evasion. *Anim. Behav.*, **23**, 779–800.

Trewick, S. A. (1997). Flightlessness and phylogeny amongst endemic rails (Aves: Rallidae) of the New Zealand region. *Phil. Trans. R. Soc. London B*, **352**, 429–446.

Trivers, R. L. (1972). Parental investment and sexual selection. In: *Sexual Selection and the Descent of Man*, ed. B. Campbell. Chicago: Aldine. pp. 136–179.

(1974). Parent–offspring conflict. *Am. Zool.*, **14**, 249–264.

Trivers, R. L. and Hare, H. (1976). Haplo-diploidy and the evolution of the social insects. *Science*, **191**, 249–263.

Trivers, R. L. and Willard, D. E. (1973). Natural selection of parental ability to vary the sex ratio of offspring. *Science*, **179**, 90–92.

Tsuji, K. and Tsuji, N. (1998). Indices of reproductive skew depend on average reproductive success. *Evol. Ecol.*, **12**, 141–152.

Tuttle, E. M. and Pruett-Jones, S. (2003). Estimates of extreme sperm production. Morphological and experimental evidence from reproductively promiscuous fairy wrens (*Malurus*). *Anim. Behav.*, in press.

Tuttle, E. M., Pruett-Jones, S. and Webster, M. S. (1996). Cloacal protuberances and extreme sperm production in Australian fairy-wrens. *Proc. R. Soc. London Ser. B*, **263**, 1359–1364.

Urban, E. K., Fry, C. H. and Keith, S. (1997). *The Birds of Africa*, vol. 5. San Diego, CA: Academic Press.

van der Jeugd, H. (1999). Life history decisions in a changing environment. Ph.D. dissertation, Uppsala University, Sweden.

van Hoof, J. A. R. A. M. and van Schaik, C. P. (1994). Male bonds: affiliative relationships among non-human primate males. *Behaviour*, **130**, 309–337.

Vander Wall, S. B. and Balda, R. P. (1981). Ecology and evolution of food-storage behaviour in conifer-seed-caching corvids. *Z. Tierpsychol.*, **56**, 217–242.

Vehrencamp, S. L. (1977). Relative fecundity and parental effort in communally nesting anis, *Crotophaga sulcirostris*. *Science*, **197**, 403–405.

(1978). The adaptive significance of communal nesting in groove-billed anis (*Crotophaga sulcirostris*). *Behav. Ecol. Sociobiol.*, **4**, 1–33.

(1979). The roles of individual, kin, and group selection in the evolution of sociality. In: *Handbook of Behavioral Neurobiology: Social Behavior and Communication*, ed. P. Marler and J. G. Vendenbergh. New York, NY: Plenum Press. pp. 351–394.

(1980). To skew or not to skew? *Proc. Int. Ornithol. Congr.*, **17**, 869–874.

(1982). Body temperatures of incubating versus non-incubating roadrunners. *Condor*, **84**, 203–207.

(1983a). A model for the evolution of despotic versus egalitarian societies. *Anim. Behav.*, **31**, 667–682.

(1983b). Optimal degree of skew in cooperative societies. *Am. Zool.*, **23**, 327–335.

(2000). Evolutionary routes to joint-female nesting in birds. *Behav. Ecol.*, **11**, 334–344.

Vehrencamp, S. L. and Bradbury, J. W. (1984). Mating systems and ecology. In: *Behavioural Ecology: an Evolutionary Approach*, 2nd ed., ed. J. R. Krebs and N. B. Davies. Oxford: Blackwell. pp. 251–278.

Vehrencamp, S. L., Bowen, B. S. and Koford, R. R. (1986). Breeding roles and pairing patterns within communal groups of groove-billed anis. *Anim. Behav.*, **34**, 347–366.

Vehrencamp, S. L., Koford, R. L. and Bowen, B. S. (1988). Effects of breeding unit size on fitness components in groove-billed anis. In: *Reproductive Success*, ed. T. H. Clutton-Brock. Chicago, IL: University of Chicago Press. pp. 291–304.

Veltman, C. J. (1989). Flock, pair and group living lifestyles without cooperative breeding by Australian magpies *Gymnorhina tibicen*. *Ibis*, **131**, 601–608.

Verbeek, N. A. M. (1973). The exploitation system of the yellow-billed magpie. *Univ. Calif. Publ. Zool.*, **99**, 1–58.

Vleck, C. M. and Brown, J. L. (1999). Testosterone and social and reproductive behaviour in *Aphelocoma* jays. *Anim. Behav.*, **58**, 943–951.

Vleck, C. M. and Patrick, D. J. (1999). Effects of vasoactive intestinal peptide on prolactin secretion in three species of passerine birds. *Gen. Comp. Endocrinol.*, **113**, 146–154.

Vleck, C. M., Mays, N. A., Dawson, J. W. and Goldsmith, A. R. (1991). Hormonal correlates of parental and helping behavior in cooperatively breeding Harris' hawks (*Parabuteo unicinctus*). *Auk*, **108**, 638–648.

Wagner, R. H. (1998). Hidden leks: sexual selection and the clustering of avian territories. *Ornithol. Monogr.*, **49**, 123–145.

Waite, T. A. and Strickland, D. (1997). Cooperative breeding in gray jays: philopatric offspring provision juvenile siblings. *Condor*, **99**, 523–525.

Walls, S. S. and Kenward, R. E. (1998). Movements of radio-tagged buzzards *Buteo buteo* in early life. *Ibis*, **140**, 561–568.

Walters, J. R. (1990). Red-cockaded woodpeckers: a "primitive" cooperative breeder. In *Cooperative Breeding in Birds*, ed. P. B. Stacey and W. D. Koenig. Cambridge: Cambridge University Press. pp. 67–102.

(1998). The ecological basis of avian sensitivity to habitat fragmentation. In: *Avian Conservation: Research and Management*, ed. J. M. Marzluff and R. Sallabanks. Washington, DC: Island Press. pp. 181–192.

Walters, J. R., Doerr, P. D. and Carter, J. H., III. (1988). The cooperative breeding system of the red-cockaded woodpecker. *Ethology*, **78**, 275–305.

Walters, J. R., Copeyon, C. K. and Carter, J. H., III. (1992a). Test of the ecological basis of cooperative breeding in red-cockaded woodpeckers. *Auk*, **109**, 90–97.

Walters, J. R., Doerr, P. D. and Carter, J. H., III. (1992b). Delayed dispersal and reproduction as a life-history tactic in cooperative breeders: fitness calculations from red-cockaded woodpeckers. *Am. Nat.*, **139**, 623–643.

Walters, J. R., Ford, H. A. and Cooper, C. B. (1999). Variation in population structure and ecology of brown treecreepers between contiguous and fragmented woodland: a preliminary assessment. *Biol. Conserv.*, **90**, 13–20.

Walters, J. R., Crowder, L. B. and Priddy, J. A. (2002). Population viability analysis for red-cockaded woodpeckers using an individual-based model. *Ecol. Appl.*, **12**, 249–260.

Warkentin, I. G., Curzon, A. D., Carter, R. E., Wetton, J. H., James, P. C., Oliphant, L. W. and Parkin, D. T. (1994).

No evidence for extrapair fertilizations in the merlin revealed by DNA fingerprinting. *Mol. Ecol.*, **3**, 229–234.

Waser, P. M. (1981). Sociality of territorial defense? The influence of resource renewal. *Behav. Ecol. Sociobiol.*, **8**, 231–237.

Waser, P. M. and Waser, M. S. (1985). *Ichneumia albicauda* and the evolution of viverrid gregariousness. *Z. Tierpsychol.*, **68**, 137–151.

Waser, P. M., Austad, S. N. and Keane, B. (1986). When should animals tolerate inbreeding? *Am. Nat.*, **128**, 529–537.

Waser, P. M., Creel, S. R. and Lucas, J. R. (1994). Death and disappearance: estimating mortality risks associated with philopatry and dispersal. *Behav. Ecol.*, **5**, 135–141.

Waser, P. M., Elliot, L. F., Creel, N. M. and Creel, S. R. (1995). Habitat variation and mongoose demography. In: *Serengeti II. Dynamics, Management and Conservation of an Ecosystem*, ed. A. R. E. Sinclair and P. Arcese. Chicago, IL: University of Chicago Press. pp. 421–448.

Washabaugh, K. F., Snowdon, C. T. and Ziegler, T. E. (2002). Variations in care for cottontop tamarin, *Saguinus oedipus*, infants as a function of parental experience and group size. *Anim. Behav.*, **63**, 1163–1174.

Watson, A., Moss, R., Parr, R., Mountford, M. D. and Rothery, P. (1994). Kin land ownership, differential aggression between kin and non-kin, and population fluctuations in red grouse. *J. Anim. Ecol.*, **63**, 39–50.

Weatherhead, P. J. and Forbes, M. R. L. (1994). Natal philopatry in passerine birds: genetic or ecological influences? *Behav. Ecol.*, **5**, 426–433.

Weathers, W. W., Koenig, W. D. and Stanback, M. T. (1990). Breeding energetics and thermal ecology of the acorn woodpecker in central coastal California. *Condor*, **92**, 341–359.

Webber, T. and Cox, J. A. (1987). Breeding and behavior of scrub jays *Aphelocoma coerulescens* in captivity. *Avicult. Mag.*, **93**, 6–14.

Webster, M. S., Pruett-Jones, S., Westneat, D. F. and Arnold, S. J. (1995). The effects of pairing success, extra-pair copulations and mate quality on the opportunity for sexual selection. *Evolution*, **49**, 1147–1157.

Webster, M. S., Tarvin, K. A., Tuttle, E. M. and Pruett-Jones, S. (2003). Reproductive promiscuity in the splendid fairy-wren: effects of group size and reproduction by helpers. *Behav. Ecol.*, in press.

Wegner, W. A. (1976). Extra-parental assistance by male American kestrel. *Wilson Bull.*, **88**, 670.

Welty, J. C. and Baptista, L. (1988). *The Life of Birds*, 4th edn. New York, NY: Saunders.

Westerdahl, H., Bensch, S., Hansson, B., Hasselquist, D. and von Schantz, T. (1997). Sex ratio variation among broods of great reed warblers *Acrocephalus arundinaceus*. *Mol. Ecol.*, **6**, 543–548.

White, F. N., Bartholomew, G. A. and Howell, T. R. (1975). The thermal significance of the nest of the sociable weaver *Philetairus socius*: winter observations. *Ibis*, **117**, 171–179.

Whitehead, H. (1996). Babysitting, dive synchrony, and indications of alloparental care in sperm whales. *Behav. Ecol. Sociobiol.*, **38**, 237–244.

Whitehead, P. J. (1999). Aspects of the nesting biology of the magpie goose *Anseranas semipalmata*: incubation period, hatching synchrony and patterns of nest attendance and defence. *Emu*, **99**, 121–134.

Whitehead, P. J. and Saalfeld, K. (2000). Nesting phenology of magpie geese (*Anseranas semipalmata*) in monsoonal northern Australia: responses to antecedent rainfall. *J. Zool.*, **251**, 495–508.

Whitehead, P. J. and Tschirner, K. (1990a). Eggs and hatchlings of the magpie goose *Anseranas semipalmata*. *Emu*, **90**, 154–160.

  (1990b). Magpie goose, *Anseranas semipalmata*, nesting on the Mary River floodplain, Northern territory, Australia: extent and frequency of flooding losses. *Aust. Wildl. Res.*, **17**, 147–157.

  (1991). Patterns of egg-laying and variation in egg size in the magpie goose *Anseranas semipalmata*: evidence for intra-specific nest parasitism. *Emu*, **91**, 26–31.

Whittingham, L. A. and Dunn, P. O. (1998). Male parental effort and paternity in a variable mating system. *Anim. Behav.*, **55**, 629–640.

Whittingham, L. A., Dunn, P. O. and Magrath, R. D. (1997). Relatedness, polyandry and extra-group paternity in the cooperatively-breeding white-browed scrubwren (*Sericornis frontalis*). *Behav. Ecol. Sociobiol.*, **40**, 261–270.

Widowski, T. M., Ziegler, T. E., Elowson, A. M. and Snowdon, C. T. (1990). The role of males in the stimulation of reproductive function in female cotton-top tamarins, *Saguinus o. oedipus*. *Anim. Behav.*, **40**, 731–741.

Wiens, J. J. and Servedio, M. R. (1998). Phylogenetic analysis and intraspecific variation: performance of parsimony, likelihood, and distance methods. *Syst. Biol.*, **47**, 228–253.

Wiley, R. H. and Rabenold, K. N. (1984). The evolution of cooperative breeding by delayed reciprocity and queuing for favorable social positions. *Evolution*, **38**, 609–621.

Wilkinson, R. (1988). Long-tailed glossy starlings *Lamrotornis caudatus* in field and aviary with observations on co-operative breeding in captivity. *Avicult. Mag.*, **94**, 143–154.

Williams, D. A., Lawton, M. F. and Lawton, R. O. (1994). Population growth, range expansion, and competition in the cooperatively breeding brown jay, *Cyanocorax morio*. *Anim. Behav.*, **48**, 309–322.

Williams, G. C. (1966). *Adaptation and Natural Selection*. Princeton, NJ: Princeton University Press.

(1979). On the question of adaptive sex ratio in outcrossed vertebrates. *Proc. R. Soc. London Ser. B*, **205**, 567–580.

(1992). *Natural Selection: Domains, Levels, and Challenges*. Oxford: Oxford University Press.

Williams, J. B. and Du Plessis, M. A. (1996). Field metabolism and water flux of sociable weavers *Philetairus socius* in the Kalahari Desert. *Ibis*, **138**, 168–171.

Williams, J. B., Du Plessis, M. A. and Siegfried, W. R. (1991). Green woodhoopoes (*Phoeniculus purpureus*) and obligate cavity roosting provide a test of the thermoregulatory insufficiency hypothesis. *Auk*, **108**, 285–293.

Wilson, F. E. and Follett, B. K. (1975). Corticosterone-induced gonadosuppression in photostimulated tree sparrows. *Life Sci.*, **17**, 1451–1456.

Wilson, K. and Hardy, I. C. W. (2002). Statistical analysis of sex ratios: an introduction. In: *The Sex Ratio Handbook*, ed. I. C. W. Hardy. Cambridge: Cambridge University Press. pp. 48–92.

Wingfield, J. C. (1988). Changes in reproductive function of free-living birds in direct response to environmental perturbations. In: *Processing of Environmental Information in Vertebrates*, ed. M. H. Stetson. Berlin: Springer-Verlag. pp. 121–148.

Wingfield, J. C. and Lewis, D. M. (1993). Hormonal and behavioural responses to simulated territorial intrusion in the cooperatively breeding white-browed sparrow weaver, *Plocepasser mahali*. *Anim. Behav.*, **45**, 1–11.

Wingfield, J. C. and Soma, K. K. (2002). Spring and autumn territoriality in song sparrows: same behavior different mechanisms? *Int. Comp. Biol.*, **42**, 11–20.

Wingfield, J. C., Hegner, R. E., Dufty, A. M. and Ball, G. F. (1990). The "challenge hypothesis": theoretical implications for patterns of testosterone secretion, mating systems, and breeding strategies. *Am. Nat.*, **136**, 829–845.

Wingfield, J. C., Hegner, R. E. and Lewis, D. M. (1991). Circulating levels of luteinizing hormone and steroid hormones in relation to social status in the cooperatively breeding white-browed sparrow weaver, *Plocepasser mahali*. *J. Zool.*, **225**, 43–58.

(1992). Hormonal responses to removal of a breeding male in the cooperatively breeding white-browed sparrow weaver, *Plocepasser mahali*. *Horm. Behav.*, **26**, 145–155.

Winkler, H., Christie, D. A. and Nurney, D. (1995). *Woodpeckers*. Sussex: Pica Press.

Winterbottom, M., Burke, T. and Birkhead, T. R. (1999). A stimulatory phalloid organ in a weaver bird. *Nature*, **399**, 28.

(2001). The phalloid organ, orgasm and sperm competition in a polygynandrous bird: the red-billed buffalo weaver (*Bubalornis niger*). *Behav. Ecol. Sociobiol.*, **50**, 474–482.

With, K. A. and King, A. W. (1999). Extinction thresholds for species in fractal landscapes. *Conserv. Biol.*, **13**, 314–326.

Witmer, M. C. (1993). Cooperative breeding by rufous hornbills on Mindanao Island, Philippines. *Auk*, **110**, 933–936.

Witter, M. S. and Cuthill, I. E. (1993). The ecological costs of avian fat storage. *Proc. R. Soc. London Ser. B*, **340**, 73–90.

Woolfenden, G. E. (1975). Florida scrub jay helpers at the nest. *Auk*, **92**, 1–15.

(1978). Growth and survival of young Florida scrub jays. *Wilson Bull.*, **90**, 1–18.

Woolfenden, G. E. and Fitzpatrick, J. W. (1984). *The Florida Scrub Jay: Demography of a Cooperative-Breeding Bird*. Princeton, NJ: Princeton University Press.

(1986). Sexual asymmetries in the life history of the Florida scrub jay. In: *Ecological Aspects of Social Evolution*, ed. D. Rubenstein and R. W. Wrangham. Princeton, NJ: Princeton University Press, pp. 87–107.

(1990). Florida scrub jays: a synopsis after 18 years of study. In: *Cooperative Breeding in Birds*, ed. P. B. Stacey and W. D. Koenig. Cambridge: Cambridge University Press. pp. 241–266.

(1996). Florida scrub-jay (*Aphelocoma coerulescens*). In: *The Birds of North America*, ed. A. Poole and F. Gill. Philadelphia, PA and Washington, DC: Academy of Natural Sciences and American Ornithologists' Union.

Wrangham, R. W. (1980). An ecological model of female-bonded primate groups. *Behaviour*, **75**, 262–300.

Wright, J. (1997). Helping-at-the-nest in Arabian babblers: signalling social status or sensible investment in chicks. *Anim. Behav.*, **54**, 1439–1448.

Wright, J., Parker, P. G. and Lundy, K. J. (1999). Relatedness and chick-feeding effort in the cooperatively breeding Arabian babbler. *Anim. Behav.*, **58**, 779–785.

Wright, J., Malkakov, A. A. and Khazin, V. (2000). State-dependent sentinels: an experimental study in the Arabian babbler. *Proc. R. Soc. London Ser. B*, **268**, 821–826.

Wright, J., Berg, E., De Kort, S., Khazin, V. and Malkakov, A. A. (2001). Cooperative sentinel behaviour in the Arabian babbler. *Anim. Behav.*, **62**, 973–979.

Yamashina, M. (1938). A sociable breeding unit habit among timaliine birds. *Proc. Int. Ornithol. Congr.*, **9**, 453–456.

Yates, C. J. and Hobbs, R. J. (1996). Woodland restoration in the Western Australian wheatbelt: a conceptual framework using a state and transition model. *Restor. Ecol.*, **5**, 28–35.

Yezerinac, S. M. (1999). Sex allocation in co-operatively breeding birds. *Proc. Int. Ornithol. Congr.*, **22**, 467–482.

Yom-Tov, Y. (1980). Intraspecific nest parasitism in birds. *Biol. Rev.*, **55**, 93–108.

  (1987). The reproductive rates of Australian passerines. *Aust. Wildl. Res.*, **14**, 319–330.

Yom-Tov, Y. and Ollason, J. G. (1976). Sexual dimorphism and sex ratios in wild birds. *Oikos*, **27**, 81–85.

Young, A. J. (2003). Subordinate tactics in cooperative meerkats: helping, breeding and dispersal. Ph.D. dissertation, University of Cambridge.

Young, E. C. (1999). *The Millennium Bird: Cooperative Breeding in Chatham Island Skua*s. Privately published.

Zack, S. W. (1990). Coupling delayed breeding with short-distance dispersal in cooperatively breeding birds. *Ethology*, **86**, 265–289.

  (1995). Cooperative breeding in *Lanius* shrikes. III. A reply in hindsight to Zack and Ligon I, II. *Proc. West. Found. Vert. Zool.*, **6**, 34–38.

Zack, S. W. and Ligon, J. D. (1985). Cooperative breeding in *Lanius* shrikes.I. Habitat and demography of two sympatric species. *Auk*, **102**, 754–765.

Zack, S. W. and Rabenold, K. N. (1989). Assessment, age and proximity in dispersal contests among cooperative wrens: field experiments. *Anim. Behav.*, **38**, 235–247.

Zahavi, A. (1974). Communal nesting by the Arabian babbler: a case of individual selection. *Ibis*, **116**, 84–87.

  (1990). Arabian babblers: the quest for social status in a cooperative breeder. In: *Cooperative Breeding in Birds*, ed. P. B. Stacey and W. D. Koenig. Cambridge: Cambridge University Press. pp. 103–130.

  (1995). Altruism as a handicap: the limitations of kin selection and reciprocity. *J. Avian Biol.*, **26**, 1–3.

Zahavi, A. and Zahavi, A. (1997). *The Handicap Principle: a Missing Piece of Darwin's Puzzle*. Oxford: Oxford University Press.

Zanette, L. and Jenkins, B. (2000). Nesting success and nest predators in forest fragments: a study using real and artificial nests. *Auk*, **117**, 445–454.

Zhang, D. Y., Jiang, X. H. and Zhao, S. L. (1996). Evolutionary stable reproductive strategies in sexual organisms. II. Dioecy and optimal resource allocation. *Am. Nat.*, **147**, 1115–1123.

Zimmer, K. J. and Whittaker, A. (2000). The rufous cacholote (Furnariidae: Pseudoseisura) is two species. *Condor*, **102**, 409–422.

Zink, A. G. (2000). The evolution of intraspecific brood parasitism in birds and insects. *Am. Nat.*, **155**, 395–405.

Taxonomic index

Acanthisitta chloris, see rifleman
Acanthisittidae, 25, 207
Acanthiza chrysorrhoa, 27, 207
   lineata, 27, 207
   murina, 27
   nana, 27
   pusilla, 40, 239
   reguloides, 27, 207
   uropygialis, 27
Acanthizinae, 27, 33
Acanthizini, 27
Accentor, alpine, 32, 84, 93, 99, 209, 239
Accipiter cooperii, see hawk, Cooper's
Accipitridae, 25, 33, 34, 207
Accipitrinae, 25, 33, 34
Aceros comatus, 23
Acrocephalinae, 31, 33
Acrocephalus melanopogon, see warbler, moustached
   sechellensis, see warbler, Seychelles
   vaughani, 31
Aegithalidae, 31, 209
Aegithalos caudatus, see tit, long-tailed
Agelaius phoeniceus, see blackbird, red-winged
Aimophila ruficauda, 34
Alauahio, Maui, 209, 239
Alaudidae, 34
Amaurornis flavirostra, 25
Amytornis barbatus, 26
   dorotheae, 26
   goyderi, 26
   housei, 26
   purnelli, 26
   striatus, 26
   textiles, 26
   woodwardi, 26
Amytornithinae, 26
Ani, greater, 25
   groove-billed, 25, 173, 188–191, 195, 207, 232, 239
   smooth-billed, 25, 188–190, 195, 207, 240
Anorrhinus austeni, 23
   galeritus, 23
   tickelli, 23
Anseranas semipalmata, see goose, magpie
Anseranatidae, 33, 181, 206
Anseriformes, 181, 192, 193
Anteater-chat, northern, 30

Anthochaera carunculata, 27
   lunulata, 27
Aphelocephala leucopsis, 27
   nigricincta, 27
Aphelocoma, 18, 19, 21, 134, 139, 145, 234
   insularis, see scrub-jay, island
   californica, see scrub-jay, western
   coerulescens, see scrub-jay, Florida
   ultramarina, see jay, Mexican
   unicolor, see jay, unicolored
Apodidae, 25
Apostlebird, 28, 208, 230, 239
Apteryx mantelli, see kiwi
Apus horus, 25
Aracari, collared, 23
   fiery-billed, 23
Artamini, 29
Artamus cinereus, 29, 208
   cyanopterus, 29
   leucorhynchus, 29, 208
   maximus, 29
   minor, 29

Babbler, Arabian, 32, 44, 86, 95, 96, 100, 101, 121, 124, 127, 190, 201, 209, 222, 239
   arrowmarked, 32, 209
   bare-cheeked, 209
   blackcap, 32, 209
   black-lored, 32, 209
   brown, 32, 209
   chestnut-crowned, 28, 207
   common, 32, 209, 239
   grey-crowned, 28, 57, 208
   Hall's, 28, 207
   jungle, 32, 44, 209, 239
   large grey, 32, 209
   New Guinea, 28, 207
   striated, 32
   white-browed, 28, 208
   yellow-billed, 31
   yellow-eyed, 31
Baeolophus bicolor, 34
   wollweberi, 31
Barbet, Anchieta's, 22
   bearded, 23
   black-backed, 23

Barbet, Anchieta's (*cont.*)
  black-billed, 23
  black-breasted, 23
  black-collared, 23
  black-throated, 22
  bristle-nosed, 22
  brown-breasted, 23
  Chaplin's, 23
  D'Arnaud's, 23, 206
  double-toothed, 23
  green, 22
  grey-throated, 22
  naked-faced, 22
  pied, 22
  red-and-yellow, 23, 206
  red-faced, 23
  spot-flanked, 22
  toucan, 23, 206
  Vieillot's, 23
  white-eared, 22
  white-headed, 23
  Whyte's, 22
  yellow-breasted, 23
Bateleur, 25
Bee-eater, black-headed, 24
  blue-cheeked, 24
  blue-headed, 24
  blue-tailed, 24
  blue-throated, 24
  carmine, 24, 207
  chestnut-headed, 24
  cinnamon-chested, 24
  European, 24, 58, 86, 97, 207, 239
  little, 24
  little green, 24, 207
  rainbow, 24, 207
  red-throated, 24, 78, 122, 207, 240
  rosy, 24
  swallow-tailed, 24
  white-fronted, 24, 53, 58, 61, 78, 173, 193, 205, 207, 230, 240
  white-throated, 24
Blackbird, Austral, 32
  Bolivian, 32
  red-winged, 60, 240
Blue-flycatcher, African, 29
Bluebird, eastern, 30
  western, 30, 38, 41, 45, 51, 55, 57, 58, 60, 61, 77, 87, 97, 99, 100, 107–109, 111, 208, 240
Bobolink, 32
Booby, blue-footed, 105, 239
Bradornis mariquensis, 34
  pallidus, 30
Brushrunner, lark-like, 26
Bubalornis niger, *see* buffalo-weaver, red-browed
Bubo lacteus, 25
Bucconidae, 23, 206

Bucephala clangola, *see* goldeneye
Buceros hydrocorax, 23
  rhinoceros, 23
Bucerotidae, 23
Bucerotiformes, 10
Bucorvidae, 24, 206
Bucorvus leadbeateri, 24, 206
Buffalo-weaver, red-billed, 91, 99
  red-browed, 84, 91, 99, 100, 240
Bunting, chestnut-eared, 34
Buphagus africanus, 30
  erythrorhynchus, 30
Bush baby, 106, 241
Bush-crow, Stresemann's, 29
Bushtit, 31, 44, 209, 239
Butcherbird, grey, 29
  hooded, 29
  pied, 29
Buteo galapagoensis, *see* hawk, Galápagos
  jamaicensis, 34
  swainsoni, 34

Cacholote, rufous, 26
Cactus-finch, common, 32, 209
Calcarius pictus, *see* longspur, Smith's
Callithrix jacchus, *see* marmoset, common
  kuhli, 241
Calocitta colliei, 28
  formosa, 28, 208, 240
Campylorhynchus, 10, 14, 16–21, 52, 88, 95, 234
  albobrunneus, 31
  brunneicapillus, *see* wren, cactus
  chiapensis, 31
  fasciatus, 30, 208
  griseus, *see* wren, bicolored
  gularis, 31, 209
  jocosus, 31
  megalopterus, 31, 208
  nuchalis, *see* wren, stripe-backed
  rufinucha, 31
  turdinus, 31
  yucatanicus, 31
  zonatus, 31, 209
Canis aureus, 241
  familiaris, *see* dingo
  lupus, *see* wolf, gray
  mesomelas, 241
  simensis, *see* wolf, Ethiopian
Capitoninae, 23
Cardinal, northern, 34
Cardinalini, 32, 34
Cardinalis cardinalis, 34
Caryothraustes poliogaster, 32
Cassowary, 178, 179, 239
Casuarius casuarius, *see* cassowary
Catharacta lonnbergi, *see* skua, brown
  maccormicki, 34

Ceratogymna atrata, 23
  bucinator, 23
  subcylindricus, 23
Certhia familiaris, *see* treecreeper
Certhiidae, 30, 34, 208
Cervus elaphus, *see* deer, red
Ceryle rudis, *see* kingfisher, pied
Cerylidae, 24, 207
Chaetops frenatus, 30, 208
Chaetura andrei, 25
  brachyura, 25
  pelagica, 25
  vauxi, 25
Charadriidae, 33
Charadriinae, 33
Chat, anteater, 208
  white-fronted, 27
Chersomanes albofasciata, 34
Chough, white-winged, 2, 11, 28, 69–70, 78–79, 83, 86, 88, 96, 99–101, 119–120, 122, 170–172, 190, 193, 195, 208, 230, 233, 240
Chrysomma sinense, 31
Chthonicola sagittatus, 27
Ciconiiformes, 7, 10
Cinclosoma cinnamomeum, 28
Cinclosomatinae, 28
Cinnycerthia peruana, 31
Cinnyricinclus leucogaster, 30
Climacteridae, 26, 207
Climacteris erythrops, 26, 207
  melanura, 26, 207
  picumnus, *see* treecreeper, brown
  rufa, 26, 207
Clytomyias insignis, 26
Coati, 218, 241
Coccyzus americanus, *see* cuckoo, yellow-billed
Colaptes campestris, 22
Coliidae, 24, 207
Coliiformes, 10, 119–121
Coliinae, 24
Colius, 117
  castanotus, *see* mousebird, red-backed
  colius, *see* mousebird, white-backed
  leucocephalus, 24
  striatus, *see* mousebird, speckled
Collocalia spodiopygius, 25
Columbiformes, 192
Conopophilia rufogularis, 27
Coot, American, 192, 239
  giant, 25
  red-knobbed, 25
Coraciae, 7
Coraciiformes, 10
Coracina maxima, 29, 208
Corcoracinae, 28
Corcorax melanorhamphos, *see* chough, white-winged
Corvida, 13, 14, 21, 48, 234, 237

Corvidae, 28, 197, 208
Corvinae, 28, 234
Corvinella corvina, 28, 208, 240
  melanoleuca, 28
Corvini, 28
Corvus brachyrhynchos, 28, 208, 239
  caurinus, 28, 208
  corone, 28, 239
Coryphistera alaudina, 26
Corythaixoides concolor, 25
  leucogaster, 34
Cosmopsarus regius, 30, 208
Cotinginae, 25
Coturnix chinensis, *see* quail, button
Cowbird, bay-winged, 32, 209
Cracticus cassicus, 29
  nigrogularis, 29
  torquatus, 29
Crake, black, 25
Criniferinae, 25, 34
Crocuta crocuta, 241
Crotophaga ani, *see* ani, smooth-billed
  major, 25
  sulcirostris, *see* ani, groove-billed
Crotophagidae, 25, 207
Crotophaginae, 188–190
Crotophagini, 25
Crow, American, 28, 56, 198, 199, 208, 239
  carrion, 28, 36, 40, 239
  northwestern, 28, 208
Cryptomys anselli, *see* mole-rat, common
  damarensis, *see* mole-rat, Damaraland
  darlingi, *see* mole-rat, Mashona
  mechowi, *see* mole-rat, Zambian
Crypturellus boucardii, 240
Cuckoo, guira, 25, 84, 91, 188–190, 195, 207, 232, 239
  yellow-billed, 192, 240
Cuckoo-shrike, ground, 29, 208
Cuculiformes, 188, 192
Curaeus curaeus, 32
Cyanocorax affinis, 29
  beecheii, 29, 208, 239
  chrysops, 208
  cristatellus, 28
  dickeyi, 29, 208
  melanocyaneus, 29, 208
  sanblasianus, 29, 208
  violaceus, 28
  yncas, 29, 208, 239
  yucatanicus, 29, 208
Cyanopica cyana, 28, 208
Cynomys ludovicianus, 241
Cyphorhinus ardus, 31

Dacelo leachii, 24, 206
  novaeguineae, 24, 206, 239

Daphoenositta, 10
   chrysoptera, 28, 208
   miranda, 28
Deer, red, 102–103, 241
Delichon urbica, 31
Dendrocopos medius, 33
Dicrurinae, 29
Dicrurini, 29
Dicrurus macrocercus, 29
Dingo, 182, 241
Dioptrornis chocolatinus, 30
Dog, African wild, 127, 211, 214, 216, 220, 223, 241
Dolichonyx oryzivorus, 32
Donacobius atricapillus, 31, 209
Donacobius, black-capped, 31, 209
Dromaius novaehollandiae, 239
Drongo, black, 29
Dryoscopus cubla, 29
Dunnock, 5, 33, 81, 84, 88, 92, 99, 164, 173, 209, 226, 229, 230, 239

Eagle, bald, 34
Eagle-owl, Verreaux's, 25
Eclectus roratus, 33, 207, 239
Elephant, African, 210, 241
Elk, *see* deer, red
Elminia longicauda, 29
Emberiza fucata, 34
Emberizinae, 32–34
Emberizini, 33, 34
Emu, 178, 179, 239
Entomyzon cyanotis, 27
Eopsaltria australis, 28
   georgiana, 28
   griseogularis, 28
Ephthianura albifrons, 27
Eremomela, greencap, 31
   Senegal, 31
Eremomela pusilla, 31
   scotops, 31
Erithacus rubecula, 34
Erythrocercus mccalli, 29
Eucalyptus, 199, 200, 206
Euphonia laniirostris, 32
Euphonia, thick-billed, 32
Eurocephalus anguitimens, 28, 208
   rueppelli, 28

Fairy-wren, blue-breasted, 26, 207
   broad-billed, 26
   Campbell's, 26
   emperor, 26
   lovely, 26
   orange-crowned, 26
   purple-crowned, 26, 207
   red-backed, 26
   red-winged, 26, 201, 202, 207
   splendid, 1, 26, 77, 83, 86, 96, 146, 152, 201, 202, 207, 235, 240
   superb, 26, 51, 82, 83, 86, 96, 99, 100, 138–140, 146, 149, 153, 201, 202, 207, 233, 235–236, 240
   variegated, 26
   Wallace's, 26
   white-shouldered, 26
   white-winged, 26, 230, 240
Falco columbarius, 33
   peregrinus, 25
   sparverius, 34
Falcon, peregrine, 25
Falconidae, 25, 33, 34
Falcunculini, 28
Falcunculus frontatus, 28
Figbird, green, 29
Finch, zebra, 105, 240
Flicker, campo, 22
Flycatcher, chestnut-capped, 29
   Mariqua, 34
   pale, 30
   rusty-margined, 25
   white-bearded, 25, 207
Forest-flycatcher, African, 30
Fraseria ocreata, 30
Friarbird, little, 27
Fringillidae, 32–34, 209
Fruitcrow, purple-throated, 25
Fulica americana see coot, American
   cristata, 25
   gigantea, 25
Furnariidae, 26
Furnariinae, 26

Galago crassicaudatus, *see* bush baby
Galbalcyrhynchus purusianus, 23
Galbula ruficauda, 23
Galbulidae, 23
Galliformes, 192
Gallinula chloropus, *see* moorhen, common
   mortierii, *see* native hen, Tasmanian
   tenebrosa, 25, 207
Gallinule, purple, 25, 207
Geococcyx californianus, *see* roadrunner, greater
Geocolaptes olivaceus, 22
Geospiza fortis, 32, 209
   scandens, 32, 209
Gerbil, Mongolian, 211, 220, 222, 224, 241
Gerygone, brown, 27
   yellow-bellied, 27
Gerygone chrysogaster, 27
   mouki, 27
Globicephala melas, *see* whale, pilot
Glossy-starling, long-tailed, 30
   red-shouldered, 30
Go-away-bird, grey, 25
   white-bellied, 34
Goldeneye, 90, 239
Goose, magpie, 33, 90, 181–182, 193–195, 206, 239
Goshawk, pale chanting, 33, 95, 118, 127, 240

Grallina cyanoleuca, 29
Grasswren, black, 26
  carpentarian, 26
  dusky, 26
  Eyrean, 26
  grey, 26
  striated, 26
  thick-billed, 26
  white-throated, 26
Grebe, Australasian, 25
  horned, 34
Greenbul, spotted, 31
  swamp, 31
Grosbeak, black-faced, 32
Ground-finch, medium, 32, 209
Ground-hornbill, southern, 24, 206
Ground squirrel, Belding's, 211, 220, 222, 223, 241
Grouse, red, 240
Gruiformes, 7, 182
Guira guira, *see* cuckoo, guira
Guirini, 25
Gymnobucco bonapartei, 22
  calvus, 22
  peli, 22
Gymnorhina tibicen, *see* magpie, Australian
Gymnorhinus cyanocephalus, *see* jay, pinyon

Halcyon chelicuti, 24
  cinnamomina, 207
Halcyonidae, 24, 206
Haliaeetus leucocephalus, 34
Hammerkop, 34
Hawk, Cooper's, 34, 105, 239
  Galápagos, 33, 85, 94, 95, 99, 101, 127, 200, 207, 239
  Harris's, 25, 107, 108, 113, 118, 135–136, 139, 140, 207, 239
  red-tailed, 34
  Swainson's, 34
Helmetshrike, chestnut-fronted, 30, 208
  Retz's, 30, 208
  white, 30, 208
  yellow-crested, 29
Helogale parvula, *see* mongoose, dwarf
Heterocephalus glaber, *see* mole-rat, naked
Hihi, *see* stitchbird
Hirundinidae, 31, 34
Hirundininae, 31, 34
Hirundo atrocaerulea, 34
  rustica, 31
Hoatzin, 24, 117–118, 201, 202, 207, 239
Honeyeater, black-chinned, 27
  black-headed, 27
  blue-faced, 27
  brown-headed, 27
  crescent, 107, 239
  golden-backed, 27
  New Holland, 27
  rufous-throated, 27
  striped, 27
  strong-billed, 27
  varied, 27
  white-lined, 27
  white-naped, 27
  white-plumed, 27
  white-throated, 27
  yellow-faced, 107, 240
  yellow-tufted, 27
Hoopoe, 87, 98, 239
  African, 24
Hornbill, Assam, 23
  black-and-white-casqued, 23
  black-casqued, 23
  brown, 23
  bushy-crested, 23
  Luzon, 24
  red-billed dwarf, 23
  rhinoceros, 23
  rufous, 23
  Sulawesi, 23
  Tarictic, 24
  trumpeter, 23
  white-crowned, 23
House-martin, northern, 31
Hyaena brunnea, *see* hyena, brown
Hyena, brown, 211, 220, 222, 241
  spotted, 210, 211, 215, 217, 218, 224, 241

Icterini, 32
Ictinia mississippiensis, 25
Ixonotus guttatus, 31

Jacamar, chestnut, 23
  rufous-tailed, 23
  three-toed, 23
Jacamaralcyon tridactyla, 23
Jackal, black-backed, 241
  golden, 241
Jay, Beechey, 208, 239
  black-chested, 29
  brown, 29, 208, 230, 239
  bushy-crested, 29, 208
  curl-crested, 28
  gray, 29, 59, 239
  green, 29, 59, 208, 239
  Mexican, 18, 28, 58, 82, 85, 99, 133–134, 139, 140, 144–146, 148, 201, 205, 208, 235, 236, 240
  pinyon, 18, 29, 56, 58, 119, 208, 240
  plush-capped, 208
  purplish-backed, 29
  San Blas, 29, 208
  Siberian, 2, 29, 40, 44, 45, 52, 58, 82, 229, 240
  tufted, 29, 208
  unicolored, 18, 28, 208, 240
  violaceous, 28
  Yucatan, 29, 208

Kestrel, American, 34
Kingfisher, forest, 24
  Micronesian, 207
  pied, 5, 24, 53, 55, 69, 77, 78, 95, 107, 108, 122–123, 125, 130–131, 140, 173, 207, 222, 240
  striped, 24
Kite, Mississippi, 25
Kiwi, 178, 179, 239
Kookaburra, blue-winged, 24, 206
  laughing, 24, 56, 78, 87, 97–99, 107, 108, 110, 112–113, 198, 206, 239

Lagopus scoticus, 240
Lamprologus brichardi, 69, 122
Lamprotornis caudatus, 30
  nitens, 30
  pulcher, 30, 208
  superbus, 30
Laniidae, 20, 28, 208
Lanius, 16–21
  cabanisi, 28
  excubitoroides, 28, 208
Lapwing, southern, 33
Laridae, 25, 34, 192, 207
Larinae, 25, 34
Lark, spike-heeled, 34
Leontopithecus rosalia, *see* tamarin, golden lion
Leptopterus chabert, 29
Lichenostomus chrysops, *see* honeyeater, yellow-faced
  melanops, 27
  penicillatus, 27
  versicolor, 27
Lion, African, 95, 211, 214, 215, 217, 220, 222, 241
Logrunner, 28
Longspur, Smith's, 33, 84, 99, 240
Loxodonta africana, *see* elephant, African
Lybiidae, 22, 206
Lybius bidentatus, 23
  chaplini, 23
  dubius, 23
  guifsobalito, 23
  leucocephalus, 23
  melanopterus, 23
  minor, 23
  rolleti, 23
  rubrifacies, 23
  torquatus, 23
  vieilloti, 23
Lycaon pictus, *see* dog, African wild

Magpie, Australian, 29, 59, 135, 208, 239
  azure-winged, 28, 208
  Formosan, 28, 208
Magpie-jay, black-throated, 28
  white-throated, 28, 78, 88, 190, 195, 201, 202, 208, 223, 240
Magpie-lark, 29

Malaconotinae, 29
Malaconotini, 29
Maluridae, 26, 207
Malurinae, 26
Malurini, 26
Malurus, 40, 96, 144, 146, 234, 235
  alboscapulatus, 26
  amabilis, 26
  campbelli, 26
  coronatus, 26, 207
  cyaneus, *see* fairy-wren, superb
  cyanocephalus, 26
  elegans, *see* fairy-wren, red-winged
  grayi, 26
  lamberti, 26
  leucopterus, *see* fairy-wren, white-winged
  melanocephalus, 26
  pulcherrimus, 26, 207
  splendens, *see* fairy-wren, splendid
Manorina, 82, 88, 94, 98, 99
  flavigula, 27
  melanocephala, *see* miner, noisy
  melanophrys, *see* miner, bell
  melanotis, *see* miner, black-eared
Marmoset, common, 211, 216–218, 224, 241
  Wied's black tufted-ear, 241
Marmot, alpine, 211, 220, 222, 241
Marmota marmota, *see* marmot, alpine
Marshbird, brown-and-yellow, 32, 209
Martin, brown-chested, 34
Meerkat, 70, 125–127, 152–154, 211, 213, 216–218, 220–222, 224, 225, 241
Melanerpes, 124
  cactorum, 22
  candidus, 22
  chrysauchen, 22
  cruentatus, 22
  flavifrons, 22
  formicivorus, *see* woodpecker, acorn
  lewis, *see* woodpecker, Lewis'
  striatus, 22
Melanodryas cucullata, *see* robin, hooded
Melierax canorus, *see* goshawk, pale chanting
Meliphagidae, 27, 94, 207
Melithreptus affinis, 27
  albilineata, 27
  albogularis, 27
  brevirostris, 27
  gularis, 27
  laetior, 27
  lunatus, 27
  validirostris, 27
Meriones unguiculatus, *see* gerbil, Mongolian
Merlin, 33
Meropidae, 15–16, 24, 207, 234

Merops, 15, 17
   albicollis, 24
   apiaster, 24, 207, 239
   breweri, 24
   bullockoides, see bee-eater, white-fronted
   bulocki, see bee-eater, red-throated
   hirundineus, 24
   leschenaulti, 24
   malimbicus, 24
   muelleri, 24
   nubicus, 24, 207
   oreobates, 24
   orientalis, 24, 207
   ornatus, 24, 207
   persicus, 24
   philippinus, 24
   pusillus, 24
   viridis, 24
Messelirrisoridae, 15
Microtus ochrogaster, 241
   pinetorum, 241
Mimini, 30
Mimus gilvus, 30
   longicaudatus, 30
   saturninus, 30
Miner, bell, 2, 27, 61, 65, 85, 99, 107–109, 111–112, 114, 136–137, 140, 207, 234, 239
   black-eared, 27, 107, 108, 239
   noisy, 2, 27, 85, 99, 107, 108, 113, 206, 207, 240
   yellow-throated, 27
Mistletoe, 38, 41
Mitrospingus cassinii, 32
Mockingbird, chalk-browed, 30
   Charles, 30
   Galápagos, 30, 55, 58, 83, 85, 100, 201, 202, 208, 239
   hood, 30
   long-tailed, 30
   tropical, 30
Mohoua albicilla, 28
   ochrocephala, 28
Mohouini, 28
Mole-rat, common, 153, 211, 216, 217, 241
   Damaraland, 152, 153, 211, 213, 216, 217, 241
   Mashona, 144, 153, 211, 216, 217, 241
   naked, 144, 147–148, 153, 211, 213, 216–218, 221, 224, 241
   Zambian, 211, 216, 217, 241
Molothrus badius, 32, 209
Monarchini, 29
Monasa morphoeus, 23, 206
Mongoose, banded, 211, 213, 216, 217, 220, 221, 241
   dwarf, 144, 147–148, 211, 213, 216, 217, 220, 241
Moorhen, common, 25, 57, 77, 84, 92, 99, 144, 149, 156, 183–185, 192, 193, 195, 207, 232, 236, 239
   dusky, 25, 207
Motacilla capensis, 32
Motacillinae, 32

Mousebird, blue-naped, 24
   red-backed, 24, 121, 240
   red-faced, 24
   speckled, 10, 24, 121, 207, 240
   white-backed, 24, 121, 240
   white-headed, 24
Mungos mungo, see mongoose, banded
Muscicapidae, 30, 34, 208
Muscicapinae, 30, 34
Muscicapini, 30, 34
Muskox, 210, 241
Musophagidae, 25, 34
Myiozetetes cayanensis, 25
Myrmecocichla aethiops, 30, 208

Nasua narica, see coati
Native hen, Tasmanian, 33, 84, 183–185, 193–195, 207, 230, 240
Neafrapus boehmi, 25
   cassini, 25
Neoaves, 7
Neosittini, 28
Nesomimus macdonaldi, 30
   parvulus, see mockingbird, Galápagos
   trifasciatus, 30
Nothoprocta cinerascens, see tinamou, brushland
   ornata, see tinamou, ornate
Notiomystis cincta, see stitchbird
Nunbird, white-fronted, 23, 206
Nuthatch, brown-headed, 18, 30, 208, 239
   pygmy, 18, 30, 208, 240

Oenanthe lugubris, 30
Opisthocomidae, 24, 207
Opisthocomus hoazin, see hoatzin
Orcinus orca, see whale, killer
Oreospar bolivianus, 32
Oriolini, 29
Orthonychidae, 28
Orthonyx temminckii, 28
Ostrich, 33, 117, 178–181, 195, 206, 240
Ovenbird, 12, 34, 240
Ovibos moschatus, see muskox
Oxpecker, red-billed, 30
   yellow-billed, 30

Pachycephalinae, 28
Panthera leo, see lion, African
Parabuteo unicinctus, see hawk, Harris's
Paradise-flycatcher, African, 29
Paradise-kingfisher, buff-breasted, 24
Pardalote, striated, 34
Pardalotidae, 27, 33, 34, 207
Pardalotinae, 34
Pardalotus striatus, 34
Paridae, 31, 34, 209
Parinae, 31, 34
Paroreomyza montana, 209, 239

Parrot, eclectus, 2, 33, 82–83, 85, 88, 93, 107, 108, 114, 207, 239
Parulini, 34
Parus bicolor, *see* titmouse, tufted
   caeruleus, *see* tit, blue
   major, *see* tit, great
   niger, 31, 209
Passer domesticus, 32
Passeri, 234
Passerida, 13, 14, 21
Passeridae, 32, 33, 209
Passeriformes, 192, 234
Passerinae, 32
Penelopides exarhatus, 23
   manilloe, 24
   panini, 24
Perisoreus canadensis, *see* jay, gray
   infaustus, *see* jay, Siberian
Petroicidae, 28, 33, 207
Phacellodomus rufifrons, 26
Phaeoprogne tapera, 34
Phelpsia inornata, 25, 207
Philemon citreogularis, 27
Philetairus socius, *see* weaver, sociable
Phoeniculidae, 15–16, 24, 206, 234
Phoeniculus, 15
   bollei, *see* woodhoopoe, white-headed
   castaneiceps, *see* woodhoopoe, forest
   damarensis, 24
   purpureus, *see* woodhoopoe, green
   somaliensis, *see* woodhoopoe, black-billed
Pholidornis rushiae, 31
Phoradendron villosum, *see* mistletoe
Phylidonyris lanceolata, 27
   novaehollandiae, 27
   pyrrhoptera, *see* honeyeater, crescent
Piapiac, 29
Picathartidae, 30, 208
Picidae, 22, 33, 206
Piciformes, 10, 185
Picoides, 16
   borealis, *see* woodpecker, red-cockaded
   pubescens, *see* woodpecker, downy
   villosus, *see* woodpecker, hairy
Pied-babbler, southern, 32
Pine, pinyon, 119
Platysteira peltata, 29
   tonsa, 29
Ploceinae, 32
Plocepasser mahali, *see* sparrow-weaver, white-browed
Podiceps auritus, 34
Podicipedidae, 25, 34
Pomatostomidae, 10, 28, 207
Pomatostomus halli, 28, 207
   isidorei, 28, 207
   ruficeps, 28, 207
   superciliosus, 28, 208
   temporalis, *see* babbler, grey-crowned

Porphyrio porphyrio, *see* pukeko
Porphyrula martinica, 25, 207
Prairie dog, black-tailed, 211, 220, 222, 241
Primates
Prionops alberti, 29
   plumatus, 30, 208
   retzii, 30, 208
   scopifrons, 30, 208
Prunella, 88
   collaris, *see* accentor, alpine
   modularis, *see* dunnock
Prunellinae, 32, 33
Psaltriparus minimus, *see* bushtit
Pseudoleistes virescens, 32, 209
Pseudonigrita arnaudi, 32, 209
Pseudoseisura cristata, 26
Psilorhinus morio, *see* jay, brown
Psittacidae, 33, 207
Psophia leucoptera, *see* trumpeter, pale-winged
Psophiidae, 25, 207
Psophocichla litsipsirupa, 34
Pteroglossus frantzii, 23
   torquatus, 23
Ptilostomus afer, 29
Puffback, black-backed, 29
Pukeko, 25, 84, 91, 99, 144–146, 170, 182–185, 192–195, 207, 230, 233, 240
Pycnonotidae, 31

Quail, button, 182, 239
Quail-thrush, cinnamon, 28
Querula purpurata, 25

Rallidae, 25, 33, 192, 207
Ramphastidae, 23, 206
Ramphastinae, 23
Reed-warbler, Pitcairn, 31
Remiz pendulatus, *see* tit, penduline
Remizinae, 31
Rhea americana, *see* rhea, greater
Rhea, greater, 33, 178–181, 195, 239
Rheidae, 33
Rhinopomastidae, 15, 16
Rhinopomastus, 15
Rifleman, 25, 83, 87, 98, 207, 240
Roadrunner, greater, 192, 239
Robin, European, 34
   grey-breasted, 28
   hooded, 33, 206, 207, 239
   white-breasted, 28
   yellow, 28
Rockjumper, rufous, 30, 208

Saguinus fuscicollis *see* tamarin, saddle-back
   oedipus, *see* tamarin, cotton-top
Saxicolini, 30, 34
Schetba rufa, 30

Scopidae, 34
Scopus umbretta, 34
Scrub-jay, Florida, 1, 18, 28, 41, 57, 77, 82, 83, 87, 97, 98, 119, 124, 131–133, 138–140, 152, 153, 173, 190, 199, 201, 202, 206, 208, 239
    island, 18, 239
    western, 18–20, 28, 119, 133, 139, 240
Scrubwren, large-billed, 27
    white-browed, 2, 33, 56, 58, 78, 86, 95, 99, 100, 153, 167–169, 207, 240
Seiurus aurocapillus, *see* ovenbird
Semnornis ramphastinus, 206
Sericornis frontalis, *see* scrubwren, white-browed
    magnirostris, 27
Sericornithini, 27, 33
Shrike, grey-backed fiscal, 28, 208
    long-tailed fiscal, 28
    magpie, 28
    white-crowned, 28, 208
    white-rumped, 28
    yellow-billed, 28, 208, 230, 240
Shrike-tit, crested, 28
Sialia mexicana, *see* bluebird, western
    sialis, 30
Sipodotus wallacii, 26
Sitella, black, 28, 208
    varied, 28
Sitta, 18
    pusilla, *see* nuthatch, brown-headed
    pygmaea, *see* nuthatch, pygmy
Sittidae, 30, 208
Sittinae, 30
Skua, brown, 85, 94, 99, 207, 239
    south polar, 34
Slaty-flycatcher, Abyssinian, 30
Smicrornis brevirostris, 27
Social-weaver, grey-headed, 32, 209
Sparrow, house, 32
    stripe-headed, 34
    white-throated, 240
Sparrow-weaver, white-browed, 32, 77, 134–135, 140, 208, 240
Spermophilus beldingi, *see* ground squirrel, Belding's
Sphecotheres viridis, 29
Spinetail, bat-like, 25
    Cassin's, 25
Spreo bicolor, 30, 208
    fischeri, 30
Stactolaema anchietae, 22
    leucotis, 22
    olivacea, 22
    whytii, 22
Starling, African pied, 30, 208
    chestnut-bellied, 30, 208
    Fischer's, 30
    golden-breasted, 30, 208
    superb, 30
    violet-backed, 30

Stercorariini, 34
Sterna paradisaea, 25
Sternini, 25
Stitchbird, 27, 85, 93, 94, 99, 100, 235, 240
Strigidae, 25
Strigimorphae, 7
Struthidea cinerea, *see* apostlebird
Struthio camelus, *see* ostrich
Struthionidae, 33, 206
Struthioniformes, 178
Stubtail, Asian, 31
Sturnidae, 30, 208
Sturnini, 30
Sula nebouxii, *see* booby, blue-footed
Suricata suricatta, *see* meerkat
Swallow, barn, 31
    blue, 34
    tree, 31, 235, 240
Swamphen, purple, *see* pukeko
Swift, alpine, 25
    ashy-tailed, 25
    chimney, 25
    horus, 25
    mottled, 25
    short-tailed, 25
    Vaux's, 25
Swiftlet, white-rumped, 25
Sylviidae, 31, 33, 209
Sylviinae, 14, 31

Tachybaptus novaehollandiae, 25
Tachycineta bicolor, *see* swallow, tree
Tachymarptis aequatorialis, 25
    melba, 25
Taeniopygia guttata, *see* finch, zebra
Tamarin, cotton-top, 211, 216, 217, 220, 221, 224, 241
    golden lion, 211, 216–218, 220–222, 241
    saddle-back, 224, 241
Tanager, dusky-faced, 32
    golden-hooded, 32
    plain-colored, 32
    speckled, 32
    turquoise, 32
Tangara guttata, 32
    inornata, 32
    larvata, 32
    mexicana, 32
Tanysiptera sylvia, 24
Terathopius ecaudatus, 25
Tern, Arctic, 25
Terpsiphone viridis, 29
Thescelocichla leucopleura, 31
Thornbill, brown, 40, 239
    buff-rumped, 27, 207
    chestnut-rumped, 27
    Papuan, 27

Thornbill (*cont.*)
    striated, 27, 207
    yellow, 27
    yellow-rumped, 27, 207
Thornbird, plain, 26
Thraupini, 32
Thrush, groundscraper, 34
Thryothorus pleurostictus, 31
Timaliini, 14, 31
Tinamiformes, 178
Tinamou, brushland, 178, 239
    ornate, 178, 240
    slaty-breasted, 179, 240
Tit, black, 31, 209
    blue, 62, 239
    great, 62, 115, 239
    long-tailed, 2, 31, 40, 44, 55, 58, 59, 61, 63, 86, 97, 150, 153, 154, 209, 239
    penduline, 82, 240
Tit-hylia, 31
Titmouse, bridled, 31
    tufted, 34, 46, 240
Tockus camurus, 23
Todidae, 24, 206
Todirhamphus macleayii, 24
Todus mexicanus, 24, 206
Tody, Puerto Rican, 24, 206
Trachyphonus darnaudii, 23, 206
    erythrocephalus, 23, 206
    margaritatus, 23
Treecreeper, 62, 240
    black-tailed, 26, 207
    brown, 26, 197, 201–203, 205–207, 230, 239
    red-browed, 26, 207
    rufous, 26, 207
Tricholaema lacrymosa, 22
    leucomelas, 22
    melanocephala, 22
Troglodytes troglodytes, *see* wren
Troglodytinae, 14, 30, 34
Trumpeter, pale-winged, 25, 88, 207, 240
Turdinae, 30, 34
Turdoides, 10, 14, 44
    affinis, 31
    bicolor, 32
    caudatus, 32, 209, 239
    earlei, 32
    gymnogens, 209
    jardineii, 32, 209
    malcolmi, 32, 209
    melanops, 32, 209
    plebejus, 32, 209
    reinwardtii, 32, 209
    squamiceps, *see* babbler, Arabian
    striatus, *see* babbler, jungle
Tyrannidae, 25, 207
Tyranninae, 25

Upupa africana, 24
    epops, *see* hoopoe
Upupidae, 15, 16, 24
Upupiformes, 10, 15
Urocissa caerulea, 28, 208
Urocoliinae, 24
Urocolius indicus, 24
    macrourus, 24
Urosphena squameiceps, 31

Vanellus chilensis, 33
Vanga, Chabert's, 29
    rufous, 30
Vangini, 29
Varanus, 182
Vole, pine, 211, 218–220, 241
    prairie, 211, 217–221, 241

Wagtail, Cape, 32
Warbler, hooded, 12, 34, 239
    moustached, 33, 83, 87, 98, 240
    Seychelles, 2, 31, 41, 51, 53, 55, 57, 67–68, 78, 79, 84, 89, 90, 100, 105, 107–114, 144, 146, 148–149, 155, 156, 190–196, 205, 209, 215, 223, 235, 236, 240
    speckled, 27
Wattlebird, little, 27
    red, 27
Wattle-eye, black-throated, 29
    white-spotted, 29
Weaver, sociable, 32, 91, 107, 108, 111, 121, 240
Weebill, 27
Whale, killer, 210, 241
    pilot, 146, 153, 154, 241
Wheatear, Schalow's, 30
White-eye, Mascarene grey, 31
    Seychelles grey, 31
Whiteface, banded, 27
    southern, 27
Whitehead, 28
Wilsonia citrina, *see* warbler, hooded
Wolf, Ethiopian, 146, 153, 154, 211, 213, 241
    gray, 211, 216, 217, 241
Woodhoopoe, black-billed, 15, 24, 239
    forest, 15, 239
    green, 10, 24, 44, 58, 65, 107, 108, 110, 112, 120, 125, 127, 144, 146–148, 201, 206, 236, 239
    violet, 24
    white-headed, 15, 24, 240
Woodpecker, acorn, 1, 2, 22, 41, 44, 51, 57, 64, 65, 77, 82, 84, 91, 99, 107, 108, 110, 112, 118–120, 123–127, 144, 149–155, 169–170, 173, 185–188, 193–195, 201, 204, 205, 206, 232, 233, 235–236, 239
    downy, 18, 239
    golden-naped, 22
    ground, 22
    hairy, 18, 239
    Hispaniolan, 22

Lewis', 119, 239
  middle-spotted, 33
  red-cockaded, 16–18, 22, 52, 53, 65, 82, 87, 97, 107–110, 137–139, 144, 150–151, 153–155, 197, 199–206, 234, 236, 240
  white, 22
  white-fronted, 22
  yellow-fronted, 22
  yellow-tufted, 22
Woodswallow, black-faced, 29, 208
  dusky, 29
  great, 29
  little, 29
  white-breasted, 29, 208
Wren, 44, 62, 240
  band-backed, 31, 209
  banded, 31
  bicolored, 31, 57, 77, 85, 153, 201, 208, 239
  Boucard's, 31
  cactus, 16–20, 34, 239
  fasciated, 30, 208
  giant, 31
  gray-barred, 31, 208
  musician, 31
  rufous-naped, 31
  sepia-brown, 31
  spotted, 31, 209
  stripe-backed, 31, 53, 70, 85, 99, 153, 201, 209, 240
  thrush-like, 31
  white-headed, 31
  yucatan, 31

Yellowhead, 28
Yuhina brunneiceps, *see* yuhina, Taiwan
Yuhina, Taiwan, 32, 192–193, 195, 196, 240

Zavattariornis stresemanni, 29
Zonotrichia albicollis, 240
Zosteropidae, 31
Zosterops borbonicus, 31
  modesta, 31

Subject index

acorn storage, 118, 120, 186
alarm calling, 43, 44, 222
Allee effect, 63
alliances, *see* coalitions
allogrooming, 210
alloparental behavior, *see* helpers and helping
altriciality, 7–10, 13–14
    origins of, 6
Australia, cooperative breeding in, 2, 12, 39, 48, 126, 233, 237
avian radiation, 7

begging behavior, 6, 10
behavioral plasticity, 45, 62
Bergmann's rule, 20
best of a bad job, 35, 49, 50, 97, 149, 214, 215
breeding constraints, 48, 50, 60, 63
breeding failure, 11
breeding vacancies, 212
brood parasitism, 11, 90, 149, 178, 180, 181, 183, 185–186, 188–190, 192–194, 196
brood reduction, 113

challenge hypothesis, 131, 138
clans, 61, 205
climatic factors, 5, 12
cloaca pecking, 92
coalitions, 171, 172, 219
    formation of, 68
coercion, 68
coloniality, 17, 83
communal nesting, *see* joint nesting
communal nests, 91–92, 121, 185
communal roosting, 20, 44, 120–121, 125, 127
comparative analyses, 228
confidence of parentage, 80
conservation biology, 197
conspecific brood parasitism, *see* brood parasitism
cooperative breeding
    age of, 21
    climate and, 39
    definition of, 3, 5–6, 82, 210, 228–229
    diet and, 117–118, 233
    frequency and occurrence of, 13, 22–34
    genetic basis of, 21
    in Australia, 2, 12, 39, 48, 126, 233, 237
    in South Africa, 12, 39, 126, 233
    maintenance of, 11–12
    obligate and opportunistic, 49, 170, 171, 230–231
    paradox of, 67
    phylogeny and, 234
    regular vs. irregular, 11–12
    route to, 35
    savanna habitat and, 199
    year-round residency and, 38–39
cooperative defense, 48, 210
cooperative hunting, 118
cooperative polyandry, 177
cooperative polygamy, 49, 54, 65
copulation, intrasexual mounting, 91
corticosterone, 132, 134, 136, 140, 224–225
coteries, 65
current functional utility, 54

declining-population paradigm, 198
delayed breeding, 215–219
delayed dispersal, 3, 18, 35, 51, 212–215, 225
    benefits of, 39–45
    costs for offspring, 45–46
    costs for parents, 46
    diet and, 212–214
    uncoupled from helping, 36, 58–59
    where to wait, 37–38
    without helping, 52
delayed reciprocity, 55
demographic factors, 13, 150
demographic stochasticity, 204
depart-and-search, 200–204
despotic societies, 158
digestion, solar-enhanced, 119
dispersal
    coalitions, 171, 223
    decisions, 38–39
    constraints on, 231
    costs of, 52, 61, 212
    habitat fragmentation and, 200
    sex-biased, 59
    strategies, 200–202
    timing of, 63
divorce, 40, 98
    incest avoidance and, 150, 154, 236

DNA fingerprinting, 1, 142, 178, 189, 192, *see also* parentage analysis
DNA–DNA hybridization, 182
dominance hierarchies, 45
    reproductive skew and, 166, 170, 171, 173
doubly labeled water technique, 122, 125

ecological constraints, 1, 3, 38, 50, 153, 177, 182, 194, 195, 236
    reproductive skew and, 169
ecological factors, 1, 5, 7, 11–13, 21, 213, 231, 233–234
egalitarian societies, 90–92, 95, 158, 170, 215, 219
egg destruction, 91–92, 180, 186, 194, 196, 232
egg dumping, 90, 190
egg hatchability, 149, 181, 236
El Niño Southern Oscillation, 126
environmental factors, 13, 21
environmental stochasticity, 19, 126, 204
estrogen, 132, 136, 140, 224
evolutionary origins, 5–11, 36, 49
evolutionary stasis, 15, 22
exaptation, 7
experience hypothesis, 50, 53–55, 68, 78, 119–120, 222, 231
extended families, 205
extra-group fertilizations, 40, 49, 53, 93, 115
extra-group matings, 148, 236
extra-group paternity, 148
extrinsic constraints, 52, 53

familiar neighbor hypothesis, 60
Fisher's theorem, 102, 115
fitness benefits, 68, 79, 232
    delayed, 161
    direct, 3, 43, 52, 53, 55, 57, 222–223, 231
    future direct, 94
    inclusive, 36–37, 51, 60, 90, 106, 115, 143
    indirect, 35–39, 53, 55, 57, 222
    of delayed dispersal, 39–45
    paired comparisons, 56
    reproductive skew and, 158
fitness consequences, 48
floaters, 37, 45, 50, 52, 67, 204, 212
food storage, 118–119
food supplementation, 69
forced copulation, 93

Gondwanaland, 126
group augmentation, 55, 223
group living
    predation and, 213
    resource distribution and, 213
group size effects, 68

habitat
    degradation, 199
    fragmentation, 4, 199–200, 202–204
    loss, 4, 198–199
    quality, variation in, 41, 51, 52

    saturation, 14, 38, 50–52, 80, 109, 145, 190, 193, 212, 231, 236
    specialization, 198
Hamilton's rule, 43, 69, 142, 226
haplodiploidy, 102, 143
helpers and helping
    addition experiments, 57
    additive, 72–73
    advertisement and, 222
    age effects and, 68
    benefits of, 219–225
    casual, 98
    compensatory, 72–76
    costs and benefits of, 52, 69–71, 221, 223–224, 232
    courtship and, 98
    defection of, 96
    lactation and, 219
    model of, 71–75
    origins of, 10–11
    parental aggression and, 59
    quality of, 70
    redirected, 50, 58, 62, 63, 97
    regular, 98
    removal experiments, 56–57
    sex bias in, 79, 227
    unattached, 83, 98
    uncoupled from delayed dispersal, 58–59
    unrelated, 55
heterogamety, 102
historical perspective, 49–58
homology, 6, 21
hormones
    binding proteins
    deactivation of, 130
    dominance and, 216–218
    helpers vs. breeders, 131
    parental, 129, 134
    precursors to, 130
    regulation of, 130
    relatedness and, 140
    reproductive, 128–129
    stress, 129
    target tissues, 128
    testis size and, 131

inbreeding depression, 143–145, 149, 150, 152, 155, 184, 205
incest and incest avoidance, 2–3, 40, 46, 59, 78, 82, 88, 90, 92, 96, 100, 142, 186, 196, 202, 205, 236
    demographic consequences of, 155
    dispersal and, 150–151
    extra-pair matings and, 142, 146, 147
    hatchability and, 149, 236
    in humans, 155
    mechanisms of, 150–154
    reproductive skew and, 154–155, 165–166, 169, 171, 172, 175
    sex-biased dispersal and, 152
inclusive fitness, *see* fitness benefits

incubation
    communal nesting and, 177
    nocturnal, 188, 189, 192
    length, 123
indicator species, 205
individual quality, 41
individual reproductive decisions, 60
infanticide, 189, 190, 219, 233
information sharing, 48
interspecific feeding, 11
intrinsic benefits, 52
island endemics, 198

joint nesting, 3, 83, 89–91, 149, 151, 154, 177, 232
    male parental care and, 194
    phylogenetic origins of, 177
    sexual conflict in, 177
juvenile helpers, 57

keystone species, 205
kidnapping, 55
kin discrimination, 213, 222, 236
    reproductive skew and, 174
kin neighborhoods, 62, 63, 213, 224
kin preferences, 55, 63
kin selection, 3, 68, 75, 76, 79, 143, 226, 231

leks, 94, 230
    hidden, 83, 96–97, 100
    paradox, 88
levels of analysis, 3, 6, 234
life-history traits, 1, 4, 6, 12, 36, 41, 63, 79, 80, 113, 213, 230, 233–234, 237
lifetime reproductive success, 44
load-lightening, 98, 221
local mate competition, 45, 106, 109
local resource competition, 106, 107, 112, 115
local resource enhancement, 106, 107, 109, 112, 115
luteinizing hormone, 131, 134–136, 140, 216

marginal-habitat hypothesis, 50, 52
mate choice, 88–89, 98–99
    genetic benefits and, 96, 100
    reproductive skew and, 164–165, 172
mate guarding, 147, 186, 233
mate limitation, 51
mating systems, 81, 89–98, 229–230
microsatellite markers, 65
molecular sexing, 103
monogamy, 82
    exclusive, 83, 97–98
multiple mating, 92, 219

natal philopatry, 35–36, 167
nepotism, 36–37, 40, 41, 43, 46, 212–214, 219
    concession of food and, 42–43
    defensive behavior and, 43–44

    roosting behavior and, 44
    survival and, 42
    vs. territory quality, 44–45

optimal outbreeding, 143
outbreeding depression, 143

paired comparisons, 62
parentage analysis, 1, 3, 228, *see also* DNA fingerprinting
parental facilitation, 161, 175
parental investment, extended, 36, 44
parsimony reconstruction, 14
paternity, 91
    guards, 89, 99
    shared, 95
pay-to-stay, 49, 53, 68, 76, 100, 222
peace incentives, 160
phenotype matching, 154
philopatry, benefits of, 3, 51–52, 59, 67, 212, 236
phylogenetic analyses, 8, 16, 17, 19, 36, 38, 228
    limitations of, 21–22
phylogenetic history, 6, 13–15
physiological constraints, 3, 10, 20
    helping and, 69
physiological ecology, 117
plural breeding, 83, 92, 93
polyandry, 82, 98
    cooperative, 2
    egalitarian, 94–95, 100
    flexible, 95–96
polygynandry, 5, 82, 92, 169
    egalitarian, 83, 90
    flexible, 83, 92, 100
    nest defense and, 93
    opportunistic, 185
polygyny, 82
    resource defense, 94
population bottleneck, 155, 190
population structure, 65, 142
power struggles, 50, 186
predation, sociality and, 12
prolactin, 130, 132–133, 136–140, 224–225
promiscuity, 138
psychological castration, 69, 122, 135

queuing, 41, 52, 93, 161

RAPDs, 90
rent, payment of, *see* pay-to-stay
repayment model, 106
reproduction, costs of, 160, 216
reproductive competition, 150–152, 155
reproductive skew, 2, 3, 81, 100, 157, 177, 183, 194–196, 215–219, 226, 228, 233
    age and, 168
    assessment and, 173
    bidding wars, 161

bribery and, 160
competitive ability and, 162
compromise and synthetic models, 162–164
concession models, 158–159, 215, 218–219
confounding variables and, 165–167
dynamic models, 161
ecological constraints and, 196
eviction and, 173
female control and, 167, 175
group productivity effect, 158
group size and, 162, 173–175
group stability and, 159
incest avoidance and, 165–166, 169, 171, 172, 175
incomplete control and, 162, 173, 219
manipulation and, 161
mate choice and, 164–165, 172
mechanisms of, 216–218
multiple paternity and, 168
null models, 169
relatedness and, 159–161, 163, 167
restraint models, 158–159
sexual conflict and, 164–165
transactional models, 158–162
tug-of-war model, 162–163
work incentive model, 164
reproductive suppression, 135, 136, 154, 194, 217, 218, 226
reproductive vacancies, 151, 152
residual reproductive value, 77
resource defense, 213
resource-holding potential, 42

safe-haven hypothesis, 41, 50
selfish-herd effect, 43, 48
sentinels and sentinel behavior, 12, 121, 213
sex-biased dispersal, 59
sex determination, 102
sex ratio, 2, 102
 adult, 63, 94
 avian life-cycles and, 105
 conflicts of interest and, 115
 evolutionarily stable strategy and, 104
 facultative adjustment of, 106
 frequency-dependent selection and, 104, 106
 genetic conflict and, 104
 kin competition and, 104
 laying order and, 113–114
 mate competition and, 107
 mechanisms of adjustment, 104, 114
 parental condition and, 104
 primary, 103, 105, 107, 115

pseudoreplication and, 107, 114
repayment and, 106
secondary, 103, 115
sexual dimorphism and, 105, 106
sibling competition and, 105, 106
sexual conflict, 81, 88–89, 100, 188
sexual dimorphism, reverse size, 95
sexual monomorphism, 81
sexual selection, 81, 234–235
shortage of mates, 63
siblicide, 105, 113
sibling rivalry, 40
simultaneous breeder–helpers, 58
singular nesting, 83
skew, see reproductive skew
skills hypothesis, see experience hypothesis
small-population paradigm, 198, 204
social conflict, 42
social prestige, 68
sociality, origins of, 10
South Africa, cooperative breeding in, 12, 39, 126, 233
spatial autocorrelation, 52
 of territory quality, 41
sperm competition, 99
sperm depletion, 89
stay-and-foray, 52, 59, 63, 200–204
staying incentives, 158, 160, 164
supersaturation hypothesis, 63, 65, 194

tannins, 118, 124
territory
 abdication, 150
 budding, 40, 52, 90, 201
 inheritance, 40, 51–52, 153, 202, 204, 212
 variation in quality, 41, 52
testis size, 136
testosterone, 129–132, 134–141, 216, 224
thermoneutral zone, 124
thermoregulation, 120

umbrella species, 205
unselected hypothesis, 2, 3, 6, 139

vigilance, 43, 48, 121, 127, 210

why delay?, 54
why help?, 52–58, 219–225
why stay?, 54
winter social behavior, 55, 59–62